STATA BASE REFERENCE MANUAL
VOLUME 3
Q–Z
RELEASE 10

A Stata Press Publication
StataCorp LP
College Station, Texas

Stata Press, 4905 Lakeway Drive, College Station, Texas 77845

The suggested citation for this software is

StataCorp. 2007. *Stata Statistical Software: Release 10*. College Station, TX: StataCorp LP.

Title

qc — Quality control charts

Syntax

Draw a c chart

    cchart *defect_var unit_var* [, *cchart_options*]

Draw a p (fraction-defective) chart

    pchart *reject_var unit_var ssize_var* [, *pchart_options*]

Draw an R (range or dispersion) chart

    rchart *varlist* [*if*] [*in*] [, *rchart_options*]

Draw an $\overline{X}$ (control line) chart

    xchart *varlist* [*if*] [*in*] [, *xchart_options*]

Draw vertically aligned $\overline{X}$ and R charts

    shewhart *varlist* [*if*] [*in*] [, *shewhart_options*]

cchart_options	description
Plot	
connect_options	affect rendition of the plotted points
marker_options	change look of markers (color, size, etc.)
marker_label_options	add marker labels; change look or position
Control limits	
clopts(*cline_options*)	affect rendition of the control limits
Add plots	
addplot(*plot*)	add other plots to the generated graph
Y axis, X axis, Titles, Legend, Overall	
twoway_options	any options other than by() documented in [G] ***twoway_options***

(Continued on next page)

1

pchart_options	description
Main	
stabilized	stabilize the p chart when sample sizes are unequal
Plot	
connect_options	affect rendition of the plotted points
marker_options	change look of markers (color, size, etc.)
marker_label_options	add marker labels; change look or position
Control limits	
clopts(*cline_options*)	affect rendition of the control limits
Add plots	
addplot(*plot*)	add other plots to the generated graph
Y axis, X axis, Titles, Legend, Overall	
twoway_options	any options other than by() documented in [G] *twoway_options*

rchart_options	description
Main	
std(*#*)	user-specified standard deviation
Plot	
connect_options	affect rendition of the plotted points
marker_options	change look of markers (color, size, etc.)
marker_label_options	add marker labels; change look or position
Control limits	
clopts(*cline_options*)	affect rendition of the control limits
Add plots	
addplot(*plot*)	add other plots to the generated graph
Y axis, X axis, Titles, Legend, Overall	
twoway_options	any options other than by() documented in [G] *twoway_options*

xchart_options	description
Main	
<u>std</u>(#)	user-specified standard deviation
<u>mean</u>(#)	user-specified mean
<u>lower</u>(#) <u>upper</u>(#)	lower and upper limits of the X-bar limits
Plot	
connect_options	affect rendition of the plotted points
marker_options	change look of markers (color, size, etc.)
marker_label_options	add marker labels; change look or position
Control limits	
<u>clop</u>ts(cline_options)	affect rendition of the control limits
Add plots	
addplot(plot)	add other plots to the generated graph
Y axis, X axis, Titles, Legend, Overall	
twoway_options	any options other than by() documented in [G] *twoway_options*

shewhart_options	description
Main	
<u>std</u>(#)	user-specified standard deviation
<u>mean</u>(#)	user-specified mean
Plot	
connect_options	affect rendition of the plotted points
marker_options	change look of markers (color, size, etc.)
marker_label_options	add marker labels; change look or position
Control limits	
<u>clop</u>ts(cline_options)	affect rendition of the control limits
Y axis, X axis, Titles, Legend, Overall	
combine_options	any options documented in [G] **graph combine**

Description

These commands provide standard quality-control charts. cchart draws a c chart; pchart, a p (fraction-defective) chart; rchart, an R (range or dispersion) chart; xchart, an $\overline{X}$ (control line) chart; and shewhart, vertically aligned $\overline{X}$ and R charts.

Options

┌─ Main ┐

stabilized stabilizes the p chart when sample sizes are unequal.

std(#) specifies the standard deviation of the process. The R chart is calculated (based on the range) if this option is not specified.

mean(#) specifies the grand mean, which is calculated if not specified.

lower(#) and upper(#) must be specified together or not at all. They specify the lower and upper limits of the $\overline{X}$ chart. Calculations based on the mean and standard deviation (whether specified by option or calculated) are used otherwise.

┌─ Plot ┐

connect_options affect whether lines connect the plotted points and the rendition of those lines; see [G] ***connect_options***.

marker_options affect the rendition of markers drawn at the plotted points, including their shape, size, color, and outline; see [G] ***marker_options***.

marker_label_options specify if and how the markers are to be labeled; see [G] ***marker_label_options***.

┌─ Control limits ┐

clopts(*cline_options*) affects the rendition of the control limits; see [G] ***cline_options***.

┌─ Add plots ┐

addplot(*plot*) provides a way to add other plots to the generated graph. See [G] ***addplot_option***.

┌─ Y axis, X axis, Titles, Legend, Overall ┐

twoway_options are any of the options documented in [G] ***twoway_options***, excluding by(). These include options for titling the graph (see [G] ***title_options***) and for saving the graph to disk (see [G] ***saving_option***).

combine_options (shewhart only) are any of the options documented in [G] **graph combine**. These include options for titling the graph (see [G] ***title_options***) and for saving the graph to disk (see [G] ***saving_option***).

Remarks

Control charts may be used to define the goal of a repetitive process, to control that process, and to determine if the goal has been achieved. Walter A. Shewhart of Bell Telephone Laboratories devised the first control chart in 1924. In 1931, Shewhart published *Economic Control of Quality of Manufactured Product*. According to Burr, "Few fields of knowledge have ever been so completely explored and charted in the first exposition" (1976, 29). Shewhart states that "a phenomenon will be said to be controlled when, through the use of past experience, we can predict, at least within limits, how the phenomenon may be expected to vary in the future. Here it is understood that prediction within limits means that we can state, at least approximately, the probability that the observed phenomenon will fall within given limits" (1931, 6).

For more information on quality-control charts, see Burr (1976), Duncan (1986), Harris (1999), or Ryan (2000).

▷ Example 1: cchart

cchart graphs a c chart showing the number of nonconformities in a unit, where *defect_var* records the number of defects in each inspection unit and *unit_var* records the unit number. The unit numbers need not be in order. For instance, consider the following example dataset from Ryan (2000, 156):

```
. use http://www.stata-press.com/data/r10/ncu

. describe

Contains data from http://www.stata-press.com/data/r10/ncu.dta
  obs:            30
  vars:            2                          31 Mar 2007 03:56
  size:          360 (99.9% of memory free)
```

	storage	display	value	
variable name	type	format	label	variable label
day	float	%9.0g		Days in April
defects	float	%9.0g		Numbers of Nonconforming Units

```
Sorted by:

. list in 1/5
```

	day	defects
1.	1	7
2.	2	5
3.	3	11
4.	4	13
5.	5	9

```
. cchart defects day, title(c Chart for Nonconforming Transistors)
```

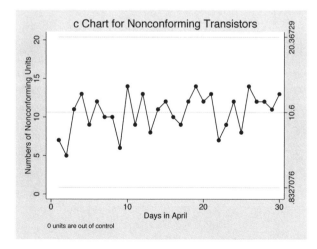

The expected number of defects is 10.6, with lower- and upper-control limits of .8327 and 20.37, respectively. No units are out of control.

◁

▷ Example 2: pchart

pchart graphs a p chart, which shows the fraction of nonconforming items in a subgroup, where *reject_var* records the number rejected in each inspection unit, *unit_var* records the inspection unit number, and *ssize_var* records the number inspected in each unit.

Consider the example dataset from Ryan (2000, 156) of the number of nonconforming transistors out of 1,000 inspected each day during the month of April:

```
. use http://www.stata-press.com/data/r10/ncu2

. describe

Contains data from http://www.stata-press.com/data/r10/ncu2.dta
  obs:            30
  vars:            3                          31 Mar 2007 14:13
  size:          480 (99.9% of memory free)
```

variable name	storage type	display format	value label	variable label
day	float	%9.0g		Days in April
rejects	float	%9.0g		Numbers of Nonconforming Units
ssize	float	%9.0g		Sample size

```
Sorted by:

. list in 1/5
```

	day	rejects	ssize
1.	1	7	1000
2.	2	5	1000
3.	3	11	1000
4.	4	13	1000
5.	5	9	1000

```
. pchart rejects day ssize
```

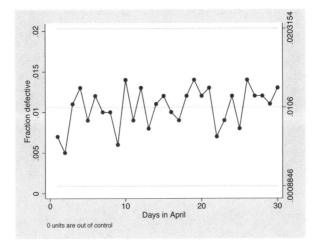

All the points are within the control limits, which are .0009 for the lower limit and .0203 for the upper limit.

Here the sample sizes are fixed at 1,000, so the `ssize` variable contains 1,000 for each observation. Sample sizes need not be fixed, however. Say that our data were slightly different:

```
. use http://www.stata-press.com/data/r10/ncu3
. list in 1/5
```

	day	rejects	ssize
1.	1	7	920
2.	2	5	920
3.	3	11	920
4.	4	13	950
5.	5	9	950

```
. pchart rejects day ssize
```

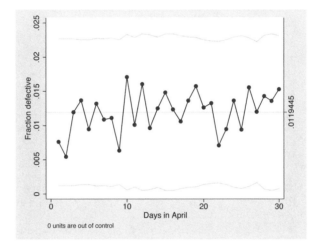

In this case, the control limits are, like the sample size, no longer constant. The `stabilize` option will stabilize the control chart:

```
. pchart rejects day ssize, stabilize
```

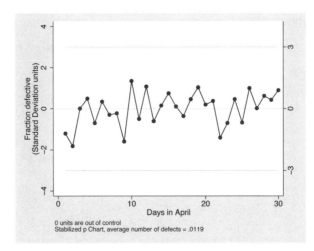

◁

▷ Example 3: rchart

`rchart` displays an R chart showing the range for repeated measurements at various times. Variables within observations record measurements. Observations represent different samples.

For instance, say that we take five samples of 5 observations each. In our first sample, our measurements are 10, 11, 10, 11, and 12. The data are

. list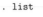

	m1	m2	m3	m4	m5
1.	10	11	10	11	12
2.	12	10	9	10	9
3.	10	11	10	12	10
4.	9	9	9	10	11
5.	12	12	12	12	13

. rchart m1-m5, connect(l)

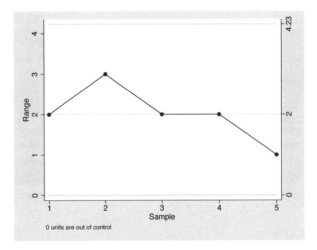

The expected range in each sample is 2 with lower- and upper-control limits of 0 and 4.23, respectively. If we knew that the process standard deviation is 0.3, we could specify

. rchart m1-m5, connect(l) std(.3)

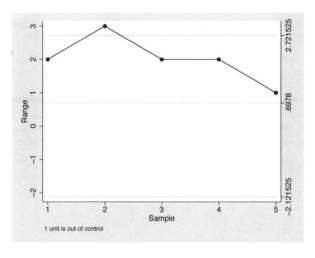

▷ Example 4: xchart

xchart graphs an $\overline{X}$ chart for repeated measurements at various times. Variables within observations record measurements, and observations represent different samples. Using the same data as in the previous example, we type

. xchart m1-m5, connect(l)

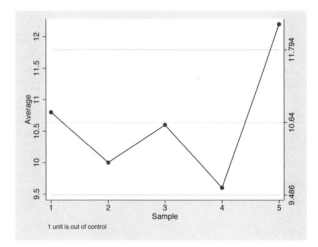

The average measurement in the sample is 10.64, and the lower- and upper-control limits are 9.486 and 11.794, respectively. Suppose that we knew from prior information that the mean of the process is 11. Then we would type

```
. xchart m1-m5, connect(l) mean(11)
```

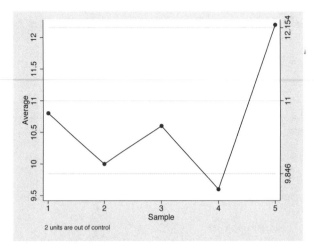

If we also knew that the standard deviation of the process is 0.3, we could type

```
. xchart m1-m5, connect(l) mean(11) std(.3)
```

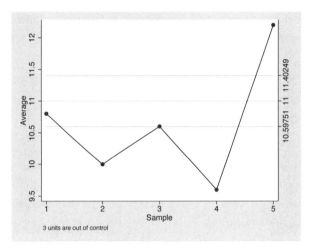

Finally, xchart allows us to specify our own control limits:

```
. xchart m1-m5, connect(l) mean(11) lower(10) upper(12)
```

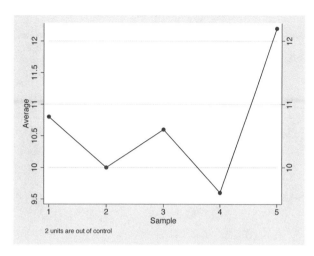

◁

Walter Andrew Shewhart (1891–1967) was born in Illinois and educated as a physicist, with degrees from the Universities of Illinois and California. After a brief period teaching physics, he worked for the Western Electric Company and (from 1925) the Bell Telephone Laboratories. His name is most associated with control charts used in quality controls, but his many other interests ranged from quality assurance generally to the philosophy of science.

▷ Example 5: shewhart

shewhart displays a vertically aligned $\overline{X}$ and R chart in the same image. To produce the best-looking combined image possible, you will want to use the xchart and rchart commands separately and then combine the graphs. shewhart, however, is more convenient.

Using the same data as previously, but realizing that the standard deviation should have been 0.4, we type

(Continued on next page)

. shewhart m1-m5, connect(l) mean(11) std(.4)

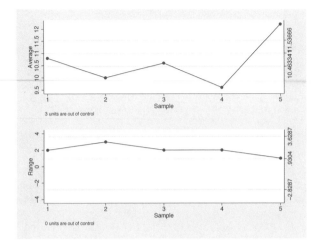

Methods and Formulas

cchart, pchart, rchart, xchart, and shewhart are implemented as ado-files.

For the c chart, the number of defects per unit, C, is taken to be a value of a random variable having a Poisson distribution. If k is the number of units available for estimating λ, the parameter of the Poisson distribution, and if C_i is the number of defects in the ith unit, then λ is estimated by $\overline{C} = \sum_i C_i/k$. Then

$$\text{central line} = \overline{C}$$

$$\text{UCL} = \overline{C} + 3\sqrt{\overline{C}}$$

$$\text{LCL} = \overline{C} - 3\sqrt{\overline{C}}$$

Control limits for the p chart are based on the sampling theory for proportions, using the normal approximation to the binomial. If k samples are taken, the estimator of p is given by $\overline{p} = \sum_i \widehat{p}_i/k$, where $\widehat{p}_i = x_i/n_i$, and x_i is the number of defects in the ith sample of size n_i. The central line and the control limits are given by

$$\text{central line} = \overline{p}$$

$$\text{UCL} = \overline{p} + 3\sqrt{\overline{p}(1 - \overline{p})/n_i}$$

$$\text{LCL} = \overline{p} - 3\sqrt{\overline{p}(1 - \overline{p})/n_i}$$

Control limits for the R chart are based on the distribution of the range of samples of size n from a normal population. If the standard deviation of the process σ is known,

$$\text{central line} = d_2\sigma$$

$$\text{UCL} = D_2\sigma$$

$$\text{LCL} = D_1\sigma$$

where d_2, D_1, and D_2 are functions of the number of observations in the sample and are obtained from the table published in Beyer (1976).

When σ is unknown,

$$\text{central line} = \overline{R}$$
$$\text{UCL} = (D_2/d_2)\overline{R}$$
$$\text{LCL} = (D_1/d_2)\overline{R}$$

where $\overline{R} = \sum_i R_i/k$ is the range of the k sample ranges R_i.

Control limits for the $\overline{X}$ chart are given by

$$\text{central line} = \overline{x}$$
$$\text{UCL} = \overline{x} + (3/\sqrt{n})\sigma$$
$$\text{LCL} = \overline{x} - (3/\sqrt{n})\sigma$$

if σ is known. If σ is unknown,

$$\text{central line} = \overline{x}$$
$$\text{UCL} = \overline{x} + A_2\overline{R}$$
$$\text{LCL} = \overline{x} - A_2\overline{R}$$

where $\overline{R}$ is the average range as defined above and A_2 is a function (op. cit.) of the number of observations in the sample.

References

Bayart, D. 2001. Walter Andrew Shewhart. In *Statisticians of the Centuries*, ed. C. C. Heyde and E. Seneta, 398–401. New York: Springer.

Beyer, W. H., ed. 1976. Factors for computing control limits. In *Handbook of Tables for Probability and Statistics.* 2nd ed. 451–465. Cleveland, OH: The Chemical Rubber Company.

Burr, I. W. 1976. *Statistical Quality Control Methods.* New York: Dekker.

Caulcutt, R. 2004. Control charts in practice. *Significance* 1: 81–84.

Duncan, A. J. 1986. *Quality Control and Industrial Statistics.* 5th ed. Homewood, IL: Irwin.

Harris, R. L. 1999. *Information Graphics: A Comprehensive Illustrated Reference.* New York: Oxford University Press.

Ryan, T. P. 2000. *Statistical Methods for Quality Improvement.* 2nd ed. New York: Wiley.

Saw, S. L. C., and T. W. Soon. 1994. sqc1: Estimating process capability indices with Stata. *Stata Technical Bulletin* 17: 18–19. Reprinted in *Stata Technical Bulletin Reprints*, vol. 3, pp. 174–175.

Shewhart, W. A. 1931. *Economic Control of Quality of Manufactured Product.* New York: Van Nostrand.

Also See

[R] **serrbar** — Graph standard error bar chart

Title

qreg — Quantile regression

Syntax

Quantile regression

    qreg *depvar* [*indepvars*] [*if*] [*in*] [*weight*] [, *qreg_options*]

Interquantile range regression

    iqreg *depvar* [*indepvars*] [*if*] [*in*] [, *iqreg_options*]

Simultaneous-quantile regression

    sqreg *depvar* [*indepvars*] [*if*] [*in*] [, *sqreg_options*]

Quantile regression with bootstrap standard errors

    bsqreg *depvar* [*indepvars*] [*if*] [*in*] [, *bsqreg_options*]

Internal estimation command for quantile regression

    _qreg [*depvar* [*indepvars*] [*if*] [*in*] [*weight*]] [, *_qreg_options*]

qreg_options	description
Model	
<u>q</u>uantile(*#*)	estimate # quantile; default is quantile(.5)
Reporting	
<u>l</u>evel(*#*)	set confidence level; default is level(95)
Opt options	
optimization_options	control the optimization process; seldom used
<u>wls</u>iter(*#*)	attempt # weighted least-squares iterations before doing linear programming iterations

iqreg_options	description
Model	
<u>q</u>uantiles(*# #*)	interquantile range; default is quantiles(.25 .75)
<u>r</u>eps(*#*)	perform # bootstrap replications; default is reps(20)
Reporting	
<u>l</u>evel(*#*)	set confidence level; default is level(95)
<u>nod</u>ots	suppress display of the replication dots

14

sqreg_options	description
Model	
quantiles(# [# [# …]])	estimate # quantiles; default is quantiles(.5)
reps(#)	perform # bootstrap replications; default is reps(20)
Reporting	
level(#)	set confidence level; default is level(95)
nodots	suppress display of the replication dots

bsqreg_options	description
Model	
quantile(#)	estimate # quantile; default is quantile(.5)
reps(#)	perform # bootstrap replications; default is reps(20)
Reporting	
level(#)	set confidence level; default is level(95)

_qreg_options	description
quantile(#)	estimate # quantile; default is quantile(.5)
level(#)	set confidence level; default is level(95)
accuracy(#)	relative accuracy required for linear programming algorithm; should not be specified
optimization_options	control the optimization process; seldom used

by, rolling, statsby, and xi are allowed by qreg, iqreg, sqreg, and bsqreg; nestreg and stepwise are allowed only with qreg; see [U] **11.1.10 Prefix commands**.

qreg and _qreg allow aweights and fweights; see [U] **11.1.6 weight**.

See [U] **20 Estimation and postestimation commands** for more capabilities of estimation commands.

Description

qreg fits quantile (including median) regression models, also known as least–absolute value models (LAV or MAD) and minimum L1-norm models.

iqreg estimates interquantile range regressions, regressions of the difference in quantiles. The estimated variance–covariance matrix of the estimators (VCE) is obtained via bootstrapping.

sqreg estimates simultaneous-quantile regression. It produces the same coefficients as qreg for each quantile. Reported standard errors will be similar, but sqreg obtains an estimate of the VCE via bootstrapping, and the VCE includes between-quantile blocks. Thus you can test and construct confidence intervals comparing coefficients describing different quantiles.

bsqreg is equivalent to sqreg with one quantile.

_qreg is the internal estimation command for quantile regression. _qreg is not intended to be used directly; see *Methods and Formulas* below.

Options for qreg

⌐ Model ⌐

quantile(#) specifies the quantile to be estimated and should be a number between 0 and 1, exclusive. Numbers larger than 1 are interpreted as percentages. The default value of 0.5 corresponds to the median.

⌐ Reporting ⌐

level(#); see [R] **estimation options**.

⌐ Opt options ⌐

optimization_options: <u>iter</u>ate(#), [<u>no</u>]<u>log</u>, <u>trace</u>. iterate() specifies the maximum number of iterations; log/nolog specifies whether to show the iteration log; and trace specifies that the iteration log should include the current parameter vector. These options are seldom used.

wlsiter(#) specifies the number of weighted least-squares iterations that will be attempted before the linear programming iterations are started. The default value is 1. If there are convergence problems, increasing this number should help.

Options for iqreg

⌐ Model ⌐

quantiles(# #) specifies the quantiles to be compared. The first number must be less than the second, and both should be between 0 and 1, exclusive. Numbers larger than 1 are interpreted as percentages. Not specifying this option is equivalent to specifying quantiles(.25 .75), meaning the interquartile range.

reps(#) specifies the number of bootstrap replications to be used to obtain an estimate of the variance–covariance matrix of the estimators (standard errors). reps(20) is the default and is arguably too small. reps(100) would perform 100 bootstrap replications. reps(1000) would perform 1,000 replications.

⌐ Reporting ⌐

level(#); see [R] **estimation options**.

nodots suppresses display of the replication dots.

Options for sqreg

⌐ Model ⌐

quantiles(# [# [# ...]]) specifies the quantiles to be estimated and should contain numbers between 0 and 1, exclusive. Numbers larger than 1 are interpreted as percentages. The default value of 0.5 corresponds to the median.

reps(#) specifies the number of bootstrap replications to be used to obtain an estimate of the variance–covariance matrix of the estimators (standard errors). reps(20) is the default and is arguably too small. reps(100) would perform 100 bootstrap replications. reps(1000) would perform 1,000 replications.

⌐‾‾‾ Reporting ‾‾‾

level(*#*); see [R] **estimation options**.

nodots suppresses display of the replication dots.

Options for bsqreg

⌐‾‾‾ Model ‾‾

quantile(*#*) specifies the quantile to be estimated and should be a number between 0 and 1, exclusive. Numbers larger than 1 are interpreted as percentages. The default value of 0.5 corresponds to the median.

reps(*#*) specifies the number of bootstrap replications to be used to obtain an estimate of the variance–covariance matrix of the estimators (standard errors). reps(20) is the default and is arguably too small. reps(100) would perform 100 bootstrap replications. reps(1000) would perform 1,000 replications.

⌐‾‾‾ Reporting ‾‾

level(*#*); see [R] **estimation options**.

Options for _qreg

quantile(*#*) specifies the quantile to be estimated and should be a number between 0 and 1, exclusive. The default value of 0.5 corresponds to the median.

level(*#*); see [R] **estimation options**.

accuracy(*#*) should not be specified; it specifies the relative accuracy required for the linear programming algorithm. If the potential for improving the sum of weighted deviations by deleting an observation from the basis is less than this on a percentage basis, the algorithm will be said to have converged. The default value is 10^{-10}.

optimization_options: <u>iter</u>ate(*#*), [<u>no</u>]<u>log</u>, <u>trace</u>. iterate() specifies the maximum number of iterations; log/nolog specifies whether to show the iteration log; and trace specifies that the iteration log should include the current parameter vector. These options are seldom used.

Remarks

Remarks are presented under the following headings:

> *Median regression*
> *Generalized quantile regression*
> *Estimated standard errors*
> *Interquantile and simultaneous-quantile regression*

Median regression

qreg without options fits quantile regression models. The most common form is median regression, where the object is to estimate the median of the dependent variable, conditional on the values of the independent variables. This method is similar to ordinary regression, where the objective is to estimate the mean of the dependent variable. Simply put, median regression finds a line through the data that minimizes the sum of the *absolute* residuals rather than the sum of the *squares* of the residuals, as in ordinary regression.

▷ Example 1

Consider a two-group experimental design with 5 observations per group:

```
. use http://www.stata-press.com/data/r10/twogrp
. list
```

	x	y
1.	0	0
2.	0	1
3.	0	3
4.	0	4
5.	0	95
6.	1	14
7.	1	19
8.	1	20
9.	1	22
10.	1	23

```
. qreg y x
Iteration  1:  WLS sum of weighted deviations =  121.88268

Iteration  1: sum of abs. weighted deviations =      111
Iteration  2: sum of abs. weighted deviations =      110
```

Median regression   Number of obs =   10
    Raw sum of deviations   157 (about 14)
    Min sum of deviations   110   Pseudo R2   =   0.2994

y	Coef.	Std. Err.	t	P>|t|	[95% Conf. Interval]	
x	17	3.924233	4.33	0.003	7.950702	26.0493
_cons	3	2.774852	1.08	0.311	-3.39882	9.39882

We have estimated the equation

$$y_{\text{median}} = 3 + 17\,x$$

We look back at our data. x takes on the values 0 and 1, so the median for the x==0 group is 3, whereas for x==1 it is $3 + 17 = 20$. The output reports that the raw sum of absolute deviations about 14 is 157; that is, the sum of $|y - 14|$ is 157. Fourteen is the unconditional median of y, although in these data, any value between 14 and 19 could also be considered an unconditional median (we have an even number of observations, so the median is bracketed by those two values). In any case, the raw sum of deviations of y about the median would be the same no matter what number we choose between 14 and 19. (With a "median" of 14, the raw sum of deviations is 157. Now think of choosing a slightly larger number for the median and recalculating the sum. Half the observations will have larger negative residuals, but the other half will have smaller positive residuals, resulting in no net change.)

We turn now to the actual estimated equation. The sum of the absolute deviations about the solution $y_{median} = 3 + 17x$ is 110. The pseudo-R^2 is calculated as $1 - 110/157 \approx 0.2994$. This result is based on the idea that the median regression is the maximum likelihood estimate for the double-exponential distribution.

◁

❑ Technical Note

qreg is an alternative to regular regression or robust regression—see [R] **regress** and [R] **rreg**. Let's compare the results:

```
. regress y x
```

Source	SS	df	MS				
				Number of obs =		10	
				F(1, 8) =		0.00	
Model	2.5	1	2.5	Prob > F =		0.9586	
Residual	6978.4	8	872.3	R-squared =		0.0004	
				Adj R-squared =		-0.1246	
Total	6980.9	9	775.655556	Root MSE =		29.535	

y	Coef.	Std. Err.	t	P>|t|	[95% Conf. Interval]	
x	-1	18.6794	-0.05	0.959	-44.07477	42.07477
_cons	20.6	13.20833	1.56	0.157	-9.858465	51.05847

Unlike qreg, regress fits ordinary linear regression and is concerned with predicting the mean rather than the median, so both results are, in a technical sense, correct. Putting aside those technicalities, however, we tend to use either regression to describe the central tendency of the data, of which the mean is one measure and the median another. Thus we can ask: which method better describes the central tendency of these data?

Means—and therefore ordinary linear regression—are sensitive to outliers, and our data were purposely designed to contain two such outliers: 95 for x==0 and 14 for x==1. These two outliers dominated the ordinary regression and produced results that do not reflect the central tendency well—you are invited to enter the data and graph y against x.

Robust regression attempts to correct the outlier-sensitivity deficiency in ordinary regression:

```
. rreg y x, genwt(wt)

     Huber iteration 1:  maximum difference in weights = .7311828
     Huber iteration 2:  maximum difference in weights = .17695779
     Huber iteration 3:  maximum difference in weights = .03149585
  Biweight iteration 4:  maximum difference in weights = .1979335
  Biweight iteration 5:  maximum difference in weights = .23332905
  Biweight iteration 6:  maximum difference in weights = .09960067
  Biweight iteration 7:  maximum difference in weights = .02691458
  Biweight iteration 8:  maximum difference in weights = .0009113
```

	Robust regression				Number of obs =	10
					F(1, 8) =	80.63
					Prob > F =	0.0000

y	Coef.	Std. Err.	t	P>|t|	[95% Conf. Interval]	
x	18.16597	2.023114	8.98	0.000	13.50066	22.83128
_cons	2.000003	1.430558	1.40	0.200	-1.298869	5.298875

Here rreg discarded the first outlier completely. (We know this because we included the genwt() option on rreg and, after fitting the robust regression, examined the weights.) For the other "outlier", rreg produced a weight of 0.47.

In any case, the answers produced by qreg and rreg to describe the central tendency are similar, but the standard errors are different. In general, robust regression will have smaller standard errors since it is not as sensitive to the exact placement of observations near the median. Also, some authors (Rousseeuw and Leroy 1987, 11) have noted that quantile regression, unlike the median, may be sensitive to even one outlier, if its leverage is high enough.   ❑

▷ Example 2

Let us now consider a less artificial example using the automobile data described in [U] **1.2.1 Sample datasets**. Using median regression, we will regress each car's price on its weight and length and whether it is of foreign manufacture:

```
. use http://www.stata-press.com/data/r10/auto, clear
(1978 Automobile Data)

. qreg price weight length foreign
Iteration  1:  WLS sum of weighted deviations =  112795.66

Iteration  1: sum of abs. weighted deviations =    111901
Iteration  2: sum of abs. weighted deviations =  110529.43
 (output omitted )
Iteration  8: sum of abs. weighted deviations =  108822.59
```

Median regression					Number of obs =	74
Raw sum of deviations 142205 (about 4934)						
Min sum of deviations 108822.6					Pseudo R2 =	0.2347

price	Coef.	Std. Err.	t	P>|t|	[95% Conf. Interval]	
weight	3.933588	.8602183	4.57	0.000	2.217937	5.649239
length	-41.25191	28.8693	-1.43	0.157	-98.82991	16.32609
foreign	3377.771	577.3391	5.85	0.000	2226.305	4529.237
_cons	344.6494	3260.244	0.11	0.916	-6157.702	6847.001

The estimated equation is

$$\text{price}_{\text{median}} = 3.93\,\text{weight} - 41.25\,\text{length} + 3377.8\,\text{foreign} + 344.65$$

The output may be interpreted in the same way as linear regression output; see [R] **regress**. The variables weight and foreign are significant, but length is not significant. The median price of the cars in these data is $4,934. This value is a median (one of the two center observations), not *the* median, which would typically be defined as the midpoint of the two center observations.

                                                                                                        ◁

Generalized quantile regression

Generalized quantile regression is similar to median regression in that it estimates an equation describing a quantile other than the .5 (median) quantile. For example, specifying `quant(.25)` estimates the 25th percentile or the first quartile.

▷ Example 3

Again we will begin with the 10-observation artificial dataset we used at the beginning of *Median regression* above. We will estimate the .6667 quantile:

```
. use http://www.stata-press.com/data/r10/twogrp

. qreg y x, quant(0.6667)
Iteration  1:  WLS sum of weighted deviations =   152.32472

Iteration  1: sum of abs. weighted deviations =   138.0054
Iteration  2: sum of abs. weighted deviations =   136.6714

.6667 Quantile regression                        Number of obs =        10
  Raw sum of deviations 159.3334 (about 20)
  Min sum of deviations 136.6714                  Pseudo R2     =    0.1422
```

y	Coef.	Std. Err.	t	P>|t|	[95% Conf. Interval]	
x	18	54.21918	0.33	0.748	-107.0297	143.0297
_cons	4	38.33875	0.10	0.919	-84.40932	92.40932

The 0.6667 quantile in the data is 20. The estimated values are 4 for x==0 and 22 for x==1. These values are appropriate since the usual convention is to "count in" $(n + 1) \times$ quantile observations.

◁

▷ Example 4

Returning to real data, the equation for the 25th percentile of `price` based on `weight`, `length`, and `foreign` in our automobile data is

```
. use http://www.stata-press.com/data/r10/auto
(1978 Automobile Data)

. qreg price weight length foreign, quant(.25)
Iteration  1:  WLS sum of weighted deviations =   98938.466

Iteration  1: sum of abs. weighted deviations =   99457.766
Iteration  2: sum of abs. weighted deviations =   91339.779
 (output omitted )
Iteration 10: sum of abs. weighted deviations =   69603.554

.25 Quantile regression                          Number of obs =        74
  Raw sum of deviations   83825.5 (about 4187)
  Min sum of deviations 69603.55                  Pseudo R2     =    0.1697
```

price	Coef.	Std. Err.	t	P>|t|	[95% Conf. Interval]	
weight	1.831789	.668093	2.74	0.008	.4993194	3.164258
length	2.845558	24.78057	0.11	0.909	-46.57773	52.26885
foreign	2209.925	434.631	5.08	0.000	1343.081	3076.769
_cons	-1879.775	2808.067	-0.67	0.505	-7480.287	3720.737

Compared with our previous median regression, the coefficient on `length` now has a positive sign, and the coefficients on `foreign` and `weight` are reduced. The actual lower quantile is $4,187, substantially less than the median $4,934. It appears that the factors are weaker in this part of the distribution.

We can also estimate the upper quartile as a function of the same three variables:

```
. qreg price weight length foreign, quant(.75)
Iteration  1:  WLS sum of weighted deviations =  110931.48

Iteration  1: sum of abs. weighted deviations =  111305.91
Iteration  2: sum of abs. weighted deviations =  105989.57
 (output omitted )
Iteration  7: sum of abs. weighted deviations =  98395.935

.75 Quantile regression                          Number of obs =        74
   Raw sum of deviations 159721.5 (about 6342)
   Min sum of deviations 98395.94                 Pseudo R2     =    0.3840
```

price	Coef.	Std. Err.	t	P>|t|	[95% Conf. Interval]	
weight	9.22291	2.653579	3.48	0.001	3.930515	14.51531
length	-220.7833	80.13907	-2.76	0.007	-380.6156	-60.95096
foreign	3595.133	1727.704	2.08	0.041	149.3355	7040.931
_cons	20242.9	8534.529	2.37	0.020	3221.323	37264.49

This result tells a different story: `weight` is much more important, and `length` is now significant—with a negative coefficient! The prices of high-priced cars seem to be determined by factors different from those affecting the prices of low-priced cars.

◁

❑ Technical Note

One explanation for having substantially different regression functions for different quantiles is that the data are heteroskedastic, as we will demonstrate below. The following statements create a sharply heteroskedastic set of data:

```
. drop _all
. set obs 10000
obs was 0, now 10000
. set seed 50550
. gen x = .1 + .9 * uniform()
. gen y = x * uniform()^2
```

Let us now fit the regressions for the 5th and 95th quantiles:

```
. qreg y x, quant(.05)
Iteration  1:  WLS sum of weighted deviations =  1080.7273

Iteration  1: sum of abs. weighted deviations =  1078.3192
Iteration  2: sum of abs. weighted deviations =  282.73545
 (output omitted )
Iteration  9: sum of abs. weighted deviations =  182.25244
```

```
.05 Quantile regression                          Number of obs =      10000
   Raw sum of deviations  182.357 (about .0009234)
   Min sum of deviations 182.2524                 Pseudo R2     =     0.0006
```

y	Coef.	Std. Err.	t	P>|t|	[95% Conf. Interval]
x	.002601	.0002737	9.50	0.000	.0020646	.0031374
_cons	-.0001393	.0001666	-0.84	0.403	-.000466	.0001874

```
. qreg y x, quant(.95)
Iteration  1:  WLS sum of weighted deviations =  1237.5569

Iteration  1: sum of abs. weighted deviations =  1238.0014
Iteration  2: sum of abs. weighted deviations =  456.65044
 (output omitted )
Iteration  5: sum of abs. weighted deviations =   338.4389
```

```
.95 Quantile regression                          Number of obs =      10000
   Raw sum of deviations 554.6889 (about .61326343)
   Min sum of deviations 338.4389                 Pseudo R2     =     0.3899
```

y	Coef.	Std. Err.	t	P>|t|	[95% Conf. Interval]
x	.8898259	.006085	146.23	0.000	.8778981	.9017537
_cons	.0021514	.0036897	0.58	0.560	-.0050812	.0093839

The coefficient on x, in particular, differs markedly between the two estimates. For the mathematically inclined, it is not too difficult to show that the theoretical lines are $y = .0025\,x$ for the 5th percentile and $y = .9025\,x$ for the 95th, numbers in close agreement with our numerical results.

❏

Estimated standard errors

qreg estimates the variance–covariance matrix of the coefficients by using a method of Koenker and Bassett (1982) and Rogers (1993). This approach is described in *Methods and Formulas* below. Rogers (1992) reports that, although this method seems adequate for homoskedastic errors, it appears to understate the standard errors for heteroskedastic errors. The irony is that exploring heteroskedastic errors is one of the major benefits of quantile regression. Gould (1992, 1997) introduced generalized versions of qreg that obtain estimates of the standard errors by using bootstrap resampling (see Efron and Tibshirani 1993 or Wu 1986 for an introduction to bootstrapped standard errors). The iqreg, sqreg, and bsqreg commands provide a bootstrapped estimate of the entire variance–covariance matrix of the estimators.

▷ Example 5

The first example of qreg on real data above was a median regression of price on weight, length, and foreign using the automobile data. Here is the result of repeating the estimation using bootstrapped standard errors:

```
. set seed 1001

. use http://www.stata-press.com/data/r10/auto, clear
(1978 Automobile Data)

. bsqreg price weight length foreign
(fitting base model)
(bootstrapping ...................)
Median regression, bootstrap(20) SEs             Number of obs =       74
   Raw sum of deviations   142205 (about 4934)
   Min sum of deviations 108822.6                Pseudo R2      =    0.2347
```

price	Coef.	Std. Err.	t	P>|t|	[95% Conf. Interval]	
weight	3.933588	3.12446	1.26	0.212	-2.297951	10.16513
length	-41.25191	83.71266	-0.49	0.624	-208.2116	125.7077
foreign	3377.771	1057.281	3.19	0.002	1269.09	5486.452
_cons	344.6494	7053.301	0.05	0.961	-13722.72	14412.01

The coefficient estimates are the same—indeed, they are obtained using the same technique. Only the standard errors differ. Therefore, the t statistics, significance levels, and confidence intervals also differ.

Since bsqreg (as well as sqreg and iqreg) obtains standard errors by randomly resampling the data, the standard errors it produces will not be the same from run to run, unless we first set the random-number seed to the same number; see [D] **generate**.

By default, bsqreg, sqreg, and iqreg use 20 replications. We can control the number of replications by specifying the reps() option:

```
. set seed 1001

. bsqreg price weight length foreign, reps(1000)
(fitting base model)
(bootstrapping ..................(output omitted )...)
Median regression, bootstrap(1000) SEs           Number of obs =       74
   Raw sum of deviations   142205 (about 4934)
   Min sum of deviations 108822.6                Pseudo R2      =    0.2347
```

price	Coef.	Std. Err.	t	P>|t|	[95% Conf. Interval]	
weight	3.933588	2.670394	1.47	0.145	-1.392346	9.259521
length	-41.25191	69.64598	-0.59	0.556	-180.1564	97.65261
foreign	3377.771	1095.481	3.08	0.003	1192.904	5562.638
_cons	344.6494	5944.884	0.06	0.954	-11512.05	12201.35

A comparison of the standard errors is informative:

variable	qreg	bsqreg reps(20)	bsqreg reps(1000)
weight	.8602	3.124	2.670
length	28.87	83.71	69.65
foreign	577.3	1057.	1095.
_cons	3260.	7053.	5945.

The results shown above are typical for models with heteroskedastic errors. (Our dependent variable is price; if our model had been in terms of ln(price), the standard errors estimated by qreg and bsqreg would have been nearly identical.) Also, even for heteroskedastic errors, 20 replications is generally sufficient for hypothesis tests against 0.

◁

Interquantile and simultaneous-quantile regression

Consider a quantile-regression model where the qth quantile is given by

$$Q_q(y) = a_q + b_{q,1}x_1 + b_{q,2}x_2$$

For instance, the 75th and 25th quantiles are given by

$$Q_{.75}(y) = a_{.75} + b_{.75,1}x_1 + b_{.75,2}x_2$$
$$Q_{.25}(y) = a_{.25} + b_{.25,1}x_1 + b_{.25,2}x_2$$

The difference in the quantiles is then

$$Q_{.75}(y) - Q_{.25}(y) = (a_{.75} - a_{.25}) + (b_{.75,1} - b_{.25,1})x_1 + (b_{.75,2} - b_{.25,2})x_2$$

qreg fits models such as $Q_{.75}(y)$ and $Q_{.25}(y)$. iqreg fits interquantile models, such as $Q_{.75}(y) - Q_{.25}(y)$. The relationships of the coefficients estimated by qreg and iqreg are exactly as shown: iqreg reports coefficients that are the difference in coefficients of two qreg models, and, of course, iqreg reports the appropriate standard errors, which it obtains by bootstrapping.

sqreg is like qreg in that it estimates the equations for the quantiles

$$Q_{.75}(y) = a_{.75} + b_{.75,1}x_1 + b_{.75,2}x_2$$
$$Q_{.25}(y) = a_{.25} + b_{.25,1}x_1 + b_{.25,2}x_2$$

The coefficients it obtains are the same that would be obtained by estimating each equation separately using qreg. sqreg differs from qreg in that it estimates the equations simultaneously and obtains an estimate of the entire variance–covariance matrix of the estimators by bootstrapping. Thus you can perform hypothesis tests concerning coefficients both within and across equations.

For example, to fit the above model, you could type

```
. qreg y x1 x2, q(.25)
. qreg y x1 x2, q(.75)
```

Doing this, you would obtain estimates of the parameters, but you could not test whether $b_{.25,1} = b_{.75,1}$ or, equivalently, $b_{.75,1} - b_{.25,1} = 0$. If your interest really is in the difference of coefficients, you could type

```
. iqreg y x1 x2, q(.25 .75)
```

The "coefficients" reported would be the difference in quantile coefficients. You could also estimate both quantiles simultaneously and then test the equality of the coefficients:

```
. sqreg y x1 x2, q(.25 .75)
. test [q25]x1 = [q75]x1
```

Whether you use iqreg or sqreg makes no difference for this test. sqreg, however, because it estimates the quantiles simultaneously, allows you to test other hypotheses. iqreg, by focusing on quantile differences, presents results in a way that is easier to read.

Finally, sqreg can estimate quantiles singly,

```
. sqreg y x1 x2, q(.5)
```

and can thereby be used as a substitute for the slower bsqreg. (Gould [1997] presents timings demonstrating that sqreg is faster than bsqreg.) sqreg can also estimate more than two quantiles simultaneously:

```
. sqreg y x1 x2, q(.25 .5 .75)
```

▷ Example 6

In demonstrating qreg, we performed quantile regressions using the automobile data. We discovered that the regression of price on weight, length, and foreign produced vastly different coefficients for the .25, .5, and .75 quantile regressions. Here are the coefficients that we obtained:

Variable	25th percentile	50th percentile	75th percentile
weight	1.83	3.93	9.22
length	2.85	−41.25	−220.8
foreign	2209.9	3377.8	3595.1
_cons	−1879.8	344.6	20242.9

All we can say, having estimated these equations separately, is that price seems to depend differently on the weight, length, and foreign variables depending on the portion of the price distribution we examine. We cannot be more precise because the estimates have been made separately. With sqreg, however, we can estimate all the effects simultaneously:

```
. set seed 1001

. sqreg price weight length foreign, q(.25 .5 .75) reps(100)
(fitting base model)
(bootstrapping ................  (output omitted ) .....)
```

Simultaneous quantile regression					Number of obs =	74
bootstrap(100) SEs					.25 Pseudo R2 =	0.1697
					.50 Pseudo R2 =	0.2347
					.75 Pseudo R2 =	0.3840

price	Coef.	Bootstrap Std. Err.	t	P>|t|	[95% Conf. Interval]	
q25						
weight	1.831789	1.574777	1.16	0.249	−1.309005	4.972583
length	2.845558	38.63523	0.07	0.941	−74.20999	79.9011
foreign	2209.925	1008.521	2.19	0.032	198.494	4221.357
_cons	−1879.775	3665.184	−0.51	0.610	−9189.753	5430.204
q50						
weight	3.933588	2.529541	1.56	0.124	−1.111423	8.978599
length	−41.25191	68.62258	−0.60	0.550	−178.1153	95.6115
foreign	3377.771	1017.422	3.32	0.001	1348.586	5406.956
_cons	344.6494	6199.257	0.06	0.956	−12019.38	12708.68
q75						
weight	9.22291	2.483676	3.71	0.000	4.269374	14.17645
length	−220.7833	86.17422	−2.56	0.013	−392.6524	−48.91422
foreign	3595.133	1147.216	3.13	0.003	1307.083	5883.184
_cons	20242.9	9414.242	2.15	0.035	1466.79	39019.02

The coefficient estimates above are the same as those previously estimated, although the standard error estimates are a little different. sqreg obtains estimates of variance by bootstrapping. Rogers (1992) provides evidence that, for quantile regression, the bootstrapped standard errors are better than those calculated analytically by Stata.

The important thing here, however, is that the full covariance matrix of the estimators has been estimated and stored, and thus it is now possible to perform hypothesis tests. Are the effects of weight the same at the 25th and 75th percentiles?

```
. test [q25]weight = [q75]weight

 ( 1)  [q25]weight - [q75]weight = 0

       F(  1,    70) =     8.97
             Prob > F =    0.0038
```

It appears that they are not. We can obtain a confidence interval for the difference by using `lincom`:

```
. lincom [q75]weight-[q25]weight

 ( 1)  - [q25]weight + [q75]weight = 0
```

price	Coef.	Std. Err.	t	P>|t|	[95% Conf. Interval]
(1)	7.391121	2.467548	3.00	0.004	2.469752 12.31249

Indeed, we could test whether the `weight` and `length` sets of coefficients are equal at the three quantiles estimated:

```
. quietly test [q25]weight = [q50]weight

. quietly test [q25]weight = [q75]weight, accum

. quietly test [q25]length = [q50]length, accum

. test [q25]length = [q75]length, accum

 ( 1)  [q25]weight - [q50]weight = 0
 ( 2)  [q25]weight - [q75]weight = 0
 ( 3)  [q25]length - [q50]length = 0
 ( 4)  [q25]length - [q75]length = 0

       F(  4,    70) =     2.43
             Prob > F =    0.0553
```

`iqreg` focuses on one quantile comparison but presents results that are more easily interpreted:

```
. set seed 1001

. iqreg price weight length foreign, q(.25 .75) reps(100) nodots
.75-.25 Interquantile regression                    Number of obs =        74
  bootstrap(100) SEs                                 .75 Pseudo R2 =    0.3840
                                                     .25 Pseudo R2 =    0.1697
```

price	Coef.	Bootstrap Std. Err.	t	P>|t|	[95% Conf. Interval]
weight	7.391121	2.467548	3.00	0.004	2.469752 12.31249
length	-223.6288	83.09868	-2.69	0.009	-389.3639 -57.89376
foreign	1385.208	1193.557	1.16	0.250	-995.2672 3765.683
_cons	22122.68	9009.159	2.46	0.017	4154.478 40090.88

Looking only at the .25 and .75 quantiles (the interquartile range), the `iqreg` command output is easily interpreted. Increases in `weight` correspond significantly to increases in `price` dispersion. Increases in `length` correspond to decreases in `price` dispersion. The `foreign` variable does not significantly change `price` dispersion.

Do not make too much of these results; the purpose of this example is simply to illustrate the `sqreg` and `iqreg` commands and to do so in a context that suggests why analyzing dispersion might be of interest.

lincom after sqreg produced the same t statistic for the interquartile range of weight, as did the iqreg command above. In general, they will not agree exactly because of the randomness of bootstrapping, unless the random-number seed is set to the same value before estimation (as was done here).

◁

Gould (1997) presents simulation results showing that the coverage—the actual percentage of confidence intervals containing the true value—for iqreg is appropriate.

Saved Results

qreg saves the following in e():

Scalars

e(N)	number of observations	e(sum_adev)	sum of absolute deviations
e(df_m)	model degrees of freedom	e(sum_rdev)	sum of raw deviations
e(df_r)	residual degrees of freedom	e(f_r)	residual density estimate
e(q)	quantile requested	e(convcode)	0 if converged; otherwise,
e(q_v)	value of the quantile		return code for why nonconvergence

Macros

e(cmd)	qreg	e(predict)	program used to implement predict
e(cmdline)	command as typed	e(properties)	b V
e(depvar)	name of dependent variable		

Matrices

e(b)	coefficient vector	e(V)	variance–covariance matrix of the estimators

Functions

e(sample)	marks estimation sample

iqreg saves the following in e():

Scalars

e(N)	number of observations	e(sumrdev0)	lower quantile sum of raw deviations
e(df_r)	residual degrees of freedom	e(sumrdev1)	upper quantile sum of raw deviations
e(reps)	number of replications	e(sumadev0)	lower quantile sum of absolute deviations
e(q0)	lower quantile requested	e(sumadev1)	upper quantile sum of absolute deviations
e(q1)	upper quantile requested	e(convcode)	0 if converged; otherwise, return code for why nonconvergence

Macros

e(cmd)	iqreg	e(vcetype)	title used to label Std. Err.
e(cmdline)	command as typed	e(properties)	b V
e(depvar)	name of dependent variable	e(predict)	program used to implement predict

Matrices

e(b)	coefficient vector	e(V)	variance–covariance matrix of the estimators

Functions

e(sample)	marks estimation sample

sqreg saves the following in e():

Scalars

e(N)	number of observations	e(reps)	number of replications
e(n_q)	number of quantiles requested	e(sumrdv#)	sum of raw deviations for q#
e(q#)	the quantiles requested	e(sumadv#)	sum of absolute deviations for q#
e(df_r)	residual degrees of freedom	e(convcode)	0 if converged; otherwise, return code for why nonconvergence

Macros

e(cmd)	sqreg	e(vcetype)	title used to label Std. Err.
e(cmdline)	command as typed	e(properties)	b V
e(depvar)	name of dependent variable	e(predict)	program used to implement predict
e(eqnames)	names of equations		

Matrices

e(b)	coefficient vector	e(V)	variance–covariance matrix of the estimators

Functions

e(sample)	marks estimation sample

bsqreg saves the following in e():

Scalars

e(N)	number of observations	e(q_v)	value of the quantile
e(df_r)	residual degrees of freedom	e(sum_adev)	sum of absolute deviations
e(reps)	number of replications	e(sum_rdev)	sum of raw deviations
e(q)	quantile requested	e(convcode)	0 if converged; otherwise, return code for why nonconvergence

Macros

e(cmd)	bsqreg	e(properties)	b V
e(cmdline)	command as typed	e(predict)	program used to implement predict
e(depvar)	name of dependent variable		

Matrices

e(b)	coefficient vector	e(V)	variance–covariance matrix of the estimators

Functions

e(sample)	marks estimation sample

_qreg saves the following in r():

Scalars

r(N)	number of observations	r(ic)	number of iterations
r(df_m)	model degrees of freedom	r(f_r)	residual density estimate
r(sum_w)	sum of the weights	r(q)	quantile requested
r(sum_adev)	sum of absolute deviations	r(q_v)	value of the quantile
r(sum_rdev)	sum of raw deviations	r(convcode)	1 if converged; 0 otherwise

Methods and Formulas

qreg, iqreg, sqreg, and bsqreg are implemented as ado-files.

According to Stuart and Ord (1991, 1084), the method of minimum absolute deviations was first proposed by Boscovich in 1757 and was later developed by Laplace; Stigler (1986, 39–55) and Hald (1998, 97–103, 112–116) provide historical details. According to Bloomfield and Steiger (1980), Harris (1950) later observed that the problem of minimum absolute deviations could be turned into the linear programming problem that was first implemented by Wagner (1959). Interest has grown in this method because of interest in robust methods. Statistical and computational properties of minimum absolute deviation estimators are surveyed by Narula and Wellington (1982).

Define q as the quantile to be estimated; the median is $q = .5$. For each observation i, let r_i be the residual

$$r_i = y_i - \sum_j \beta_j x_{ij}$$

Define the multiplier h_i

$$h_i = \begin{cases} 2q & \text{if } r_i > 0 \\ 2(1-q) & \text{otherwise} \end{cases}$$

The quantity being minimized with respect to β_j is $\sum_i |r_i| h_i$, so quantiles other than the median are estimated by weighting the residuals. For example, if we want to estimate the 75th percentile, we weight the negative residuals by 0.50 and the positive residuals by 1.50. It can be shown that the criterion is minimized when 75% of the residuals are negative.

This is set up as a linear programming problem and is solved via linear programming techniques, as suggested by Armstrong, Frome, and Kung (1979) and used by courtesy of Marcel Dekker, Inc. The definition of convergence is exact in the sense that no amount of added iterations could improve the solution. Each step is described by a set of observations through which the regression plane passes, called the *basis*. A step is taken by replacing a point in the basis if the sum of weighted absolute deviations can be improved. If this occurs, a line is printed in the iteration log. The linear programming method is started by doing a weighted least-squares (WLS) regression to identify a good set of observations to use as a starting basis. The WLS algorithm for $q = 0.5$ is taken from Schlossmacher (1973) with a generalization for $0 < q < 1$ implied from Hunter and Lange (2000).

The variances are estimated using a method suggested by Koenker and Bassett (1982). This method can be put into a form recommended by Huber (1967) for M estimates, where

$$\text{cov}(\boldsymbol{\beta}) = \mathbf{R}_2^{-1} \mathbf{R}_1 \mathbf{R}_2^{-1}$$

$\mathbf{R}_1 = \mathbf{X}'\mathbf{W}\mathbf{W}'\mathbf{X}$ (in the Huber formulation), $\mathbf{W}$ is a diagonal matrix with elements

$$W_{ii} = \begin{cases} q/f_{\text{residuals}}(0) & \text{if } r > 0 \\ (1-q)/f_{\text{residuals}}(0) & \text{if } r < 0 \\ 0 & \text{otherwise} \end{cases}$$

and $\mathbf{R}_2$ is the design matrix $\mathbf{X}'\mathbf{X}$. This is derived from formula 3.11 in Koenker and Bassett, although their notation is much different. $f_{\text{residuals}}()$ refers to the density of the true residuals. Koenker and Bassett leave much unspecified, including how to obtain a density estimate for the errors in real data. At this point, we offer our contribution (Rogers 1993).

We first sort the residuals and locate the observation in the residuals corresponding to the quantile in question, taking into account weights if they are applied. We then calculate w_n, the square root of the sum of the weights. Unweighted data are equivalent to weighted data in which each observation has weight 1, resulting in $w_n = \sqrt{n}$. For analytically weighted data, the weights are rescaled so that the sum of the weights is the number of observations, resulting in $\sqrt{n}$ again. For frequency-weighted data, w_n literally is the square root of the sum of the weights.

We locate the closest observation in each direction, such that the sum of weights for all closer observations is w_n. If we run off the end of the dataset, we stop. We calculate w_s, the sum of weights for all observations in this middle space. Typically, w_s is slightly greater than w_n.

If there are k parameters, then exactly k of the residuals must be zero. Thus we calculate an adjusted weight $w_a = w_s - k$. The density estimate is the distance spanned by these observations divided by w_a. Because the distance spanned by this mechanism converges toward zero, this estimate of density converges in probability to the true density.

The pseudo-R^2 is calculated as

$$1 - \frac{\text{sum of weighted deviations about estimated quantile}}{\text{sum of weighted deviations about raw quantile}}$$

This is based on the likelihood for a double-exponential distribution $e^{h_i|r_i|}$.

References

Armstrong, R. D., E. L. Frome, and D. S. Kung. 1979. Algorithm 79-01: A revised simplex algorithm for the absolute deviation curve fitting problem. In *Communications in Statistics, Simulation and Computation* B8(2), 175–190. New York: Dekker.

Bloomfield, P., and W. Steiger. 1980. Least absolute deviations curve-fitting. *SIAM Journal on Scientific and Statistical Computing* 1: 290–301.

Efron, B., and R. J. Tibshirani. 1993. *An Introduction to the Bootstrap.* New York: Chapman & Hall/CRC.

Gould, W. W. 1992. sg11.1: Quantile regression with bootstrapped standard errors. *Stata Technical Bulletin* 9: 19–21. Reprinted in *Stata Technical Bulletin Reprints*, vol. 2, pp. 137–139.

——.1997. sg70: Interquantile and simultaneous-quantile regression. *Stata Technical Bulletin* 38: 14–22. Reprinted in *Stata Technical Bulletin Reprints*, vol. 7, pp. 167–176.

Gould, W. W., and W. H. Rogers. 1994. Quantile regression as an alternative to robust regression. *1994 Proceedings of the Statistical Computing Section.* Alexandria, VA: American Statistical Association.

Hald, A. 1998. *A History of Mathematical Statistics from 1750 to 1930.* New York: Wiley.

Harris, T. 1950. Regression using minimum absolute deviations. *American Statistician* 4: 14–15.

Huber, P. J. 1967. The behavior of maximum likelihood estimates under nonstandard conditions. In *Proceedings of the Fifth Symposium on Mathematical Statistics and Probability*, vol. 1, 221–233. Berkeley: University of California Press.

——.1981. *Robust Statistics.* New York: Wiley.

Hunter, D. R., and K. Lange. 2000. Quantile regression via an MM algorithm. *Journal of Computational and Graphical Statistics* 9: 60–77.

Jolliffe, D., B. Krushelnytskyy, and A. Semykina. 2000. sg153: Censored least absolute deviations estimator: CLAD. *Stata Technical Bulletin* 58: 13–16. Reprinted in *Stata Technical Bulletin Reprints*, vol. 10, pp. 240–244.

Koenker, R., and G. Bassett, Jr. 1982. Robust tests for heteroscedasticity based on regression quantiles. *Econometrica* 50: 43–61.

Koenker, R., and K. Hallock. 2001. Quantile regression. *Journal of Economic Perspectives* 15: 143–156.

Narula, S. C., and J. F. Wellington. 1982. The minimum sum of absolute errors regression: A state of the art survey. *International Statistical Review* 50: 317–326.

Rogers, W. H. 1992. sg11: Quantile regression standard errors. *Stata Technical Bulletin* 9: 16–19. Reprinted in *Stata Technical Bulletin Reprints*, vol. 2, pp. 133–137.

——.1993. sg11.2: Calculation of quantile regression standard errors. *Stata Technical Bulletin* 13: 18–19. Reprinted in *Stata Technical Bulletin Reprints*, vol. 3, pp. 77–78.

Rousseeuw, P. J., and A. M. Leroy. 1987. *Robust Regression and Outlier Detection.* New York: Wiley.

Schlossmacher, E. J. 1973. An iterative technique for absolute deviations curve fitting. *Journal of the American Statistical Association* 344: 857–859.

Stigler, S. M. 1986. *The History of Statistics: The Measurement of Uncertainty before 1900.* Cambridge, MA: Belknap Press.

Stuart, A., and J. K. Ord. 1991. *Kendall's Advanced Theory of Statistics, Vol. 2.* 5th ed. New York: Oxford University Press.

Wagner, H. M. 1959. Linear programming techniques for regression analysis. *Journal of the American Statistical Association* 54: 206–212.

Wu, C. F. J. 1986. Jackknife, bootstrap and other resampling methods in regression analysis. *Annals of Statistics* 14: 1261–1350 (including comments and reply).

Also See

[R] **qreg postestimation** — Postestimation tools for qreg, iqreg, sqreg, and bsqreg

[R] **bootstrap** — Bootstrap sampling and estimation

[R] **regress** — Linear regression

[R] **rreg** — Robust regression

[U] **20 Estimation and postestimation commands**

Title

> **qreg postestimation** — Postestimation tools for qreg, iqreg, sqreg, and bsqreg

Description

The following postestimation commands are available for `qreg`, `iqreg`, `bsqreg`, and `sqreg`:

command	description
adjust	adjusted predictions of $\mathbf{x}\beta$
estat	VCE and estimation sample summary
estimates	cataloging estimation results
lincom	point estimates, standard errors, testing, and inference for linear combinations of coefficients
linktest	link test for model specification
mfx	marginal effects or elasticities
nlcom	point estimates, standard errors, testing, and inference for nonlinear combinations of coefficients
predict	predictions, residuals, influence statistics, and other diagnostic measures
predictnl	point estimates, standard errors, testing, and inference for generalized predictions
test	Wald tests for simple and composite linear hypotheses
testnl	Wald tests of nonlinear hypotheses

See the corresponding entries in the *Stata Base Reference Manual* for details.

Syntax for predict

For qreg, iqreg, *and* bsqreg

> predict $\big[$ *type* $\big]$ *newvar* $\big[$ *if* $\big]$ $\big[$ *in* $\big]$ $\big[$, $\big[$ xb | stdp | <u>r</u>esiduals $\big]\big]$

For sqreg

> predict $\big[$ *type* $\big]$ *newvar* $\big[$ *if* $\big]$ $\big[$ *in* $\big]$ $\big[$, <u>eq</u>uation(*eqno* $\big[$,*eqno* $\big]$) *statistic* $\big]$

statistic	description
Main	
xb	linear prediction; the default
stdp	standard error of the linear prediction
stddp	standard error of the difference in linear predictions
<u>r</u>esiduals	residuals

These statistics are available both in and out of sample; type `predict ... if e(sample) ...` if wanted only for the estimation sample.

Options for predict

⌐ Main ⌐

xb, the default, calculates the linear prediction.

stdp calculates the standard error of the linear prediction.

stddp is allowed only after you have fitted a model using sqreg. The standard error of the difference in linear predictions $(\mathbf{x}_{1j}\mathbf{b} - \mathbf{x}_{2j}\mathbf{b})$ between equations 1 and 2 is calculated.

residuals calculates the residuals, that is, $y_j - \mathbf{x}_j\mathbf{b}$.

equation(*eqno* [,*eqno*]) specifies the equation to which you are making the calculation.

> equation() is filled in with one *eqno* for options xb, stdp, and residuals. equation(#1) would mean that the calculation is to be made for the first equation, equation(#2) would mean the second, and so on. You could also refer to the equations by their names. equation(income) would refer to the equation named income and equation(hours) to the equation named hours.

> If you do not specify equation(), results are the same as if you had specified equation(#1).

> To use stddp, you must specify two equations. You might specify equation(#1, #2) or equation(q80, q20) to indicate the 80th and 20th quantiles.

Methods and Formulas

All postestimation commands listed above are implemented as ado-files.

Also See

[R] **qreg** — Quantile regression

[U] **20 Estimation and postestimation commands**

Title

query — Display system parameters

Syntax

query [memory | output | interface | graphics | efficiency | network |

update | trace | mata | other]

Description

query displays the settings of various Stata parameters.

Remarks

query provides more system information than you will ever want to know. You do not need to understand every line of output that query produces if all you need is one piece of information. Here is what happens when you type query:

```
. query
```

```
      Memory settings (also see query memory for a more complete report)
          set memory        10M
          set maxvar        5000
          set matsize       400

      Output settings
          set more          on
          set rmsg          off
          set dp            period      may be period or comma
          set linesize      79          characters
          set pagesize      23          lines

          set logtype       smcl        may be smcl or text

      Interface settings
          set dockable      on
          set dockingguides on
          set floatresults  on
          set floatwindows  on
          set locksplitters off
          set pinnable      on
          set persistfv     off
          set persistvtopic off
          set doublebuffer  on

          set linegap       1           pixels
          set scrollbufsize 32000       characters
          set varlabelpos   32          column number
          set reventries    5000        lines

          set maxdb         50          dialog boxes
```

```
        Graphics settings
             set graphics       on
             set autotabgraphs  off
             set scheme         s2color
             set printcolor     automatic  may be automatic, asis, gs1, gs2, gs3
             set copycolor      automatic  may be automatic, asis, gs1, gs2, gs3
        ─────────────────────────────────────────────────────────────────────────
        Efficiency settings
             set adosize        500        kilobytes
             set virtual        off
        ─────────────────────────────────────────────────────────────────────────
        Network settings
             set checksum       off
             set timeout1       120        seconds
             set timeout2       300        seconds
             ────────────────────────────────────────────────────
             set httpproxy      off
             set httpproxyhost
             set httpproxyport  80
             ────────────────────────────────────────────────────
             set httpproxyauth  off
             set httpproxyuser
             set httpproxypw
        ─────────────────────────────────────────────────────────────────────────
        Update settings
             set update_query   on
             set update_interval 7
             set update_prompt  on
        ─────────────────────────────────────────────────────────────────────────
        Trace (programming debugging) settings
             set trace          off
             set tracedepth     32000
             set traceexpand    on
             set tracesep       on
             set traceindent    on
             set tracenumber    off
             set tracehilite
        ─────────────────────────────────────────────────────────────────────────
        Mata settings
             set matastrict     off
             set matalnum       off
             set mataoptimize   on
             set matafavor      space      may be space or speed
             set matacache      400        kilobytes
             set matalibs       lmatabase;lmataopt;lmataado
             set matamofirst    off
        ─────────────────────────────────────────────────────────────────────────
        Other settings
             set type           float      may be float or double
             set level          95         percent confidence intervals
             set maxiter        16000      max iterations for estimation commands
             set searchdefault  local      may be local, net, or all
             set seed           X075bcd151f123bb5159a55e50022865746ad
             set varabbrev      on
             set processors     1
        ─────────────────────────────────────────────────────────────────────────
```

The output is broken into several divisions: memory, output, interface, graphics, efficiency, network, update, trace, mata, and other settings, and we will discuss each one in turn.

We generated the output above using Stata/SE for Windows. Here is what happens when we type query and we are running Stata/IC for Linux(GUI):

```
. query
```

```
Memory settings
    set memory          1M
    set maxvar          2048        (not settable in this version of Stata)
    set matsize         200
```

```
Output settings
    set more            on
    set rmsg            off
    set dp              period      may be period or comma
    set linesize        83          characters
    set pagesize        24          lines

    set logtype         smcl        may be smcl or text
```

```
Interface settings

    set linegap         1           pixels
    set scrollbufsize   32000       characters
    set fastscroll      on
    set varlabelpos     32          column number

    set maxdb           50          dialog boxes
```

```
Graphics settings
    set graphics        on
    set scheme          s2color
    set printcolor      automatic   may be automatic, asis, gs1, gs2, gs3
```

```
Efficiency settings
    set adosize         500         kilobytes
    set virtual         off
```

```
Network settings
    set checksum        off
    set timeout1        120         seconds
    set timeout2        300         seconds

    set httpproxy       off
    set httpproxyhost
    set httpproxyport   8080

    set httpproxyauth   off
    set httpproxyuser
    set httpproxypw
```

```
Update settings
    set update_query                (not relevant)
    set update_interval             (not relevant)
    set update_prompt               (not relevant)
```

```
Trace (programming debugging) settings
    set trace           off
    set tracedepth      32000
    set traceexpand     on
    set tracesep        on
    set traceindent     on
    set tracenumber     off
    set tracehilite
```

```
Mata settings
    set matastrict     off
    set matalnum       off
    set mataoptimize   on
    set matafavor      space         may be space or speed
    set matacache      400           kilobytes
    set matalibs       lmatabase;lmataopt;lmataado
    set matamofirst    off

Other settings
    set type           float         may be float or double
    set level          95            percent confidence intervals
    set maxiter        16000         max iterations for estimation commands
    set searchdefault  local         may be local, net, or all
    set seed           X075bcd151f123bb5159a55e50022865746ad
    set varabbrev      on
    set processors     1
    set odbcmgr        off
```

Memory settings

Memory settings indicate how much memory is allocated to Stata, the maximum number of variables, and the maximum size of a matrix.

For more information, see

```
memory     [D] memory
maxvar     [D] memory
matsize    [R] matsize
```

Output settings

Output settings show how Stata displays output on the screen and in log files.

For more information, see

```
     more   [R] more
     rmsg   [P] rmsg
       dp   [D] format
 linesize   [R] log
 pagesize   [R] more
  logtype   [R] log
  eolchar   [R] set
```

Interface settings

Interface settings control how Stata's interface works.

For more information, see

dockable	[R] set
dockingguides	[R] set
floatresults	[R] set
floatwindows	[R] set
locksplitters	[R] set
pinnable	[R] set
persistfv	[R] set
persistvtopic	[R] set
doublebuffer	[R] set
smoothfonts	[R] set
smoothsize	[R] set
linegap	[R] set
revwindow	[R] set
varwindow	[R] set
scrollbufsize	[R] set
fastscroll	[R] set
varlabelpos	[R] set
reventries	[R] set
maxdb	[R] **db**

Graphics settings

Graphics settings indicate how Stata's graphics are displayed.

For more information, see

graphics	[G] **set graphics**
autotabgraphs	[R] set
scheme	[G] **set scheme**
printcolor	[G] **set printcolor**
copycolor	[G] **set printcolor**
macgphengine	[R] set
piccomments	[R] set

Efficiency settings

The efficiency settings set the maximum amount of memory allocated to automatically loaded do-files, the maximum number of remembered-contents dialog boxes, and the use of virtual memory.

For more information, see

adosize	[P] **sysdir**
virtual	[D] **memory**

Network settings

Network settings determine how Stata interacts with the Internet.

For more information, see [R] **netio**.

Update settings

Update settings determine how Stata performs updates.

For more information, see [R] **update**.

Trace settings

Trace settings adjust Stata's behavior and are particularly useful in debugging code.

For more information, see [P] **trace**.

Mata settings

Mata settings affect Mata's system parameters.

For more information, see [M-3] **mata set**.

Other settings

The other settings are a miscellaneous collection.

For more information, see

type	[D]	**generate**
level	[R]	**level**
maxiter	[R]	**maximize**
searchdefault	[R]	**search**
seed	[D]	**generate**
varabbrev	[R]	**set**
processors	[R]	**set**
odbcmgr	[D]	**odbc**

In general, the parameters displayed by query can be changed by set; see [R] **set**.

Also See

[R] **set** — Overview of system parameters

[P] **creturn** — Return c-class values

[M-3] **mata set** — Set and display Mata system parameters

Title

ranksum — Equality tests on unmatched data

Syntax

Wilcoxon rank-sum test

> **ranksum** *varname* [*if*] [*in*] , **by**(*groupvar*) [**porder**]

Nonparametric equality-of-medians test

> **median** *varname* [*if*] [*in*] [*weight*] , **by**(*groupvar*) [*median_options*]

ranksum options	description
Main	
* **by**(*groupvar*)	grouping variable
porder	probability that variable for first group is larger than variable for second group

median_options	description
Main	
* **by**(*groupvar*)	grouping variable
exact	perform Fisher's exact test
medianties(**below**)	assign values equal to the median to below group
medianties(**above**)	assign values equal to the median to above group
medianties(**drop**)	drop values equal to the median from the analysis
medianties(**split**)	split values equal to the median equally between the two groups

* **by**(*groupvar*) is required.

by is allowed with ranksum and median; see [D] **by**.

fweights are allowed with median; see [U] **11.1.6 weight**.

Description

ranksum tests the hypothesis that two independent samples (i.e., *unmatched* data) are from populations with the same distribution using the Wilcoxon rank-sum test, which is also known as the Mann–Whitney two-sample statistic (Wilcoxon 1945; Mann and Whitney 1947).

median performs a nonparametric k-sample test on the equality of medians. It tests the null hypothesis that the k samples were drawn from populations with the same median. For two samples, the chi-squared test statistic is computed both with and without a continuity correction.

ranksum and median are for use with *unmatched* data. For equality tests on matched data, see [R] **signrank**.

Options for ranksum

⌐ Main ⌐

by(*groupvar*) is required. It specifies the name of the grouping variable.

porder displays an estimate of the probability that a random draw from the first population is larger than a random draw from the second population.

Options for median

⌐ Main ⌐

by(*groupvar*) is required. It specifies the name of the grouping variable.

exact displays the significance calculated by Fisher's exact test. For two samples, both one- and two-sided probabilities are displayed.

medianties(below | above | drop | split) specifies how values equal to the overall median are to be handled. The median test computes the median for *varname* by using all observations and then divides the observations into those falling above the median and those falling below the median. When values for an observation are equal to the sample median, they can be dropped from the analysis by specifying medianties(drop); added to the group above or below the median by specifying medianties(above) or medianties(below), respectively; or if there is more than 1 observation with values equal to the median, they can be equally divided into the two groups by specifying medianties(split). If this option is not specified, medianties(below) is assumed.

Remarks

▷ Example 1

We are testing the effectiveness of a new fuel additive. We run an experiment with 24 cars: 12 cars with the fuel treatment and 12 cars without. We input these data by creating a dataset with 24 observations. mpg records the mileage rating, and treat records 0 if the mileage corresponds to untreated fuel and 1 if it corresponds to treated fuel.

```
. use http://www.stata-press.com/data/r10/fuel2
. ranksum mpg, by(treat)
Two-sample Wilcoxon rank-sum (Mann-Whitney) test
       treat |     obs    rank sum    expected
-------------+----------------------------------
           0 |      12         128         150
           1 |      12         172         150
-------------+----------------------------------
    combined |      24         300         300

unadjusted variance      300.00
adjustment for ties       -4.04
                       ----------
adjusted variance        295.96

Ho: mpg(treat==0) = mpg(treat==1)
           z =  -1.279
   Prob > |z| =   0.2010
```

These results indicate that the medians are not statistically different at any level smaller than 20.1%. Similarly, the median test

```
. median mpg, by(treat) exact

Median test
```

Greater than the median	treat 0	1	Total
no	7	5	12
yes	5	7	12
Total	12	12	24

```
              Pearson chi2(1) =   0.6667   Pr = 0.414
              Fisher's exact =              0.684
      1-sided Fisher's exact =              0.342

Continuity corrected:
              Pearson chi2(1) =   0.1667   Pr = 0.683
```

fails to reject the null hypothesis that there is no difference between the two fuel additives.

Compare these results from these two tests with those obtained from the `signrank` and `signtest` where we found significant differences; see [R] **signrank**. An experiment run on 24 different cars is not as powerful as a before-and-after comparison using the same 12 cars.

◁

Saved Results

`ranksum` saves the following in `r()`:

Scalars

r(N_1)	sample size n_1	r(group1)	value of variable for first group
r(N_2)	sample size n_2	r(sum_obs)	actual sum of ranks for first group
r(z)	z statistic	r(sum_exp)	expected sum of ranks for first group
r(Var_a)	adjusted variance		

`median` saves the following in `r()`:

Scalars

r(N)	sample size	r(groups)	number of groups compared
r(chi2)	Pearson's χ^2	r(chi2_cc)	continuity-corrected Pearson's χ^2
r(p)	significance of Pearson's χ^2	r(p_cc)	continuity-corrected significance
r(p_exact)	Fisher's exact p	r(p1_exact)	one-sided Fisher's exact p

Methods and Formulas

`ranksum` and `median` are implemented as ado-files.

For a practical introduction to these techniques with an emphasis on examples rather than theory, see Acock (2006), Bland (2000), or Sprent and Smeeton (2007). For a summary of these tests, see Snedecor and Cochran (1989).

ranksum

For the Wilcoxon rank-sum test, there are two independent random variables X_1 and X_2, and we test the null hypothesis that $X_1 \sim X_2$. We have a sample of size n_1 from X_1 and another of size n_2 from X_2.

The data are then ranked without regard to the sample to which they belong. If the data are tied, averaged ranks are used. Wilcoxon's test statistic (1945) is the sum of the ranks for the observations in the first sample:

$$T = \sum_{i=1}^{n_1} R_{1i}$$

Mann and Whitney's U statistic (1947) is the number of pairs (X_{1i}, X_{2j}) such that $X_{1i} > X_{2j}$. These statistics differ only by a constant:

$$U = T - \frac{n_1(n_1 + 1)}{2}$$

Again Fisher's Principle of Randomization provides a method for calculating the distribution of the test statistic, ties or not. The randomization distribution consists of the $\binom{n}{n_1}$ ways to choose n_1 ranks from the set of all $n = n_1 + n_2$ ranks and assign them to the first sample.

It is a straightforward exercise to verify that

$$E(T) = \frac{n_1(n + 1)}{2} \qquad \text{and} \qquad \text{Var}(T) = \frac{n_1 n_2 s^2}{n}$$

where s is the standard deviation of the combined ranks r_i for both groups:

$$s^2 = \frac{1}{n - 1} \sum_{i=1}^{n} (r_i - \bar{r})^2$$

This formula for the variance is exact and holds both when there are no ties and when there are ties and we use averaged ranks. (Indeed, the variance formula holds for the randomization distribution of choosing n_1 numbers from any set of n numbers.)

Using a normal approximation, we calculate

$$z = \frac{T - E(T)}{\sqrt{\text{Var}(T)}}$$

When the option `porder` is specified, the probability

$$p = \frac{U}{n_1 n_2}$$

is computed.

median

The median test examines whether it is likely that two or more samples came from populations with the same median. The null hypothesis is that the samples were drawn from populations with the same median. The alternative hypothesis is that at least one sample was drawn from a population with a different median. The test should be used only with ordinal or interval data.

Assume that there are score values for k independent samples to be compared. The median test is performed by first computing the median score for all observations combined, regardless of the sample group. Each score is compared with this computed grand median and is classified as being above the grand median, below the grand median, or equal to the grand median. Observations with scores equal to the grand median can be dropped, added to the "above" group, added to the "below" group, or split between the two groups.

Once all observations are classified, the data are cast into a $2 \times k$ contingency table, and a Pearson's chi-squared test or Fisher's exact test is performed.

Henry Berthold Mann (1905–2000) was born in Vienna, where he completed a doctorate in algebraic number theory. He moved to the United States in 1938 and for several years earned a living by tutoring in New York. During this time he proved a celebrated conjecture in number theory and studied statistics at Columbia with Abraham Wald, with whom he wrote three papers. After the war he taught at Ohio State and the Universities of Wisconsin and Arizona. In addition to his work in number theory and statistics, he made major contributions to algebra and combinatorics.

Donald Ransom Whitney (1915–) studied at Princeton and Ohio State Universities and worked at the latter throughout his career. His Ph.D. thesis under Henry Mann was on nonparametric statistics and produced the test that bears his name.

References

Acock, A. C. 2006. *A Gentle Introduction to Stata*. College Station, TX: Stata Press.

Bland, M. 2000. *An Introduction to Medical Statistics*. 3rd ed. Oxford: Oxford University Press.

Feiveson, A. H. 2002. Power by simulation. *Stata Journal* 2: 107–124.

Fisher, R. A. 1935. *Design of Experiments*. Edinburgh: Oliver & Boyd.

Goldstein, R. 1997. sg69: Immediate Mann–Whitney and binomial effect-size display. *Stata Technical Bulletin* 36: 29–31. Reprinted in *Stata Technical Bulletin Reprints*, vol. 6, pp. 187–189.

Kruskal, W. H. 1957. Historical notes on the Wilcoxon unpaired two-sample test. *Journal of the American Statistical Association* 52: 356–360.

Mann, H. B., and D. R. Whitney. 1947. On a test of whether one of two random variables is stochastically larger than the other. *Annals of Mathematical Statistics* 18: 50–60.

Newson, R. 2000a. snp15: somersd—Confidence intervals for nonparametric statistics and their differences. *Stata Technical Bulletin* 55: 47–55. Reprinted in *Stata Technical Bulletin Reprints*, vol. 10, pp. 312–322.

——. 2000b. snp15.1: Update to somersd. *Stata Technical Bulletin* 57: 35. Reprinted in *Stata Technical Bulletin Reprints*, vol. 10, pp. 322–323.

——. 2000c. snp15.2: Update to somersd. *Stata Technical Bulletin* 58: 30. Reprinted in *Stata Technical Bulletin Reprints*, vol. 10, p. 323.

——. 2001. snp15.3: Update to somersd. *Stata Technical Bulletin* 61: 22. Reprinted in *Stata Technical Bulletin Reprints*, vol. 10, p. 324.

——. 2003. snp15_4: Software updates for somersd. *Stata Journal* 3: 325.

——. 2005. snp15_5: Software updates for somersd. *Stata Journal* 5: 470.

——. 2006. snp15_6: Confidence intervals for rank statistics: Somers' D and extensions. *Stata Journal* 6: 309–334.

Perkins, A. M. 1998. snp14: A two-sample multivariate nonparametric test. *Stata Technical Bulletin* 42: 47–49. Reprinted in *Stata Technical Bulletin Reprints*, vol. 7, pp. 243–245.

Snedecor, G. W., and W. G. Cochran. 1989. *Statistical Methods*. 8th ed. Ames, IA: Iowa State University Press.

Sprent, P., and N. C. Smeeton. 2007. *Applied Nonparametric Statistical Methods.* 4th ed. Boca Raton, FL: Chapman & Hall/CRC.

Sribney, W. M. 1995. crc40: Correcting for ties and zeros in sign and rank tests. *Stata Technical Bulletin* 26: 2–4. Reprinted in *Stata Technical Bulletin Reprints*, vol. 5, pp. 5–8.

Wilcoxon, F. 1945. Individual comparisons by ranking methods. *Biometrics* 1: 80–83.

Also See

[R] **signrank** — Equality tests on matched data

[R] **ttest** — Mean-comparison tests

Title

ratio — Estimate ratios

Syntax

Basic syntax

> ratio $[$ *name*: $]$ *varname* $[$ / $]$ *varname*

Full syntax

> ratio ($[$ *name*: $]$ *varname* $[$ / $]$ *varname*)
>
> > $[$ ($[$ *name*: $]$ *varname* $[$ / $]$ *varname*) ... $]$ $[$ *if* $]$ $[$ *in* $]$ $[$ *weight* $]$ $[$, *options* $]$

options	description
Model	
<u>std</u>ize(*varname*)	variable identifying strata for standardization
stdweight(*varname*)	weight variable for standardization
<u>nostdr</u>escale	do not rescale the standard weight variable
if/in/over	
over(*varlist*$[$, <u>nolab</u>el $]$)	group over subpopulations defined by *varlist*; optionally, suppress group labels
SE/Cluster	
vce(*vcetype*)	*vcetype* may be <u>linear</u>ized, <u>c</u>luster *clustvar*, <u>boot</u>strap, or jackknife
Reporting	
<u>level</u>(#)	set confidence level; default is level(95)
<u>nohe</u>ader	suppress table header
<u>nole</u>gend	suppress table legend

bootstrap, jackknife, rolling, statsby, and svy are allowed; see [U] **11.1.10 Prefix commands**.

Weights are not allowed with the bootstrap prefix.

vce() and weights are not allowed with the svy prefix.

fweights, iweights, and pweights are allowed; see [U] **11.1.6 weight**.

See [U] **20 Estimation and postestimation commands** for more capabilities of estimation commands.

Description

ratio produces estimates of ratios, along with standard errors.

Options

Model

stdize(*varname*) specifies that the point estimates be adjusted by direct standardization across the strata identified by *varname*. This option requires the stdweight() option.

stdweight(*varname*) specifies the weight variable associated with the standard strata identified in the stdize() option. The standardization weights must be constant within the standard strata.

nostdrescale prevents the standardization weights from being rescaled within the over() groups. This option requires stdize() but is ignored if the over() option is not specified.

if/in/over

over(*varlist* [, nolabel]) specifies that estimates be computed for multiple subpopulations, which are identified by the different values of the variables in *varlist*.

When this option is supplied with one variable name, such as over(*varname*), the value labels of *varname* are used to identify the subpopulations. If *varname* does not have labeled values (or there are unlabeled values), the values themselves are used, provided that they are nonnegative integers. Noninteger values, negative values, and labels that are not valid Stata names are substituted with a default identifier.

When over() is supplied with multiple variable names, each subpopulation is assigned a unique default identifier.

nolabel requests that value labels attached to the variables identifying the subpopulations be ignored.

SE/Cluster

vce(*vcetype*) specifies the type of standard error reported, which includes types that are derived from asymptotic theory, that allow for intragroup correlation, and that use bootstrap or jackknife methods; see [R] *vce_option*.

vce(linearized), the default, uses the linearized or sandwich estimator of variance.

Reporting

level(*#*); see [R] **estimation options**.

noheader prevents the table header from being displayed. This option implies nolegend.

nolegend prevents the table legend identifying the subpopulations from being displayed.

Remarks

▷ Example 1

Using the fuel data from [R] **ttest**, we estimate the ratio of mileage for the cars without the fuel treatment (mpg1) to those with the fuel treatment (mpg2).

```
. use http://www.stata-press.com/data/r10/fuel
. ratio myratio: mpg1/mpg2
Ratio estimation                    Number of obs   =      12
      myratio: mpg1/mpg2
```

		Linearized		
	Ratio	Std. Err.	[95% Conf. Interval]	
myratio	.9230769	.032493	.8515603	.9945936

Using these results, we can test to see if this ratio is significantly different from one.

```
. test _b[myratio] = 1
 ( 1)  myratio = 1
       F(  1,    11) =     5.60
            Prob > F =    0.0373
```

We find that the ratio is different from one at the 5% significance level but not at the 1% significance level.

◁

▷ Example 2

Using state-level census data, we want to test whether the marriage rate is equal to the death rate.

```
. use http://www.stata-press.com/data/r10/census2
(1980 Census data by state)
. ratio (deathrate: death/pop) (marrate: marriage/pop)
Ratio estimation                    Number of obs   =      50
    deathrate: death/pop
      marrate: marriage/pop
```

		Linearized		
	Ratio	Std. Err.	[95% Conf. Interval]	
deathrate	.0087368	.0002052	.0083244	.0091492
marrate	.0105577	.0006184	.009315	.0118005

```
. test _b[deathrate]  =  _b[marrate]
 ( 1)  deathrate - marrate = 0
       F(  1,    49) =     6.93
            Prob > F =    0.0113
```

◁

(Continued on next page)

Saved Results

ratio saves the following in e():

Scalars

e(N)	number of observations	e(N_clust)	number of clusters
e(N_over)	number of subpopulations	e(k_eq)	number of equations in e(b)
e(N_stdize)	number of standard strata	e(df_r)	sample degrees of freedom

Macros

e(cmd)	ratio	e(over)	*varlist* from over()
e(cmdline)	command as typed	e(over_labels)	labels from over() variables
e(varlist)	*varlist*	e(over_namelist)	names from e(over_labels)
e(stdize)	*varname* from stdize()	e(vce)	*vcetype* specified in vce()
e(stdweight)	*varname* from stdweight()	e(vcetype)	title used to label Std. Err.
e(wtype)	weight type	e(estat_cmd)	program used to implement estat
e(wexp)	weight expression	e(namelist)	ratio identifiers
e(title)	title in estimation output	e(properties)	b V
e(cluster)	name of cluster variable		

Matrices

e(b)	vector of mean estimates
e(V)	(co)variance estimates
e(_N)	vector of numbers of nonmissing observations
e(_N_stdsum)	number of nonmissing observations within the standard strata
e(_p_stdize)	standardizing proportions
e(error)	error code corresponding to e(b)

Functions

e(sample)	marks estimation sample

Methods and Formulas

ratio is implemented as an ado-file.

The ratio estimator

Let $R = Y/X$ be the ratio to be estimated, where Y and X are totals; see [R] **total**. The estimate for R is $\widehat{R} = \widehat{Y}/\widehat{X}$ (the ratio of the sample totals). From the delta method (i.e., a first-order Taylor expansion), the approximate variance of the sampling distribution of the linearized $\widehat{R}$ is

$$V(\widehat{R}) \approx \frac{1}{X^2}\left\{ V(\widehat{Y}) - 2R\text{Cov}(\widehat{Y}, \widehat{X}) + R^2 V(\widehat{X}) \right\}$$

Direct substitution of $\widehat{X}$, $\widehat{R}$, and the estimated variances and covariance of $\widehat{X}$ and $\widehat{Y}$ leads to the following variance estimator:

$$\widehat{V}(\widehat{R}) = \frac{1}{\widehat{X}^2}\left\{ \widehat{V}(\widehat{Y}) - 2\widehat{R}\widehat{\text{Cov}}(\widehat{Y}, \widehat{X}) + \widehat{R}^2\widehat{V}(\widehat{X}) \right\} \tag{1}$$

Survey data

See [SVY] **variance estimation**, [SVY] **direct standardization**, and [SVY] **poststratification** for discussions that provide background information for the following formulas.

The survey ratio estimator

Let Y_j and X_j be survey items for the jth individual in the population, where $j = 1, \ldots, M$ and M is the size of the population. The associated population ratio for the items of interest is $R = Y/X$ where

$$Y = \sum_{j=1}^{M} Y_j \quad \text{and} \quad X = \sum_{j=1}^{M} X_j$$

Let y_j and x_j be the corresponding survey items for the jth sampled individual from the population, where $j = 1, \ldots, m$ and m is the number of observations in the sample.

The estimator $\widehat{R}$ for the population ratio R is $\widehat{R} = \widehat{Y}/\widehat{X}$, where

$$\widehat{Y} = \sum_{j=1}^{m} w_j y_j \quad \text{and} \quad \widehat{X} = \sum_{j=1}^{m} w_j x_j$$

and w_j is a sampling weight. The score variable for the ratio estimator is

$$z_j(\widehat{R}) = \frac{y_j - \widehat{R} x_j}{\widehat{X}} = \frac{\widehat{X} y_j - \widehat{Y} x_j}{\widehat{X}^2}$$

The standardized ratio estimator

Let D_g denote the set of sampled observations that belong to the gth standard stratum and define $I_{D_g}(j)$ to indicate if the jth observation is a member of the gth standard stratum; where $g = 1, \ldots, L_D$ and L_D is the number of standard strata. Also let π_g denote the fraction of the population that belongs to the gth standard stratum, thus $\pi_1 + \cdots + \pi_{L_D} = 1$. Note that π_g is derived from the `stdweight()` option.

The estimator for the standardized ratio is

$$\widehat{R}^D = \sum_{g=1}^{L_D} \pi_g \frac{\widehat{Y}_g}{\widehat{X}_g}$$

where

$$\widehat{Y}_g = \sum_{j=1}^{m} I_{D_g}(j) \, w_j y_j$$

and $\widehat{X}_g$ is similarly defined. The score variable for the standardized ratio is

$$z_j(\widehat{R}^D) = \sum_{g=1}^{L_D} \pi_g I_{D_g}(j) \frac{\widehat{X}_g y_j - \widehat{Y}_g x_j}{\widehat{X}_g^2}$$

The poststratified ratio estimator

Let P_k denote the set of sampled observations that belong to poststratum k and define $I_{P_k}(j)$ to indicate if the jth observation is a member of poststratum k; where $k = 1, \ldots, L_P$ and L_P is the number of poststrata. Also let M_k denote the population size for poststratum k. P_k and M_k are identified by specifying the `poststrata()` and `postweight()` options on `svyset`.

The estimator for the poststratified ratio is

$$\widehat{R}^P = \frac{\widehat{Y}^P}{\widehat{X}^P}$$

where

$$\widehat{Y}^P = \sum_{k=1}^{L_P} \frac{M_k}{\widehat{M}_k} \widehat{Y}_k = \sum_{k=1}^{L_P} \frac{M_k}{\widehat{M}_k} \sum_{j=1}^{m} I_{P_k}(j)\, w_j y_j$$

and $\widehat{X}^P$ is similarly defined. The score variable for the poststratified ratio is

$$z_j(\widehat{R}^P) = \frac{z_j(\widehat{Y}^P) - \widehat{R}^P z_j(\widehat{X}^P)}{\widehat{X}^P} = \frac{\widehat{X}^P z_j(\widehat{Y}^P) - \widehat{Y}^P z_j(\widehat{X}^P)}{(\widehat{X}^P)^2}$$

where

$$z_j(\widehat{Y}^P) = \sum_{k=1}^{L_P} I_{P_k}(j) \frac{M_k}{\widehat{M}_k} \left(y_j - \frac{\widehat{Y}_k}{\widehat{M}_k} \right)$$

and $z_j(\widehat{X}^P)$ is similarly defined.

The standardized poststratified ratio estimator

The estimator for the standardized poststratified ratio is

$$\widehat{R}^{DP} = \sum_{g=1}^{L_D} \pi_g \frac{\widehat{Y}_g^P}{\widehat{X}_g^P}$$

where

$$\widehat{Y}_g^P = \sum_{k=1}^{L_p} \frac{M_k}{\widehat{M}_k} \widehat{Y}_{g,k} = \sum_{k=1}^{L_p} \frac{M_k}{\widehat{M}_k} \sum_{j=1}^{m} I_{D_g}(j) I_{P_k}(j)\, w_j y_j$$

and $\widehat{X}_g^P$ is similarly defined. The score variable for the standardized poststratified ratio is

$$z_j(\widehat{R}^{DP}) = \sum_{g=1}^{L_D} \pi_g \frac{\widehat{X}_g^P z_j(\widehat{Y}_g^P) - \widehat{Y}_g^P z_j(\widehat{X}_g^P)}{(\widehat{X}_g^P)^2}$$

where

$$z_j(\widehat{Y}_g^P) = \sum_{k=1}^{L_P} I_{P_k}(j) \frac{M_k}{\widehat{M}_k} \left\{ I_{D_g}(j) y_j - \frac{\widehat{Y}_{g,k}}{\widehat{M}_k} \right\}$$

and $z_j(\widehat{X}_g^P)$ is similarly defined.

Subpopulation estimation

Let S denote the set of sampled observations that belong to the subpopulation of interest, and define $I_S(j)$ to indicate if the jth observation falls within the subpopulation.

The estimator for the subpopulation ratio is $\widehat{R}^S = \widehat{Y}^S/\widehat{X}^S$, where

$$\widehat{Y}^S = \sum_{j=1}^{m} I_S(j)\, w_j y_j \qquad \text{and} \qquad \widehat{X}^S = \sum_{j=1}^{m} I_S(j)\, w_j x_j$$

Its score variable is

$$z_j(\widehat{R}^S) = I_S(j)\frac{y_j - \widehat{R}^S x_j}{\widehat{X}^S} = I_S(j)\frac{\widehat{X}^S y_j - \widehat{Y}^S x_j}{(\widehat{X}^S)^2}$$

The estimator for the standardized subpopulation ratio is

$$\widehat{R}^{DS} = \sum_{g=1}^{L_D} \pi_g \frac{\widehat{Y}_g^S}{\widehat{X}_g^S}$$

where

$$\widehat{Y}_g^S = \sum_{j=1}^{m} I_{D_g}(j) I_S(j)\, w_j y_j$$

and $\widehat{X}_g^S$ is similarly defined. Its score variable is

$$z_j(\widehat{R}^{DS}) = \sum_{g=1}^{L_D} \pi_g I_{D_g}(j) I_S(j)\frac{\widehat{X}_g^S y_j - \widehat{Y}_g^S x_j}{(\widehat{X}_g^S)^2}$$

The estimator for the poststratified subpopulation ratio is

$$\widehat{R}^{PS} = \frac{\widehat{Y}^{PS}}{\widehat{X}^{PS}}$$

where

$$\widehat{Y}^{PS} = \sum_{k=1}^{L_P} \frac{M_k}{\widehat{M}_k} \widehat{Y}_k^S = \sum_{k=1}^{L_P} \frac{M_k}{\widehat{M}_k} \sum_{j=1}^{m} I_{P_k}(j) I_S(j)\, w_j y_j$$

and $\widehat{X}^{PS}$ is similarly defined. Its score variable is

$$z_j(\widehat{R}^{PS}) = \frac{\widehat{X}^{PS} z_j(\widehat{Y}^{PS}) - \widehat{Y}^{PS} z_j(\widehat{X}^{PS})}{(\widehat{X}^{PS})^2}$$

where

$$z_j(\widehat{Y}^{PS}) = \sum_{k=1}^{L_P} I_{P_k}(j) \frac{M_k}{\widehat{M}_k} \left\{ I_S(j)\, y_j - \frac{\widehat{Y}_k^S}{\widehat{M}_k} \right\}$$

and $z_j(\widehat{X}^{PS})$ is similarly defined.

The estimator for the standardized poststratified subpopulation ratio is

$$\widehat{R}^{DPS} = \sum_{g=1}^{L_D} \pi_g \frac{\widehat{Y}_g^{PS}}{\widehat{X}_g^{PS}}$$

where

$$\widehat{Y}_g^{PS} = \sum_{k=1}^{L_p} \frac{M_k}{\widehat{M}_k} \widehat{Y}_{g,k}^S = \sum_{k=1}^{L_p} \frac{M_k}{\widehat{M}_k} \sum_{j=1}^{m} I_{D_g}(j) I_{P_k}(j) I_S(j)\, w_j y_j$$

and $\widehat{X}_g^{PS}$ is similarly defined. Its score variable is

$$z_j(\widehat{R}^{DPS}) = \sum_{g=1}^{L_D} \pi_g \frac{\widehat{X}_g^{PS} z_j(\widehat{Y}_g^{PS}) - \widehat{Y}_g^{PS} z_j(\widehat{X}_g^{PS})}{(\widehat{X}_g^{PS})^2}$$

where

$$z_j(\widehat{Y}_g^{PS}) = \sum_{k=1}^{L_P} I_{P_k}(j) \frac{M_k}{\widehat{M}_k} \left\{ I_{D_g}(j) I_S(j)\, y_j - \frac{\widehat{Y}_{g,k}^S}{\widehat{M}_k} \right\}$$

and $z_j(\widehat{X}_g^{PS})$ is similarly defined.

References

Cochran, W. G. 1977. *Sampling Techniques.* 3rd ed. New York: Wiley.

Stuart, A., and J. K. Ord. 1994. *Kendall's Advanced Theory of Statistics: Distribution Theory, Vol. I.* 6th ed. London: Arnold.

Also See

Title

> **ratio postestimation** — Postestimation tools for ratio

Description

The following postestimation commands are available for `ratio`:

command	description
estat	VCE
estat (svy)	postestimation statistics for survey data
estimates	cataloging estimation results
lincom	point estimates, standard errors, testing, and inference for linear combinations of coefficients
nlcom	point estimates, standard errors, testing, and inference for nonlinear combinations of coefficients
test	Wald tests for simple and composite linear hypotheses
testnl	Wald tests of nonlinear hypotheses

See the corresponding entries in the *Stata Base Reference Manual* for details, but see [SVY] **estat** for details about `estat` (svy).

Remarks

For examples of the use of `test` after `ratio`, see [R] **ratio**.

Methods and Formulas

All postestimation commands listed above are implemented as ado-files.

Also See

[R] **ratio** — Estimate ratios

[SVY] **svy postestimation** — Postestimation tools for svy

[U] **20 Estimation and postestimation commands**

Title

> **reg3** — Three-stage estimation for systems of simultaneous equations

Syntax

Basic syntax

> reg3 (*depvar₁ varlist₁*) (*depvar₂ varlist₂*) ... (*depvarₙ varlistₙ*) $\left[\,if\,\right]$ $\left[\,in\,\right]$ $\left[\,weight\,\right]$

Full syntax

> reg3 ($\left[\,eqname_1:\right]depvar_{1a}\left[\,depvar_{1b}\,...\,=\,\right]varlist_1$ $\left[\,,\,\underline{no}constant\,\right]$)
>
> ($\left[\,eqname_2:\right]depvar_{2a}\left[\,depvar_{2b}\,...\,=\,\right]varlist_2$ $\left[\,,\,\underline{no}constant\,\right]$)
>
> ...
>
> ($\left[\,eqname_N:\right]depvar_{Na}\left[\,depvar_{Nb}\,...\,=\,\right]varlist_N$ $\left[\,,\,\underline{no}constant\,\right]$)
>
> $\left[\,if\,\right]$ $\left[\,in\,\right]$ $\left[\,weight\,\right]$ $\left[\,,\,options\,\right]$

options	description
Model	
<u>i</u>reg3	iterate until estimates converge
<u>c</u>onstraints(*constraints*)	apply specified linear constraints
Model 2	
<u>ex</u>og(*varlist*)	exogenous variables not specified in system equations
<u>en</u>dog(*varlist*)	additional right-hand-side (RHS) endogenous variables
<u>i</u>nst(*varlist*)	full list of exogenous variables
allexog	all RHS variables are exogenous
<u>no</u>constant	suppress constant from instrument list
Est. method	
3sls	three-stage least squares; the default
2sls	two-stage least squares
<u>o</u>ls	ordinary least squares (OLS)
<u>su</u>re	seemingly unrelated regression
<u>m</u>vreg	sure with OLS degrees-of-freedom adjustment
corr(*correlation*)	<u>un</u>structured or <u>in</u>dependent correlation structure; default is unstructured
df adj.	
<u>sm</u>all	report small-sample statistics
dfk	use small-sample adjustment
dfk2	use alternate adjustment

Reporting

level(#)	set confidence level; default is level(95)
first	report first-stage regression

Opt options

optimization_options	control the optimization process; seldom used
† noheader	suppress display of header
† notable	suppress display of coefficient table
† nofooter	suppress display of footer

† noheader, notable, and nofooter do not appear in the dialog box.

depvar and *varlist* may contain time-series operators; see [U] **11.4.3 Time-series varlists**.

bootstrap, by, jackknife, rolling, statsby, and xi are allowed; see [U] **11.1.10 Prefix commands**.

Weights are not allowed with the bootstrap prefix.

aweights are not allowed with the jackknife prefix.

aweights and fweights are allowed; see [U] **11.1.6 weight**.

See [U] **20 Estimation and postestimation commands** for more capabilities of estimation commands.

Explicit equation naming (*eqname*:) cannot be combined with multiple dependent variables in an equation specification.

Description

reg3 estimates a system of structural equations, where some equations contain endogenous variables among the explanatory variables. Estimation is via three-stage least squares (3SLS); see Zellner and Theil (1962). Typically, the endogenous explanatory variables are dependent variables from other equations in the system. reg3 supports iterated GLS estimation and linear constraints.

reg3 can also estimate systems of equations by seemingly unrelated regression (SURE), multivariate regression (MVREG), and equation-by-equation ordinary least squares (OLS) or two-stage least squares (2SLS).

Nomenclature

Under 3SLS or 2SLS estimation, a *structural equation* is defined as one of the equations specified in the system. A *dependent variable* will have its usual interpretation as the left-hand-side variable in an equation with an associated disturbance term. All dependent variables are explicitly taken to be *endogenous* to the system and are treated as correlated with the disturbances in the system's equations. Unless specified in an endog() option, all other variables in the system are treated as *exogenous* to the system and uncorrelated with the disturbances. The exogenous variables are taken to be *instruments* for the endogenous variables.

Options

    Model

ireg3 causes reg3 to iterate over the estimated disturbance covariance matrix and parameter estimates until the parameter estimates converge. Although the iteration is usually successful, there is no guarantee that it will converge to a stable point. Under seemingly unrelated regression, this iteration converges to the maximum likelihood estimates.

constraints(*constraints*); see [R] **estimation options**.

exog(*varlist*) specifies additional exogenous variables that are included in none of the system equations. This can occur when the system contains identities that are not estimated. If implicitly exogenous variables from the equations are listed here, reg3 will just ignore the additional information. Specified variables will be added to the exogenous variables in the system and used in the first stage as instruments for the endogenous variables. By specifying dependent variables from the structural equations, you can use exog() to override their endogeneity.

endog(*varlist*) identifies variables in the system that are not dependent variables but are endogenous to the system. These variables must appear in the variable list of at least one equation in the system. Again the need for this identification often occurs when the system contains identities. For example, a variable that is the sum of an exogenous variable and a dependent variable may appear as an explanatory variable in some equations.

inst(*varlist*) specifies a full list of all exogenous variables and may not be used with the endog() or exog() options. It must contain a full list of variables to be used as instruments for the endogenous regressors. Like exog(), the list may contain variables not specified in the system of equations. This option can be used to achieve the same results as the endog() and exog() options, and the choice is a matter of convenience. Any variable not specified in the *varlist* of the inst() option is assumed to be endogenous to the system. As with exog(), including the dependent variables from the structural equations will override their endogeneity.

allexog indicates that all right-hand-side variables are to be treated as exogenous—even if they appear as the dependent variable of another equation in the system. This option can be used to enforce a seemingly unrelated regression or multivariate regression estimation even when some dependent variables appear as regressors.

noconstant; see [R] **estimation options**.

3sls specifies the full three-stage least-squares estimation of the system and is the default for reg3.

2sls causes reg3 to perform equation-by-equation two-stage least squares on the full system of equations. This option implies dfk, small, and corr(independent).

Cross-equation testing should not be performed after estimation with this option. With 2sls, no covariance is estimated between the parameters of the equations. For cross-equation testing, use full 3sls.

ols causes reg3 to perform equation-by-equation OLS on the system—even if dependent variables appear as regressors or the regressors differ for each equation; see [R] **mvreg**. ols implies allexog, dfk, small, and corr(independent); nodfk and nosmall may be specified to override dfk and small.

The covariance of the coefficients between equations is not estimated under this option, and cross-equation tests should not be performed after estimation with ols. For cross-equation testing, use sureg or 3sls (the default).

sure causes reg3 to perform a seemingly unrelated regression estimation of the system—even if dependent variables from some equations appear as regressors in other equations; see [R] **sureg**. sure is a synonym for allexog.

mvreg is identical to sure, except that the disturbance covariance matrix is estimated with an OLS degrees-of-freedom adjustment—the dfk option. If the regressors are identical for all equations, the parameter point estimates will be the standard multivariate regression results. If any of the regressors differ, the point estimates are those for seemingly unrelated regression with an OLS degrees-of-freedom adjustment in computing the covariance matrix. nodfk and nosmall may be specified to override dfk and small.

corr(*correlation*) specifies the assumed form of the correlation structure of the equation disturbances and is rarely requested explicitly. For the family of models fitted by reg3, the only two allowable correlation structures are <u>un</u>structured and <u>in</u>dependent. The default is unstructured.

This option is used almost exclusively to estimate a system of equations by two-stage least squares or to perform OLS regression with reg3 on multiple equations. In these cases, the correlation is set to independent, forcing reg3 to treat the covariance matrix of equation disturbances as diagonal in estimating model parameters. Thus a set of two-stage coefficient estimates can be obtained if the system contains endogenous right-hand-side variables, or OLS regression can be imposed, even if the regressors differ across equations. Without imposing independent disturbances, reg3 would estimate the former by three-stage least squares and the latter by seemingly unrelated regression.

Any tests performed after estimation with the independent option will treat coefficients in different equations as having no covariance; cross-equation tests should not be used after specifying corr(independent).

    ⌐ df adj. ⌐

small specifies that small-sample statistics be computed. It shifts the test statistics from χ^2 and Z statistics to F statistics and t statistics. This option is intended primarily to support multivariate regression. Although the standard errors from each equation are computed using the degrees of freedom for the equation, the degrees of freedom for the t statistics are all taken to be those for the first equation. This approach poses no problem under multivariate regression because the regressors are the same across equations.

dfk specifies the use of an alternative divisor in computing the covariance matrix for the equation residuals. As an asymptotically justified estimator, reg3 by default uses the number of sample observations n as a divisor. When the dfk option is set, a small-sample adjustment is made, and the divisor is taken to be $\sqrt{(n - k_i)(n - k_j)}$, where k_i and k_j are the numbers of parameters in equations i and j, respectively.

dfk2 specifies the use of an alternative divisor in computing the covariance matrix for the equation errors. When the dfk2 option is set, the divisor is taken to be the mean of the residual degrees of freedom from the individual equations.

    ⌐ Reporting ⌐

level(*#*); see [R] **estimation options**.

first requests that the first-stage regression results be displayed during estimation.

    ⌐ Opt options ⌐

optimization_options control the iterative process that minimizes the sum of squared errors when ireg3 is specified. These options are seldom used.

iterate(*#*) specifies the maximum number of iterations. When the number of iterations equals *#*, the optimizer stops and presents the current results, even if the convergence tolerance has not been reached. The default value of iterate() is the current value of set maxiter, which is iterate(16000) if maxiter has not been changed.

trace adds to the iteration log a display of the current parameter vector nolog suppresses the display of the iteration log.

tolerance(#) specifies the tolerance for the coefficient vector. When the relative change in the coefficient vector from one iteration to the next is less than or equal to #, the optimization process is stopped. tolerance(1e-6) is the default.

The following options are available with reg3 but are not shown in the dialog box:

noheader suppresses display of the header reporting the estimation method and the table of equation summary statistics.

notable suppresses display of the coefficient table.

nofooter suppresses display of the footer reporting the list of endogenous variables in the model.

Remarks

reg3 estimates systems of structural equations where some equations contain endogenous variables among the explanatory variables. Generally, these endogenous variables are the dependent variables of other equations in the system, though not always. The disturbance is correlated with the endogenous variables—violating the assumptions of ordinary least squares. Further, since some of the explanatory variables are the dependent variables of other equations in the system, the error terms among the equations are expected to be correlated. reg3 uses an instrumental variable approach to produce consistent estimates and generalized least squares (GLS) to account for the correlation structure in the disturbances across the equations. Good general references on three-stage estimation include Kmenta (1997) and Greene (2003, 405–407).

Three-stage least squares can be thought of as producing estimates from a three-step process.

Stage 1. Develop instrumented values for all endogenous variables. These instrumented values can simply be considered as the predicted values resulting from a regression of each endogenous variable on all exogenous variables in the system. This stage is identical to the first step in two-stage least squares and is critical for the consistency of the parameter estimates.

Stage 2. Obtain a consistent estimate for the covariance matrix of the equation disturbances. These estimates are based on the residuals from a two-stage least-squares estimation of each structural equation.

Stage 3. Perform a GLS-type estimation using the covariance matrix estimated in the second stage and with the instrumented values in place of the right-hand-side endogenous variables.

❏ Technical Note

The estimation and use of the covariance matrix of disturbances in three-stage estimation is almost identical to the seemingly unrelated regression estimation (SURE) method—sureg. As with SURE, using this covariance matrix improves the efficiency of the three-stage estimator. Even without the covariance matrix, the estimates would be consistent. (They would be two-stage least-squares estimates.) This improvement in efficiency comes with a caveat. All the parameter estimates now depend on the consistency of the covariance matrix estimates. If one equation in the system is misspecified, the disturbance covariance estimates will be inconsistent, and the resulting coefficients will be biased and inconsistent. Alternatively, if each equation is estimated separately by two-stage least squares ([R] **regress**), only the coefficients in the misspecified equation are affected.

❏

❏ Technical Note

If an equation is just identified, the three-stage least-squares point estimates for that equation are identical to the two-stage least-squares estimates. However, as with sureg, even if all equations are just identified, fitting the model via reg3 has at least one advantage over fitting each equation separately via ivregress; by using reg3, tests involving coefficients in different equations can be performed easily using test or testnl.

❏

▷ Example 1

A simple macroeconomic model relates consumption (consump) to private and government wages paid (wagepriv and wagegovt). Simultaneously, private wages depend on consumption, total government expenditures (govt), and the lagged stock of capital in the economy (capital1). Although this is not a plausible model, it does meet the criterion of being simple. This model could be written as

$$consump = \beta_0 + \beta_1 \, wagepriv + \beta_2 \, wagegovt + \epsilon_1$$

$$wagepriv = \beta_3 + \beta_4 \, consump + \beta_5 \, govt + \beta_6 \, capital1 + \epsilon_2$$

If we assume that this is the full system, consump and wagepriv will be endogenous variables, with wagegovt, govt, and capital1 exogenous. Data for the U.S. economy on these variables are taken from Klein (1950). This model can be fitted with reg3 by typing

```
. use http://www.stata-press.com/data/r10/klein

. reg3 (consump wagepriv wagegovt) (wagepriv consump govt capital1)
Three-stage least-squares regression
```

Equation	Obs	Parms	RMSE	"R-sq"	chi2	P
consump	22	2	1.776297	0.9388	208.02	0.0000
wagepriv	22	3	2.372443	0.8542	80.04	0.0000

	Coef.	Std. Err.	z	P>|z|	[95% Conf. Interval]	
consump						
wagepriv	.8012754	.1279329	6.26	0.000	.5505314	1.052019
wagegovt	1.029531	.3048424	3.38	0.001	.432051	1.627011
_cons	19.3559	3.583772	5.40	0.000	12.33184	26.37996
wagepriv						
consump	.4026076	.2567312	1.57	0.117	-.1005764	.9057916
govt	1.177792	.5421253	2.17	0.030	.1152461	2.240338
capital1	-.0281145	.0572111	-0.49	0.623	-.1402462	.0840173
_cons	14.63026	10.26693	1.42	0.154	-5.492552	34.75306

Endogenous variables: consump wagepriv
Exogenous variables: wagegovt govt capital1

Without showing the two-stage least-squares results, we note that the consumption function in this system falls under the conditions noted earlier. That is, the two-stage and three-stage least-squares coefficients for the equation are identical.

◁

▷ Example 2

Some of the most common simultaneous systems encountered are supply-and-demand models. A simple system could be specified as

$$\texttt{qDemand} = \beta_0 + \beta_1\,\texttt{price} + \beta_2\,\texttt{pcompete} + \beta_3\,\texttt{income} + \epsilon_1$$

$$\texttt{qSupply} = \beta_4 + \beta_5\,\texttt{price} + \beta_6\,\texttt{praw} + \epsilon_2$$

$$\texttt{Equilibrium condition: quantity} = \texttt{qDemand} = \texttt{qSupply}$$

where

quantity is the quantity of a product produced and sold

price is the price of the product

pcompete is the price of a competing product

income is the average income level of consumers

praw is the price of raw materials used to produce the product

In this system, price is assumed to be determined simultaneously with demand. The important statistical implications are that price is not a predetermined variable and that it is correlated with the disturbances of both equations. The system is somewhat unusual: quantity is associated with two disturbances. This fact really poses no problem because the disturbances are specified on the behavioral demand and supply equations—two separate entities. Often one of the two equations is rewritten to place price on the left-hand side, making this endogeneity explicit in the specification.

To provide a concrete illustration of the effects of simultaneous equations, we can simulate data for the above system by using known coefficients and disturbance properties. Specifically, we will simulate the data as

$$\texttt{qDemand} = 40 - 1.0\,\texttt{price} + 0.25\,\texttt{pcompete} + 0.5\,\texttt{income} + \epsilon_1$$

$$\texttt{qSupply} = 0.5\,\texttt{price} - 0.75\,\texttt{praw} + \epsilon_2$$

where

$$\epsilon_1 \sim N(0, 2.4)$$

$$\epsilon_2 \sim N(0, 3.8)$$

For comparison, we can estimate the supply and demand equations separately by OLS. The estimates for the demand equation are

```
. use http://www.stata-press.com/data/r10/supDem
. regress quantity price pcompete income
```

Source	SS	df	MS		Number of obs =	49
					F(3, 45) =	1.00
Model	23.1579302	3	7.71931008		Prob > F =	0.4004
Residual	346.459313	45	7.69909584		R-squared =	0.0627
					Adj R-squared =	0.0002
Total	369.617243	48	7.70035923		Root MSE =	2.7747

quantity	Coef.	Std. Err.	t	P>|t|	[95% Conf. Interval]	
price	.1186265	.1716014	0.69	0.493	-.2269965	.4642496
pcompete	.0946416	.1200815	0.79	0.435	-.1472149	.3364981
income	.0785339	.1159867	0.68	0.502	-.1550754	.3121432
_cons	7.563261	5.019479	1.51	0.139	-2.54649	17.67301

The OLS estimates for the supply equation are

```
. regress quantity price praw
```

Source	SS	df	MS
Model	224.819549	2	112.409774
Residual	144.797694	46	3.14777596
Total	369.617243	48	7.70035923

Number of obs = 49
F(2, 46) = 35.71
Prob > F = 0.0000
R-squared = 0.6082
Adj R-squared = 0.5912
Root MSE = 1.7742

quantity	Coef.	Std. Err.	t	P>|t|	[95% Conf. Interval]	
price	.724675	.1095657	6.61	0.000	.5041307	.9452192
praw	-.8674796	.1066114	-8.14	0.000	-1.082077	-.652882
_cons	-6.97291	3.323105	-2.10	0.041	-13.66197	-.283847

Examining the coefficients from these regressions, we note that they are not close to the known parameters used to generate the simulated data. In particular, the positive coefficient on `price` in the demand equation stands out. We constructed our simulated data to be consistent with economic theory—people demand less of a product if its price rises and more if their personal income rises. Although the `price` coefficient is statistically insignificant, the positive value contrasts starkly with what is predicted from economic price theory and the -1.0 value that we used in the simulation. Likewise, we are disappointed with the insignificance and level of the coefficient on average `income`. The supply equation has correct signs on the two main parameters, but their levels are different from the known values. In fact, the coefficient on `price` (.724675) is different from the simulated parameter (0.5) at the 5% level of significance.

All these problems are to be expected. We explicitly constructed a simultaneous system of equations that violated one of the assumptions of least squares. Specifically, the disturbances were correlated with one of the regressors—`price`.

Two-stage least squares can be used to address the correlation between regressors and disturbances. Using instruments for the endogenous variable, `price`, two-stage least squares will produce consistent estimates of the parameters in the system. Let's use `ivregress` to see how our simulated system behaves when fitted using two-stage least squares.

```
. ivregress 2sls quantity (price = praw) pcompete income
```

Instrumental variables (2SLS) regression

Number of obs = 49
Wald chi2(3) = 8.77
Prob > chi2 = 0.0326
R-squared = .
Root MSE = 3.7333

quantity	Coef.	Std. Err.	z	P>|z|	[95% Conf. Interval]	
price	-1.015817	.374209	-2.71	0.007	-1.749253	-.282381
pcompete	.3319504	.172912	1.92	0.055	-.0069508	.6708517
income	.5090607	.1919482	2.65	0.008	.1328491	.8852723
_cons	39.89988	10.77378	3.70	0.000	18.78366	61.01611

Instrumented: price
Instruments: pcompete income praw

```
. ivregress 2sls quantity (price = pcompete income) praw
```

Instrumental variables (2SLS) regression

Number of obs =	49
Wald chi2(2) =	39.25
Prob > chi2 =	0.0000
R-squared =	0.5928
Root MSE =	1.7525

quantity	Coef.	Std. Err.	z	P>|z|	[95% Conf. Interval]	
price	.5773133	.1749974	3.30	0.001	.2343247	.9203019
praw	-.7835496	.1312414	-5.97	0.000	-1.040778	-.5263213
_cons	-2.550694	5.273067	-0.48	0.629	-12.88571	7.784327

Instrumented: price
Instruments: praw pcompete income

We are now much happier with the estimation results. All the coefficients from both equations are close to the true parameter values for the system. In particular, the coefficients are all well within 95% confidence intervals for the parameters. The missing R-squared in the demand equation seems unusual; we will discuss that more later.

Finally, this system could be estimated using three-stage least squares. To demonstrate how large systems might be handled and to avoid multiline commands, we will use global macros (see [P] **macro**) to hold the specifications for our equations.

```
. global demand "(qDemand: quantity price pcompete income)"
. global supply "(qSupply: quantity price praw)"
. reg3 $demand $supply, endog(price)
```

We must specify `price` as endogenous since it does not appear as a dependent variable in either equation. Without this option, `reg3` would assume that there are no endogenous variables in the system and produce seemingly unrelated regression (`sureg`) estimates. The `reg3` output from our series of commands is

Three-stage least-squares regression

Equation	Obs	Parms	RMSE	"R-sq"	chi2	P
qDemand	49	3	3.739686	-0.8540	8.68	0.0338
qSupply	49	2	1.752501	0.5928	39.25	0.0000

	Coef.	Std. Err.	z	P>|z|	[95% Conf. Interval]	
qDemand						
price	-1.014345	.3742036	-2.71	0.007	-1.74777	-.2809194
pcompete	.2647206	.1464194	1.81	0.071	-.0222561	.5516973
income	.5299146	.1898161	2.79	0.005	.1578819	.9019472
_cons	40.08749	10.77072	3.72	0.000	18.97726	61.19772
qSupply						
price	.5773133	.1749974	3.30	0.001	.2343247	.9203019
praw	-.7835496	.1312414	-5.97	0.000	-1.040778	-.5263213
_cons	-2.550694	5.273067	-0.48	0.629	-12.88571	7.784327

Endogenous variables: quantity price
Exogenous variables: pcompete income praw

The use of three-stage least squares over two-stage least squares is essentially an efficiency issue. The coefficients of the demand equation from three-stage least squares are close to the coefficients from two-stage least squares, and those of the supply equation are identical. The latter case was mentioned earlier for systems with some exactly identified equations. However, even for the demand equation, we do not expect the coefficients to change systematically. What we do expect from three-stage least squares are more precise estimates of the parameters given the validity of our specification and reg3's use of the covariances among the disturbances.

Let's summarize the results. With OLS, we got obviously biased estimates of the parameters. No amount of data would have improved the OLS estimates—they are inconsistent in the face of the violated OLS assumptions. With two-stage least squares, we obtained consistent estimates of the parameters, and these would have improved with more data. With three-stage least squares, we obtained consistent estimates of the parameters that are more efficient than those obtained by two-stage least squares.

◁

❑ Technical Note

We noted earlier that the R-squared was missing from the two-stage estimates of the demand equation. Now we see that the R-squared is negative for the three-stage estimates of the same equation. How can we have a negative R-squared?

In most estimators, other than least squares, the R-squared is no more than a summary measure of the overall in-sample predictive power of the estimator. The computational formula for R-squared is $R\text{-squared} = 1 - RSS/TSS$, where RSS is the residual sum of squares (sum of squared residuals) and TSS is the total sum of squared deviations about the mean of the dependent variable. In a standard linear model with a constant, the model from which the TSS is computed is nested within the full model from which RSS is computed—they both have a constant term based on the same data. Thus it must be that $TSS \geq RSS$ and R-squared is constrained between 0 and 1.

For two- and three-stage least squares, some of the regressors enter the model as instruments when the parameters are estimated. However, since our goal is to fit the structural model, the actual values, not the instruments for the endogenous right-hand-side variables, are used to determine R-squared. The model residuals are computed over a different set of regressors from those used to fit the model. The two- and/or three-stage estimates are no longer nested within a constant-only model of the dependent variable, and the residual sum of squares is no longer constrained to be smaller than the total sum of squares.

A negative R-squared in three-stage least squares should be taken for exactly what it is—an indication that the structural model predicts the dependent variable worse than a constant-only model. Is this a problem? It depends on the application. Three-stage least squares applied to our contrived supply-and-demand example produced good estimates of the known true parameters. Still, the demand equation produced an R-squared of -0.854. How do we feel about our parameter estimates? This should be determined by the estimates themselves, their associated standard errors, and the overall model significance. On this basis, negative R-squared and all, we feel pretty good about all the parameter estimates for both the supply and demand equations. Would we want to make predictions about equilibrium quantity by using the demand equation alone? Probably not. Would we want to make these quantity predictions by using the supply equation? Possibly, because based on in-sample predictions, they seem better than those from the demand equations. However, both the supply and demand estimates are based on limited information. If we are interested in predicting quantity, a reduced-form equation containing all our independent variables would usually be preferred.

❑

❑ Technical Note

As a matter of syntax, we could have specified the supply-and-demand model on one line without using global macros.

```
. reg3 (quantity price pcompete income) (quantity price praw), endog(price)
Three-stage least-squares regression
```

Equation	Obs	Parms	RMSE	"R-sq"	chi2	P
quantity	49	3	3.739686	-0.8540	8.68	0.0338
2quantity	49	2	1.752501	0.5928	39.25	0.0000

	Coef.	Std. Err.	z	P>|z|	[95% Conf. Interval]	
quantity						
price	-1.014345	.3742036	-2.71	0.007	-1.74777	-.2809194
pcompete	.2647206	.1464194	1.81	0.071	-.0222561	.5516973
income	.5299146	.1898161	2.79	0.005	.1578819	.9019472
_cons	40.08749	10.77072	3.72	0.000	18.97726	61.19772
2quantity						
price	.5773133	.1749974	3.30	0.001	.2343247	.9203019
praw	-.7835496	.1312414	-5.97	0.000	-1.040778	-.5263213
_cons	-2.550694	5.273067	-0.48	0.629	-12.88571	7.784327

```
Endogenous variables:   quantity price
Exogenous variables:    pcompete income praw
```

However, here reg3 has been forced to create a unique equation name for the supply equation—2quantity. Both the supply and demand equations could not be designated as quantity, so a number was prefixed to the name for the supply equation.

We could have specified

```
. reg3 (qDemand: quantity price pcompete income) (qSupply: quantity price praw),
> endog(price)
```

and obtained the same results and equation labeling as when we used global macros to hold the equation specifications.

Without explicit equation names, reg3 always assumes that the dependent variable should be used to name equations. When each equation has a different dependent variable, this rule causes no problems and produces easily interpreted result tables. If the same dependent variable appears in more than one equation, however, reg3 will create a unique equation name based on the dependent variable name. Since equation names must be used for cross-equation tests, you have more control in this situation if explicit names are placed on the equations.

❑

▷ Example 3: Using the full syntax of reg3

Klein's (1950) model of the U.S. economy is often used to demonstrate system estimators. It contains several common features that will serve to demonstrate the full syntax of reg3. The Klein model is defined by the following seven relationships.

$$c = \beta_0 + \beta_1 p + \beta_2 p1 + \beta_3 w + \epsilon_1 \tag{1}$$

$$i = \beta_4 + \beta_5 p + \beta_6 p1 + \beta_7 k1 + \epsilon_2 \tag{2}$$

$$wp = \beta_8 + \beta_9 y + \beta_{10} y1 + \beta_{11} yr + \epsilon_3 \tag{3}$$

$$y = c + i + g \tag{4}$$

$$p = y - t - wp \tag{5}$$

$$k = k1 + i \tag{6}$$

$$w = wg + wp \tag{7}$$

The variables in the model are listed below. Two sets of variable names are shown. The concise first name uses traditional economics mnemonics, whereas the second name provides more guidance for everyone else. The concise names serve to keep the specification of the model small (and quite understandable to economists).

Short name	Long name	Variable definition	Type
c	consump	Consumption	endogenous
p	profits	Private industry profits	endogenous
p1	profits1	Last year's private industry profits	exogenous
wp	wagepriv	Private wage bill	endogenous
wg	wagegovt	Government wage bill	exogenous
w	wagetot	Total wage bill	endogenous
i	invest	Investment	endogenous
k1	capital1	Last year's level of capital stock	exogenous
y	totinc	Total income/demand	endogenous
y1	totinc1	Last year's total income	exogenous
g	govt	Government spending	exogenous
t	taxnetx	Indirect bus. taxes + net exports	exogenous
yr	year	Year—1931	exogenous

Equations (1)–(3) are behavioral and contain explicit disturbances (ϵ_1, ϵ_2, and ϵ_3). The remaining equations are identities that specify additional variables in the system and their accounting relationships with the variables in the behavioral equations. Some variables are explicitly endogenous by appearing as dependent variables in (1)–(3). Others are implicitly endogenous as linear combinations that contain other endogenous variables (e.g., w and p). Still other variables are implicitly exogenous by appearing in the identities but not in the behavioral equations (e.g., wg and g).

Using the concise names, we can fit Klein's model with the following command:

```
. use http://www.stata-press.com/data/r10/kleinAbr
. reg3 (c p p1 w) (i p p1 k1) (wp y y1 yr), endog(w p y) exog(t wg g)
Three-stage least-squares regression
```

Equation	Obs	Parms	RMSE	"R-sq"	chi2	P
c	21	3	.9443305	0.9801	864.59	0.0000
i	21	3	1.446736	0.8258	162.98	0.0000
wp	21	3	.7211282	0.9863	1594.75	0.0000

	Coef.	Std. Err.	z	P>|z|	[95% Conf. Interval]	
c						
p	.1248904	.1081291	1.16	0.248	-.0870387	.3368194
p1	.1631439	.1004382	1.62	0.104	-.0337113	.3599992
w	.790081	.0379379	20.83	0.000	.715724	.8644379
_cons	16.44079	1.304549	12.60	0.000	13.88392	18.99766
i						
p	-.0130791	.1618962	-0.08	0.936	-.3303898	.3042316
p1	.7557238	.1529331	4.94	0.000	.4559805	1.055467
k1	-.1948482	.0325307	-5.99	0.000	-.2586072	-.1310893
_cons	28.17785	6.793768	4.15	0.000	14.86231	41.49339
wp						
y	.4004919	.0318134	12.59	0.000	.3381388	.462845
y1	.181291	.0341588	5.31	0.000	.1143411	.2482409
yr	.149674	.0279352	5.36	0.000	.094922	.2044261
_cons	1.797216	1.115854	1.61	0.107	-.3898181	3.984251

```
Endogenous variables:  c i wp w p y
Exogenous variables:   p1 k1 y1 yr t wg g
```

We used the `exog()` option to identify t, wg, and g as exogenous variables in the system. These variables must be identified because they are part of the system but appear directly in none of the behavioral equations. Without this option, `reg3` would not know they were part of the system. The `endog()` option specifying w, p, and y is also required. Without this information, `reg3` would be unaware that these variables are linear combinations that include endogenous variables.

❑ Technical Note

Rather than listing additional endogenous and exogenous variables, we could specify the full list of exogenous variables in an `inst()` option,

```
. reg3 (c p p1 w) (i p p1 k1) (wp y y1 yr), inst(g t wg yr p1 k1 y1)
```

or equivalently,

```
. global conseqn "(c p p1 w)"
. global inveqn  "(i p p1 k1)"
. global wageqn  "(wp y y1 yr)"
. global inlist  "g t wg yr p1 k1 y1"
. reg3 $conseqn $inveqn $wageqn, inst($inlist)
```

Macros and explicit equations can also be mixed in the specification

```
. reg3 $conseqn (i p p1 k1) $wageqn, endog(w p y) exog(t wg g)
```

or

```
. reg3 (c p p1 w) $inveqn (wp y y1 yr), endog(w p y) exog(t wg g)
```

Placing the equation-binding parentheses in the global macros was also arbitrary. We could have used

```
. global consump  "c p p1 w"
. global invest   "i p p1 k1"
. global wagepriv "wp y y1 yr"
. reg3 ($consump) ($invest) ($wagepriv), endog(w p y) exog(t wg g)
```

reg3 is tolerant of all combinations, and these commands will produce identical output.

❑

Switching to the full variable names, we can fit Klein's model with the commands below. We will use global macros to store the lists of endogenous and exogenous variables. Again this is not necessary: these lists could have been typed directly on the command line. However, assigning the lists to local macros makes additional processing easier if alternative models are to be fitted. We will also use the ireg3 option to produce the iterated estimates.

```
. use http://www.stata-press.com/data/r10/klein

. global conseqn "(consump profits profits1 wagetot)"

. global inveqn  "(invest profits profits1 capital1)"

. global wageqn "(wagepriv totinc totinc1 year)"

. global enlist "wagetot profits totinc"

. global exlist "taxnetx wagegovt govt"

. reg3 $conseqn $inveqn $wageqn, endog($enlist) exog($exlist) ireg3
Iteration 1:   tolerance = .3712549
Iteration 2:   tolerance = .1894712
Iteration 3:   tolerance = .1076401
 (output omitted )
Iteration 24:   tolerance = 7.049e-07

Three-stage least-squares regression, iterated
```

Equation	Obs	Parms	RMSE	"R-sq"	chi2	P
consump	21	3	.9565088	0.9796	970.31	0.0000
invest	21	3	2.134327	0.6209	56.78	0.0000
wagepriv	21	3	.7782334	0.9840	1312.19	0.0000

	Coef.	Std. Err.	z	P>|z|	[95% Conf. Interval]	
consump						
profits	.1645096	.0961979	1.71	0.087	-.0240348	.3530539
profits1	.1765639	.0901001	1.96	0.050	-.0000291	.3531569
wagetot	.7658011	.0347599	22.03	0.000	.6976729	.8339294
_cons	16.55899	1.224401	13.52	0.000	14.15921	18.95877
invest						
profits	-.3565316	.2601568	-1.37	0.171	-.8664296	.1533664
profits1	1.011299	.2487745	4.07	0.000	.5237098	1.498888
capital1	-.2602	.0508694	-5.12	0.000	-.3599022	-.1604978
_cons	42.89629	10.59386	4.05	0.000	22.13271	63.65987
wagepriv						
totinc	.3747792	.0311027	12.05	0.000	.3138191	.4357394
totinc1	.1936506	.0324018	5.98	0.000	.1301443	.257157
year	.1679262	.0289291	5.80	0.000	.1112263	.2246261
_cons	2.624766	1.195559	2.20	0.028	.2815124	4.968019

```
Endogenous variables:    consump invest wagepriv wagetot profits totinc
Exogenous variables:     profits1 capital1 totinc1 year taxnetx wagegovt govt
```

◁

▷ Example 4: Constraints with reg3

As a simple example of constraints, (1) above may be rewritten with both wages explicitly appearing (rather than as a variable containing the sum). Using the longer variable names, we have

$$\text{consump} = \beta_0 + \beta_1 \, \text{profits} + \beta_2 \, \text{profits1} + \beta_3 \, \text{wagepriv} + \beta_{12} \, \text{wagegovt} + \epsilon_1$$

To retain the effect of the identity in (7), we need $\beta_3 = \beta_{12}$ as a constraint on the system. We obtain this result by defining the constraint in the usual way and then specifying its use in reg3. Since reg3 is a system estimator, we will need to use the full equation syntax of constraint. The assumption that the following commands are entered after the model above has been estimated. We are simply changing the definition of the consumption equation (consump) and adding a constraint on two of its parameters. The rest of the model definition is carried forward.

```
. global conseqn "(consump profits profits1 wagepriv wagegovt)"

. constraint 1 [consump]wagepriv = [consump]wagegovt

. reg3 $conseqn $inveqn $wageqn, endog($enlist) exog($exlist) constr(1) ireg3
note:  additional endogenous variables not in the system have no effect
       and are ignored:  wagetot
Iteration 1:   tolerance =   .3712547
Iteration 2:   tolerance =   .189471
Iteration 3:   tolerance =   .10764
 (output omitted )
Iteration 24:   tolerance =  7.049e-07

Three-stage least-squares regression, iterated

Constraints:
 ( 1)   [consump]wagepriv - [consump]wagegovt = 0
```

Equation	Obs	Parms	RMSE	"R-sq"	chi2	P
consump	21	3	.9565086	0.9796	970.31	0.0000
invest	21	3	2.134326	0.6209	56.78	0.0000
wagepriv	21	3	.7782334	0.9840	1312.19	0.0000

	Coef.	Std. Err.	z	P>|z|	[95% Conf. Interval]	
consump						
profits	.1645097	.0961978	1.71	0.087	-.0240346	.353054
profits1	.1765639	.0901001	1.96	0.050	-.0000291	.3531568
wagepriv	.7658012	.0347599	22.03	0.000	.6976729	.8339294
wagegovt	.7658012	.0347599	22.03	0.000	.6976729	.8339294
_cons	16.55899	1.224401	13.52	0.000	14.1592	18.95877
invest						
profits	-.3565311	.2601567	-1.37	0.171	-.8664288	.1533666
profits1	1.011298	.2487744	4.07	0.000	.5237096	1.498887
capital1	-.2601999	.0508694	-5.12	0.000	-.359902	-.1604977
_cons	42.89626	10.59386	4.05	0.000	22.13269	63.65984
wagepriv						
totinc	.3747792	.0311027	12.05	0.000	.313819	.4357394
totinc1	.1936506	.0324018	5.98	0.000	.1301443	.257157
year	.1679262	.0289291	5.80	0.000	.1112263	.2246261
_cons	2.624766	1.195559	2.20	0.028	.281512	4.968019

```
Endogenous variables:   consump invest wagepriv wagetot profits totinc
Exogenous variables:    profits1 wagegovt capital1 totinc1 year taxnetx govt
```

As expected, none of the parameter or standard error estimates has changed from the previous estimates (before the seventh significant digit). We have simply decomposed the total wage variable into its two parts and constrained the coefficients on these parts. The warning about additional endogenous variables was just `reg3`'s way of letting us know that we had specified some information that was irrelevant to the estimation of the system. We had left the variable `wagetot` in our `endog` macro. It does not mean anything to the system to specify `wagetot` as endogenous since it is no longer in the system. That's fine with `reg3` and fine for our current purposes.

We can also impose constraints across the equations. For example, the admittedly meaningless constraint of requiring `profits` to have the same effect in both the consumption and investment equations could be imposed. Retaining the constraint on the wage coefficients, we would estimate this constrained system.

```
. constraint 2 [consump]profits = [invest]profits

. reg3 $conseqn $inveqn $wageqn, endog($enlist) exog($exlist) constr(1 2) ireg3
note:  additional endogenous variables not in the system have no effect
       and are ignored:  wagetot
Iteration 1:   tolerance =    .1427927
Iteration 2:   tolerance =     .032539
Iteration 3:   tolerance =   .00307811
Iteration 4:   tolerance =   .00016903
Iteration 5:   tolerance =   .00003409
Iteration 6:   tolerance =   7.763e-06
Iteration 7:   tolerance =   9.240e-07

Three-stage least-squares regression, iterated

Constraints:
 ( 1)  [consump]wagepriv - [consump]wagegovt = 0
 ( 2)  [consump]profits - [invest]profits = 0
```

Equation	Obs	Parms	RMSE	"R-sq"	chi2	P
consump	21	3	.9504669	0.9798	1019.54	0.0000
invest	21	3	1.247066	0.8706	144.57	0.0000
wagepriv	21	3	.7225276	0.9862	1537.45	0.0000

	Coef.	Std. Err.	z	P>|z|	[95% Conf. Interval]	
consump						
profits	.1075413	.0957767	1.12	0.262	-.0801777	.2952602
profits1	.1712756	.0912613	1.88	0.061	-.0075932	.3501444
wagepriv	.798484	.0340876	23.42	0.000	.7316734	.8652946
wagegovt	.798484	.0340876	23.42	0.000	.7316734	.8652946
_cons	16.2521	1.212157	13.41	0.000	13.87631	18.62788
invest						
profits	.1075413	.0957767	1.12	0.262	-.0801777	.2952602
profits1	.6443378	.1058682	6.09	0.000	.43684	.8518356
capital1	-.1766669	.0261889	-6.75	0.000	-.2279962	-.1253375
_cons	24.31931	5.284325	4.60	0.000	13.96222	34.6764
wagepriv						
totinc	.4014106	.0300552	13.36	0.000	.3425035	.4603177
totinc1	.1775359	.0321583	5.52	0.000	.1145068	.240565
year	.1549211	.0282291	5.49	0.000	.099593	.2102492
_cons	1.959788	1.14467	1.71	0.087	-.2837242	4.203299

```
Endogenous variables:   consump invest wagepriv wagetot profits totinc
Exogenous variables:    profits1 wagegovt capital1 totinc1 year taxnetx govt
```

◁

❑ Technical Note

Identification in a system of simultaneous equations involves the notion that there is enough information to estimate the parameters of the model given the specified functional form. Under-identification usually manifests itself as one matrix in the three-stage least-squares computations. The most commonly violated order condition for two- or three-stage least squares involves the number of endogenous and exogenous variables. There must be at least as many noncollinear exogenous variables in the remaining system as there are endogenous right-hand-side variables in an equation. This condition must hold for each structural equation in the system.

Put as a set of rules the following:

1. Count the number of right-hand-side endogenous variables in an equation and call this m_i.

2. Count the number of exogenous variables in the same equation and call this k_i.

3. Count the total number of exogenous variables in all the structural equations plus any additional variables specified in an `exog()` or `inst()` option and call this K.

4. If $m_i > (K - k_i)$ for any structural equation (i), then the system is underidentified and cannot be estimated by three-stage least squares.

We are also possibly in trouble if any of the exogenous variables are linearly dependent. We must have m_i linearly independent variables among the exogenous variables represented by $(K - k_i)$.

The complete conditions for identification involve rank-order conditions on several matrices. For a full treatment, see Theil (1971) or Greene (2003, 405–407).

❑

Saved Results

`reg3` saves the following in `e()`:

Scalars

`e(N)`	number of observations	`e(rmse_#)`	root mean squared error for eqn. #
`e(k)`	number of parameters	`e(dfk2_adj)`	divisor used with VCE when `dfk2`
`e(k_eq)`	number of equations		specified
`e(mss_#)`	model sum of squares for eqn. #	`e(ll)`	log likelihood
`e(df_m#)`	model degrees of freedom for eqn. #	`e(chi2_#)`	χ^2 for equation #
`e(rss_#)`	residual sum of squares for eqn. #	`e(p_#)`	significance for equation #
`e(df_r)`	residual degrees of freedom (`small`)	`e(ic)`	number of iterations
`e(r2_#)`	R-squared for equation #	`e(cons_#)`	1 when equation # has a constant;
`e(F_#)`	F statistic for eqn. # (`small`)		0 otherwise

Macros

`e(cmd)`	reg3	`e(wtype)`	weight type
`e(cmdline)`	command as typed	`e(wexp)`	weight expression
`e(depvar)`	names of dependent variables	`e(method)`	3sls, 2sls, ols, sure, or mvreg
`e(exog)`	names of exogenous variables	`e(small)`	small
`e(endog)`	names of endogenous variables	`e(properties)`	b V
`e(eqnames)`	names of equations	`e(predict)`	program used to implement `predict`
`e(corr)`	correlation structure		

Matrices

`e(b)`	coefficient vector	`e(V)`	variance–covariance matrix of the
`e(Sigma)`	$\widehat{\Sigma}$ matrix		estimators

Functions

`e(sample)`	marks estimation sample

Methods and Formulas

reg3 is implemented as an ado-file.

The most concise way to represent a system of equations for three-stage least squares requires thinking of the individual equations and their associated data as being stacked. reg3 does not expect the data in this format, but it is a convenient shorthand. The system could then be formulated as

$$
\begin{bmatrix} \mathbf{y}_1 \\ \mathbf{y}_2 \\ \vdots \\ \mathbf{y}_M \end{bmatrix} = \begin{bmatrix} \mathbf{Z}_1 & 0 & \dots & 0 \\ 0 & \mathbf{Z}_2 & \dots & 0 \\ \vdots & \vdots & \ddots & \vdots \\ 0 & 0 & \dots & \mathbf{Z}_M \end{bmatrix} \begin{bmatrix} \beta_1 \\ \beta_2 \\ \vdots \\ \beta_M \end{bmatrix} + \begin{bmatrix} \epsilon_1 \\ \epsilon_2 \\ \vdots \\ \epsilon_M \end{bmatrix}
$$

In full matrix notation, this is just

$$
\mathbf{y} = \mathbf{Z}\,\mathbf{B} + \epsilon
$$

The $\mathbf{Z}$ elements in these matrices represent both the endogenous and the exogenous right-hand-side variables in the equations.

Also assume that there will be correlation between the disturbances of the equations so that

$$
E(\epsilon\epsilon') = \boldsymbol{\Sigma}
$$

where the disturbances are further assumed to have an expected value of 0; $E(\epsilon) = 0$.

The first stage of three-stage least-squares regression requires developing instrumented values for the endogenous variables in the system. These values can be derived as the predictions from a linear regression of each endogenous regressor on all exogenous variables in the system, or, more succinctly, as the projection of each regressor through the projection matrix of all exogenous variables onto the regressors. Designating the set of all exogenous variables as $\mathbf{X}$ results in

$$
\widehat{\mathbf{z}}_i = \mathbf{X}(\mathbf{X}'\mathbf{X})^{-1}\mathbf{X}'\mathbf{z}_i \quad \text{for each } i
$$

Taken collectively, these $\widehat{\mathbf{Z}}$ contain the instrumented values for all the regressors. They take on the actual values for the exogenous variables and first-stage predictions for the endogenous variables. Given these instrumented variables, a generalized least squares (GLS) or Aitken (1935) estimator can be formed for the parameters of the system

$$
\widehat{\mathbf{B}} = \left\{ \widehat{\mathbf{Z}}'(\boldsymbol{\Sigma}^{-1} \otimes \mathbf{I})\widehat{\mathbf{Z}} \right\}^{-1} \widehat{\mathbf{Z}}'(\boldsymbol{\Sigma}^{-1} \otimes \mathbf{I})\mathbf{y}
$$

All that remains is to obtain a consistent estimator for $\boldsymbol{\Sigma}$. This estimate can be formed from the residuals of two-stage least squares estimates of each equation in the system. Alternately, and identically, the residuals can be computed from the estimates formed by taking $\boldsymbol{\Sigma}$ to be an identity matrix. This maintains the full system of coefficients and allows constraints to be applied when the residuals are computed.

If we take $\mathbf{E}$ to be the matrix of residuals from these estimates, a consistent estimate of $\boldsymbol{\Sigma}$ is

$$
\widehat{\boldsymbol{\Sigma}} = \frac{\mathbf{E}'\mathbf{E}}{n}
$$

where n is the number of observations in the sample. An alternative divisor for this estimate can be obtained with the dfk option as outlined under options.

With the estimate of $\widehat{\Sigma}$ placed into the GLS estimating equation

$$\widehat{\mathbf{B}} = \left\{ \widehat{\mathbf{Z}}'(\widehat{\Sigma}^{-1} \otimes \mathbf{I})\widehat{\mathbf{Z}} \right\}^{-1} \widehat{\mathbf{Z}}'(\widehat{\Sigma}^{-1} \otimes \mathbf{I})\mathbf{y}$$

is the three-stage least-squares estimates of the system parameters.

The asymptotic variance–covariance matrix of the estimator is just the standard formulation for a GLS estimator

$$\mathbf{V}_{\widehat{\mathbf{B}}} = \left\{ \widehat{\mathbf{Z}}'(\widehat{\Sigma}^{-1} \otimes \mathbf{I})\widehat{\mathbf{Z}} \right\}^{-1}$$

Iterated three-stage least-squares estimates can be obtained by computing the residuals from the three-stage parameter estimates, using these to formulate a new $\widehat{\Sigma}$, and recomputing the parameter estimates. This process is repeated until the estimates $\widehat{\mathbf{B}}$ converge—if they converge. Convergence is not guaranteed. When estimating a system by SURE, these iterated estimates will be the maximum likelihood estimates for the system. The iterated solution can also be used to produce estimates that are invariant to choice of system and restriction parameterization for many linear systems under full three-stage least squares.

The exposition above follows the parallel developments in Greene (2003) and Kmenta (1997).

Henri Theil (1924–2000) was born in Amsterdam and awarded a Ph.D. in 1951 by the University of Amsterdam. He researched and taught econometric theory, statistics, microeconomics, macroeconomic modeling, and economic forecasting, and policy at what is now Erasmus University in Rotterdam, the University of Chicago, and the University of Florida. Theil's many specific contributions include work on two- and three-stage least squares, inequality and concentration, and consumer demand.

References

Aitken, A. C. 1935. On least squares and linear combination of observations. *Proceedings, Royal Society of Edinburgh* 55: 42–48.

Bewley, R. 2000. Mr. Henri Theil: An interview with the *International Journal of Forecasting*. *International Journal of Forecasting* 16: 1–16.

Greene, W. H. 2003. *Econometric Analysis*. 5th ed. Upper Saddle River, NJ: Prentice Hall.

Klein, L. R. 1950. *Economic Fluctuations in the United States 1921–1941*. New York: Wiley.

Kmenta, J. 1997. *Elements of Econometrics*. 2nd ed. Ann Arbor: University of Michigan Press.

Poi, B. P. 2006. Jackknife instrumental variables estimation in Stata. *Stata Journal* 6: 364–376.

Theil, H. 1971. *Principles of Econometrics*. New York: Wiley.

Weesie, J. 1999. sg121: Seemingly unrelated estimation and the cluster-adjusted sandwich estimator. *Stata Technical Bulletin* 52: 34–47. Reprinted in *Stata Technical Bulletin Reprints*, vol. 9, pp. 231–248.

Zellner, A., and H. Theil. 1962. Three stage least squares: Simultaneous estimate of simultaneous equations. *Econometrica* 29: 63–68.

Also See

[R] **reg3 postestimation** — Postestimation tools for reg3

[R] **ivregress** — Single-equation instrumental-variables regression

[R] **mvreg** — Multivariate regression

[R] **nlsur** — Estimation of nonlinear systems of equations

[R] **regress** — Linear regression

[R] **sureg** — Zellner's seemingly unrelated regression

[U] **20 Estimation and postestimation commands**

Title

reg3 postestimation — Postestimation tools for reg3

Description

The following postestimation commands are available for reg3:

command	description
adjust	adjusted predictions of $\mathbf{x}\beta$
*estat	AIC, BIC, VCE, and estimation sample summary
estimates	cataloging estimation results
hausman	Hausman's specification test
lincom	point estimates, standard errors, testing, and inference for linear combinations of coefficients
mfx	marginal effects or elasticities
nlcom	point estimates, standard errors, testing, and inference for nonlinear combinations of coefficients
predict	predictions, residuals, influence statistics, and other diagnostic measures
predictnl	point estimates, standard errors, testing, and inference for generalized predictions
test	Wald tests for simple and composite linear hypotheses
testnl	Wald tests of nonlinear hypotheses

*estat ic is not appropriate after reg3, 2sls.

See the corresponding entries in the *Stata Base Reference Manual* for details.

Syntax for predict

predict [*type*] *newvar* [*if*] [*in*] [, <u>equation</u>(*eqno*[,*eqno*]) *statistic*]

statistic	description
Main	
xb	linear prediction; the default
stdp	standard error of the linear prediction
<u>residuals</u>	residuals
<u>difference</u>	difference between the linear predictions of two equations
stddp	standard error of the difference in linear predictions

These statistics are available both in and out of sample; type predict ... if e(sample) ... if wanted only for the estimation sample.

Options for predict

⌐ Main ⌐

equation(*eqno*[,*eqno*]) specifies to which equation you are referring.

> equation() is filled in with one *eqno* for options xb, stdp, and residuals. equation(#1) would mean the calculation is to be made for the first equation, equation(#2) would mean the second, and so on. You could also refer to the equations by their names. equation(income) would refer to the equation named income and equation(hours) to the equation named hours.

> If you do not specify equation(), results are the same as if you specified equation(#1).

> difference and stddp refer to between-equation concepts. To use these options, you must specify two equations, e.g., equation(#1,#2) or equation(income,hours). When two equations must be specified, equation() is required.

xb, the default, calculates the linear prediction (fitted values)—the prediction of $x_j b$ for the specified equation.

stdp calculates the standard error of the prediction for the specified equation. It can be thought of as the standard error of the predicted expected value or mean for the observation's covariate pattern. The standard error of the prediction is also referred to as the standard error of the fitted value.

residuals calculates the residuals.

difference calculates the difference between the linear predictions of two equations in the system. With equation(#1,#2), difference computes the prediction of equation(#1) minus the prediction of equation(#2).

stddp is allowed only after you have previously fitted a multiple-equation model. The standard error of the difference in linear predictions $(x_{1j} b - x_{2j} b)$ between equations 1 and 2 is calculated.

For more information on using predict after multiple-equation estimation commands, see [R] **predict**.

Remarks

▷ Example 1

> In example 2 of [R] **reg3**, we fitted a simple supply-and-demand. Here we obtain the obtain the fitted supply and demand curves assuming that the exogenous regressors equal their sample means. We first replace each of the three exogenous regressors with their sample means, then we call predict to obtain the predictions.

```
. use http://www.stata-press.com/data/r10/supDem
. global demand "(qDemand: quantity price pcompete income)"
. global supply "(qSupply: quantity price praw)"
. reg3 $demand $supply, endog(price)
  (output omitted )
. summarize pcompete, meanonly
. replace pcompete = r(mean)
(49 real changes made)
. summarize income, meanonly
. replace income = r(mean)
(49 real changes made)
. summarize praw, meanonly
. replace praw = r(mean)
(49 real changes made)
```

```
. predict demand, equation(qDemand)
(option xb assumed; fitted values)
. predict supply, equation(qSupply)
(option xb assumed; fitted values)
. graph twoway line demand price, sort || line supply price
```

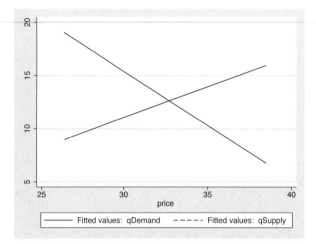

As we would expect based on economic theory, the demand curve slopes downward while the supply curve slopes upward. With the exogenous variables at their mean levels, the equilibrium price and quantity are slightly less than 33 and 13, respectively.

◁

Methods and Formulas

All postestimation commands listed above are implemented as ado-files.

The computational formulas for the statistics produced by predict can be found in [R] **predict** and [R] **regress postestimation**.

Also See

[R] **reg3** — Three-stage estimation for systems of simultaneous equations

[U] **20 Estimation and postestimation commands**

Title

regress — Linear regression

Syntax

regress *depvar* [*indepvars*] [*if*] [*in*] [*weight*] [, *options*]

options	description
Model	
<u>noc</u>onstant	suppress constant term
<u>hasc</u>ons	has user-supplied constant
tsscons	compute total sum of squares with constant; seldom used
SE/Robust	
vce(*vcetype*)	*vcetype* may be ols, <u>r</u>obust, <u>c</u>luster *clustvar*, <u>boot</u>strap, <u>jack</u>knife, hc2, or hc3
Reporting	
<u>l</u>evel(*#*)	set confidence level; default is level(95)
beta	report standardized beta coefficients
<u>ef</u>orm(*string*)	report exponentiated coefficients and label as *string*
<u>noh</u>eader	suppress table header
plus	make table extendable
<u>depn</u>ame(*varname*)	substitute dependent variable name; programmer's option
† <u>mse1</u>	force mean squared error to 1

depvar and *indepvars* may contain time-series operators; see [U] **11.4.3 Time-series varlists**.

bootstrap, by, jackknife, nestreg, rolling, statsby, stepwise, svy, and xi are allowed; see [U] **11.1.10 Prefix commands**.

Weights are not allowed with the bootstrap prefix.

aweights are not allowed with the jackknife prefix.

hascons, tsscons, vce(), beta, noheader, plus, depname(), mse1, and weights are not allowed with the svy prefix.

†mse1 does not appear in the dialog box.

aweights, fweights, iweights, and pweights are allowed; see [U] **11.1.6 weight**.

See [U] **20 Estimation and postestimation commands** for more capabilities of estimation commands.

Description

regress fits a model of *depvar* on *indepvars* using linear regression.

Here is a short list of other regression commands that may be of interest. See [I] **estimation commands** for a complete list.

areg	[R] **areg**	an easier way to fit regressions with many dummy variables
arch	[TS] **arch**	regression models with ARCH errors
arima	[TS] **arima**	ARIMA models
boxcox	[R] **boxcox**	Box–Cox regression models
cnreg	[R] **cnreg**	censored-normal regression
cnsreg	[R] **cnsreg**	constrained linear regression
eivreg	[R] **eivreg**	errors-in-variables regression
frontier	[R] **frontier**	stochastic frontier models
heckman	[R] **heckman**	Heckman selection model
intreg	[R] **intreg**	interval regression
ivregress	[R] **ivregress**	single-equation instrumental-variables regression
ivtobit	[R] **ivtobit**	tobit regression with endogenous variables
newey	[TS] **newey**	regression with Newey–West standard errors
qreg	[R] **qreg**	quantile (including median) regression
reg3	[R] **reg3**	three-stage least-squares (3SLS) regression
rreg	[R] **rreg**	a type of robust regression
sureg	[R] **sureg**	seemingly unrelated regression
tobit	[R] **tobit**	tobit regression
treatreg	[R] **treatreg**	treatment-effects model
truncreg	[R] **truncreg**	truncated regression
xtabond	[XT] **xtabond**	Arellano–Bond linear, dynamic panel-data estimation
xtdpd	[XT] **xtdpd**	linear, dynamic panel-data estimation
xtfrontier	[XT] **xtfrontier**	panel-data stochastic frontier models
xtgls	[XT] **xtgls**	panel-data GLS models
xthtaylor	[XT] **xthtaylor**	Hausman–Taylor estimator for error-components models
xtintreg	[XT] **xtintreg**	panel-data interval regression models
xtivreg	[XT] **xtivreg**	panel-data instrumental variables (2SLS) regression
xtpcse	[XT] **xtpcse**	linear regression with panel-corrected standard errors
xtreg	[XT] **xtreg**	fixed- and random-effects linear models
xtregar	[XT] **xtregar**	fixed- and random-effects linear models with an AR(1) disturbance
xttobit	[XT] **xttobit**	panel-data tobit models

Options

＿＿＿＿ Model

noconstant; see [R] **estimation options**.

hascons indicates that a user-defined constant or its equivalent is specified among the independent variables in *indepvars*. Some caution is recommended when specifying this option, as resulting estimates may not be as accurate as they otherwise would be. Use of this option requires "sweeping" the constant last, so the moment matrix must be accumulated in absolute rather than deviation form.

This option may be safely specified when the means of the dependent and independent variables are all reasonable and there is not much collinearity between the independent variables. The best procedure is to view `hascons` as a reporting option—estimate with and without `hascons` and verify that the coefficients and standard errors of the variables not affected by the identity of the constant are unchanged.

`tsscons` forces the total sum of squares to be computed as though the model has a constant, that is, as deviations from the mean of the dependent variable. This is a rarely used option that has an effect only when specified with `noconstant`. It affects the total sum of squares and all results derived from the total sum of squares.

<hr>

⌐SE/Robust⌐

`vce(`*vcetype*`)` specifies the type of standard error reported, which includes types that are derived from asymptotic theory, that are robust to some kinds of misspecification, that allow for intragroup correlation, and that use bootstrap or jackknife methods; see [R] **vce_option**.

`vce(ols)`, the default, uses the standard variance estimator for ordinary least-squares regression.

`regress` also allows the following:

`vce(hc2)` and `vce(hc3)` specify an alternative bias correction for the robust variance calculation. `vce(hc2)` and `vce(hc3)` may not be specified with `svy` prefix. In the unclustered case, `vce(robust)` uses $\hat{\sigma}_j^2 = \{n/(n-k)\}u_j^2$ as an estimate of the variance of the jth observation, where u_j is the calculated residual and $n/(n-k)$ is included to improve the overall estimate's small-sample properties.

> `vce(hc2)` instead uses $u_j^2/(1-h_{jj})$ as the observation's variance estimate, where h_{jj} is the diagonal element of the hat (projection) matrix. This estimate is unbiased if the model really is homoskedastic. `vce(hc2)` tends to produce slightly more conservative confidence intervals.

> `vce(hc3)` uses $u_j^2/(1-h_{jj})^2$ as suggested by Davidson and MacKinnon (1993), who report that this method tends to produce better results when the model really is heteroskedastic. `vce(hc3)` produces confidence intervals that tend to be even more conservative.

> See Davidson and MacKinnon (1993, 554–556) for more discussion on these two bias corrections.

<hr>

⌐Reporting⌐

`level(#)`; see [R] **estimation options**.

`beta` asks that standardized beta coefficients be reported instead of confidence intervals. The beta coefficients are the regression coefficients obtained by first standardizing all variables to have a mean of 0 and a standard deviation of 1. `beta` may not be specified with `vce(cluster` *clustvar*`)` or the `svy` prefix.

`eform(`*string*`)` is used only in programs and ado-files that use `regress` to fit models other than linear regression. `eform()` specifies that the coefficient table be displayed in exponentiated form as defined in [R] **maximize** and that *string* be used to label the exponentiated coefficients in the table.

`noheader` suppresses the display of the ANOVA table and summary statistics at the top of the output; only the coefficient table is displayed. This option is often used in programs and ado-files.

`plus` specifies that the output table be made extendable. This option is often used in programs and ado-files.

depname(*varname*) is used only in programs and ado-files that use `regress` to fit models other than linear regression. `depname()` may be specified only at estimation time. *varname* is recorded as the identity of the dependent variable, even though the estimates are calculated using *depvar*. This method affects the labeling of the output—not the results calculated—but could affect subsequent calculations made by `predict`, where the residual would be calculated as deviations from *varname* rather than *depvar*. `depname()` is most typically used when *depvar* is a temporary variable (see [P] **macro**) used as a proxy for *varname*.

The following option is available with `probit` but is not shown in the dialog box:

mse1 is used only in programs and ado-files that use `regress` to fit models other than linear regression and is not allowed with the `svy` prefix. `mse1` sets the mean squared error to 1, forcing the variance–covariance matrix of the estimators to be $(\mathbf{X'DX})^{-1}$ (see *Methods and Formulas* below) and affecting calculated standard errors. Degrees of freedom for t statistics are calculated as n rather than $n - k$.

Remarks

Remarks are presented under the following headings:

Ordinary least squares
Treatment of the constant
Robust standard errors
Weighted regression
Instrumental variables and two-stage least-squares regression

`regress` performs linear regression, including ordinary least squares and weighted least squares. For a general discussion of linear regression, see Draper and Smith (1998), Johnston and DiNardo (1997), or Kmenta (1997).

See Wooldridge (2006) for an excellent treatment of estimation, inference, interpretation, and specification testing in linear regression models. This presentation stands out for its clarification of the statistical issues, as opposed to the algebraic issues. See Wooldridge (2002, chap. 4) for a more advanced discussion along the same lines.

See Hamilton (2006, chap. 6) for an introduction to linear regression using Stata. Dohoo, Martin, and Stryhn (2003) discuss linear regression using examples from epidemiology, and Stata datasets and do-files used in the text are available.

Chatterjee and Hadi (2006) explain regression analysis by using examples containing typical problems that you might encounter when performing exploratory data analysis. We also recommend Weisberg (2005), who emphasizes the importance of the assumptions of linear regression and problems resulting from these assumptions. For a discussion of model-selection techniques and exploratory data analysis, see Mosteller and Tukey (1977). For a mathematically rigorous treatment, see Peracchi (2001, chap. 6). Finally, see Plackett (1972) if you are interested in the history of regression. Least squares, which dates back to the 1790s, was discovered independently by Legendre and Gauss.

Ordinary least squares

▷ Example 1

Suppose that we have data on the mileage rating and weight of 74 automobiles. The variables in our data are `mpg`, `weight`, and `foreign`. The last variable assumes the value 1 for foreign and 0 for domestic automobiles. We wish to fit the model

$$\text{mpg} = \beta_0 + \beta_1\text{weight} + \beta_2\text{weight}^2 + \beta_3\text{foreign} + \epsilon$$

We do this by creating a new variable called `weightsq` and then typing `regress mpg weight weightsq foreign`:

```
. use http://www.stata-press.com/data/r10/auto
(1978 Automobile Data)

. generate weightsq=weight^2

. regress mpg weight weightsq foreign
```

Source	SS	df	MS		
Model	1689.15372	3	563.05124		
Residual	754.30574	70	10.7757963		
Total	2443.45946	73	33.4720474		

	Number of obs =	74
	F(3, 70) =	52.25
	Prob > F =	0.0000
	R-squared =	0.6913
	Adj R-squared =	0.6781
	Root MSE =	3.2827

mpg	Coef.	Std. Err.	t	P>|t|	[95% Conf. Interval]	
weight	-.0165729	.0039692	-4.18	0.000	-.0244892	-.0086567
weightsq	1.59e-06	6.25e-07	2.55	0.013	3.45e-07	2.84e-06
foreign	-2.2035	1.059246	-2.08	0.041	-4.3161	-.0909002
_cons	56.53884	6.197383	9.12	0.000	44.17855	68.89913

◁

`regress` produces a variety of summary statistics along with the table of regression coefficients. At the upper left, `regress` reports an analysis-of-variance (ANOVA) table. The column headings SS, df, and MS stand for "sum of squares", "degrees of freedom", and "mean square", respectively. In the previous example, the total sum of squares is 2,443.5: 1,689.2 accounted for by the model and 754.3 left unexplained. Since the regression included a constant, the total sum reflects the sum after removal of means, as does the sum of squares due to the model. The table also reveals that there are 73 total degrees of freedom (counted as 74 observations less 1 for the mean removal), of which 3 are consumed by the model, leaving 70 for the residual.

To the right of the ANOVA table are presented other summary statistics. The F statistic associated with the ANOVA table is 52.25. The statistic has 3 numerator and 70 denominator degrees of freedom. The F statistic tests the hypothesis that all coefficients *excluding the constant* are zero. The chance of observing an F statistic that large or larger is reported as 0.0000, which is Stata's way of indicating a number smaller than 0.00005. The R-squared (R^2) for the regression is 0.6913, and the R-squared adjusted for degrees of freedom (R_a^2) is 0.6781. The root mean squared error, labeled Root MSE, is 3.2827. It is the square root of the mean squared error reported for the residual in the ANOVA table.

Finally, Stata produces a table of the estimated coefficients. The first line of the table indicates that the left-hand-side variable is `mpg`. Thereafter follow the four estimated coefficients. Our fitted model is

$$\text{mpg\_hat} = 56.54 - 0.0166\,\text{weight} + 1.59 \cdot 10^{-6}\,\text{weightsq} - 2.20\,\text{foreign}$$

Reported to the right of the coefficients in the output are the standard errors. For instance, the standard error for the coefficient on `weight` is 0.0039692. The corresponding t statistic is -4.18, which has a two-sided significance level of 0.000. This number indicates that the significance is less than 0.0005. The 95% confidence interval for the coefficient is $[-.024, -.009]$.

▷ Example 2

regress shares the features of all estimation commands. Among other things, this means that after running a regression, we can use test to test hypotheses about the coefficients, estat vce to examine the covariance matrix of the estimators, and predict to obtain predicted values, residuals, and influence statistics. See [U] **20 Estimation and postestimation commands**. Options that affect how estimates are displayed, such as beta or level(), can be used when replaying results.

Suppose that we meant to specify the beta option to obtain beta coefficients (regression coefficients normalized by the ratio of the standard deviation of the regressor to the standard deviation of the dependent variable). Even though we forgot, we can specify the option now:

```
. regress, beta
```

Source	SS	df	MS			
Model	1689.15372	3	563.05124			
Residual	754.30574	70	10.7757963			
Total	2443.45946	73	33.4720474			

		Number of obs	=	74
		F(3, 70)	=	52.25
		Prob > F	=	0.0000
		R-squared	=	0.6913
		Adj R-squared	=	0.6781
		Root MSE	=	3.2827

mpg	Coef.	Std. Err.	t	P>|t|	Beta
weight	-.0165729	.0039692	-4.18	0.000	-2.226321
weightsq	1.59e-06	6.25e-07	2.55	0.013	1.32654
foreign	-2.2035	1.059246	-2.08	0.041	-.17527
_cons	56.53884	6.197383	9.12	0.000	.

◁

Treatment of the constant

By default, regress includes an intercept (constant) term in the model. The noconstant option suppresses it, and the hascons option tells regress that the model already has one.

▷ Example 3

We wish to fit a regression of the weight of an automobile against its length, and we wish to impose the constraint that the weight is zero when the length is zero.

If we simply type regress weight length, we are fitting the model

$$\text{weight} = \beta_0 + \beta_1 \, \text{length} + \epsilon$$

Here a length of zero corresponds to a weight of β_0. We want to force β_0 to be zero or, equivalently, estimate an equation that does not include an intercept:

$$\text{weight} = \beta_1 \, \text{length} + \epsilon$$

We do this by specifying the noconstant option:

```
. regress weight length, noconstant
```

Source	SS	df	MS
Model	703869302	1	703869302
Residual	14892897.8	73	204012.299
Total	718762200	74	9713002.7

Number of obs =   74
F(1,   73) = 3450.13
Prob > F   = 0.0000
R-squared   = 0.9793
Adj R-squared = 0.9790
Root MSE   = 451.68

weight	Coef.	Std. Err.	t	P>|t|	[95% Conf. Interval]
length	16.29829	.2774752	58.74	0.000	15.74528 16.8513

In our data, `length` is measured in inches and `weight` in pounds. We discover that each inch of length adds 16 pounds to the weight.

◁

Sometimes there is no need for Stata to include a constant term in the model. Most commonly, this occurs when the model contains a set of mutually exclusive indicator variables. `hascons` is a variation of the `noconstant` option—it tells Stata not to add a constant to the regression because the regression specification already has one, either directly or indirectly.

For instance, we now refit our model of `weight` as a function of `length` and include separate constants for foreign and domestic cars. Our dataset already has an indicator variable called `foreign` marking foreign-made automobiles, so we create a corresponding `domestic` variable and fit the regression:

```
. generate domestic=!foreign
. regress weight length domestic foreign, hascons
```

Source	SS	df	MS
Model	39647744.7	2	19823872.3
Residual	4446433.7	71	62625.8268
Total	44094178.4	73	604029.841

Number of obs =   74
F(2,   71) = 316.54
Prob > F   = 0.0000
R-squared   = 0.8992
Adj R-squared = 0.8963
Root MSE   = 250.25

weight	Coef.	Std. Err.	t	P>|t|	[95% Conf. Interval]
length	31.44455	1.601234	19.64	0.000	28.25178 34.63732
domestic	-2850.25	315.9691	-9.02	0.000	-3480.274 -2220.225
foreign	-2983.927	275.1041	-10.85	0.000	-3532.469 -2435.385

❑ Technical Note

There is a subtle distinction between the `hascons` and `noconstant` options. We can most easily reveal it by refitting the last regression, specifying `noconstant` rather than `hascons`:

(Continued on next page)

```
. regress weight length domestic foreign, noconstant
```

Source	SS	df	MS
Model	714315766	3	238105255
Residual	4446433.7	71	62625.8268
Total	718762200	74	9713002.7

```
Number of obs =      74
F(  3,     71) = 3802.03
Prob > F      =  0.0000
R-squared     =  0.9938
Adj R-squared =  0.9936
Root MSE      =  250.25
```

weight	Coef.	Std. Err.	t	P>|t|	[95% Conf. Interval]	
length	31.44455	1.601234	19.64	0.000	28.25178	34.63732
domestic	-2850.25	315.9691	-9.02	0.000	-3480.274	-2220.225
foreign	-2983.927	275.1041	-10.85	0.000	-3532.469	-2435.385

Comparing this output with that produced by the previous **regress** command, we see that they are almost, but not quite, identical. The parameter estimates and their associated statistics—the second half of the output—are identical. The overall summary statistics and the ANOVA table—the first half of the output—are different, however.

In the first case, the R^2 is shown as 0.8992; here it is shown as 0.9938. In the first case, the F statistic is 316.54; now it is 3802.03. The numerator degrees of freedom are different as well. In the first case, the numerator degrees of freedom are 2; now they are 3. Which is correct?

Both are. Specifying the **hascons** option causes **regress** to adjust the ANOVA table and its associated statistics for the explanatory power of the constant. The regression in effect has a constant; it is just written in such a way that a separate constant is unnecessary. No such adjustment is made with the **noconstant** option.

❏

❏ Technical Note

When the **hascons** option is specified, **regress** checks to make sure that the model does in fact have a constant term. If **regress** cannot find a constant term, it automatically adds one. Fitting a model of **weight** on **length** and specifying the **hascons** option, we obtain

```
. regress weight length, hascons
(note: hascons false)
```

Source	SS	df	MS
Model	39461306.8	1	39461306.8
Residual	4632871.55	72	64345.4382
Total	44094178.4	73	604029.841

```
Number of obs =      74
F(  1,     72) =  613.27
Prob > F      =  0.0000
R-squared     =  0.8949
Adj R-squared =  0.8935
Root MSE      =  253.66
```

weight	Coef.	Std. Err.	t	P>|t|	[95% Conf. Interval]	
length	33.01988	1.333364	24.76	0.000	30.36187	35.67789
_cons	-3186.047	252.3113	-12.63	0.000	-3689.02	-2683.073

Even though we specified **hascons**, **regress** included a constant, anyway. It also added a note to our output: "note: hascons false".

❏

❑ Technical Note

Even if the model specification effectively includes a constant term, we need not specify the hascons option. regress is always on the lookout for collinear variables and drops them as necessary. For instance,

```
. regress weight length domestic foreign
```

Source	SS	df	MS
Model	39647744.7	2	19823872.3
Residual	4446433.7	71	62625.8268
Total	44094178.4	73	604029.841

Number of obs =	74
F(2, 71) =	316.54
Prob > F =	0.0000
R-squared =	0.8992
Adj R-squared =	0.8963
Root MSE =	250.25

weight	Coef.	Std. Err.	t	P>|t|	[95% Conf. Interval]	
length	31.44455	1.601234	19.64	0.000	28.25178	34.63732
domestic	(dropped)					
foreign	-133.6775	77.47615	-1.73	0.089	-288.1605	20.80555
_cons	-2850.25	315.9691	-9.02	0.000	-3480.274	-2220.225

❑

Robust standard errors

regress with the vce(robust) option substitutes a robust variance matrix calculation for the conventional calculation, or if vce(cluster *clustvar*) is specified, allows relaxing the assumption of independence within groups. How this method works is explained in [U] **20.15 Obtaining robust variance estimates**. Below we show how well this approach works.

▷ Example 4

Specifying the vce(robust) option is equivalent to requesting White-corrected standard errors in the presence of heteroskedasticity. We use the automobile data and, in the process of looking at the energy efficiency of cars, analyze a variable with considerable heteroskedasticity.

We will examine the amount of energy—measured in gallons of gasoline—that the cars in the data need to move 1,000 pounds of their weight 100 miles. We are going to examine the relative efficiency of foreign and domestic cars.

```
. gen gpmw = ((1/mpg)/weight)*100*1000
```

```
. summarize gpmw
```

Variable	Obs	Mean	Std. Dev.	Min	Max
gpmw	74	1.682184	.2426311	1.09553	2.30521

In these data, the engines consume between 1.10 and 2.31 gallons of gas to move 1,000 pounds of the car's weight 100 miles. If we ran a regression with conventional standard errors of gpmw on foreign, we would obtain

(Continued on next page)

```
. regress gpmw foreign
```

Source	SS	df	MS
Model	.936705572	1	.936705572
Residual	3.36079459	72	.046677703
Total	4.29750017	73	.058869865

Number of obs =   74
F(1,   72) =   20.07
Prob > F   = 0.0000
R-squared   = 0.2180
Adj R-squared = 0.2071
Root MSE   = .21605

gpmw	Coef.	Std. Err.	t	P>|t|	[95% Conf. Interval]
foreign	.2461526	.0549487	4.48	0.000	.1366143 .3556909
_cons	1.609004	.0299608	53.70	0.000	1.549278 1.66873

regress with the vce(robust) option, on the other hand, reports

```
. regress gpmw foreign, vce(robust)
```
Linear regression

Number of obs =   74
F(1,   72) =   13.13
Prob > F   = 0.0005
R-squared   = 0.2180
Root MSE   = .21605

gpmw	Coef.	Robust Std. Err.	t	P>|t|	[95% Conf. Interval]
foreign	.2461526	.0679238	3.62	0.001	.1107489 .3815563
_cons	1.609004	.0234535	68.60	0.000	1.56225 1.655758

The point estimates are the same (foreign cars need one-quarter gallon more gas), but the standard errors differ by roughly 20%. Conventional regression reports the 95% confidence interval as $[.14, .36]$, whereas the robust standard errors make the interval $[.11, .38]$.

Which is right? Notice that gpmw is a variable with considerable heteroskedasticity:

```
. tabulate foreign, summarize(gpmw)
```

Car type	Summary of gpmw Mean	Std. Dev.	Freq.
Domestic	1.6090039	.16845182	52
Foreign	1.8551565	.30186861	22
Total	1.6821844	.24263113	74

Thus here we favor the robust standard errors. In [U] **20.15 Obtaining robust variance estimates**, we show another example using linear regression where it makes little difference whether we specify vce(robust). The linear-regression assumptions were true, and we obtained nearly linear-regression results. The advantage of the robust estimate is that in neither case did we have to check assumptions.

◁

❑ Technical Note

regress purposefully suppresses displaying the ANOVA table when vce(robust) is specified, as it is no longer appropriate in a statistical sense, even though, mechanically, the numbers would be unchanged. That is, sums of squares remain unchanged, but the meaning of those sums is no longer relevant. The F statistic, for instance, is no longer based on sums of squares; it becomes a Wald test based on the robustly estimated variance matrix. Nevertheless, `regress` continues to report the R^2 and the root MSE even though both numbers are based on sums of squares and are, strictly speaking, irrelevant. In this, the root MSE is more in violation of the spirit of the robust estimator than is R^2. As a goodness-of-fit statistic, R^2 is still fine; just do not use it in formulas to obtain F statistics because those formulas no longer apply. The root MSE is valid in a literal sense—it is the square root of the mean squared error, but it is no longer an estimate of σ because there is no single σ; the variance of the residual varies observation by observation.

❑

▷ Example 5

Options `vce(hc2)` and `vce(hc3)` modify the robust variance calculation. In the context of linear regression without clustering, the idea behind the robust calculation is somehow to measure σ_j^2, the variance of the residual associated with the jth observation, and then to use that estimate to improve the estimated variance of $\widehat{\beta}$. Since residuals have (theoretically and practically) mean 0, one estimate of σ_j^2 is the observation's squared residual itself—u_j^2. A finite-sample correction could improve that by multiplying u_j^2 by $n/(n-k)$, and, as a matter of fact, `vce(robust)` uses $\{n/(n-k)\}u_j^2$ as its estimate of the residual's variance.

`vce(hc2)` and `vce(hc3)` use alternative estimators of the observation-specific variances. For instance, if the residuals are homoskedastic, we can show that the expected value of u_j^2 is $\sigma^2(1-h_{jj})$, where h_{jj} is the jth diagonal element of the projection (hat) matrix. h_{jj} has average value k/n, so $1-h_{jj}$ has average value $1-k/n = (n-k)/n$. Thus the default robust estimator $\widehat{\sigma}_j = \{n/(n-k)\}u_j^2$ amounts to dividing u_j^2 by the average of the expectation.

`vce(hc2)` divides u_j^2 by $1-h_{jj}$ itself, so it should yield better estimates if the residuals really are homoskedastic. `vce(hc3)` divides u_j^2 by $(1-h_{jj})^2$ and has no such clean interpretation. Davidson and MacKinnon (1993) show that $u_j^2/(1-h_{jj})^2$ approximates a more complicated estimator that they obtain by jackknifing (MacKinnon and White 1985).

Here are the results of refitting our efficiency model using `vce(hc2)` and `vce(hc3)`:

```
. regress gpmw foreign, vce(hc2)
Linear regression                               Number of obs =       74
                                                F(  1,     72) =    12.93
                                                Prob > F      =   0.0006
                                                R-squared     =   0.2180
                                                Root MSE      =  .21605
```

gpmw	Coef.	Robust HC2 Std. Err.	t	P>|t|	[95% Conf. Interval]
foreign	.2461526	.0684669	3.60	0.001	.1096662 .3826389
_cons	1.609004	.0233601	68.88	0.000	1.562437 1.655571

(Continued on next page)

```
. regress gpmw foreign, vce(hc3)
```

Linear regression

```
                                        Number of obs =      74
                                        F(  1,     72) =   12.38
                                        Prob > F      =  0.0008
                                        R-squared     =  0.2180
                                        Root MSE      =  .21605
```

gpmw	Coef.	Robust HC3 Std. Err.	t	P>|t|	[95% Conf. Interval]	
foreign	.2461526	.069969	3.52	0.001	.1066719	.3856332
_cons	1.609004	.023588	68.21	0.000	1.561982	1.656026

◁

▷ Example 6

The vce(cluster *clustvar*) option relaxes the assumption of independence. Below we have 28,534 observations on 4,711 women aged 14–46 years. Data were collected on these women between 1968 and 1988. We are going to fit a classic earnings model, and we begin by ignoring that each woman appears an average of 6.056 times in the data.

```
. use http://www.stata-press.com/data/r10/regsmpl
(NLS Women 14-26 in 1968)

. regress ln_wage age age2 tenure
```

Source	SS	df	MS			
Model	1054.52501	3	351.508335			
Residual	5360.43962	28097	.190783344			
Total	6414.96462	28100	.228290556			

```
                                        Number of obs =    28101
                                        F(  3, 28097) = 1842.45
                                        Prob > F      =  0.0000
                                        R-squared     =  0.1644
                                        Adj R-squared =  0.1643
                                        Root MSE      =  .43679
```

ln_wage	Coef.	Std. Err.	t	P>|t|	[95% Conf. Interval]	
age	.0752172	.0034736	21.65	0.000	.0684088	.0820257
age2	-.0010851	.0000575	-18.86	0.000	-.0011979	-.0009724
tenure	.0390877	.0007743	50.48	0.000	.0375699	.0406054
_cons	.3339821	.0504413	6.62	0.000	.2351148	.4328495

The number of observations in our model is 28,101 because Stata drops observations that have a missing value for one or more of the variables in the model. We can be reasonably certain that the standard errors reported above are meaningless. Without a doubt, a woman with higher-than-average wages in one year typically has higher-than-average wages in other years, and so the residuals are not independent. One way to deal with this would be to fit a random-effects model—and we are going to do that—but first we fit the model using regress specifying vce(cluster id), which treats only observations with different person ids as truly independent:

```
. regress ln_wage age age2 tenure, vce(cluster id)
```

Linear regression

Number of obs =   28101
F(3,   4698) = 748.82
Prob > F   = 0.0000
R-squared   = 0.1644
Root MSE   = .43679

(Std. Err. adjusted for 4699 clusters in idcode)

ln_wage	Coef.	Robust Std. Err.	t	P>|t|	[95% Conf. Interval]	
age	.0752172	.0045711	16.45	0.000	.0662557	.0841788
age2	-.0010851	.0000778	-13.94	0.000	-.0012377	-.0009325
tenure	.0390877	.0014425	27.10	0.000	.0362596	.0419157
_cons	.3339821	.0641918	5.20	0.000	.208136	.4598282

For comparison, we focus on the tenure coefficient, which in economics jargon can be interpreted as the rate of return for keeping your job. The 95% confidence interval we previously estimated—an interval we do not believe—is $[.038, .041]$. The robust interval is twice as wide, being $[.036, .042]$.

As we said, one correct way to fit this model is by random-effects regression. Here is the random-effects result:

```
. xtreg ln_wage age age2 tenure, re
```

Random-effects GLS regression
Group variable: idcode

Number of obs   =   28101
Number of groups   =   4699

R-sq: within = 0.1370
       between = 0.2154
       overall = 0.1608

Obs per group: min =   1
                     avg =   6.0
                     max =   15

Random effects u_i ~ Gaussian
corr(u_i, X)   = 0 (assumed)

Wald chi2(3)   = 4717.05
Prob > chi2   =   0.0000

ln_wage	Coef.	Std. Err.	z	P>|z|	[95% Conf. Interval]	
age	.0568296	.0026958	21.08	0.000	.0515459	.0621132
age2	-.0007566	.0000447	-16.93	0.000	-.0008441	-.000669
tenure	.0260135	.0007477	34.79	0.000	.0245481	.0274789
_cons	.6136792	.0394611	15.55	0.000	.5363368	.6910216
sigma_u	.33542449					
sigma_e	.29674679					
rho	.56095413	(fraction of variance due to u_i)				

Robust regression estimated the 95% interval $[.036, .042]$, and xtreg estimates $[.025, .027]$. Which is better? The random-effects regression estimator assumes a lot. We can check some of these assumptions by performing a Hausman test. Using estimates, we save the random-effects estimation results, and then we run the required fixed-effects regression to perform the test.

(Continued on next page)

```
. estimates store random

. xtreg ln_wage age age2 tenure, fe
Fixed-effects (within) regression              Number of obs      =      28101
Group variable: idcode                         Number of groups   =       4699

R-sq:  within  = 0.1375                         Obs per group: min =          1
       between = 0.2066                                        avg =        6.0
       overall = 0.1568                                        max =         15

                                               F(3,23399)         =    1243.00
corr(u_i, Xb)  = 0.1380                         Prob > F           =     0.0000
```

ln_wage	Coef.	Std. Err.	t	P>|t|	[95% Conf. Interval]	
age	.0522751	.002783	18.78	0.000	.0468202	.05773
age2	-.0006717	.0000461	-14.56	0.000	-.0007621	-.0005813
tenure	.021738	.000799	27.21	0.000	.020172	.023304
_cons	.687178	.0405944	16.93	0.000	.6076103	.7667456

sigma_u	.38743138		
sigma_e	.29674679		
rho	.6302569	(fraction of variance due to u_i)	

```
F test that all u_i=0:     F(4698, 23399) =      7.98          Prob > F = 0.0000
. hausman . random
```

	─── Coefficients ───			
	(b)	(B)	(b-B)	sqrt(diag(V_b-V_B))
	.	random	Difference	S.E.
age	.0522751	.0568296	-.0045545	.0006913
age2	-.0006717	-.0007566	.0000849	.0000115
tenure	.021738	.0260135	-.0042756	.0002816

```
                        b = consistent under Ho and Ha; obtained from xtreg
            B = inconsistent under Ha, efficient under Ho; obtained from xtreg
    Test:  Ho:  difference in coefficients not systematic
                chi2(3) = (b-B)'[(V_b-V_B)^(-1)](b-B)
                        =      336.62
              Prob>chi2 =      0.0000
```

The Hausman test casts grave suspicions on the random-effects model we just fitted, so we should be careful in interpreting those results.

Meanwhile, our robust regression results still stand, as long as we are careful about the interpretation. The correct interpretation is that, if the data collection were repeated (on women sampled the same way as in the original sample), and if we were to refit the model, 95% of the time we would expect the estimated coefficient on tenure to be in the range $[.036, .042]$.

Even with robust regression, we must be careful about going beyond that statement. Here the Hausman test is probably picking up something that differs within and between person, which would cast doubt on our robust regression model in terms of interpreting $[.036, .042]$ to contain the rate of return for keeping a job, economywide, for all women, without exception.

◁

Weighted regression

regress can perform weighted and unweighted regression. We indicate the weight by specifying the [*weight*] qualifier. By default, regress assumes analytic weights; see the technical note below.

▷ Example 7

We have census data recording the death rate (drate) and median age (medage) for each state. The data also record the region of the country in which each state is located and the overall population of the state:

```
. use http://www.stata-press.com/data/r10/census9
(1980 Census data by state)
. describe
Contains data from http://www.stata-press.com/data/r10/census9.dta
  obs:            50                          1980 Census data by state
  vars:            5                          13 Jan 2007 22:03
  size:         1,550 (99.9% of memory free)
```

variable name	storage type	display format	value label	variable label
state	str14	%-14s		State
drate	float	%9.0g		Death Rate
pop	long	%12.0gc		Population
medage	float	%9.2f		Median age
region	byte	%-8.0g	cenreg	Census region

Sorted by:

We can use the xi command to automatically create and include dummy variables for region. Since the variables in the regression reflect means rather than individual observations, the appropriate method of estimation is analytically weighted least squares (Johnston and DiNardo 1997), where the weight is total population:

```
. xi: regress drate medage I.region [w=pop]
I.region          _Iregion_1-4       (naturally coded; _Iregion_1 omitted)
(analytic weights assumed)
(sum of wgt is    2.2591e+08)
```

Source	SS	df	MS
Model	4096.6093	4	1024.15232
Residual	1238.40987	45	27.5202192
Total	5335.01916	49	108.877942

Number of obs = 50
F(4,   45) =   37.21
Prob > F   =   0.0000
R-squared   =   0.7679
Adj R-squared =   0.7472
Root MSE   =   5.246

drate	Coef.	Std. Err.	t	P>|t|	[95% Conf. Interval]
medage	4.283183	.5393329	7.94	0.000	3.196911 5.369455
_Iregion_2	.3138738	2.456431	0.13	0.899	-4.633632 5.26138
_Iregion_3	-1.438452	2.320244	-0.62	0.538	-6.111663 3.234758
_Iregion_4	-10.90629	2.681349	-4.07	0.000	-16.30681 -5.505777
_cons	-39.14727	17.23613	-2.27	0.028	-73.86262 -4.431915

To weight the regression by population, we added the qualifier [w=pop] to the end of the regress command. Our qualifier was vague (we did not say [aweight=pop]), but unless told otherwise, Stata assumes analytic weights for regress. Stata informed us that the sum of the weight is $2.2591 \cdot 10^8$; there were approximately 226 million people residing in the United States according to our 1980 data.

xi provides one way to include dummy variables and can be used with any estimation command. In the special case of linear regression, another alternative would be to use anova with the regress option. This would probably be better, but only because anova has special logic to fit models with many dummy variables, which uses less memory and computer time.

◁

❑ Technical Note

Once we fit a weighted regression, we can obtain the appropriately weighted variance–covariance matrix of the estimators using estat vce and perform appropriately weighted hypothesis tests using test.

In the weighted regression in the previous example, we see that _Iregion_4 is statistically significant but that _Iregion_2 and _Iregion_3 are not. We use test to test the joint significance of the region variables:

```
. test _Iregion_2 _Iregion_3 _Iregion_4
 ( 1)  _Iregion_2 = 0
 ( 2)  _Iregion_3 = 0
 ( 3)  _Iregion_4 = 0
       F(  3,    45) =    9.84
            Prob > F =    0.0000
```

The results indicate that the region variables are jointly significant.

❑

regress also accepts frequency weights (fweights). Frequency weights are appropriate when the data do not reflect cell means, but instead represent replicated observations. Specifying aweights or fweights will not change the parameter estimates, but it will change the corresponding significance levels.

For instance, if we specified [fweight=pop] in the weighted regression example above—which would be statistically incorrect—Stata would treat the data as if the data represented 226 million independent observations on death rates and median age. The data most certainly do not represent that—they represent 50 observations on state averages.

With aweights, Stata treats the number of observations on the process as the number of observations in the data. When we specify fweights, Stata treats the number of observations as if it were equal to the sum of the weights; see *Methods and Formulas* below.

❑ Technical Note

A popular request on the help line is to describe the effect of specifying [aweight=*exp*] with regress in terms of transformation of the dependent and independent variables. The mechanical answer is that typing

```
. regress y x1 x2 [aweight=n]
```

is equivalent to fitting the model

$$y_j \sqrt{n_j} = \beta_0 \sqrt{n_j} + \beta_1 x_{1j} \sqrt{n_j} + \beta_2 x_{2j} \sqrt{n_j} + u_j \sqrt{n_j}$$

This regression will reproduce the coefficients and covariance matrix produced by the aweighted regression. The mean squared errors (estimates of the variance of the residuals) will, however, be different. The transformed regression reports s_t^2, an estimate of $\mathrm{Var}(u_j \sqrt{n_j})$. The aweighted regression reports s_a^2, an estimate of $\mathrm{Var}(u_j \sqrt{n_j} \sqrt{N / \sum_k n_k})$, where N is the number of observations. Thus

$$s_a^2 = \frac{N}{\sum_k n_k} s_t^2 = \frac{s_t^2}{\overline{n}} \tag{1}$$

The logic for this adjustment is as follows: consider the model

$$y = \beta_0 + \beta_1 x_1 + \beta_2 x_2 + u$$

Assume that, were this model fitted on individuals, $\mathrm{Var}(u) = \sigma_u^2$, a constant. Assume that individual data are not available; what is available are averages $(\overline{y}_j, \overline{x}_{1j}, \overline{x}_{2j})$ for $j = 1, \ldots, N$, and each average is calculated over n_j observations. Then it is still true that

$$\overline{y}_j = \beta_0 + \beta_1 \overline{x}_{1j} + \beta_2 \overline{x}_{2j} + \overline{u}_j$$

where $\overline{u}_j$ is the average of n_j mean 0, variance σ_u^2 deviates and has variance $\sigma_{\overline{u}}^2 = \sigma_u^2/n_j$. Thus multiplying through by $\sqrt{n_j}$ produces

$$\overline{y}_j \sqrt{n_j} = \beta_0 \sqrt{n_j} + \beta_1 \overline{x}_{1j} \sqrt{n_j} + \beta_2 \overline{x}_{2j} \sqrt{n_j} + \overline{u}_j \sqrt{n_j}$$

and $\mathrm{Var}(\overline{u}_j \sqrt{n_j}) = \sigma_u^2$. The mean squared error, s_t^2, reported by fitting this transformed regression is an estimate of σ_u^2. The coefficients and covariance matrix could also be obtained by `aweighted regress`. The only difference would be in the reported mean squared error, which from (1) is $\sigma_u^2/\overline{n}$. On average, each observation in the data reflects the averages calculated over $\overline{n} = \sum_k n_k/N$ individuals, and thus this reported mean squared error is the average variance of an observation in the dataset. We can retrieve the estimate of σ_u^2 by multiplying the reported mean squared error by $\overline{n}$.

More generally, `aweight`s are used to solve general heteroskedasticity problems. In these cases, we have the model

$$y_j = \beta_0 + \beta_1 x_{1j} + \beta_2 x_{2j} + u_j$$

and the variance of u_j is thought to be proportional to a_j. If the variance is proportional to a_j, it is also proportional to αa_j, where α is any positive constant. Not quite arbitrarily, but with no loss of generality, we could choose $\alpha = \sum_k (1/a_k)/N$, the average value of the inverse of a_j. We can then write $\mathrm{Var}(u_j) = k\alpha a_j \sigma^2$, where k is the constant of proportionality that is no longer a function of the scale of the weights.

Dividing this regression through by the $\sqrt{a_j}$,

$$y_j/\sqrt{a_j} = \beta_0/\sqrt{a_j} + \beta_1 x_{1j}/\sqrt{a_j} + \beta_2 x_{2j}/\sqrt{a_j} + u_j/\sqrt{a_j}$$

produces a model with $\mathrm{Var}(u_j/\sqrt{a_j}) = k\alpha\sigma^2$, which is the constant part of $\mathrm{Var}(u_j)$. This variance is a function of α, the average of the reciprocal weights; if the weights are scaled arbitrarily, then so is this variance.

We can also fit this model by typing

```
. regress y x1 x2 [aweight=1/a]
```

This input will produce the same estimates of the coefficients and covariance matrix; the reported mean squared error is, from (1), $\left\{N/\sum_k(1/a_k)\right\}k\alpha\sigma^2 = k\sigma^2$. This variance is independent of the scale of a_j.

❑

Instrumental variables and two-stage least-squares regression

An alternate syntax for `regress` can be used to produce instrumental variables (two-stage least squares) estimates.

$$\underline{\text{regress }} depvar \left[varlist_1 \left[(varlist_2) \right] \right] \left[if \right] \left[in \right] \left[weight \right] \left[, regress\_options \right]$$

This syntax is used mainly by programmers developing estimators using the instrumental variables estimates as intermediate results. `ivregress` is normally used to directly fit these models; see [R] **ivregress**.

With this syntax, `regress` fits a structural equation of *depvar* on *varlist₁* using instrumental variables regression; (*varlist₂*) indicates the list of instrumental variables. With the exception of `vce(hc2)` and `vce(hc3)`, all standard `regress` options are allowed.

Saved Results

`regress` saves the following in e():

Scalars

e(N)	number of observations	e(F)	F statistic
e(mss)	model sum of squares	e(rmse)	root mean squared error
e(df_m)	model degrees of freedom	e(ll)	log likelihood under additional assumption
e(rss)	residual sum of squares		of i.i.d. normal errors
e(df_r)	residual degrees of freedom	e(ll_0)	log likelihood, constant-only model
e(r2)	R-squared	e(N_clust)	number of clusters
e(r2_a)	adjusted R-squared		

Macros

e(cmd)	regress	e(clustvar)	name of cluster variable
e(cmdline)	command as typed	e(vce)	*vcetype* specified in vce()
e(depvar)	name of dependent variable	e(vcetype)	title used to label Std. Err.
e(model)	ols or iv	e(properties)	b V
e(wtype)	weight type	e(estat_cmd)	program used to implement estat
e(wexp)	weight expression	e(predict)	program used to implement predict
e(title)	title in estimation output		
	when vce() is not ols		

Matrices

e(b)	coefficient vector	e(V)	variance–covariance matrix of the estimators

Functions

e(sample)	marks estimation sample

Methods and Formulas

Variables printed in lowercase and not boldfaced (e.g., x) are scalars. Variables printed in lowercase and boldfaced (e.g., **x**) are column vectors. Variables printed in uppercase and boldfaced (e.g., **X**) are matrices.

Let **v** be a column vector of weights specified by the user. If no weights are specified, $\mathbf{v} = \mathbf{1}$. Let **w** be a column vector of normalized weights. If no weights are specified or if the user specified `fweights` or `iweights`, $\mathbf{w} = \mathbf{v}$. Otherwise, $\mathbf{w} = \{\mathbf{v}/(\mathbf{1'v})\}(\mathbf{1'1})$.

The *number of observations*, n, is defined as $\mathbf{1}'\mathbf{w}$. For iweights, this is truncated to an integer. The *sum of the weights* is $\mathbf{1}'\mathbf{v}$. Define $c = 1$ if there is a constant in the regression and zero otherwise. Define k as the number of RHS variables (including the constant).

Let $\mathbf{X}$ denote the matrix of observations on the RHS variables, $\mathbf{y}$ the vector of observations on the left-hand-side variable, and $\mathbf{Z}$ the matrix of observations on the instruments. If the user specifies no instruments, then $\mathbf{Z} = \mathbf{X}$. In the following formulas, if the user specifies weights, then $\mathbf{X}'\mathbf{X}$, $\mathbf{X}'\mathbf{y}$, $\mathbf{y}'\mathbf{y}$, $\mathbf{Z}'\mathbf{Z}$, $\mathbf{Z}'\mathbf{X}$, and $\mathbf{Z}'\mathbf{y}$ are replaced by $\mathbf{X}'\mathbf{DX}$, $\mathbf{X}'\mathbf{Dy}$, $\mathbf{y}'\mathbf{Dy}$, $\mathbf{Z}'\mathbf{DZ}$, $\mathbf{Z}'\mathbf{DX}$, and $\mathbf{Z}'\mathbf{Dy}$, respectively, where $\mathbf{D}$ is a diagonal matrix whose diagonal elements are the elements of $\mathbf{w}$. We suppress the $\mathbf{D}$ below to simplify the notation.

If no instruments are specified, define $\mathbf{A}$ as $\mathbf{X}'\mathbf{X}$ and $\mathbf{a}$ as $\mathbf{X}'\mathbf{y}$. Otherwise, define $\mathbf{A}$ as $\mathbf{X}'\mathbf{Z}(\mathbf{Z}'\mathbf{Z})^{-1}(\mathbf{X}'\mathbf{Z})'$ and $\mathbf{a}$ as $\mathbf{X}'\mathbf{Z}(\mathbf{Z}'\mathbf{Z})^{-1}\mathbf{Z}'\mathbf{y}$.

The coefficient vector $\mathbf{b}$ is defined as $\mathbf{A}^{-1}\mathbf{a}$. Although not shown in the notation, unless hascons is specified, $\mathbf{A}$ and $\mathbf{a}$ are accumulated in deviation form and the constant is calculated separately. This comment applies to all statistics listed below.

The *total sum of squares*, TSS, equals $\mathbf{y}'\mathbf{y}$ if there is no intercept and $\mathbf{y}'\mathbf{y} - \left\{(\mathbf{1}'\mathbf{y})^2/n\right\}$ otherwise. The *degrees of freedom* are $n - c$.

The *error sum of squares*, ESS, is defined as $\mathbf{y}'\mathbf{y} - 2\mathbf{b}\mathbf{X}'\mathbf{y} + \mathbf{b}'\mathbf{X}'\mathbf{Xb}$ if there are instruments and as $\mathbf{y}'\mathbf{y} - \mathbf{b}'\mathbf{X}'\mathbf{y}$ otherwise. The *degrees of freedom* are $n - k$.

The *model sum of squares*, MSS, equals TSS − ESS. The *degrees of freedom* are $k - c$.

The *mean squared error*, s^2, is defined as $\text{ESS}/(n - k)$. The *root mean squared error* is s, its square root.

The F statistic with $k - c$ and $n - k$ degrees of freedom is defined as

$$F = \frac{\text{MSS}}{(k-c)s^2}$$

if no instruments are specified. If instruments are specified and $c = 1$, then F is defined as

$$F = \frac{(\mathbf{b} - \mathbf{c})'\mathbf{A}(\mathbf{b} - \mathbf{c})}{(k-1)s^2}$$

where $\mathbf{c}$ is a vector of $k - 1$ zeros and kth element $\mathbf{1}'\mathbf{y}/n$. Otherwise, F is defined as *missing*. (Here you may use the test command to construct any F test that you wish.)

The *R-squared*, R^2, is defined as $R^2 = 1 - \text{ESS}/\text{TSS}$.

The *adjusted R-squared*, R_a^2, is $1 - (1 - R^2)(n - c)/(n - k)$.

If vce(robust) is not specified, the conventional estimate of variance is $s^2\mathbf{A}^{-1}$. The handling of vce(robust) is described below.

A general notation for the robust variance calculation

Put aside all context of linear regression and the notation that goes with it—we will return to it. First, we are going to establish a notation for describing robust variance calculations.

The calculation formula for the robust variance calculation is

$$\widehat{\mathcal{V}} = q_c \widehat{\mathbf{V}}\left(\sum_{k=1}^{M} \mathbf{u}_k^{(G)\prime}\mathbf{u}_k^{(G)}\right)\widehat{\mathbf{V}}$$

where

$$\mathbf{u}_k^{(G)} = \sum_{j \in G_k} w_j \mathbf{u}_j$$

$G_1, G_2, \ldots, G_M$ are the clusters specified by vce(cluster *clustvar*), and w_j are the user-specified weights, normalized if aweights or pweights are specified and equal to 1 if no weights are specified.

For fweights without clusters, the variance formula is

$$\widehat{\mathcal{V}} = q_c \widehat{\mathbf{V}} \left(\sum_{j=1}^{N} w_j \mathbf{u}_j' \mathbf{u}_j \right) \widehat{\mathbf{V}}$$

which is the same as expanding the dataset and making the calculation on the unweighted data.

If vce(cluster *clustvar*) is not specified, $M = N$, and each cluster contains 1 observation. The inputs into this calculation are

1. $\widehat{\mathbf{V}}$, which is typically a conventionally calculated variance matrix;

2. $\mathbf{u}_j$, $j = 1, \ldots, N$, a row vector of scores; and

3. q_c, a constant finite-sample adjustment.

Thus we can now describe how estimators apply the robust calculation formula by defining $\widehat{\mathbf{V}}$, $\mathbf{u}_j$, and q_c.

Two definitions are popular enough for q_c to deserve a name. The regression-like formula for q_c (Fuller et al. 1986) is

$$q_c = \frac{N-1}{N-k} \frac{M}{M-1}$$

where M is the number of clusters and N is the number of observations. For weights, N refers to the sum of the weights if weights are frequency weights and the number of observations in the dataset (ignoring weights) in all other cases. Also note that, weighted or not, $M = N$ when vce(cluster *clustvar*) is not specified, and, then $q_c = N/(N-k)$.

The asymptotic-like formula for q_c is

$$q_c = \frac{M}{M-1}$$

where $M = N$ if vce(cluster *clustvar*) is not specified.

See [U] **20.15 Obtaining robust variance estimates** and [P] **_robust** for a discussion of the robust variance estimator and a development of these formulas.

Robust calculation for regress

For regress, $\widehat{\mathbf{V}} = \mathbf{A}^{-1}$. The other terms are

No instruments, vce(robust), but not vce(hc2) or vce(hc3),

$$\mathbf{u}_j = (y_j - \mathbf{x}_j \mathbf{b}) \mathbf{x}_j$$

and q_c is given by its regression-like definition.

No instruments, vce(hc2),

$$\mathbf{u}_j = \frac{1}{\sqrt{1 - h_{jj}}}(y_j - \mathbf{x}_j\mathbf{b})\mathbf{x}_j$$

where $q_c = 1$ and $h_{jj} = \mathbf{x}_j(\mathbf{X}'\mathbf{X})^{-1}\mathbf{x}_j'$.

No instruments, vce(hc3),

$$\mathbf{u}_j = \frac{1}{1 - h_{jj}}(y_j - \mathbf{x}_j\mathbf{b})\mathbf{x}_j$$

where $q_c = 1$ and $h_{jj} = \mathbf{x}_j(\mathbf{X}'\mathbf{X})^{-1}\mathbf{x}_j'$.

Instrumental variables,

$$\mathbf{u}_j = (y_j - \mathbf{x}_j\mathbf{b})\widehat{\mathbf{x}}_j$$

where q_c is given by its regression-like definition, and

$$\widehat{\mathbf{x}}_j' = \mathbf{P}\mathbf{z}_j'$$

where $\mathbf{P} = (\mathbf{X}'\mathbf{Z})(\mathbf{Z}'\mathbf{Z})^{-1}$.

Acknowledgments

The robust estimate of variance was first implemented in Stata by Mead Over, World Bank; Dean Jolliffe, U.S. Department of Agriculture; and Andrew Foster, Department of Economics, Brown University (Over, Jolliffe, and Foster 1996).

References

Alexandersson, A. 1998. gr32: Confidence ellipses. *Stata Technical Bulletin* 46: 10–13. Reprinted in *Stata Technical Bulletin Reprints*, vol. 8, pp. 54–57.

Chatterjee, S., and A. S. Hadi. 2006. *Regression Analysis by Example*. 4th ed. New York: Wiley.

Davidson, R., and J. G. MacKinnon. 1993. *Estimation and Inference in Econometrics*. New York: Oxford University Press.

Dohoo, I., W. Martin, and H. Stryhn. 2003. *Veterinary Epidemiologic Research*. Charlottetown, Prince Edward Island: AVC.

Draper, N., and H. Smith. 1998. *Applied Regression Analysis*. 3rd ed. New York: Wiley.

Dunnington, G. W. 1955. *Gauss: Titan of Science*. New York: Hafner Publishing.

Fuller, W. A., W. J. Kennedy, Jr., D. Schnell, G. Sullivan, H. J. Park. 1986. *PC Carp*. Ames, IA: Iowa State University.

Gillham, N. W. 2001. *A Life of Sir Francis Galton: From African Exploration to the Birth of Eugenics*. New York: Oxford University Press.

Gillispie, C. C. 1997. *Pierre-Simon Laplace, 1749–1827: A Life in Exact Science*. Princeton: Princeton University Press.

Hald, A. 1998. *A History of Mathematical Statistics from 1750 to 1930*. New York: Wiley.

Hamilton, L. C. 2006. *Statistics with Stata (Updated for Version 9)*. Belmont, CA: Duxbury.

Johnston, J., and J. DiNardo. 1997. *Econometric Methods*. 4th ed. New York: McGraw–Hill.

Kmenta, J. 1997. *Elements of Econometrics.* 2nd ed. Ann Arbor: University of Michigan Press.

Kohler, U., and F. Kreuter. 2005. *Data Analysis Using Stata.* College Station, TX: Stata Press.

Long, J. S., and J. Freese. 2000. sg152: Listing and interpreting transformed coefficients from certain regression models. *Stata Technical Bulletin* 57: 27–34. Reprinted in *Stata Technical Bulletin Reprints,* vol. 10, pp. 231–240.

MacKinnon, J. G., and H. White. 1985. Some heteroskedasticity consistent covariance matrix estimators with improved finite sample properties. *Journal of Econometrics* 29: 305–325.

Mosteller, F., and J. W. Tukey. 1977. *Data Analysis and Regression: A Second Course in Statistics.* Reading, MA: Addison–Wesley.

Over, M., D. Jolliffe, and A. Foster. 1996. sg46: Huber correction for two-stage least-squares estimates. *Stata Technical Bulletin* 29: 24–25. Reprinted in *Stata Technical Bulletin Reprints,* vol. 5, pp. 140–142.

Peracchi, F. 2001. *Econometrics.* Chichester, UK: Wiley.

Plackett, R. L. 1972. The discovery of the method of least squares. *Biometrika* 59: 239–251.

Rogers, W. H. 1991. smv2: Analyzing repeated measurements—some practical alternatives. *Stata Technical Bulletin* 4: 10–16. Reprinted in *Stata Technical Bulletin Reprints,* vol. 1, pp. 123–131.

Royston, J. P., and G. Ambler. 1998. sg79: Generalized additive models. *Stata Technical Bulletin* 42: 38–43. Reprinted in *Stata Technical Bulletin Reprints,* vol. 7, pp. 217–224.

Schonlau, M. 2005. Boosted regression (boosting): An introductory tutorial and a Stata plugin. *Stata Journal* 5: 330–354.

Stigler, S. M. 1986. *The History of Statistics: The Measurement of Uncertainty before 1900.* Cambridge, MA: Belknap Press.

Tyler, J. H. 1997. sg73: Table making program. *Stata Technical Bulletin* 40: 18–23. Reprinted in *Stata Technical Bulletin Reprints,* vol. 7, pp. 186–192.

Weesie, J. 1998. sg77: Regression analysis with multiplicative heteroskedasticity. *Stata Technical Bulletin* 42: 28–32. Reprinted in *Stata Technical Bulletin Reprints,* vol. 7, pp. 204–210.

Weisberg, S. 2005. *Applied Linear Regression.* 3rd ed. New York: Wiley.

Wooldridge, J. M. 2002. *Econometric Analysis of Cross Section and Panel Data.* Cambridge, MA: MIT Press.

——. 2006. *Introductory Econometrics: A Modern Approach.* 3rd ed. Cincinnati, OH: South-Western.

Zimmerman, F. 1998. sg93: Switching regressions. *Stata Technical Bulletin* 45: 30–33. Reprinted in *Stata Technical Bulletin Reprints,* vol. 8, pp. 183–186.

Also See

[R] **regress postestimation** — Postestimation tools for regress

[R] **regress postestimation time series** — Postestimation tools for regress with time series

[R] **anova** — Analysis of variance and covariance

[U] **20 Estimation and postestimation commands**

The history of regression is long and complicated: the books by Stigler (1986) and Hald (1998) are devoted largely to the story. Legendre published first on least squares in 1805. Gauss published later in 1809, but he had the idea earlier. Gauss, and especially Laplace, tied least squares to a normal errors assumption. The idea of the normal distribution can itself be traced back to De Moivre in 1733. Laplace discussed a variety of other estimation methods and error assumptions over his long career, while linear models long predate either innovation. Most of this work was linked to problems in astronomy and geodesy.

A second wave of ideas started when Galton used graphical and descriptive methods on data bearing on heredity to develop what he called regression. His term reflects the common phenomenon that characteristics of offspring are positively correlated with those of parents but with regression slope such that offspring "regress toward the mean". Galton's work was rather intuitive: contributions from Pearson, Edgeworth, Yule, and others introduced more formal machinery, developed related ideas on correlation, and extended application into the biological and social sciences. So most of the elements of regression as we know it were in place by 1900.

Pierre-Simon Laplace (1749–1827) was born in Normandy and was early recognized as a remarkable mathematician. He weathered a changing political climate well enough to rise to Minister of the Interior under Napoleon in 1799 (although only for 6 weeks) and to be made a Marquis by Louis XVIII in 1817. He made many contributions to mathematics and physics, his two main interests being theoretical astronomy and probability theory (including statistics). Laplace transforms are named for him.

Adrien-Marie Legendre (1752–1833) was born in Paris (or possibly in Toulouse) and educated in mathematics and physics. He worked in number theory, geometry, differential equations, calculus, function theory, applied mathematics, and geodesy. The Legendre polynomials are named for him. His main contribution to statistics is as one of the discoverers of least squares. He died in poverty, having refused to bow to political pressures.

Johann Carl Friedrich Gauss (1777–1855) was born in Braunschweig (Brunswick), now in Germany. He studied there and at Göttingen. His doctoral dissertation at the University of Helmstedt was a discussion of the fundamental theorem of algebra. He made many fundamental contributions to geometry, number theory, algebra, real analysis, differential equations, numerical analysis, statistics, astronomy, optics, geodesy, mechanics, and magnetism. An outstanding genius, Gauss worked mostly in isolation in Göttingen.

Francis Galton (1822–1911) was born in Birmingham, England, into a well-to-do family with many connections: he and Charles Darwin were first cousins. After an unsuccessful foray into medicine, he became independently wealthy at the death of his father. Galton traveled widely in Europe, the Middle East, and Africa, and became celebrated as an explorer and geographer. His pioneering work on weather maps helped in the identification of anticyclones, which he named. From about 1865, most of his work was centered on quantitative problems in biology, anthropology, and psychology. In a sense, Galton (re)invented regression, and he certainly named it. Galton also promoted the normal distribution, correlation approaches, and the use of median and selected quantiles as descriptive statistics. He was knighted in 1909.

Title

regress postestimation — Postestimation tools for regress

Description

The following postestimation commands are of special interest after `regress`:

command	description
dfbeta	DFBETA influence statistics
estat hettest	tests for heteroskedasticity
estat imtest	information matrix test
estat ovtest	Ramsey regression specification-error test for omitted variables
estat szroeter	Szroeter's rank test for heteroskedasticity
estat vif	variance inflation factors for the independent variables
acprplot	augmented component-plus-residual plot
avplot	added-variable plot
avplots	all added-variables plots in one image
cprplot	component-plus-residual plot
lvr2plot	leverage-versus-squared-residual plot
rvfplot	residual-versus-fitted plot
rvpplot	residual-versus-predictor plot

These commands are not appropriate after the `svy` prefix.

For information about these commands, see below.

The following standard postestimation commands are also available:

command	description
adjust	adjusted predictions of $\mathbf{x}\beta$
estat	AIC, BIC, VCE, and estimation sample summary
estat (svy)	postestimation statistics for survey data
estimates	cataloging estimation results
hausman	Hausman's specification test
lincom	point estimates, standard errors, testing, and inference for linear combinations of coefficients
linktest	link test for model specification
lrtest[1]	likelihood-ratio test
mfx	marginal effects or elasticities
nlcom	point estimates, standard errors, testing, and inference for nonlinear combinations of coefficients
predict	predictions, residuals, influence statistics, and other diagnostic measures
predictnl	point estimates, standard errors, testing, and inference for generalized predictions
suest	seemingly unrelated estimation
test	Wald tests for simple and composite linear hypotheses
testnl	Wald tests of nonlinear hypotheses

[1] lrtest is not appropriate with svy estimation results.

See the corresponding entries in the *Stata Base Reference Manual* for details, but see [SVY] **estat** for details about estat (svy).

For postestimation tests specific to time series, see [R] **regress postestimation time series**.

Special-interest postestimation commands

These commands provide tools for diagnosing sensitivity to individual observations, analyzing residuals, and assessing specification.

dfbeta will calculate one, more than one, or all the DFBETAs after regress. Although predict will also calculate DFBETAs, predict can do this for only one variable at a time. dfbeta is a convenience tool for those who want to calculate DFBETAs for multiple variables. The names for the new variables created are chosen automatically and begin with the letters DF.

estat hettest performs three versions of the Breusch–Pagan (1979) and Cook–Weisberg (1983) test for heteroskedasticity. All three versions of this test present evidence against the null hypothesis that $\mathbf{t} = \mathbf{0}$ in $\mathrm{Var}(e) = \sigma^2 \exp(\mathbf{zt})$. In the **normal** version, performed by default, the null hypothesis also includes the assumption that the regression disturbances are independent-normal draws with variance σ^2. The normality assumption is dropped from the null hypothesis in the **iid** and **fstat** versions, which respectively produce the score and F tests discussed in *Methods and Formulas*. If *varlist* is not specified, the fitted values are used for $\mathbf{z}$. If *varlist* or the option rhs is specified, the variables specified are used for $\mathbf{z}$.

estat imtest performs an information matrix test for the regression model and an orthogonal decomposition into tests for heteroskedasticity, skewness, and kurtosis due to Cameron and Trivedi (1990); White's test for homoskedasticity against unrestricted forms of heteroskedasticity (1980) is available as an option. White's test is usually similar to the first term of the Cameron–Trivedi decomposition.

`estat ovtest` performs two versions of the Ramsey (1969) regression specification-error test (RESET) for omitted variables. This test amounts to fitting $y = \mathbf{xb} + \mathbf{zt} + u$ and then testing $\mathbf{t} = \mathbf{0}$. If option `rhs` is not specified, powers of the fitted values are used for $\mathbf{z}$. If `rhs` is specified, powers of the individual elements of $\mathbf{x}$ are used.

`estat szroeter` performs Szroeter's rank test for heteroskedasticity for each of the variables in *varlist* or for the explanatory variables of the regression if `rhs` is specified.

`estat vif` calculates the centered or uncentered variance inflation factors (VIFs) for the independent variables specified in a linear regression model.

`acprplot` graphs an augmented component-plus-residual plot (a.k.a. augmented partial residual plot) as described by Mallows (1986). This seems to work better than the component-plus-residual plot for identifying nonlinearities in the data.

`avplot` graphs an added-variable plot (a.k.a. partial-regression leverage plot, partial regression plot, or adjusted partial residual plot) after `regress`. *indepvar* may be an independent variable (a.k.a. predictor, carrier, or covariate) that is currently in the model or not.

`avplots` graphs all the added-variable plots in one image.

`cprplot` graphs a component-plus-residual plot (a.k.a. partial residual plot) after `regress`. *indepvar* must be an independent variable that is currently in the model.

`lvr2plot` graphs a leverage-versus-squared-residual plot (a.k.a. L-R plot).

`rvfplot` graphs a residual-versus-fitted plot, a graph of the residuals against the fitted values.

`rvpplot` graphs a residual-versus-predictor plot (a.k.a. independent variable plot or carrier plot), a graph of the residuals against the specified predictor.

Syntax for predict

predict [*type*] *newvar* [*if*] [*in*] [, *statistic*]

statistic	description
xb	linear prediction; the default
<u>r</u>esiduals	residuals
<u>sc</u>ore	score; equivalent to residuals
rstandard	standardized residuals
rstudent	studentized (jackknifed) residuals
<u>c</u>ooksd	Cook's distance
<u>leverage</u> | <u>hat</u>	leverage (diagonal elements of hat matrix)
<u>pr</u>(*a*,*b*)	$\Pr(y_j \mid a < y_j < b)$
e(*a*,*b*)	$E(y_j \mid a < y_j < b)$
<u>ystar</u>(*a*,*b*)	$E(y_j^*)$, $y_j^* = \max\{a, \min(y_j, b)\}$
* <u>dfb</u>eta(*varname*)	DFBETA for *varname*
stdp	standard error of the linear prediction
stdf	standard error of the forecast
stdr	standard error of the residual
* <u>covr</u>atio	COVRATIO
* <u>dfits</u>	DFITS
* <u>welsch</u>	Welsch distance

Unstarred statistics are available both in and out of sample; type predict ... if e(sample) ... if wanted only for the estimation sample. Starred statistics are calculated only for the estimation sample, even when if e(sample) is not specified.

rstandard, rstudent, cooksd, leverage, dfbeta(), stdf, stdr, covratio, dfits, and welsch are not available if any vce() other than vce(ols) was specified with regress.

xb, residuals, score, and stdp are the only options allowed with svy estimation results.

where *a* and *b* may be numbers or variables; *a* missing ($a \geq .$) means $-\infty$, and *b* missing ($b \geq .$) means $+\infty$; see [U] **12.2.1 Missing values**.

Options for predict

xb, the default, calculates the linear prediction.

residuals calculates the residuals.

score is equivalent to residuals in linear regression.

rstandard calculates the standardized residuals.

rstudent calculates the studentized (jackknifed) residuals.

cooksd calculates the Cook's D influence statistic (Cook 1977).

leverage or hat calculates the diagonal elements of the projection hat matrix.

pr(*a*,*b*) calculates $\Pr(a < \mathbf{x}_j\mathbf{b} + u_j < b)$, the probability that $y_j|\mathbf{x}_j$ would be observed in the interval (a, b).

a and *b* may be specified as numbers or variable names; *lb* and *ub* are variable names; pr(20,30) calculates $\Pr(20 < \mathbf{x}_j\mathbf{b} + u_j < 30)$;

pr(*lb*,*ub*) calculates $\Pr(lb < \mathbf{x}_j\mathbf{b} + u_j < ub)$;
and pr(20,*ub*) calculates $\Pr(20 < \mathbf{x}_j\mathbf{b} + u_j < ub)$.

a missing ($a \geq .$) means $-\infty$; pr(.,30) calculates $\Pr(-\infty < \mathbf{x}_j\mathbf{b} + u_j < 30)$;
pr(*lb*,30) calculates $\Pr(-\infty < \mathbf{x}_j\mathbf{b} + u_j < 30)$ in observations for which $lb \geq .$
and calculates $\Pr(lb < \mathbf{x}_j\mathbf{b} + u_j < 30)$ elsewhere.

b missing ($b \geq .$) means $+\infty$; pr(20,.) calculates $\Pr(+\infty > \mathbf{x}_j\mathbf{b} + u_j > 20)$;
pr(20,*ub*) calculates $\Pr(+\infty > \mathbf{x}_j\mathbf{b} + u_j > 20)$ in observations for which $ub \geq .$
and calculates $\Pr(20 < \mathbf{x}_j\mathbf{b} + u_j < ub)$ elsewhere.

e(*a*,*b*) calculates $E(\mathbf{x}_j\mathbf{b} + u_j \mid a < \mathbf{x}_j\mathbf{b} + u_j < b)$, the expected value of $y_j|\mathbf{x}_j$ conditional on $y_j|\mathbf{x}_j$ being in the interval (a, b), meaning that $y_j|\mathbf{x}_j$ is censored.
a and *b* are specified as they are for pr().

ystar(*a*,*b*) calculates $E(y_j^*)$, where $y_j^* = a$ if $\mathbf{x}_j\mathbf{b} + u_j \leq a$, $y_j^* = b$ if $\mathbf{x}_j\mathbf{b} + u_j \geq b$, and $y_j^* = \mathbf{x}_j\mathbf{b} + u_j$ otherwise, meaning that y_j^* is truncated. *a* and *b* are specified as they are for pr().

dfbeta(*varname*) calculates the DFBETA for *varname*, the difference between the regression coefficient when the *j*th observation is included and excluded, said difference being scaled by the estimated standard error of the coefficient. *varname* must have been included among the regressors in the previously fitted model. The calculation is automatically restricted to the estimation subsample.

stdp calculates the standard error of the prediction, which can be thought of as the standard error of the predicted expected value or mean for the observation's covariate pattern. The standard error of the prediction is also referred to as the standard error of the fitted value.

stdf calculates the standard error of the forecast, which is the standard error of the point prediction for 1 observation. It is commonly referred to as the standard error of the future or forecast value. By construction, the standard errors produced by stdf are always larger than those produced by stdp; see *Methods and Formulas*.

stdr calculates the standard error of the residuals.

covratio calculates COVRATIO (Belsley, Kuh, and Welsch 1980), a measure of the influence of the *j*th observation based on considering the effect on the variance–covariance matrix of the estimates. The calculation is automatically restricted to the estimation subsample.

dfits calculates DFITS (Welsch and Kuh 1977) and attempts to summarize the information in the leverage versus residual-squared plot into a single statistic. The calculation is automatically restricted to the estimation subsample.

welsch calculates Welsch distance (Welsch 1982) and is a variation on dfits. The calculation is automatically restricted to the estimation subsample.

Syntax for dfbeta

dfbeta $\big[$*indepvar* $\big[$*indepvar* $[\ldots]\big]\big]$

Syntax for estat hettest

estat <u>het</u>test $\big[$*varlist*$\big]$ $\big[$, <u>rhs</u> $\big[$<u>normal</u>|<u>iid</u>|<u>f</u>stat$\big]$ <u>m</u>test$\big[$(*spec*)$\big]\big]$

Options for estat hettest

rhs specifies that tests for heteroskedasticity be performed for the right-hand-side (explanatory) variables of the fitted regression model. Option rhs may be combined with a *varlist*.

normal, the default, causes estat hettest to compute the original Breusch–Pagan/Cook–Weisberg test, which assumes that the regression disturbances are normally distributed.

iid causes estat hettest to compute the $N * R^2$ version of the score test that drops the normality assumption.

fstat causes estat hettest to compute the F statistic version that drops the normality assumption.

mtest [(*spec*)] specifies that multiple testing be performed. The argument specifies how p-values are adjusted. The following specifications *spec* are supported:

bonferroni	Bonferroni's multiple testing adjustment
holm	Holm's multiple testing adjustment
sidak	Šidák's multiple testing adjustment
noadjust	no adjustment is made for multiple testing

mtest may be specified without an argument. This is equivalent to specifying mtest(noadjust), that is, tests for the individual variables should be performed with unadjusted p-values. By default, estat hettest does not perform multiple testing. mtest [(*spec*)] may not be specified with iid or fstat.

Syntax for estat imtest

> estat imtest [, preserve white]

Options for estat imtest

preserve specifies that the data in memory be preserved, all variables and cases that are not needed in the calculations be dropped, and at the conclusion the original data be restored. This option is costly for large datasets. However, as estat imtest has to perform an auxiliary regression on $k(k+1)/2$ temporary variables, where k is the number of regressors, it may not be able to perform the test otherwise.

white specifies that White's original heteroskedasticity test also be performed.

Syntax for estat ovtest

> estat ovtest [, rhs]

Option for estat ovtest

rhs specifies that powers of the right-hand-side (explanatory) variables be used in the test rather than powers of the fitted values.

Syntax for estat szroeter

estat szroeter [*varlist*] [, rhs mtest(*spec*)]

Options for estat szroeter

rhs specifies that tests for heteroskedasticity be performed for the right-hand-side (explanatory) variables of the fitted regression model. Option rhs may be combined with a *varlist*.

mtest(*spec*) specifies that multiple testing be performed. The argument specifies how p-values are adjusted. The following specifications *spec* are supported:

bonferroni	Bonferroni's multiple testing adjustment
holm	Holm's multiple testing adjustment
sidak	Šidák's multiple testing adjustment
noadjust	no adjustment is made for multiple testing

estat szroeter always performs multiple testing. By default, it does not adjust the p-values.

Syntax for estat vif

estat vif [, uncentered]

Option for estat vif

uncentered requests that the computation of the uncentered variance inflation factors. This option is often used to detect the collinearity of the regressors with the constant. estat vif, uncentered may be used after regression models fitted without the constant term.

Syntax for acprplot

acprplot *indepvar* [, *acprplot_options*]

acprplot_options	description
Plot	
marker_options	change look of markers (color, size, etc.)
marker_label_options	add marker labels; change look or position
Reference line	
<u>rl</u>opts(*cline_options*)	affect rendition of the reference line
Options	
<u>lowess</u>	add a lowess smooth of the plotted points
<u>ls</u>opts(*lowess_options*)	affect rendition of the lowess smooth
<u>mspline</u>	add median spline of the plotted points
<u>ms</u>opts(*mspline_options*)	affect rendition of the spline
Add plots	
<u>addplot</u>(*plot*)	add other plots to the generated graph
Y axis, X axis, Titles, Legend, Overall	
twoway_options	any options other than by() documented in [G] ***twoway_options***

Options for acprplot

⌐ Plot ⌐

marker_options affect the rendition of markers drawn at the plotted points, including their shape, size, color, and outline; see [G] ***marker_options***.

marker_label_options specify if and how the markers are to be labeled; see [G] ***marker_label_options***.

⌐ Reference line ⌐

rlopts(*cline_options*) affects the rendition of the reference line. See [G] ***cline_options***.

⌐ Options ⌐

lowess adds a lowess smooth of the plotted points to assist in detecting nonlinearities.

lsopts(*lowess_options*) affects the rendition of the lowess smooth. For an explanation of these options, especially the bwidth() option, see [R] **lowess**. Specifying lsopts() implies the lowess option.

mspline adds a median spline of the plotted points to assist in detecting nonlinearities.

msopts(*mspline_options*) affects the rendition of the spline. For an explanation of these options, especially the bands() option, see [G] **graph twoway mspline**. Specifying msopts() implies the mspline option.

⌐ Add plots ⌐

addplot(*plot*) provides a way to add other plots to the generated graph. See [G] *addplot_option*.

⌐ Y axis, X axis, Titles, Legend, Overall ⌐

twoway_options are any of the options documented in [G] *twoway_options*, excluding by(). These
include options for titling the graph (see [G] *title_options*) and for saving the graph to disk (see
[G] *saving_option*).

Syntax for avplot

avplot *indepvar* [, *avplot_options*]

avplot_options	description
Plot	
marker_options	change look of markers (color, size, etc.)
marker_label_options	add marker labels; change look or position
Reference line	
rlopts(*cline_options*)	affect rendition of the reference line
Add plots	
addplot(*plot*)	add other plots to the generated graph
Y axis, X axis, Titles, Legend, Overall	
twoway_options	any options other than by() documented in [G] *twoway_options*

Options for avplot

⌐ Plot ⌐

marker_options affect the rendition of markers drawn at the plotted points, including their shape,
size, color, and outline; see [G] *marker_options*.

marker_label_options specify if and how the markers are to be labeled; see [G] *marker_label_options*.

⌐ Reference line ⌐

rlopts(*cline_options*) affects the rendition of the reference line. See [G] *cline_options*.

⌐ Add plots ⌐

addplot(*plot*) provides a way to add other plots to the generated graph. See [G] *addplot_option*.

⌐ Y axis, X axis, Titles, Legend, Overall ⌐

twoway_options are any of the options documented in [G] *twoway_options*, excluding by(). These
include options for titling the graph (see [G] *title_options*) and for saving the graph to disk (see
[G] *saving_option*).

Syntax for avplots

avplots [, *avplots_options*]

avplots_options	description
Plot	
marker_options	change look of markers (color, size, etc.)
marker_label_options	add marker labels; change look or position
combine_options	any of the options documented in [G] **graph combine**
Reference line	
<u>rl</u>opts(*cline_options*)	affect rendition of the reference line
Y axis, X axis, Titles, Legend, Overall	
twoway_options	any options other than by() documented in [G] *twoway_options*

Options for avplots

▻ Plot

marker_options affect the rendition of markers drawn at the plotted points, including their shape, size, color, and outline; see [G] *marker_options*.

marker_label_options specify if and how the markers are to be labeled; see [G] *marker_label_options*.

combine_options are any of the options documented in [G] **graph combine**. These include options for titling the graph (see [G] *title_options*) and for saving the graph to disk (see [G] *saving_option*).

▻ Reference line

rlopts(*cline_options*) affects the rendition of the reference line. See [G] *cline_options*.

▻ Y axis, X axis, Titles, Legend, Overall

twoway_options are any of the options documented in [G] *twoway_options*, excluding by(). These include options for titling the graph (see [G] *title_options*) and for saving the graph to disk (see [G] *saving_option*).

Syntax for cprplot

cprplot *indepvar* [, *cprplot_options*]

(Continued on next page)

cprplot_options	description
Plot	
marker_options	change look of markers (color, size, etc.)
marker_label_options	add marker labels; change look or position
Reference line	
<u>rlopts</u>(*cline_options*)	affect rendition of the reference line
Options	
<u>lowess</u>	add a lowess smooth of the plotted points
<u>lsopts</u>(*lowess_options*)	affect rendition of the lowess smooth
<u>mspline</u>	add median spline of the plotted points
<u>msopts</u>(*mspline_options*)	affect rendition of the spline
Add plots	
<u>addplot</u>(*plot*)	add other plots to the generated graph
Y axis, X axis, Titles, Legend, Overall	
twoway_options	any options other than by() documented in [G] *twoway_options*

Options for cprplot

⌐ Plot ⌐

marker_options affect the rendition of markers drawn at the plotted points, including their shape, size, color, and outline; see [G] *marker_options*.

marker_label_options specify if and how the markers are to be labeled; see [G] *marker_label_options*.

⌐ Reference line ⌐

rlopts(*cline_options*) affects the rendition of the reference line. See [G] *cline_options*.

⌐ Options ⌐

lowess adds a lowess smooth of the plotted points to assist in detecting nonlinearities.

lsopts(*lowess_options*) affects the rendition of the lowess smooth. For an explanation of these options, especially the bwidth() option, see [R] **lowess**. Specifying lsopts() implies the lowess option.

mspline adds a median spline of the plotted points to assist in detecting nonlinearities.

msopts(*mspline_options*) affects the rendition of the spline. For an explanation of these options, especially the bands() option, see [G] **graph twoway mspline**. Specifying msopts() implies the mspline option.

⌐ Add plots ⌐

addplot(*plot*) provides a way to add other plots to the generated graph. See [G] *addplot_option*.

⌐ Y axis, X axis, Titles, Legend, Overall ⌐

twoway_options are any of the options documented in [G] *twoway_options*, excluding by(). These include options for titling the graph (see [G] *title_options*) and for saving the graph to disk (see [G] *saving_option*).

Syntax for lvr2plot

lvr2plot [, *lvr2plot_options*]

lvr2plot_options	description
Plot	
marker_options	change look of markers (color, size, etc.)
marker_label_options	add marker labels; change look or position
Add plots	
addplot(*plot*)	add other plots to the generated graph
Y axis, X axis, Titles, Legend, Overall	
twoway_options	any options other than by() documented in [G] ***twoway_options***

Options for lvr2plot

⌐ Plot ⌐

marker_options affect the rendition of markers drawn at the plotted points, including their shape, size, color, and outline; see [G] ***marker_options***.

marker_label_options specify if and how the markers are to be labeled; see [G] ***marker_label_options***.

⌐ Add plots ⌐

addplot(*plot*) provides a way to add other plots to the generated graph; see [G] ***addplot_option***.

⌐ Y axis, X axis, Titles, Legend, Overall ⌐

twoway_options are any of the options documented in [G] ***twoway_options***, excluding by(). These include options for titling the graph (see [G] ***title_options***) and for saving the graph to disk (see [G] ***saving_option***).

Syntax for rvfplot

rvfplot [, *rvfplot_options*]

rvfplot_options	description
Plot	
marker_options	change look of markers (color, size, etc.)
marker_label_options	add marker labels; change look or position
Add plots	
addplot(*plot*)	add plots to the generated graph
Y axis, X axis, Titles, Legend, Overall	
twoway_options	any options other than by() documented in [G] ***twoway_options***

Options for rvfplot

⌐ Plot ⌐

marker_options affect the rendition of markers drawn at the plotted points, including their shape, size, color, and outline; see [G] *marker_options*.

marker_label_options specify if and how the markers are to be labeled; see [G] *marker_label_options*.

⌐ Add plots ⌐

addplot(*plot*) provides a way to add plots to the generated graph. See [G] *addplot_option*.

⌐ Y axis, X axis, Titles, Legend, Overall ⌐

twoway_options are any of the options documented in [G] *twoway_options*, excluding by(). These include options for titling the graph (see [G] *title_options*) and for saving the graph to disk (see [G] *saving_option*).

Syntax for rvpplot

rvpplot *indepvar* [, *rvpplot_options*]

rvpplot_options	description
Plot	
marker_options	change look of markers (color, size, etc.)
marker_label_options	add marker labels; change look or position
Add plots	
addplot(*plot*)	add other plots to the generated graph
Y axis, X axis, Titles, Legend, Overall	
twoway_options	any options other than by() documented in [G] *twoway_options*

Options for rvpplot

⌐ Plot ⌐

marker_options affect the rendition of markers drawn at the plotted points, including their shape, size, color, and outline; see [G] *marker_options*.

marker_label_options specify if and how the markers are to be labeled; see [G] *marker_label_options*.

⌐ Add plots ⌐

addplot(*plot*) provides a way to add other plots to the generated graph; see [G] *addplot_option*.

⌐ Y axis, X axis, Titles, Legend, Overall ⌐

twoway_options are any of the options documented in [G] *twoway_options*, excluding by(). These include options for titling the graph (see [G] *title_options*) and for saving the graph to disk (see [G] *saving_option*).

Remarks

Remarks are presented under the following headings:

Fitted values and residuals
Prediction standard errors
Prediction with weighted data
Residual-versus-fitted plots
Added-variable plots
Component-plus-residual plots
Residual-versus-predictor plots
Leverage statistics
L-R plots
Standardized and studentized residuals
DFITS, Cook's Distance, and Welsch Distance
COVRATIO
DFBETAs
Formal tests for violations of assumptions
Variance inflation factors

Many of these commands concern identifying influential data in linear regression. This is, unfortunately, a field that is dominated by jargon, codified and partially begun by Belsley, Kuh, and
Welsch (1980). In the words of Chatterjee and Hadi (1986, 416), "Belsley, Kuh, and Welsch's book,
Regression Diagnostics, was a very valuable contribution to the statistical literature, but it unleashed
on an unsuspecting statistical community a computer speak (à la Orwell), the likes of which we
have never seen." Things have only gotten worse since then. Chatterjee and Hadi's (1986, 1988) own
attempts to clean up the jargon did not improve matters (see Hoaglin and Kempthorne 1986, Velleman
1986, and Welsch 1986). We apologize for the jargon, and for our contribution to the jargon in the
form of inelegant command names, we apologize most of all.

Model *sensitivity* refers to how estimates are affected by subsets of our data. Imagine data on y
and x, and assume that the data are to be fit by the regression $y_i = \alpha + \beta x_i + \epsilon_i$. The regression
estimates of α and β are a and b, respectively. Now imagine that the estimated a and b would be
different if a small portion of the dataset, perhaps even one observation, were deleted. As a data
analyst, you would like to think that you are summarizing tendencies that apply to all the data, but
you have just been told that the model you fitted is unduly influenced by one point or just a few
points and that, as a matter of fact, there is another model that applies to the rest of the data—a
model that you have ignored. The search for subsets of the data that, if deleted, would change the
results markedly is a predominant theme of this entry.

There are three key issues in identifying model sensitivity to individual observations, which go by
the names *residuals*, *leverage*, and *influence*. In our $y_i = a + bx_i + e_i$ regression, the residuals are,
of course, e_i—they reveal how much our fitted value $\widehat{y}_i = a + bx_i$ differs from the observed y_i. A
point (x_i, y_i) with a corresponding large residual is called an outlier. Say that you are interested in
outliers because you somehow think that such points will exert undue influence on your estimates.
Your feelings are generally right, but there are exceptions. A point might have a huge residual and
yet not affect the estimated b at all. Nevertheless, studying observations with large residuals almost
always pays off.

(x_i, y_i) can be an outlier in another way—just as y_i can be far from $\widehat{y}_i$, x_i can be far from
the center of mass of the other x's. Such an "outlier" should interest you just as much as the more
traditional outliers. Picture a scatterplot of y against x with thousands of points in some sort of mass
at the lower left of the graph and one point at the upper right of the graph. Now run a regression
line through the points—the regression line will come close to the point at the upper right of the
graph and may in fact, go through it. That is, this isolated point will not appear as an outlier as
measured by residuals because its residual will be small. Yet this point might have a dramatic effect
on our resulting estimates in the sense that, were you to delete the point, the estimates would change

markedly. Such a point is said to have high leverage. Just as with traditional outliers, a high leverage point does not necessarily have an undue effect on regression estimates, but if it does not, it is more the exception than the rule.

Now all this is a most unsatisfactory state of affairs. Points with large residuals may, but need not, have a large effect on our results, and points with small residuals may still have a large effect. Points with high leverage may, but need not, have a large effect on our results, and points with low leverage may still have a large effect. Can't you identify the influential points and simply have the computer list them for you? You can, but you will have to define what you mean by "influential".

"Influential" is defined with respect to some statistic. For instance, you might ask which points in your data have a large effect on your estimated a, which points have a large effect on your estimated b, which points have a large effect on your estimated standard error of b, and so on, but do not be surprised when the answers to these questions are different. In any case, obtaining such measures is not difficult—all you have to do is fit the regression excluding each observation one at a time and record the statistic of interest which, in the day of the modern computer, is not too onerous. Moreover, you can save considerable computer time by doing algebra ahead of time and working out formulas that will calculate the same answers as if you ran each of the regressions. (Ignore the question of pairs of observations that, together, exert undue influence, and triples, and so on, which remains largely unsolved and for which the brute force fit-every-possible-regression procedure is not a viable alternative.)

Fitted values and residuals

Typing predict *newvar* with no options creates *newvar* containing the fitted values. Typing predict *newvar*, resid creates *newvar* containing the residuals.

▷ Example 1

Using the example from [R] **predict**, we have data on automobiles, including the mileage rating (mpg), the car's weight (weight), and whether the car is foreign (foreign). We wish to fit the following model:

$$mpg = \beta_1 weight + \beta_2 weight^2 + \beta_3 foreign + \beta_4$$

We first create the $weight^2$ variable and then type the regress command:

```
. use http://www.stata-press.com/data/r10/auto
(1978 Automobile Data)

. generate weight2 = weight^2

. regress mpg weight weight2 foreign
```

Source	SS	df	MS		Number of obs =	74
					F(3, 70) =	52.25
Model	1689.15372	3	563.05124		Prob > F =	0.0000
Residual	754.30574	70	10.7757963		R-squared =	0.6913
					Adj R-squared =	0.6781
Total	2443.45946	73	33.4720474		Root MSE =	3.2827

mpg	Coef.	Std. Err.	t	P>|t|	[95% Conf. Interval]	
weight	-.0165729	.0039692	-4.18	0.000	-.0244892	-.0086567
weight2	1.59e-06	6.25e-07	2.55	0.013	3.45e-07	2.84e-06
foreign	-2.2035	1.059246	-2.08	0.041	-4.3161	-.0909002
_cons	56.53884	6.197383	9.12	0.000	44.17855	68.89913

That done, we can now obtain the predicted values from the regression. We will store them in a new variable called pmpg by typing `predict pmpg`. Since `predict` produces no output, we will follow that by summarizing our predicted and observed values.

```
. predict pmpg
(option xb assumed; fitted values)

. summarize pmpg mpg
```

Variable	Obs	Mean	Std. Dev.	Min	Max
pmpg	74	21.2973	4.810311	13.59953	31.86288
mpg	74	21.2973	5.785503	12	41

◁

▷ Example 2: Out-of-sample predictions

We can just as easily obtain predicted values from the model by using a wholly different dataset from the one on which the model was fitted. The only requirement is that the data have the necessary variables, which here are weight, weight2, and foreign.

Using the data on two new cars (the Pontiac Sunbird and the Volvo 260) from the newautos.dta dataset, we can obtain out-of-sample predictions (or forecasts) by typing

```
. use http://www.stata-press.com/data/r10/newautos, clear
(New Automobile Models)

. generate weight2=weight^2

. predict mpg
(option xb assumed; fitted values)

. list, divider
```

	make	weight	foreign	weight2	mpg
1.	Pont. Sunbird	2690	Domestic	7236100	23.47137
2.	Volvo 260	3170	Foreign	1.00e+07	17.78846

The Pontiac Sunbird has a predicted mileage rating of 23.5 mpg, whereas the Volvo 260 has a predicted rating of 17.8 mpg. In comparison, the actual mileage ratings are 24 for the Pontiac and 17 for the Volvo.

◁

Prediction standard errors

`predict` can calculate the standard error of the forecast (`stdf` option), the standard error of the prediction (`stdp` option), and the standard error of the residual (`stdr` option). It is easy to confuse `stdf` and `stdp` because both are often called the prediction error. Consider the prediction $\widehat{y}_j = \mathbf{x}_j \mathbf{b}$, where $\mathbf{b}$ is the estimated coefficient (column) vector and $\mathbf{x}_j$ is a (row) vector of independent variables for which you want the prediction. First, $\widehat{y}_j$ has a variance due to the variance of the estimated coefficient vector $\mathbf{b}$,

$$\mathrm{Var}(\widehat{y}_j) = \mathrm{Var}(\mathbf{x}_j \mathbf{b}) = s^2 h_j$$

where $h_j = \mathbf{x}_j (\mathbf{X}'\mathbf{X})^{-1} \mathbf{x}_j'$ and s^2 is the mean squared error of the regression. Do not panic over the algebra—just remember that $\mathrm{Var}(\widehat{y}_j) = s^2 h_j$, whatever s^2 and h_j are. `stdp` calculates this quantity. This is the error in the prediction due to the uncertainty about $\mathbf{b}$.

If you are about to hand this number out as your forecast, however, there is another error. According to your model, the true value of y_j is given by

$$y_j = \mathbf{x}_j \mathbf{b} + \epsilon_j = \widehat{y}_j + \epsilon_j$$

and thus the $\text{Var}(y_j) = \text{Var}(\widehat{y}_j) + \text{Var}(\epsilon_j) = s^2 h_j + s^2$, which is the square of stdf. stdf, then, is the sum of the error in the prediction plus the residual error.

stdr has to do with an analysis-of-variance decomposition of s^2, the estimated variance of y. The standard error of the prediction is $s^2 h_j$, and therefore $s^2 h_j + s^2(1 - h_j) = s^2$ decomposes s^2 into the prediction and residual variances.

▷ Example 3: standard error of the forecast

Returning to our model of mpg on weight, weight^2, and foreign, we previously predicted the mileage rating for the Pontiac Sunbird and Volvo 260 as 23.5 and 17.8 mpg, respectively. We now want to put a standard error around our forecast. Remember, the data for these two cars were in newautos.dta:

```
. use http://www.stata-press.com/data/r10/newautos, clear
(New Automobile Models)
. gen weight2=weight*weight
. predict mpg
(option xb assumed; fitted values)
. predict se_mpg, stdf
. list, divider
```

	make	weight	foreign	weight2	mpg	se_mpg
1.	Pont. Sunbird	2690	Domestic	7236100	23.47137	3.341823
2.	Volvo 260	3170	Foreign	1.00e+07	17.78846	3.438714

Thus an approximate 95% confidence interval for the mileage rating of the Volvo 260 is $17.8 \pm 2 \cdot 3.44 = [10.92, 24.68]$.

◁

Prediction with weighted data

predict can be used after frequency-weighted (fweight) estimation, just as it is used after unweighted estimation. The technical note below concerns the use of predict after analytically weighted (aweight) estimation.

❑ Technical Note

After analytically weighted estimation, predict is willing to calculate only the prediction (no options), residual (residual option), standard error of the prediction (stdp option), and diagonal elements of the projection matrix (hat option). Moreover, the results produced by hat need to be adjusted, as will be described. For analytically weighted estimation, the standard error of the forecast and residuals, the standardized and studentized residuals, and Cook's D are not statistically well-defined concepts.

To obtain the correct values of the diagonal elements of the hat matrix, you can use `predict` with the `hat` option to make a first, partially adjusted calculation, and then follow that by completing the adjustment. Assume that you are fitting a linear regression model weighting the data with the variable w (`[aweight=w]`). Begin by creating a new variable w0:

```
. predict resid if e(sample), resid
. summarize w if resid < . & e(sample)
. gen w0=w/r(mean)
```

Some caution is necessary at this step—the `summarize w` must be performed on the same sample that was used to fit the model, which means that you must include if `e(sample)` to restrict the prediction to the estimation sample. You created the residual and then included the modifier 'if resid < .' so that if the dependent variable or any of the independent variables is missing, the corresponding observations will be excluded from the calculation of the average value of the original weight.

To correct `predict`'s `hat` calculation, multiply the result by w0:

```
. predict myhat, hat
. replace myhat = w0 * myhat
```

❑

Residual-versus-fitted plots

▷ Example 4: rvfplot

Using the automobile dataset described in [U] **1.2.1 Sample datasets**, we will use `regress` to fit a model of `price` on `weight`, `mpg`, `foreign`, and the interaction of `foreign` with `mpg`.

```
. use http://www.stata-press.com/data/r10/auto, clear
(1978 Automobile Data)
. generate forXmpg=foreign*mpg
. regress price weight mpg forXmpg foreign
```

Source	SS	df	MS		
Model	350319665	4	87579916.3	Number of obs =	74
Residual	284745731	69	4126749.72	F(4, 69) =	21.22
				Prob > F =	0.0000
				R-squared =	0.5516
Total	635065396	73	8699525.97	Adj R-squared =	0.5256
				Root MSE =	2031.4

price	Coef.	Std. Err.	t	P>|t|	[95% Conf. Interval]	
weight	4.613589	.7254961	6.36	0.000	3.166263	6.060914
mpg	263.1875	110.7961	2.38	0.020	42.15527	484.2197
forXmpg	-307.2166	108.5307	-2.83	0.006	-523.7294	-90.70368
foreign	11240.33	2751.681	4.08	0.000	5750.878	16729.78
_cons	-14449.58	4425.72	-3.26	0.002	-23278.65	-5620.51

Once we have fitted a model, we may use any of the regression diagnostics commands. `rvfplot` (read residual-versus-fitted plot) graphs the residuals against the fitted values:

. rvfplot, yline(0)

All the diagnostic plot commands allow the options of `graph twoway` and `graph twoway scatter`; we specified a `yline(0)` to draw a line across the graph at $y = 0$; see [G] **graph twoway scatter**.

In a well-fitted model, there should be no pattern to the residuals plotted against the fitted values—something not true of our model. Ignoring the two outliers at the top center of the graph, we see curvature in the pattern of the residuals, suggesting a violation of the assumption that `price` is linear in our independent variables. We might also have seen increasing or decreasing variation in the residuals—heteroskedasticity. Any pattern whatsoever indicates a violation of the least-squares assumptions.

◁

Added-variable plots

▷ Example 5: avplot

We continue with our price model, and another diagnostic graph is provided by `avplot` (read added-variable plot, also known as the partial-regression leverage plot).

One of the wonderful features of one-regressor regressions (regressions of y on one x) is that we can graph the data and the regression line. There is no easier way to understand the regression than to examine such a graph. Unfortunately, we cannot do this when we have more than one regressor. With two regressors, it is still theoretically possible—the graph must be drawn in three dimensions, but with three or more regressors no graph is possible.

The added-variable plot is an attempt to project multidimensional data back to the two-dimensional world for each of the original regressors. This is, of course, impossible without making some concessions. Call the coordinates on an added-variable plot y and x. The added-variable plot has the following properties:

1. There is a one-to-one correspondence between (x_i, y_i) and the ith observation used in the original regression.

2. A regression of y on x has the same coefficient and standard error (up to a degree-of-freedom adjustment) as the estimated coefficient and standard error for the regressor in the original regression.

3. The "outlierness" of each observation in determining the slope is in some sense preserved.

It is equally important to note the properties that are not listed. The y and x coordinates of the added-variable plot cannot be used to identify functional form, or, at least, not well (see Mallows 1986). In the construction of the added-variable plot, the relationship between y and x is forced to be linear.

Let us examine the added-variable plot for mpg in our regression of price on weight, mpg, forXmpg, and foreign:

    . avplot mpg

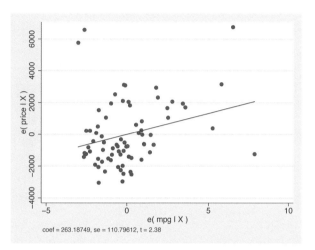

This graph suggests a problem in determining the coefficient on mpg. Were this a one-regressor regression, the two points at the top-left corner and the one at the top right would cause us concern, and so it does in our more complicated multiple-regressor case. To identify the problem points, we retyped our command, modifying it to read avplot mpg, mlabel(make), and discovered that the two cars at the top left are the Cadillac Eldorado and the Lincoln Versailles; the point at the top right is the Cadillac Seville. These three cars account for 100% of the luxury cars in our data, suggesting that our model is misspecified. By the way, the point at the lower right of the graph, also cause for concern, is the Plymouth Arrow, our data-entry error.

◁

❑ Technical Note

Stata's avplot command can be used with regressors already in the model, as we just did, or with potential regressors not yet in the model. In either case, avplot will produce the correct graph. The name "added-variable plot" is unfortunate in the case when the variable is already among the list of regressors but is, we think, still preferable to the name "partial-regression leverage plot" assigned by Belsley, Kuh, and Welsch (1980, 30) and more in the spirit of the original use of such plots by Mosteller and Tukey (1977, 271–279). Welsch (1986, 403), however, disagrees: "I am sorry to see that Chatterjee and Hadi [1986] endorse the term 'added-variable plot' when X_j is part of the original model" and goes on to suggest the name "adjusted partial residual plot".

❑

▷ Example 6: avplots

Added-variable plots are so useful that we should look at them for every regressor in the data. avplots makes this easy:

    . avplots

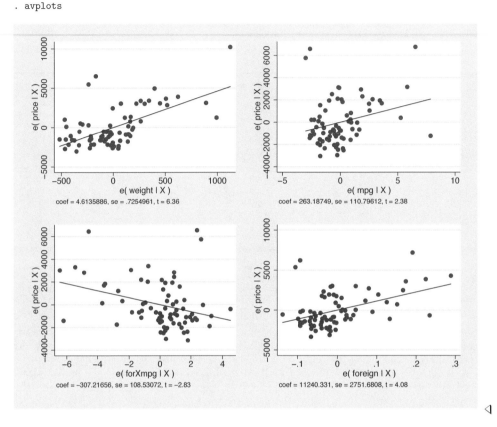

◁

Component-plus-residual plots

Added-variable plots are successful at identifying outliers, but they cannot be used to identify functional form. The component-plus-residual plot (Ezekiel 1924; Larsen and McCleary 1972) is another attempt at projecting multidimensional data into a two-dimensional form, but with different properties. Although the added-variable plot can identify outliers, the component-plus-residual plot cannot. It can, however, be used to examine the functional-form assumptions of the model. Both plots have the property that a regression line through the coordinates has a slope equal to the estimated coefficient in the regression model.

▷ Example 7: cprplot and acprplot

To illustrate these plots, we begin with a different model:

```
. use http://www.stata-press.com/data/r10/auto1, clear
(Automobile Models)

. regress price mpg weight
```

Source	SS	df	MS
Model	187716578	2	93858289
Residual	447348818	71	6300687.58
Total	635065396	73	8699525.97

```
Number of obs =      74
F(  2,     71) =   14.90
Prob > F       =  0.0000
R-squared      =  0.2956
Adj R-squared  =  0.2757
Root MSE       =  2510.1
```

price	Coef.	Std. Err.	t	P>|t|	[95% Conf. Interval]	
mpg	-55.9393	75.24136	-0.74	0.460	-205.9663	94.08771
weight	1.710992	.5861682	2.92	0.005	.5422063	2.879779
_cons	2197.9	3190.768	0.69	0.493	-4164.311	8560.11

In fact, we know that the effects of mpg in this model are nonlinear—if we added mpg squared to the model, its coefficient would have a t statistic of 2.38, the t statistic on mpg would become -2.48, and weight's effect would become about one-third of its current value and become statistically insignificant. Pretend that we do not know this.

The component-plus-residual plot for mpg is

```
. cprplot mpg, mspline msopts(bands(13))
```

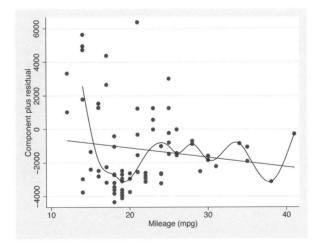

We are supposed to examine the above graph for nonlinearities, or, equivalently, ask if the regression line, which has slope equal to the estimated effect of mpg in the original model, fits the data adequately. To assist our eyes, we added a median spline. Perhaps some people may detect nonlinearity from this graph, but we assert that if we had not previously revealed the nonlinearity of mpg and if we had not added the median spline, the graph would not overly bother us.

Mallows (1986) proposed an augmented component-plus-residual plot that is often more sensitive to detecting nonlinearity:

```
. acprplot mpg, mspline msopts(bands(13))
```

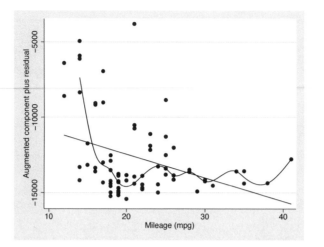

It does do somewhat better.

◁

Residual-versus-predictor plots

▷ Example 8: rvpplot

The residual-versus-predictor plot is a simple way to look for violations of the regression assumptions. If the assumptions are correct, there should be no pattern in the graph. Using our price on mpg and weight model, we type

. rvpplot mpg, yline(0)

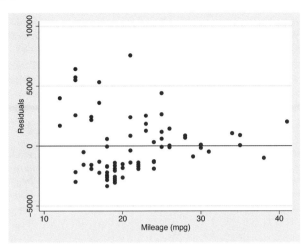

Remember, any pattern counts as a problem, and in this graph, we see that the variation in the residuals decreases as mpg increases.

◁

Leverage statistics

In addition to providing fitted values and the associated standard errors, the predict command can also be used to generate various statistics used to detect the influence of individual observations. This section provides a brief introduction to leverage (hat) statistics, and some of the following subsections discuss other influence statistics produced by predict.

▷ Example 9: diagonal elements of projection matrix

The diagonal elements of the projection matrix, obtained by the hat option, are a measure of distance in explanatory variable space. leverage is a synonym for hat.

```
. use http://www.stata-press.com/data/r10/auto
(1978 Automobile Data)
. generate weight2 = weight^2
. regress mpg weight weight2 foreign
(output omitted )
. predict xdist, hat
. summarize xdist, detail
```

		Leverage		
	Percentiles	Smallest		
1%	.0251334	.0251334		
5%	.0255623	.0251334		
10%	.0259213	.0253883	Obs	74
25%	.0278442	.0255623	Sum of Wgt.	74
50%	.04103		Mean	.0540541
		Largest	Std. Dev.	.0459218
75%	.0631279	.1593606		
90%	.0854584	.1593606	Variance	.0021088
95%	.1593606	.2326124	Skewness	3.440809
99%	.3075759	.3075759	Kurtosis	16.95135

Some 5% of our sample has an `xdist` measure in excess of 0.15. Let's force them to reveal their identities:

```
. list foreign make mpg if xdist>.15, divider
```

	foreign	make	mpg
24.	Domestic	Ford Fiesta	28
26.	Domestic	Linc. Continental	12
27.	Domestic	Linc. Mark V	12
43.	Domestic	Plym. Champ	34

To understand why these cars are on this list, we must remember that the explanatory variables in our model are `weight` and `foreign` and that `xdist` measures distance in this metric. The Ford Fiesta and the Plymouth Champ are the two lightest domestic cars in our data. The Lincolns are the two heaviest domestic cars.

◁

L-R plots

▷ Example 10: lvr2plot

One of the most useful diagnostic graphs is provided by `lvr2plot` (read leverage-versus-residual-squared plot), a graph of leverage against the (normalized) residuals squared.

```
. use http://www.stata-press.com/data/r10/auto, clear
(1978 Automobile Data)
. generate forXmpg=foreign*mpg
. regress price weight mpg forXmpg foreign
  (output omitted )
. lvr2plot
```

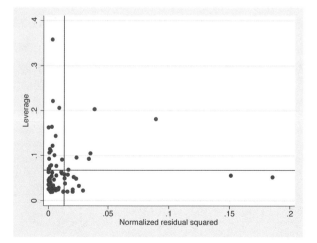

The lines on the chart show the average values of leverage and the (normalized) residuals squared. Points above the horizontal line have higher-than-average leverage; points to the right of the vertical line have larger-than-average residuals.

One point immediately catches our eye, and four more make us pause. The point at the top of the graph has high leverage and a smaller-than-average residual. The other points that bother us all have higher-than-average leverage, two with smaller-than-average residuals and two with larger-than-average residuals.

A less pretty but more useful version of the above graph specifies that make be used as the symbol (see [G] *marker_label_options*):

```
. lvr2plot, mlabel(make) mlabp(0) m(none) mlabsize(small)
```

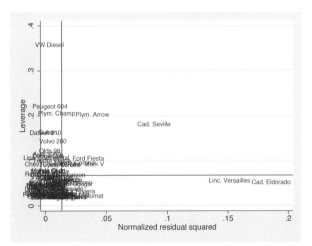

The VW Diesel, Plymouth Champ, Plymouth Arrow, and Peugeot 604 are the points that cause us the most concern. When we further examine our data, we discover that the VW Diesel is the only diesel in our data and that the data for the Plymouth Arrow were entered incorrectly into the computer. No such simple explanations were found for the Plymouth Champ and Peugeot 604.

◁

Standardized and studentized residuals

The terms standardized and studentized residuals have meant different things to different authors. In Stata, predict defines the standardized residual as $\widehat{e}_i = e_i/(s\sqrt{1-h_i})$ and the studentized residual as $r_i = e_i/(s_{(i)}\sqrt{1-h_i})$, where $s_{(i)}$ is the root mean squared error of a regression with the ith observation removed. Stata's definition of the studentized residual is the same as the one given in Bollen and Jackman (1990, 264) and is what Chatterjee and Hadi (1988, 74) call the "externally studentized" residual. Stata's "standardized" residual is the same as what Chatterjee and Hadi (1988, 74) call the "internally studentized" residual.

Standardized and studentized residuals are attempts to adjust residuals for their standard errors. Although the ϵ_i theoretical residuals are homoskedastic by assumption (i.e., they all have the same variance), the calculated e_i are not. In fact,

$$\text{Var}(e_i) = \sigma^2(1 - h_i)$$

where h_i are the leverage measures obtained from the diagonal elements of hat matrix. Thus observations with the greatest leverage have corresponding residuals with the smallest variance.

Standardized residuals use the root mean squared error of the regression for σ. Studentized residuals use the root mean squared error of a regression omitting the observation in question for σ. In general, studentized residuals are preferable to standardized residuals for purposes of outlier identification. Studentized residuals can be interpreted as the t statistic for testing the significance of a dummy variable equal to 1 in the observation in question and 0 elsewhere (Belsley, Kuh, and Welsch 1980). Such a dummy variable would effectively absorb the observation and so remove its influence in determining the other coefficients in the model. Caution must be exercised here, however, because of the simultaneous testing problem. You cannot simply list the residuals that would be individually significant at the 5% level—their joint significance would be far less (their joint significance *level* would be far greater).

▷ Example 11: standardized and studentized residuals

In the opening remarks for this entry, we distinguished residuals from leverage and speculated on the impact of an observation with a small residual but large leverage. If we had adjusted the residuals for their standard errors, however, the adjusted residual would have been (relatively) larger and perhaps large enough so that we could simply examine the adjusted residuals. Taking our `price` on `weight`, `mpg`, `forXmpg`, and `foreign` model, we can obtain the in-sample standardized and studentized residuals by typing

```
. predict esta if e(sample), rstandard
. predict estu if e(sample), rstudent
```

Under the subheading *L-R plots*, we discovered that the VW Diesel has the highest leverage in our data, but a corresponding small residual. The standardized and studentized residuals for the VW Diesel are

```
. list make price esta estu if make=="VW Diesel"
```

	make	price	esta	estu
74.	VW Diesel	5,397	.6142691	.6114758

The studentized residual of 0.611 can be interpreted as the t statistic for including a dummy variable for VW Diesel in our regression. Such a variable would not be significant.

◁

DFITS, Cook's Distance, and Welsch Distance

DFITS (Welsch and Kuh 1977), Cook's Distance (Cook 1977), and Welsch Distance (Welsch 1982) are three attempts to summarize the information in the leverage versus residual-squared plot into one statistic. That is, the goal is to create an index that is affected by the size of the residuals—outliers—and the size of h_i—leverage. Viewed mechanically, one way to write DFITS (Bollen and Jackman 1990, 265) is

$$\text{DFITS}_i = r_i \sqrt{\frac{h_i}{1 - h_i}}$$

where r_i are the studentized residuals. Thus large residuals increase the value of DFITS, as do large values of h_i. Viewed more traditionally, DFITS is a scaled difference between predicted values for the ith case when the regression is fitted with and without the ith observation, hence the name.

The mechanical relationship between DFITS and Cook's Distance D_i (Bollen and Jackman 1990, 266) is

$$D_i = \frac{1}{k} \frac{s_{(i)}^2}{s^2} \text{DFITS}_i^2$$

where k is the number of variables (including the constant) in the regression, s is the root mean squared error of the regression, and $s_{(i)}$ is the root mean squared error when the ith observation is omitted. Viewed more traditionally, D_i is a scaled measure of the distance between the coefficient vectors when the ith observation is omitted.

The mechanical relationship between DFITS and Welsch's Distance W_i (Chatterjee and Hadi 1988, 123) is

$$W_i = \text{DFITS}_i \sqrt{\frac{n-1}{1-h_i}}$$

The interpretation of W_i is more difficult, as it is based on the empirical influence curve. Although DFITS and Cook's distance are similar, the Welsch distance measure includes another normalization by leverage.

Belsley, Kuh, and Welsch (1980, 28) suggest that DFITS values greater than $2\sqrt{k/n}$ deserve more investigation, and so values of Cook's distance greater than $4/n$ should also be examined (Bollen and Jackman 1990, 265–266). Through similar logic, the cutoff for Welsch distance is approximately $3\sqrt{k}$ (Chatterjee and Hadi 1988, 124).

> ### Example 12: DFITS influence measure

Using our model of `price` on `weight`, `mpg`, `forXmpg`, and `foreign`, we can obtain the DFITS influence measure:

```
. use http://www.stata-press.com/data/r10/auto, clear
(1978 Automobile Data)
. generate forXmpg = foreign*mpg
. regress price weight mpg forXmpg foreign
  (output omitted )
. predict e if e(sample), resid
. predict dfits, dfits
```

We did not specify `if e(sample)` in computing the DFITS statistic. DFITS is only available over the estimation sample, so specifying `if e(sample)` would have been redundant. It would have done no harm, but it would not have changed the results.

Our model has $k = 5$ independent variables (k includes the constant) and $n = 74$ observations; following the $2\sqrt{k/n}$ cutoff advice, we type

```
. list make price e dfits if abs(dfits) > 2*sqrt(5/74), divider
```

	make	price	e	dfits
12.	Cad. Eldorado	14,500	7271.96	.9564455
13.	Cad. Seville	15,906	5036.348	1.356619
24.	Ford Fiesta	4,389	3164.872	.5724172
27.	Linc. Mark V	13,594	3109.193	.5200413
28.	Linc. Versailles	13,466	6560.912	.8760136
42.	Plym. Arrow	4,647	-3312.968	-.9384231

We calculate Cook's distance and list the observations greater than the suggested $4/n$ cutoff:

```
. predict cooksd if e(sample), cooksd
. list make price e cooksd if cooksd > 4/74, divider
```

	make	price	e	cooksd
40.	Cad. Eldorado	14,500	7271.96	.1492676
43.	Linc. Versailles	13,466	6560.912	.1308004
62.	Ford Fiesta	4,389	3164.872	.0638815
70.	Cad. Seville	15,906	5036.348	.3328515
71.	Plym. Arrow	4,647	-3312.968	.1700736

Here we used `if e(sample)` because Cook's distance is not restricted to the estimation sample by default. It is worth comparing this list with the preceding one.

Finally, we use Welsch distance and the suggested $3\sqrt{k}$ cutoff:

```
. predict wd, welsch
. list make price e wd if abs(wd) > 3*sqrt(5), divider
```

	make	price	e	wd
12.	Cad. Eldorado	14,500	7271.96	8.394372
13.	Cad. Seville	15,906	5036.348	12.81125
28.	Linc. Versailles	13,466	6560.912	7.703005
42.	Plym. Arrow	4,647	-3312.968	-8.981481

Here we did not need to specify `if e(sample)` since `welsch` automatically restricts the prediction to the estimation sample.

◁

COVRATIO

COVRATIO (Belsley, Kuh, and Welsch 1980) measures the influence of the ith observation by considering the effect on the variance–covariance matrix of the estimates. The measure is the ratio of the determinants of the covariances matrix, with and without the ith observation. The resulting formula is

$$\text{COVRATIO}_i = \frac{1}{1-h_i}\left(\frac{n-k-\widehat{e}_i^2}{n-k-1}\right)^k$$

where $\widehat{e}_i$ is the standardized residual.

For noninfluential observations, the value of COVRATIO is approximately 1. Large values of the residuals or large values of leverage will cause deviations from 1, although if both are large, COVRATIO may tend back toward 1 and therefore not identify such observations (Chatterjee and Hadi 1988, 139).

Belsley, Kuh, and Welsch (1980) suggest that observations for which

$$|\text{COVRATIO}_i - 1| \geq \frac{3k}{n}$$

are worthy of further examination.

▷ Example 13: COVRATIO influence measure

Using our model of `price` on `weight`, `mpg`, `forXmpg`, and `foreign`, we can obtain the COVRATIO measure and list the observations outside the suggested cutoff by typing

```
. predict covr, covratio
. list make price e covr if abs(covr-1) >= 3*5/74, divider
```

	make	price	e	covr
12.	Cad. Eldorado	14,500	7271.96	.3814242
13.	Cad. Seville	15,906	5036.348	.7386969
28.	Linc. Versailles	13,466	6560.912	.4761695
43.	Plym. Champ	4,425	1621.747	1.27782
53.	Audi 5000	9,690	591.2883	1.206842
57.	Datsun 210	4,589	19.81829	1.284801
64.	Peugeot 604	12,990	1037.184	1.348219
66.	Subaru	3,798	-909.5894	1.264677
71.	VW Diesel	5,397	999.7209	1.630653
74.	Volvo 260	11,995	1327.668	1.211888

The `covratio` option automatically restricts the prediction to the estimation sample.

◁

DFBETAs

DFBETAs are perhaps the most direct influence measure of interest to model builders. DFBETAs focus on one coefficient and measure the difference between the regression coefficient when the ith observation is included and excluded, the difference being scaled by the estimated standard error of the coefficient. Belsley, Kuh, and Welsch (1980, 28) suggest observations with $|\text{DFBETA}_i| > 2/\sqrt{n}$ as deserving special attention, but it is also common practice to use 1 (Bollen and Jackman 1990, 267), meaning that the observation shifted the estimate at least one standard error.

▷ Example 14: DFBETAs influence measure; the dfbeta() option

Using our model of `price` on `weight`, `mpg`, `forXmpg`, and `foreign`, let us first ask which observations have the greatest impact on the determination of the coefficient on `foreign`. We will use the suggested $2/\sqrt{n}$ cutoff:

```
. sort foreign make
. predict dfor, dfbeta(foreign)
. list make price foreign dfor if abs(dfor) > 2/sqrt(74), divider
```

	make	price	foreign	dfor
12.	Cad. Eldorado	14,500	Domestic	-.5290519
13.	Cad. Seville	15,906	Domestic	.8243419
28.	Linc. Versailles	13,466	Domestic	-.5283729
42.	Plym. Arrow	4,647	Domestic	-.6622424
43.	Plym. Champ	4,425	Domestic	.2371104
64.	Peugeot 604	12,990	Foreign	.2552032
69.	Toyota Corona	5,719	Foreign	-.256431

The Cadillac Seville shifted the `foreign` coefficient .82 standard deviations!

Now let us ask which observations have the greatest effect on the `mpg` coefficient:

```
. predict dmpg, dfbeta(mpg)
. list make price mpg dmpg if abs(dmpg) > 2/sqrt(74), divider
```

	make	price	mpg	dmpg
12.	Cad. Eldorado	14,500	14	-.5970351
13.	Cad. Seville	15,906	21	1.134269
28.	Linc. Versailles	13,466	14	-.6069287
42.	Plym. Arrow	4,647	28	-.8925859
43.	Plym. Champ	4,425	34	.3186909

Once again we see the Cadillac Seville heading the list, indicating that our regression results may be dominated by this one car.

◁

▷ Example 15: DFBETAs influence measure; the dfbeta command

We can use `predict, dfbeta()` or the `dfbeta` command to generate the DFBETAs. `dfbeta` makes up names for the new variables automatically and, without arguments, generates the DFBETAs for all the variables in the regression:

```
. dfbeta
             DFweight:  DFbeta(weight)
               DFmpg:  DFbeta(mpg)
            DFforXmpg:  DFbeta(forXmpg)
            DFforeign:  DFbeta(foreign)
```

`dfbeta` created four new variables in our dataset: `DFweight`, containing the DFBETAs for `weight`; `DFmpg`, containing the DFBETAs for `mpg`; and so on. Had we wanted only the DFBETAs for `mpg` and `weight`, we might have typed

```
. dfbeta mpg weight
               DFmpg:  DFbeta(mpg)
             DFweight:  DFbeta(weight)
```

In the example above, we typed `dfbeta mpg weight` instead of `dfbeta`—if we had typed `dfbeta` followed by `dfbeta mpg weight`, here is what would have happened:

```
. dfbeta
             DFweight:  DFbeta(weight)
               DFmpg:  DFbeta(mpg)
            DFforXmpg:  DFbeta(forXmpg)
            DFforeign:  DFbeta(foreign)
. dfbeta mpg weight
                 DF1:  DFbeta(mpg)
                 DF2:  DFbeta(weight)
```

`dfbeta` would have made up different names for the new variables. `dfbeta` never replaces existing variables—it instead makes up a different name, so we need to pay attention to `dfbeta`'s output.

◁

Formal tests for violations of assumptions

This section introduces some regression diagnostic commands that are designed to test for certain violations that `rvfplot` less formally attempts to detect. `estat ovtest` provides Ramsey's test for omitted variables—a pattern in the residuals. `estat hettest` provides a test for heteroskedasticity—the increasing or decreasing variation in the residuals with fitted values, with respect to the explanatory variables, or with respect to yet other variables. The score test implemented in `estat hettest` (Breusch and Pagan 1979; Cook and Weisberg 1983) performs a score test for the null hypothesis that $b = 0$ against the alternative hypothesis of multiplicative heteroskedasticity. `estat szroeter` provides a rank test for heteroskedasticity, which is an alternative to the score test computed by `estat hettest`. Finally, `estat imtest` computes an information matrix test, including an orthogonal decomposition into tests for heteroskedasticity, skewness, and kurtosis (Cameron and Trivedi 1990). The heteroskedasticity test computed by `estat imtest` is similar to the general test for heteroskedasticity that was proposed by White (1980).

▷ Example 16: estat ovtest, estat hettest, estat szroeter, and estat imtest

We run these commands just mentioned on our model:

```
. estat ovtest

Ramsey RESET test using powers of the fitted values of price
      Ho:  model has no omitted variables
                F(3, 66) =        7.77
                Prob > F =      0.0002

. estat hettest

Breusch-Pagan / Cook-Weisberg tests for heteroskedasticity
          Ho: Constant variance
          variables: fitted values of price
        chi2(1)      =      6.50
        Prob > chi2  =    0.0108
```

Testing for heteroskedasticity in the right-hand-side variables is requested by specifying the option `rhs`. By specifying the option `mtest(bonferroni)`, we request that tests be conducted for each of the variables, with a Bonferroni adjustment for the p-values to accommodate our testing multiple hypotheses.

```
. estat hettest, rhs mtest(bonf)

Breusch-Pagan / Cook-Weisberg tests for heteroskedasticity
          Ho: Constant variance
```

Variable	chi2	df	p	
weight	15.24	1	0.0004	#
mpg	9.04	1	0.0106	#
forXmpg	6.02	1	0.0566	#
foreign	6.15	1	0.0525	#
simultaneous	15.60	4	0.0036	

Bonferroni adjusted p-values

```
. estat szroeter, rhs mtest(holm)
Szroeter's test for homoskedasticity
       Ho: variance constant
       Ha: variance monotonic in variable
```

Variable	chi2	df	p
weight	17.07	1	0.0001 #
mpg	11.45	1	0.0021 #
forXmpg	6.17	1	0.0260 #
foreign	6.15	1	0.0131 #

Holm adjusted p-values

Finally, we request the information-matrix test, which is a conditional moments test with second-, third-, and fourth-order moment conditions.

```
. estat imtest
Cameron & Trivedi's decomposition of IM-test
```

Source	chi2	df	p
Heteroskedasticity	18.86	10	0.0420
Skewness	11.69	4	0.0198
Kurtosis	2.33	1	0.1273
Total	32.87	15	0.0049

We find evidence for omitted variables, heteroskedasticity, and nonnormal skewness.

So, why bother with the various graphical commands when the tests seem so much easier to interpret? In part, it is a matter of taste: both are designed to uncover the same problem, and both are, in fact, going about it in similar ways. One is based on a formal calculation, whereas the other is based on personal judgment in evaluating a graph. On the other hand, the tests are seeking evidence of specific problems, whereas judgment is more general. The careful analyst will use both.

We performed the omitted-variable test first. Omitted variables are a more serious problem than heteroskedasticity or the violations of higher moment conditions tested by estat imtest. If this were not a manual, having found evidence of omitted variables, we would never have run the estat hettest, estat szroeter, and estat imtest commands, at least not until we solved the omitted-variable problem.

◁

❑ Technical Note

estat ovtest and estat hettest both perform two flavors of their respective tests. By default, estat ovtest looks for evidence of omitted variables by fitting the original model augmented by $\widehat{y}^2$, $\widehat{y}^3$, and $\widehat{y}^4$, which are the fitted values from the original model. Under the assumption of no misspecification, the coefficients on the powers of the fitted values will be zero. With the rhs option, estat ovtest instead augments the original model with powers (second through fourth) of the explanatory variables (except for dummy variables).

estat hettest, by default, looks for heteroskedasticity by modeling the variance as a function of the fitted values. If, however, we specify a variable or variables, the variance will be modeled as a function of the specified variables. In our example, if we had, a priori, some reason to suspect heteroskedasticity and that the heteroskedasticity is a function of a car's weight, then using a test that focuses on weight would be more powerful than the more general tests such as White's test or the first term in the Cameron–Trivedi decomposition test.

`estat hettest`, by default, computes the original Breusch–Pagan/Cook–Weisberg test, which includes the assumption of normally distributed errors. Koenker (1981) derived an $N * R^2$ version of this test that drops the normality assumption. Wooldridge (2006) gives an F statistic version that does not require the normality assumption.

❑

Variance inflation factors

Problems arise in regression when the predictors are highly correlated. In this situation, there may be a significant change in the regression coefficients if you add or delete an independent variable. The estimated standard errors of the fitted coefficients are inflated, or the estimated coefficients may not be statistically significant even though a statistical relation exists between the dependent and independent variables.

Data analysts rely on these facts to check informally for the presence of multicollinearity. `estat vif`, another command for use after `regress`, calculates the variance inflation factors and tolerances for each of the independent variables.

The output shows the variance inflation factors together with their reciprocals. Some analysts compare the reciprocals with a predetermined tolerance. In the comparison, if the reciprocal of the VIF is smaller than the tolerance, the associated predictor variable is removed from the regression model. However, most analysts rely on informal rules of thumb applied to the VIF; see Chatterjee and Hadi (2006). According to these rules, there is evidence of multicollinearity if

1. The largest VIF is greater than 10 (some choose a more conservative threshold value of 30).

2. The mean of all the VIFs is considerably larger than 1.

▷ Example 17: estat vif

We examine a regression model fitted using the ubiquitous automobile dataset:

```
. regress price mpg rep78 trunk headroom length turn displ gear_ratio
```

Source	SS	df	MS		
Model	264102049	8	33012756.2	Number of obs =	69
Residual	312694909	60	5211581.82	F(8, 60) =	6.33
				Prob > F =	0.0000
				R-squared =	0.4579
				Adj R-squared =	0.3856
Total	576796959	68	8482308.22	Root MSE =	2282.9

price	Coef.	Std. Err.	t	P>|t|	[95% Conf. Interval]	
mpg	-144.84	82.12751	-1.76	0.083	-309.1195	19.43948
rep78	727.5783	337.6107	2.16	0.035	52.25638	1402.9
trunk	44.02061	108.141	0.41	0.685	-172.2935	260.3347
headroom	-807.0996	435.5802	-1.85	0.069	-1678.39	64.19061
length	-8.688914	34.89848	-0.25	0.804	-78.49626	61.11843
turn	-177.9064	137.3455	-1.30	0.200	-452.6383	96.82551
displacement	30.73146	7.576952	4.06	0.000	15.5753	45.88762
gear_ratio	1500.119	1110.959	1.35	0.182	-722.1303	3722.368
_cons	6691.976	7457.906	0.90	0.373	-8226.057	21610.01

```
. estat vif
```

Variable	VIF	1/VIF
length	8.22	0.121614
displacement	6.50	0.153860
turn	4.85	0.205997
gear_ratio	3.45	0.290068
mpg	3.03	0.330171
trunk	2.88	0.347444
headroom	1.80	0.554917
rep78	1.46	0.686147
Mean VIF	4.02	

The results are mixed. Although we have no VIFs greater than 10, the mean VIF is greater than 1, though not considerably so. We could continue the investigation of collinearity, but given that other authors advise that collinearity is a problem only when VIFs exist that are greater than 30 (contradicting our rule above), we will not do so here.

◁

▷ Example 18: estat vif, with strong evidence of multicollinearity

This example comes from a dataset described in Neter, Wasserman, and Kutner (2004) that examines body fat as modeled by caliper measurements on the triceps, midarm, and thigh.

```
. use http://www.stata-press.com/data/r10/bodyfat, clear
(Body Fat)
. regress bodyfat tricep thigh midarm
```

Source	SS	df	MS
Model	396.984607	3	132.328202
Residual	98.4049068	16	6.15030667
Total	495.389513	19	26.0731323

```
Number of obs =      20
F(  3,    16) =   21.52
Prob > F      =  0.0000
R-squared     =  0.8014
Adj R-squared =  0.7641
Root MSE      =    2.48
```

bodyfat	Coef.	Std. Err.	t	P>|t|	[95% Conf. Interval]	
triceps	4.334085	3.015511	1.44	0.170	-2.058512	10.72668
thigh	-2.856842	2.582015	-1.11	0.285	-8.330468	2.616785
midarm	-2.186056	1.595499	-1.37	0.190	-5.568362	1.19625
_cons	117.0844	99.78238	1.17	0.258	-94.44474	328.6136

```
. estat vif
```

Variable	VIF	1/VIF
triceps	708.84	0.001411
thigh	564.34	0.001772
midarm	104.61	0.009560
Mean VIF	459.26	

Here we see strong evidence of multicollinearity in our model. More investigation reveals that the measurements on the thigh and the triceps are highly correlated:

```
. corr triceps thigh midarm
(obs=20)
```

	triceps	thigh	midarm
triceps	1.0000		
thigh	0.9238	1.0000	
midarm	0.4578	0.0847	1.0000

If we remove the predictor `tricep` from the model (since it had the highest VIF), we get

```
. regress bodyfat thigh midarm
```

Source	SS	df	MS
Model	384.279748	2	192.139874
Residual	111.109765	17	6.53586854
Total	495.389513	19	26.0731323

```
Number of obs =      20
F(  2,   17) =   29.40
Prob > F      =  0.0000
R-squared     =  0.7757
Adj R-squared =  0.7493
Root MSE      =  2.5565
```

bodyfat	Coef.	Std. Err.	t	P>|t|	[95% Conf.	Interval]
thigh	.8508818	.1124482	7.57	0.000	.6136367	1.088127
midarm	.0960295	.1613927	0.60	0.560	-.2444792	.4365383
_cons	-25.99696	6.99732	-3.72	0.002	-40.76001	-11.2339

```
. estat vif
```

Variable	VIF	1/VIF
midarm	1.01	0.992831
thigh	1.01	0.992831
Mean VIF	1.01	

Note how the coefficients change and how the estimated standard errors for each of the regression coefficients become much smaller. The calculated value of R^2 for the overall regression for the subset model does not appreciably decline when we remove the correlated predictor. Removing an independent variable from the model is one way to deal with multicollinearity. Other methods include ridge regression, weighted least squares, and restricting the use of the fitted model to data that follow the same pattern of multicollinearity. In economic studies, it is sometimes possible to estimate the regression coefficients from different subsets of the data by using cross-section and time series.

◁

All examples above demonstrated the use of centered VIFs. As pointed out by Belsley (1991), the centered VIFs may fail to discover collinearity involving the constant term. One solution is to use the uncentered VIFs instead. According to the definition of the uncentered VIFs, the constant is viewed as a legitimate explanatory variable in a regression model, which allows one to obtain the VIF value for the constant term.

▷ Example 19: estat vif, with strong evidence of collinearity with the constant term

Consider the extreme example in which one of the regressors is highly correlated with the constant. We simulate the data and examine both centered and uncentered VIF diagnostics after fitted regression model as follows.

```
. use http://www.stata-press.com/data/r10/extreme_collin, clear

. summarize

(output omitted)

. regress y one x z
```

Source	SS	df	MS		
Model	223801.985	3	74600.6617		
Residual	2642.42124	96	27.5252213		
Total	226444.406	99	2287.31723		

```
                                                Number of obs =      100
                                                F(  3,     96) = 2710.27
                                                Prob > F      =   0.0000
                                                R-squared     =   0.9883
                                                Adj R-squared =   0.9880
                                                Root MSE      =   5.2464
```

y	Coef.	Std. Err.	t	P>|t|	[95% Conf. Interval]
one	-3.278582	10.5621	-0.31	0.757	-24.24419 17.68702
x	2.038696	.0242673	84.01	0.000	1.990526 2.086866
z	4.863137	.2681036	18.14	0.000	4.330956 5.395319
_cons	9.760075	10.50935	0.93	0.355	-11.10082 30.62097

```
. estat vif
```

Variable	VIF	1/VIF
z	1.03	0.968488
x	1.03	0.971307
one	1.00	0.995425
Mean VIF	1.02	

```
. estat vif, uncentered
```

Variable	VIF	1/VIF
one	402.94	0.002482
intercept	401.26	0.002492
z	2.93	0.341609
x	1.13	0.888705
Mean VIF	202.06	

According to the values of the centered VIFs (1.03, 1.03, 1.00) no harmful collinearity is detected in the model. However, by the construction of these simulated data, we know that one is highly collinear with the constant term. As such, the large values of uncentered VIFs for one (402.94) and intercept (401.26) reveal high collinearity of the variable one with the constant term.

◁

Saved Results

estat hettest saves the following results for the (multivariate) score test in r():

Scalars

r(chi2) χ^2 test statistic

r(df)   #df for the asymptotic χ^2 distribution under H_0

r(p) p-value

`estat hettest, fstat` saves results for the (multivariate) score test in `r()`:

Scalars

`r(F)`	test statistic
`r(df_m)`	#df of the test for the F distribution under H_0
`r(df_r)`	#df of the residuals for the F distribution under H_0
`r(p)`	p-value

`estat hettest` (if `mtest` is specified) and `estat szroeter` save the following in `r()`:

Matrices

   `r(mtest)`   a matrix of test results, with rows corresponding to the univariate tests

`mtest[.,1]`	χ^2 test statistic
`mtest[.,2]`	#df
`mtest[.,3]`	unadjusted p-value
`mtest[.,4]`	adjusted p-value (if an `mtest()` adjustment method is specified)

Macros

   `r(mtmethod)`   adjustment method for p-values

`estat imtest` saves the following in `r()`:

Scalars

`r(chi2_t)`	IM-test statistic $\left(= \texttt{r(chi2\_h)} + \texttt{r(chi2\_s)} + \texttt{r(chi2\_k)} \right)$
`r(df_t)`	df for limiting χ^2 distribution under H_0 $\left(= \texttt{r(df\_h)} + \texttt{r(df\_s)} + \texttt{r(df\_k)} \right)$
`r(chi2_h)`	heteroskedasticity test statistic
`r(df_h)`	df for limiting χ^2 distribution under H_0
`r(chi2_s)`	skewness test statistic
`r(df_s)`	df for limiting χ^2 distribution under H_0
`r(chi2_k)`	kurtosis test statistic
`r(df_k)`	df for limiting χ^2 distribution under H_0
`r(chi2_w)`	White's heteroskedasticity test (if `white` specified)
`r(df_w)`	df for limiting χ^2 distribution under H_0

`estat ovtest` saves the following in `r()`:

Scalars

`r(p)`	two-sided p-value	`r(df)`	degrees of freedom
`r(F)`	F statistic	`r(df_r)`	residual degrees of freedom

Methods and Formulas

All regression fit and diagnostic commands are implemented as ado-files.

See Hamilton (2006, chap. 7), Kohler and Kreuter (2005, sec. 8.3), or Baum (2006, chap. 5) for an overview of using Stata to perform regression diagnostics. See Peracchi (2001, chap. 8) for a mathematically rigorous discussion of diagnostics.

Methods and formulas for predict

Assume that you have already fitted the regression model

$$\mathbf{y} = \mathbf{Xb} + \mathbf{e}$$

where $\mathbf{X}$ is $n \times k$.

Denote the previously estimated coefficient vector by $\mathbf{b}$ and its estimated variance matrix by $\mathbf{V}$. predict works by recalling various aspects of the model, such as $\mathbf{b}$, and combining that information with the data currently in memory. Let $\mathbf{x}_j$ be the jth observation currently in memory, and let s^2 be the mean squared error of the regression.

Let $\mathbf{V} = s^2(\mathbf{X}'\mathbf{X})^{-1}$. Let k be the number of independent variables including the intercept, if any, and let y_j be the observed value of the dependent variable.

The *predicted value* (xb option) is defined $\widehat{y}_j = \mathbf{x}_j \mathbf{b}$.

Let ℓ_j represent a lower bound for an observation j and u_j represent an upper bound. The probability that $y_j|\mathbf{x}_j$ would be observed in the interval (ℓ_j, u_j) —option pr(ℓ, u) —is

$$P(\ell_j, u_j) = \Pr(\ell_j < \mathbf{x}_j\mathbf{b} + e_j < u_j) = \Phi\left(\frac{u_j - \widehat{y}_j}{s}\right) - \Phi\left(\frac{\ell_j - \widehat{y}_j}{s}\right)$$

where for the options pr(ℓ, u), e(ℓ, u), and ystar(ℓ, u), ℓ_j and u_j can be anywhere in the range $(-\infty, +\infty)$.

The option e(ℓ, u) computes the expected value of $y_j|\mathbf{x}_j$ conditional on $y_j|\mathbf{x}_j$ being in the interval (ℓ_j, u_j), that is, when $y_j|\mathbf{x}_j$ is censored. It can be expressed as

$$E(\ell_j, u_j) = E(\mathbf{x}_j\mathbf{b} + e_j \mid \ell_j < \mathbf{x}_j\mathbf{b} + e_j < u_j) = \widehat{y}_j - s\frac{\phi\left(\frac{u_j - \widehat{y}_j}{s}\right) - \phi\left(\frac{\ell_j - \widehat{y}_j}{s}\right)}{\Phi\left(\frac{u_j - \widehat{y}_j}{s}\right) - \Phi\left(\frac{\ell_j - \widehat{y}_j}{s}\right)}$$

where ϕ is the normal density and Φ is the cumulative normal.

You can also compute ystar(ℓ, u) —the expected value of $y_j|\mathbf{x}_j$, where y_j is assumed truncated at ℓ_j and u_j:

$$y_j^* = \begin{cases} \ell_j & \text{if } \mathbf{x}_j\mathbf{b} + e_j \leq \ell_j \\ \mathbf{x}_j\mathbf{b} + u & \text{if } \ell_j < \mathbf{x}_j\mathbf{b} + e_j < u_j \\ u_j & \text{if } \mathbf{x}_j\mathbf{b} + e_j \geq u_j \end{cases}$$

This computation can be expressed in several ways, but the most intuitive formulation involves a combination of the two statistics just defined:

$$y_j^* = P(-\infty, \ell_j)\ell_j + P(\ell_j, u_j)E(\ell_j, u_j) + P(u_j, +\infty)u_j$$

A diagonal element of the projection matrix (hat) or (leverage) is given by

$$h_j = \mathbf{x}_j(\mathbf{X}'\mathbf{X})^{-1}\mathbf{x}_j'$$

The *standard error of the prediction* (stdp) is defined as $s_{p_j} = \sqrt{\mathbf{x}_j\mathbf{V}\mathbf{x}_j'}$

and can also be written $s_{p_j} = s\sqrt{h_j}$.

The *standard error of the forecast* (`stdf`) is defined as $s_{f_j} = s\sqrt{1 + h_j}$.

The *standard error of the residual* (`stdr`) is defined as $s_{r_j} = s\sqrt{1 - h_j}$.

The *residuals* (`residuals`) are defined as $\widehat{e}_j = y_j - \widehat{y}_j$.

The *standardized residuals* (`rstandard`) are defined as $\widehat{e}_{s_j} = \widehat{e}_j / s_{r_j}$.

The *studentized residuals* (`rstudent`) are defined as

$$r_j = \frac{\widehat{e}_j}{s_{(j)}\sqrt{1 - h_j}}$$

where $s_{(j)}$ represents the root mean squared error with the jth observation removed, which is given by

$$s_{(j)}^2 = \frac{s^2(T - k)}{T - k - 1} - \frac{\widehat{e}_j^2}{(T - k - 1)(1 - h_j)}$$

Cook's D (`cooksd`) is given by

$$D_j = \frac{\widehat{e}_{s_j}^2 (s_{p_j}/s_{r_j})^2}{k} = \frac{h_j \widehat{e}_j^2}{ks^2(1 - h_j)^2}$$

DFITS (`dfits`) is given by

$$\text{DFITS}_j = r_j\sqrt{\frac{h_j}{1 - h_j}}$$

Welsch distance (`welsch`) is given by

$$W_j = \frac{r_j\sqrt{h_j(n - 1)}}{1 - h_j}$$

COVRATIO (`covratio`) is given by

$$\text{COVRATIO}_j = \frac{1}{1 - h_j}\left(\frac{n - k - \widehat{e}_j^2}{n - k - 1}\right)^k$$

The DFBETAs (`dfbeta`) for a particular regressor x_i are given by

$$\text{DFBETA}_j = \frac{r_j u_j}{\sqrt{U^2(1 - h_j)}}$$

where u_j are the residuals obtained from a regression of x_j on the remaining x's and $U^2 = \sum_j u_j^2$.

The omitted-variable test (Ramsey 1969) reported by `ovtest` fits the regression $y_i = \mathbf{x}_i\mathbf{b} + \mathbf{z}_i\mathbf{t} + u_i$ and then performs a standard F test of $\mathbf{t} = \mathbf{0}$. The default test uses $\mathbf{z}_i = (\widehat{y}_i^2, \widehat{y}_i^3, \widehat{y}_i^4)$. If `rhs` is specified, $\mathbf{z}_i = (x_{1i}^2, x_{1i}^3, x_{1i}^4, x_{2i}^2, \ldots, x_{mi}^4)$. In either case, the variables are normalized to have minimum 0 and maximum 1 before powers are calculated.

The test for heteroskedasticity (Breusch and Pagan 1979; Cook and Weisberg 1983) models $\text{Var}(e_i) = \sigma^2 \exp(\mathbf{zt})$, where $\mathbf{z}$ is a variable list specified by the user, the list of right-hand-side variables, or the fitted values $\mathbf{x}\widehat{\beta}$. The test is of $\mathbf{t} = \mathbf{0}$. Mechanically, `estat hettest` fits the augmented regression $\widehat{e}_i^2/\widehat{\sigma}^2 = a + \mathbf{z}_i\mathbf{t} + v_i$.

The original Breusch–Pagan/Cook–Weisberg version of the test assumes that the e_i are normally distributed under the null hypothesis which implies that the score test statistic S is equal to the model sum of squares from the augmented regression divided by 2. Under the null hypothesis, S has the χ^2 distribution with m degrees of freedom, where m is the number of columns of $\mathbf{z}$.

Koenker (1981) derived a score test of the null hypothesis that $\mathbf{t} = \mathbf{0}$ under the assumption that the e_i are independent and identically distributed (i.i.d.). Koenker showed that $S = N * R^2$ has a large-sample χ^2 distribution with m degrees of freedom, where N is the number of observations and R^2 is the R-squared in the augmented regression and m is the number of columns of $\mathbf{z}$. `estat hettest, iid` produces this version of the test.

Wooldridge (2006) showed that an F test of $\mathbf{t} = \mathbf{0}$ in the augmented regression can also be used under the assumption that the e_i are i.i.d. `estat hettest, fstat` produces this version of the test.

Szroeter's class of tests for homoskedasticity against the alternative that the residual variance increases in some variable x is defined in terms of

$$H = \frac{\sum_{i=1}^{n} h(x_i)e_i^2}{\sum_{i=1}^{n} e_i^2}$$

where $h(x)$ is some weight function that increases in x (Szroeter 1978). H is a weighted average of the $h(x)$, with the squared residuals serving as weights. Under homoskedasticity, H should be approximately equal to the unweighted average of $h(x)$. Large values of H suggest that e_i^2 tends to be large where $h(x)$ is large; i.e., the variance indeed increases in x, whereas small values of H suggest that the variance actually decreases in x. `estat szroeter` uses $h(x_i) = \text{rank}(x_i \text{ in } x_1 \ldots x_n)$; see Judge et al. 1985, 452 for details. `estat szroeter` displays a normalized version of H,

$$Q = \sqrt{\frac{6n}{n^2 - 1}} H$$

which is approximately $N(0, 1)$ distributed under the null (homoskedasticity).

`estat hettest` and `estat szroeter` provide adjustments of p-values for multiple testing. The supported methods are described in [R] **test**.

`estat imtest` performs the information matrix test for the regression model, as well as an orthogonal decomposition into tests for heteroskedasticity δ_1, nonnormal skewness δ_2, and nonnormal kurtosis δ_3 (Cameron and Trivedi 1990; Long and Trivedi 1992). The decomposition is obtained via three auxiliary regressions. Let e be the regression residuals, $\widehat{\sigma}^2$ be the maximum likelihood estimate of σ^2 in the regression, n be the number of observations, X be the set of k variables specified with `estat imtest`, and R_{un}^2 be the uncentered R^2 from a regression. δ_1 is obtained as nR_{un}^2 from a regression of $e^2 - \widehat{\sigma}^2$ on the cross-products of the variables in X. δ_2 is computed as nR_{un}^2 from a regression of $e^3 - 3\widehat{\sigma}^2 e$ on X. Finally, δ_3 is obtained as nR_{un}^2 from a regression of $e^4 - 6\widehat{\sigma}^2 e^2 - 3\widehat{\sigma}^4$ on X. δ_1, δ_2, and δ_3 are asymptotically χ^2 distributed with $\frac{1}{2}k(k+1)$, K, and 1 degrees of freedom. The information test statistic $\delta = \delta_1 + \delta_2 + \delta_3$ is asymptotically χ^2 distributed with $\frac{1}{2}k(k+3)$ degrees of freedom. White's test for heteroskedasticity is computed as nR^2 from a regression of $\widehat{u}^2$ on X and the cross-products of the variables in X. This test statistic is usually close to δ_1.

`estat vif` calculates the centered variance inflation factor (VIF_c) (Chatterjee and Hadi 2006, 235–239) for x_j is given by

$$\mathrm{VIF}_c(x_j) = \frac{1}{1 - \widehat{R}_j^2}$$

where $\widehat{R}_j^2$ is the square of the centered multiple correlation coefficient that results when x_j is regressed with intercept against all the other explanatory variables.

The uncentered variance inflation factor (VIF_{uc}) (Belsley 1991, 28–29) for x_j is given by

$$\mathrm{VIF}_{uc}(x_j) = \frac{1}{1 - \widetilde{R}_j^2}$$

where $\widetilde{R}_j^2$ is the square of the uncentered multiple correlation coefficient that results when x_j is regressed without intercept against all the other explanatory variables including the constant term.

Acknowledgments

`estat ovtest` and `estat hettest` are based on programs originally written by Richard Goldstein (1991, 1992). `estat imtest`, `estat szroeter`, and the current version of `estat hettest` were written by Jeroen Weesie, Department of Sociology, Utrecht University, The Netherlands; `estat imtest` is based in part on code written by J. Scott Long, Department of Sociology, Indiana University.

References

Baum, C. F. 2006. *An Introduction to Modern Econometrics Using Stata*. College Station, TX: Stata Press.

Baum, C. F., N. J. Cox, and V. L. Wiggins. 2000. sg137: Tests for heteroskedasticity in regression error distribution. *Stata Technical Bulletin* 55: 15–17. Reprinted in *Stata Technical Bulletin Reprints*, vol. 10, pp. 147–149.

Baum, C. F., and V. L. Wiggins. 2000a. sg135: Test for autoregressive conditional heteroskedasticity in regression error distribution. *Stata Technical Bulletin* 55: 13–14. Reprinted in *Stata Technical Bulletin Reprints*, vol. 10, pp. 143–144.

——. 2000b. sg136: Tests for serial correlation in regression error distribution. *Stata Technical Bulletin* 55: 14–15. Reprinted in *Stata Technical Bulletin Reprints*, vol. 10, pp. 145–147.

Belsley, D. 1991. *Conditional Diagnostics: Collinearity and Weak Data in Regression*. New York: Wiley.

Belsley, D. A., E. Kuh, and R. E. Welsch. 1980. *Regression Diagnostics: Identifying Influential Data and Sources of Collinearity*. New York: Wiley.

Bollen, K. A., and R. W. Jackman. 1990. Regression diagnostics: An expository treatment of outliers and influential cases. In *Modern Methods of Data Analysis*, ed. J. Fox and J. S. Long, 257–291. Newbury Park, CA: Sage.

Breusch, T., and A. Pagan. 1979. A simple test for heteroscedasticity and random coefficient variation. *Econometrica* 47: 1287–1294.

Cameron, A. C., and P. K. Trivedi. 1990. The information matrix test and its applied alternative hypotheses. Working Paper.

Chatterjee, S., and A. S. Hadi. 1986. Influential observations, high leverage points, and outliers in linear regression. *Statistical Science* 1: 379–416.

——. 1988. *Sensitivity Analysis in Linear Regression*. New York: Wiley.

——. 2006. *Regression Analysis by Example*. 4th ed. New York: Wiley.

Cook, R. D. 1977. Detection of influential observations in linear regression. *Technometrics* 19: 15–18.

Cook, R. D., and S. Weisberg. 1982. *Residuals and Influence in Regression*. New York: Chapman & Hall.

——. 1983. Diagnostics for heteroscedasticity in regression. *Biometrika* 70: 1–10.

Cox, N. J. 2004 Speaking Stata: Graphing model diagnostics. *Stata Journal* 4: 449–475.

DeMaris, A. 2004. *Regression with Social Data: Modeling Continuous and Limited Response Variables.* Hoboken, NJ: Wiley.

Ezekiel, M. 1924. A method of handling curvilinear correlation for any number of variables. *Journal of the American Statistical Association* 19: 431–453.

Garrett, J. M. 2000. sg157: Predicted values calculated from linear or logistic regression models. *Stata Technical Bulletin* 58: 27–30. Reprinted in *Stata Technical Bulletin Reprints*, vol. 10, pp. 258–261.

Goldstein, R. 1991. srd5: Ramsey test for heteroskedasticity and omitted variables. *Stata Technical Bulletin* 2: 27. Reprinted in *Stata Technical Bulletin Reprints*, vol. 1, p. 177.

——. 1992. srd14: Cook–Weisberg test of heteroskedasticity. *Stata Technical Bulletin* 10: 27–28. Reprinted in *Stata Technical Bulletin Reprints*, vol. 2, pp. 183–184.

Hamilton, L. C. 1992. *Regression with Graphics: A Second Course in Applied Statistics.* Belmont, CA: Duxbury.

——. 2006. *Statistics with Stata (Update for Version 9).* Belmont, CA: Duxbury.

Hardin, J. W. 1995. sg32: Variance inflation factors and variance-decomposition proportions. *Stata Technical Bulletin* 24: 17–22. Reprinted in *Stata Technical Bulletin Reprints*, vol. 4, pp. 154–160.

Hoaglin, D. C., and P. J. Kempthorne. 1986. Comment [on Chatterjee and Hadi 1986]. *Statistical Science* 1: 408–412.

Hoaglin, D. C., and R. E. Welsch. 1978. The hat matrix in regression and ANOVA. *American Statistician* 32: 17–22.

Judge, G. G., W. E. Griffiths, R. C. Hill, H. Lütkepohl, and T.-C. Lee. 1985. *The Theory and Practice of Econometrics.* 2nd ed. New York: Wiley.

Koenker, R. 1981. A note on studentizing a test for heteroskedasticity. *Journal of Econometrics* 17: 107–112.

Kohler, U., and F. Kreuter. 2005. *Data Analysis Using Stata.* College Station, TX: Stata Press.

Larsen, W. A., and S. J. McCleary. 1972. The use of partial residual plots in regression analysis. *Technometrics* 14: 781–790.

Long, J. S., and J. Freese. 2000. sg145: Scalar measures of fit for regression models. *Stata Technical Bulletin* 56: 34–40. Reprinted in *Stata Technical Bulletin Reprints*, vol. 10, pp. 197–205.

Long, J. S., and P. K. Trivedi. 1992. Some specification tests for the linear regression model. *Sociological Methods and Research* 21: 161–204. Reprinted in *Testing Structural Equation Models*, ed. K. A. Bollen and J. S. Long, 66–110. Newbury Park, CA: Sage.

Mallows, C. L. 1986. Augmented partial residuals. *Technometrics* 28: 313–319.

Mosteller, F., and J. W. Tukey. 1977. *Data Analysis and Regression: A Second Course in Statistics.* Reading, MA: Addison–Wesley.

Neter, J., W. Wasserman, and M. H. Kutner. 2004. *Applied Linear Regression Models.* 4th ed. Homewood, IL: Irwin.

Peracchi, F. 2001. *Econometrics.* Chichester, UK: Wiley.

Ramsey, J. B. 1969. Tests for specification errors in classical linear least-squares regression analysis. *Journal of the Royal Statistical Society, Series B* 31: 350–371.

Ramsey, J. B., and P. Schmidt. 1976. Some further results on the use of OLS and BLUS residuals in specification error tests. *Journal of the American Statistical Association* 71: 389–390.

Rousseeuw, P. J., and A. M. Leroy. 2003. *Robust Regression and Outlier Detection.* New York: Wiley.

Ryan, T. P. 1997. *Modern Regression Methods.* New York: Wiley.

Szroeter, J. 1978. A class of parametric tests for heteroscedasticity in linear econometric models. *Econometrica* 46: 1311–1328.

Velleman, P. F. 1986. Comment [on Chatterjee and Hadi 1986]. *Statistical Science* 1: 412–413.

Velleman, P. F., and R. E. Welsch. 1981. Efficient computing of regression diagnostics. *American Statistician* 35: 234–242.

Weesie, J. 2001. sg161: Analysis of the turning point of a quadratic specification. *Stata Technical Bulletin* 60: 18–20. Reprinted in *Stata Technical Bulletin Reprints*, vol. 10, pp. 273–277.

Weisberg, S. 2005. *Applied Linear Regression.* 3rd ed. New York: Wiley.

Welsch, R. E. 1982. Influence functions and regression diagnostics. In *Modern Data Analysis*, ed. R. L. Launer and A. F. Siegel, 149–169. New York: Academic Press.

——. 1986. Comment [on Chatterjee and Hadi 1986]. *Statistical Science* 1: 403–405.

Welsch, R. E., and E. Kuh. 1977. *Technical Report 923-77: Linear Regression Diagnostics.* Cambridge, MA: Sloan School of Management, Massachusetts Institute of Technology.

White, H. 1980. A heteroskedasticity-consistent covariance matrix estimator and a direct test for heteroskedasticity. *Econometrica* 48: 817–838.

Wooldridge, J. M. 2006. *Introductory Econometrics: A Modern Approach.* 3rd ed. Cincinnati, OH: South-Western.

Also See

[R] **regress** — Linear regression

[R] **regress postestimation time series** — Postestimation tools for regress with time series

[U] **20 Estimation and postestimation commands**

Title

> **regress postestimation time series** — Postestimation tools for regress with time series

Description

The following postestimation commands for time series are available for `regress`:

command	description
estat archlm	test for ARCH effects in the residuals
estat bgodfrey	Breusch–Godfrey test for higher-order serial correlation
estat durbinalt	Durbin's alternative test for serial correlation
estat dwatson	Durbin–Watson d statistic to test for first-order serial correlation

These commands provide regression diagnostic tools specific to time series. You must `tsset` your data before using these commands.

`estat archlm` tests for time-dependent volatility. `estat bgodfrey`, `estat durbinalt`, and `estat dwatson` test for serial correlation in the residuals of a linear regression. For non–time-series regression diagnostic tools, see [R] **regress postestimation**.

`estat archlm` performs Engle's LM test for the presence of autoregressive conditional heteroskedasticity.

`estat bgodfrey` performs the Breusch–Godfrey test for higher-order serial correlation in the disturbance. This test does not require that all the regressors be strictly exogenous.

`estat durbinalt` performs Durbin's alternative test for serial correlation in the disturbance. This test does not require that all the regressors be strictly exogenous.

`estat dwatson` computes the Durbin–Watson d statistic (Durbin and Watson 1950) to test for first-order serial correlation in the disturbance when all the regressors are strictly exogenous.

Syntax for estat archlm

estat archlm [, *archlm_options*]

archlm_options	description
lags(*numlist*)	test *numlist* lag orders
force	allow test after `regress, vce(robust)`

Options for estat archlm

lags(*numlist*) specifies a list of numbers, indicating the lag orders to be tested. The test will be performed separately for each order. The default is order one.

force allows the test to be run after `regress, vce(robust)`. The command will not work if the vce(cluster *clustvar*) option is specified with [R] **regress**.

Syntax for estat bgodfrey

estat <u>bg</u>odfrey [, *bgodfrey_options*]

bgodfrey_options	description
<u>lags</u>(*numlist*)	test *numlist* lag orders
<u>nomiss</u>0	do not use Davidson and MacKinnon's approach
<u>small</u>	obtain p-values by using the F or t distribution

Options for estat bgodfrey

lags(*numlist*) specifies a list of numbers, indicating the lag orders to be tested. The test will be performed separately for each order. The default is order one.

nomiss0 specifies that Davidson and MacKinnon's approach (1993, 358), which replaces the missing values in the initial observations on the lagged residuals in the auxiliary regression with zeros, not be used.

small specifies that the p-values of the test statistics be obtained using the F or t distribution instead of the default chi-squared or normal distribution.

Syntax for estat durbinalt

estat <u>dur</u>binalt [, *durbinalt_options*]

durbinalt_options	description
<u>lags</u>(*numlist*)	test *numlist* lag orders
<u>nomiss</u>0	do not use Davidson and MacKinnon's approach
<u>robust</u>	compute standard errors using the robust/sandwich estimator
<u>small</u>	obtain p-values using the F or t distribution
force	allow test after regress, vce(robust) or after newey

Options for estat durbinalt

lags(*numlist*) specifies a list of numbers, indicating the lag orders to be tested. The test will be performed separately for each order. The default is order one.

nomiss0 specifies that Davidson and MacKinnon's approach (1993, 358), which replaces the missing values in the initial observations on the lagged residuals in the auxiliary regression with zeros, not be used.

robust specifies that the Huber/White/sandwich robust estimator of the variance–covariance matrix be used in Durbin's alternative test.

small specifies that the p-values of the test statistics be obtained using the F or t distribution instead of the default chi-squared or normal distribution. This option may not be specified with robust, which always uses an F or t distribution.

force allows the test to be run after regress, vce(robust) and after newey (see [R] regress and [TS] newey). The command will not work if the vce(cluster *clustvar*) option is specified with [R] regress.

Syntax for estat dwatson

estat <u>dwa</u>tson

Remarks

The Durbin–Watson test is used to determine whether the error term in a linear regression model follows an AR(1) process. For the linear model

$$y_t = \mathbf{x}_t \boldsymbol{\beta} + u_t$$

the AR(1) process can be written as

$$u_t = \rho u_{t-1} + \epsilon_t$$

In general, an AR(1) process requires only that ϵ_t be independently and identically distributed (i.i.d.). The Durbin–Watson test, however, requires ϵ_t to be distributed $N(0, \sigma^2)$ for the statistic to have an exact distribution. Also the Durbin–Watson test can be applied only when the regressors are strictly exogenous. A regressor x is strictly exogenous if $\text{Corr}(x_s, u_t) = 0$ for all s and t, which precludes the use of the Durbin–Watson statistic with models where lagged values of the dependent variable are included as regressors.

The null hypothesis of the test is that there is no first-order autocorrelation. The Durbin–Watson d statistic can take on values between 0 and 4 and under the null d is equal to 2. Values of d less than 2 suggest positive autocorrelation ($\rho > 0$), whereas values of d greater than 2 suggest negative autocorrelation ($\rho < 0$). Calculating the exact distribution of the d statistic is difficult, but empirical upper and lower bounds have been established based on the sample size and the number of regressors. Extended tables for the d statistic have been published by Savin and White (1977). For example, suppose you have a model with 30 observations and three regressors (including the constant term). For a test of the null hypothesis of no autocorrelation versus the alternative of positive autocorrelation, the lower bound of the d statistic is 1.284, and the upper bound is 1.567 at the 5% significance level. You would reject the null if $d < 1.284$, and you would fail to reject if $d > 1.567$. A value falling within the range (1.284, 1.567) leads to no conclusion about whether or not to reject the null hypothesis.

When lagged dependent variables are included among the regressors, the past values of the error term are correlated with those lagged variables at time t, implying that they are not strictly exogenous regressors. The inclusion of covariates that are not strictly exogenous causes the d statistic to be biased toward the acceptance of the null hypothesis. Durbin (1970) suggested an alternative test for models with lagged dependent variables and extended that test to the more general AR(p) serial correlation process

$$u_t = \rho_1 u_{t-1} + \cdots + \rho_p u_{t-p} + \epsilon_t$$

where ϵ_t is independently and identically distributed with variance σ^2 but is not assumed or required to be normal for the test.

The null hypothesis of Durbin's alternative test is

$$H_0 : \rho_1 = 0, \ldots, \rho_p = 0$$

and the alternative is that at least one of the ρ's is nonzero. Although the null hypothesis was originally derived for an AR(p) process, this test turns out to have power against MA(p) processes as well. Hence, the actual null of this test is that there is no serial correlation up to order p because the MA(p) and the AR(p) models are locally equivalent alternatives under the null. See Godfrey (1988, 113–115) for a discussion of this result.

Durbin's alternative test is in fact a Lagrange multiplier (LM) test, but it is most easily computed with a Wald test on the coefficients of the lagged residuals in an auxiliary OLS regression of the residuals on their lags and all the covariates in the original regression. Consider the linear regression model

$$y_t = \beta_1 x_{1t} + \cdots + \beta_k x_{kt} + u_t \tag{1}$$

in which the covariates x_1 through x_k are not assumed to be strictly exogenous and u_t is assumed to be i.i.d. and to have finite variance. The process is also assumed to be stationary. (See Wooldridge [2006] for a discussion of stationarity.) Estimating the parameters in (1) by OLS obtains the residuals $\widehat{u}_t$. Next another OLS regression is performed of $\widehat{u}_t$ on $\widehat{u}_{t-1}, \ldots, \widehat{u}_{t-p}$ and the other regressors,

$$\widehat{u}_t = \gamma_1 \widehat{u}_{t-1} + \cdots + \gamma_p \widehat{u}_{t-p} + \beta_1 x_{1t} + \cdots + \beta_k x_{kt} + \epsilon_t \tag{2}$$

where ϵ_t stands for the random-error term in this auxiliary OLS regression. Durbin's alternative test is then obtained by performing a Wald test that $\gamma_1, \ldots, \gamma_p$ are jointly zero. The test can be made robust to an unknown form of heteroskedasticity by using a robust VCE estimator when estimating the regression in (2). When there are only strictly exogenous regressors and $p = 1$, this test is asymptotically equivalent to the Durbin–Watson test.

The Breusch–Godfrey test is also an LM test of the null hypothesis of no autocorrelation versus the alternative that u_t follows an AR(p) or MA(p) process. Like Durbin's alternative test, it is based on the auxiliary regression (2), and it is computed as NR^2, where N is the number of observations and R^2 is the simple R^2 from the regression. This test and Durbin's alternative test are asymptotically equivalent. The test statistic NR^2 has an asymptotic χ^2 distribution with p degrees of freedom. It is valid with or without the strict exogeneity assumption but is not robust to conditional heteroskedasticity, even if a robust VCE is used when fitting (2).

In fitting (2), the values of the lagged residuals will be missing in the initial periods. As noted by Davidson and MacKinnon (1993), the residuals will not be orthogonal to the other covariates in the model in this restricted sample, which implies that the R^2 from the auxiliary regression will not be zero when the lagged residuals are left out. Hence, Breusch and Godfrey's NR^2 version of the test may overreject in small samples. To correct this problem, Davidson and MacKinnon (1993) recommend setting the missing values of the lagged residuals to zero and running the auxiliary regression in (2) over the full sample used in (1). This small-sample correction has become conventional for both the Breusch–Godfrey and Durbin's alternative test, and it is the default for both commands. Specifying the `nomiss0` option overrides this default behavior and treats the initial missing values generated by regressing on the lagged residuals as missing. Hence, `nomiss0` causes these initial observations to be dropped from the sample of the auxiliary regression.

Durbin's alternative test and the Breusch–Godfrey test were originally derived for the case covered by `regress` without the `vce(robust)` option. However, after `regress`, `vce(robust)` and `newey`, Durbin's alternative test is still valid and can be invoked if the `robust` and `force` options are specified.

▷ Example 1: tests for serial correlation

Using data from Klein (1950), we first fit an OLS regression of consumption on the government wage bill:

```
. use http://www.stata-press.com/data/r10/klein
. tsset yr
        time variable:  yr, 1920 to 1941
                delta:  1 unit
. regress consump wagegovt
```

Source	SS	df	MS			
Model	532.567711	1	532.567711	Number of obs =		22
Residual	601.207167	20	30.0603584	F(1, 20) =		17.72
				Prob > F =		0.0004
				R-squared =		0.4697
				Adj R-squared =		0.4432
Total	1133.77488	21	53.9892799	Root MSE =		5.4827

consump	Coef.	Std. Err.	t	P>|t|	[95% Conf. Interval]	
wagegovt	2.50744	.5957173	4.21	0.000	1.264796	3.750085
_cons	40.84699	3.192183	12.80	0.000	34.18821	47.50577

If we assume that `wagegov` is a strictly exogenous variable, we can use the Durbin–Watson test to check for first-order serial correlation in the errors.

```
. estat dwatson
Durbin-Watson d-statistic(  2,    22) =  .3217998
```

The Durbin–Watson d statistic 0.32 is far from the center of its distribution ($d = 2.0$). Given 22 observations and two regressors (including the constant term) in the model, the lower 5% bound is about 0.997, much greater than the computed d statistic. Assuming that `wagegov` is strictly exogenous, we can reject the null of no first-order serial correlation. Rejecting the null hypothesis does not necessarily mean an AR process; other forms of misspecification may also lead to a significant test statistic. If we are willing to assume that the errors follow an AR(1) process and that `wagegov` is strictly exogenous, we could refit the model using `arima` or `prais` and model the error process explicitly; see [TS] **arima** and [TS] **prais**.

If we are not willing to assume that `wagegov` is strictly exogenous, we could instead use Durbin's alternative test or the Breusch–Godfrey to test for first-order serial correlation. Because we have only 22 observations, we will use the `small` option.

```
. estat durbinalt, small
Durbin's alternative test for autocorrelation
```

lags(p)	F	df	Prob > F
1	35.035	(1, 19)	0.0000

H0: no serial correlation

```
. estat bgodfrey, small
Breusch-Godfrey LM test for autocorrelation
```

lags(p)	F	df	Prob > F
1	14.264	(1, 19)	0.0013

H0: no serial correlation

Both tests strongly reject the null of no first-order serial correlation, so we decide to refit the model with two lags of `consump` included as regressors and then rerun `estat durbinalt` and `estat bgodfrey`. Because the revised model includes lagged values of the dependent variable, the Durbin–Watson test is not applicable.

```
. regress consump wagegovt L.consump L2.consump
```

Source	SS	df	MS			Number of obs =	20
						F(3, 16) =	44.01
Model	702.660311	3	234.220104			Prob > F =	0.0000
Residual	85.1596011	16	5.32247507			R-squared =	0.8919
						Adj R-squared =	0.8716
Total	787.819912	19	41.4642059			Root MSE =	2.307

consump	Coef.	Std. Err.	t	P>|t|	[95% Conf. Interval]	
wagegovt	.6904282	.3295485	2.10	0.052	-.0081835	1.38904
consump						
L1.	1.420536	.197024	7.21	0.000	1.002864	1.838208
L2.	-.650888	.1933351	-3.37	0.004	-1.06074	-.241036
_cons	9.209073	5.006701	1.84	0.084	-1.404659	19.82281

```
. estat durbinalt, small lags(1/2)
Durbin's alternative test for autocorrelation
```

lags(p)	F	df			Prob > F
1	0.080	(1,	15	)	0.7805
2	0.260	(2,	14	)	0.7750

H0: no serial correlation

```
. estat bgodfrey, small lags(1/2)
Breusch-Godfrey LM test for autocorrelation
```

lags(p)	F	df			Prob > F
1	0.107	(1,	15	)	0.7484
2	0.358	(2,	14	)	0.7056

H0: no serial correlation

Although `wagegov` and the constant term are no longer statistically different from zero at the 5% level, the output from `estat durbinalt` and `estat bgodfrey` indicates that including the two lags of `consump` has removed any serial correlation from the errors.

◁

Engle (1982) suggests an LM test for checking for autoregressive conditional heteroskedasticity (ARCH) in the errors. The pth-order ARCH model can be written as

$$\sigma_t^2 = E(u_t^2 | u_{t-1}, \ldots, u_{t-p})$$
$$= \gamma_0 + \gamma_1 u_{t-1}^2 + \cdots + \gamma_p u_{t-p}^2$$

To test the null hypothesis of no autoregressive conditional heteroskedasticity (i.e., $\gamma_1 = \cdots = \gamma_p = 0$), we first fit the OLS model (1), obtain the residuals $\widehat{u}_t$, and run another OLS regression on the lagged residuals:

$$\widehat{u}_t^2 = \gamma_0 + \gamma_1 \widehat{u}_{t-1}^2 + \cdots + \gamma_p \widehat{u}_{t-p}^2 + \epsilon \tag{3}$$

The test statistic is NR^2, where N is the number of observations in the sample and R^2 is the R^2 from the regression in (3). Under the null hypothesis, the test statistic follows a χ_p^2 distribution.

▷ Example 2: estat archlm

We refit the original model that does not include the two lags of consump and then use estat archlm to see if there is any evidence that the errors are autoregressive conditional heteroskedastic.

```
. regress consump wagegovt
```

Source	SS	df	MS		Number of obs =	22
					F(1, 20) =	17.72
Model	532.567711	1	532.567711		Prob > F =	0.0004
Residual	601.207167	20	30.0603584		R-squared =	0.4697
					Adj R-squared =	0.4432
Total	1133.77488	21	53.9892799		Root MSE =	5.4827

consump	Coef.	Std. Err.	t	P>|t|	[95% Conf. Interval]	
wagegovt	2.50744	.5957173	4.21	0.000	1.264796	3.750085
_cons	40.84699	3.192183	12.80	0.000	34.18821	47.50577

```
. estat archlm, lags(1 2 3)
```
LM test for autoregressive conditional heteroskedasticity (ARCH)

lags(p)	chi2	df	Prob > chi2
1	5.543	1	0.0186
2	9.431	2	0.0090
3	9.039	3	0.0288

H0: no ARCH effects   vs.   H1: ARCH(p) disturbance

estat archlm shows the results for tests of ARCH(1), ARCH(2), and ARCH(3) effects, respectively. At the 5% significance level, all three tests reject the null hypothesis that the errors are not autoregressive conditional heteroskedastic. See [TS] **arch** for information on fitting ARCH models.

◁

Saved Results

estat archlm saves the following in r():

Scalars
  r(N)   number of observations   r(N_gaps)   number of gaps
  r(k)   number of regressors
Macros
  r(lags)   lag order
Matrices
  r(arch)   test statistic for each lag order   r(p)   two-sided p-values
  r(df)   degrees of freedom

`estat bgodfrey` saves the following in `r()`:

Scalars
`r(N)`	number of observations	`r(N_gaps)`	number of gaps
`r(k)`	number of regressors		

Macros
`r(lags)`	lag order

Matrices
`r(chi2)`	χ^2 statistic for each lag order	`r(p)`	two-sided p-values
`r(F)`	F statistic for each lag order (`small` only)	`r(df)`	degrees of freedom
`r(df_r)`	residual degrees of freedom (`small` only)		

`estat durbinalt` saves the following in `r()`:

Scalars
`r(N)`	number of observations	`r(N_gaps)`	number of gaps
`r(k)`	number of regressors		

Macros
`r(lags)`	lag order

Matrices
`r(chi2)`	χ^2 statistic for each lag order	`r(p)`	two-sided p-values
`r(F)`	F statistic for each lag order (`small` only)	`r(df)`	degrees of freedom
`r(df_r)`	residual degrees of freedom (`small` only)		

`estat dwatson` saves the following in `r()`:

Scalars
`r(N)`	number of observations	`r(N_gaps)`	number of gaps
`r(k)`	number of regressors	`r(dw)`	Durbin–Watson statistic

Methods and Formulas

`estat archlm`, `estat bgodfrey`, `estat durbinalt`, and `estat dwatson` are implemented as ado-files.

Consider the regression

$$y_t = \beta_1 x_{1t} + \cdots + \beta_k x_{kt} + u_t \tag{4}$$

in which some of the covariates are not strictly exogenous. In particular, some of the x_{it} may be lags of the dependent variable. We are interested in whether the u_t are serially correlated.

The Durbin–Watson d statistic reported by `estat dwatson` is

$$d = \frac{\sum_{t=1}^{n-1} (\widehat{u}_{t+1} - \widehat{u}_t)^2}{\sum_{t=1}^{n} \widehat{u}_t^2}$$

where $\widehat{u}_t$ represents the residual of the tth observation.

To compute Durbin's alternative test and the Breusch–Godfrey test against the null hypothesis that there is no pth order serial correlation, we fit the regression in (4), compute the residuals, and then fit the following auxiliary regression of the residuals $\widehat{u}_t$ on p lags of $\widehat{u}_t$ and on all the covariates in the original regression in (4):

$$\widehat{u}_t = \gamma_1 \widehat{u}_{t-1} + \cdots + \gamma_p \widehat{u}_{t-p} + \beta_1 x_{1t} + \cdots + \beta_k x_{kt} + \epsilon \qquad (5)$$

Durbin's alternative test is computed by performing a Wald test to determine whether the coefficients of $\widehat{u}_{t-1}, \ldots, \widehat{u}_{t-p}$ are jointly different from zero. By default, the statistic is assumed to be distributed $\chi^2(p)$. When `small` is specified, the statistic is assumed to follow an $F(p, N - p - k)$ distribution. The reported p-value is a two-sided p-value. When `robust` is specified, the Wald test is performed using the Huber/White/sandwich estimator of the variance–covariance matrix, and the test is robust to an unspecified form of heteroskedasticity.

The Breusch–Godfrey test is computed as NR^2, where N is the number of observations in the auxiliary regression (5) and R^2 is the R^2 from the same regression (5). Like Durbin's alternative test, the Breusch–Godfrey test is asymptotically distributed $\chi^2(p)$, but specifying `small` causes the p-value to be computed using an $F(p, N - p - k)$.

By default, the initial missing values of the lagged residuals are replaced with zeros, and the auxiliary regression is run over the full sample used in the original regression of (4). Specifying the `nomiss0` option causes these missing values to be treated as missing values, and the observations are dropped from the sample.

Engle's LM test for ARCH(p) effects fits an OLS regression of $\widehat{u}_t^2$ on $\widehat{u}_{t-1}^2, \ldots, \widehat{u}_{t-p}^2$:

$$\widehat{u}_t^2 = \gamma_0 + \gamma_1 \widehat{u}_{t-1}^2 + \cdots + \gamma_p \widehat{u}_{t-p}^2 + \epsilon$$

The test statistic is nR^2 and is asymptotically distributed $\chi^2(p)$.

Acknowledgment

The original versions of `estat archlm`, `estat bgodfrey`, and `estat durbinalt` were written by Christopher F. Baum, Boston College.

References

Baum, C. F. 2006. *An Introduction to Modern Econometrics Using Stata*. College Station, TX: Stata Press.

Baum, C. F., and V. L. Wiggins. 2000a. sg135: Test for autoregressive conditional heteroskedasticity in regression error distribution. *Stata Technical Bulletin* 55: 13–14. Reprinted in *Stata Technical Bulletin Reprints*, vol. 10, pp. 143–144.

———. 2000b. sg136: Tests for serial correlation in regression error distribution. *Stata Technical Bulletin* 55: 14–15. Reprinted in *Stata Technical Bulletin Reprints*, vol. 10, pp. 145–147.

Beran, R. J., and N. I. Fisher. 1998. A conversation with Geoff Watson. *Statistical Science* 13: 75–93.

Breusch, T. 1978. Testing for autocorrelation in dynamic linear models. *Australian Economic Papers* 17: 334–355.

Davidson, R., and J. G. MacKinnon. 1993. *Estimation and Inference in Econometrics*. New York: Oxford University Press.

Durbin, J. 1970. Testing for serial correlation in least-squares regressions when some of the regressors are lagged dependent variables. *Econometrica* 38: 410–421.

Durbin, J., and G. S. Watson. 1950. Testing for serial correlation in least-squares regression. *Biometrika* 37: 409–428.

——. 1951. Testing for serial correlation in least-squares regression. *Biometrika* 38: 159–178.

Engle, R. F. 1982. Autoregressive conditional heteroskedasticity with estimates of the variance of United Kingdom inflation. *Econometrica* 50: 987–1007.

Fisher, N. I., and P. Hall. 1998. Geoffrey Stuart Watson: Tributes and obituary (3 December 1921—3 January 1998). *Australian and New Zealand Journal of Statistics* 40: 257–267.

Godfrey, L. G. 1978. Testing against general autoregressive and moving average error models when the regressors include lagged dependent variables. *Econometrics* 46: 1293–1301.

——. 1988. *Misspecification Tests in Econometrics: The Lagrange Multiplier Principle and Other Approaches*. Econometric Society Monographs, No. 16, Cambridge: Cambridge University Press.

Greene, W. H. 2003. *Econometric Analysis*. 5th ed. Upper Saddle River, NJ: Prentice Hall.

Johnston, J., and J. DiNardo. 1997. *Econometric Methods*. 4th ed. New York: McGraw–Hill.

Klein, L. 1950. *Economic Fluctuations in the United States 1921–1941*. New York: Wiley.

Phillips, P. C. B. 1988. The ET Interview: Professor James Durbin. *Econometric Theory* 4: 125–157.

Savin, E., and K. J. White. 1977. The Durbin–Watson test for serial correlation with extreme sample sizes or many regressors. *Econometrica* 45: 1989–1996.

Wooldridge, J. M. 2006. *Introductory Econometrics: A Modern Approach*. 3rd ed. Cincinnati, OH: South-Western.

James Durbin (1923–) is a British statistician who was born in Wigan, near Manchester. He studied mathematics at Cambridge and after military service and various research posts joined the London School of Economics in 1950. His many contributions to statistics have centered on serial correlation, time series, sample survey methodology, goodness-of-fit tests, and sample distribution functions, with emphasis on applications in the social sciences.

Geoffrey Stuart Watson (1921–1998) was born in Victoria, Australia, and earned degrees at Melbourne University and North Carolina State University. After a visit to the University of Cambridge, he returned to Australia, working at Melbourne and then the Australian National University. Following periods at Toronto and Johns Hopkins, he settled at Princeton. His best-known contributions to statistics, in a wide-ranging career, include the Durbin–Watson test for serial correlation, the Nadaraya–Watson estimator in nonparametric regression, and methods for directional data.

Leslie G. Godfrey (1946–) was born in London and earned degrees at the Universities of Exeter and London. He now holds a chair in econometrics at the University of York. His interests center on implementation and interpretation of tests of econometric models, including nonnested models.

Trevor Stanley Breusch (1949–) was born in Queensland and earned degrees at the University of Queensland and Australian National University (ANU). After a post at the University of Southampton, he returned to work at ANU. His background is in econometric methods and his recent interests include political values and social attitudes, earnings and income, and measurement of underground economic activity.

Also See

[R] **regress** — Linear regression

[TS] **tsset** — Declare data to be time-series data

[R] **regress postestimation** — Postestimation tools for regress

Title

#review — Review previous commands

Syntax

<u>#r</u>eview $\left[\ \#_1\ \left[\ \#_2\ \right]\ \right]$

Description

The **#review** command displays the last few lines typed at the terminal.

Remarks

#review (pronounced *pound-review*) is a Stata preprocessor command. *#commands* do not generate a return code or generate ordinary Stata errors. The only error message associated with *#commands* is "unrecognized #command".

The **#review** command displays the last few lines typed at the terminal. If no arguments follow **#review**, the last five lines typed at the terminal are displayed. The first argument specifies the number of lines to be reviewed, so **#review** 10 displays the last 10 lines typed. The second argument specifies the number of lines to be displayed, so **#review** 10 5 displays five lines, starting at the 10th previous line.

Stata reserves a buffer for **#review** lines and stores as many previous lines in the buffer as will fit, rolling out the oldest line to make room for the newest. Requests to **#review** lines no longer stored will be ignored. Only lines typed at the terminal are placed in the **#review** buffer. See [U] **10.5 Editing previous lines in Stata**.

▷ Example 1

Typing **#review** by itself will show the last five lines you typed at the terminal:

```
. #review
5 use mydata
4 * comments go into the #review buffer, too
3 describe
2 tabulate marriage educ [freq=number]
1 tabulate marriage educ [freq=number], chi2
. _
```

Typing **#review** 15 2 shows the 15th and 14th previous lines:

```
. #review 15 2
15 replace x=. if x<200
14 summarize x
. _
```

◁

156

Title

> **roc** — Receiver operating characteristic (ROC) analysis

Syntax

Perform nonparametric ROC analysis

> roctab *refvar classvar* [*if*] [*in*] [*weight*] [, *roctab_options*]

Test equality of ROC areas

> roccomp *refvar classvar* [*classvars*] [*if*] [*in*] [*weight*] [, *roccomp_options*]

Test equality of ROC area against a standard ROC curve

> rocgold *refvar goldvar classvar* [*classvars*] [*if*] [*in*] [*weight*] [, *rocgold_options*]

roctab_options	description
Main	
lorenz	report Gini and Pietra indices
binomial	calculate exact binomial confidence intervals
detail	show details on sensitivity/specificity for each cutpoint
table	display the raw data in a $2 \times k$ contingency table
bamber	calculate standard errors by using the Bamber method
hanley	calculate standard errors by using the Hanley method
graph	graph the ROC curve
norefline	suppress plotting the 45-degree reference line
summary	report the area under the ROC curve
specificity	graph sensitivity versus specificity
level(#)	set confidence level; default is level(95)
Plot	
plotopts(*plot_options*)	affect rendition of the ROC curve
Reference line	
rlopts(*cline_options*)	affect rendition of the reference line
Add plots	
addplot(*plot*)	add other plots to the generated graph
Y axis, X axis, Titles, Legend, Overall	
twoway_options	any options other than by() documented in [G] *twoway_options*

fweights are allowed; see [U] **11.1.6 weight**.

roccomp_options	description
Main	
by(*varname*)	split into groups by variable
test(*matname*)	use contrast matrix for comparing ROC areas
graph	graph the ROC curve
<u>norefl</u>ine	suppress plotting the 45-degree reference line
<u>separa</u>te	place each ROC curve on its own graph
<u>summa</u>ry	report the area under the ROC curve
<u>bino</u>rmal	estimate areas by using binormal distribution assumption
line#opts(*cline_options*)	affect rendition of the #th binormal fit line
<u>l</u>evel(#)	set confidence level; default is level(95)
Plots	
plot#opts(*plot_options*)	affect rendition of the #th ROC curve
Reference line	
rlopts(*cline_options*)	affect rendition of the reference line
Y axis, X axis, Titles, Legend, Overall	
twoway_options	any options documented in [G] ***twoway_options***

fweights are allowed; see [U] **11.1.6 weight**.

rocgold_options	description
Main	
<u>sidak</u>	adjust the significance probability by using Šidák's method
test(*matname*)	use contrast matrix for comparing ROC areas
graph	graph the ROC curve
<u>norefl</u>ine	suppress plotting the 45-degree reference line
<u>separa</u>te	place each ROC curve on its own graph
<u>summa</u>ry	report the area under the ROC curve
<u>bino</u>rmal	estimate areas by using binormal distribution assumption
line#opts(*cline_options*)	affect rendition of the #th binormal fit line
<u>l</u>evel(#)	set confidence level; default is level(95)
Plots	
plot#opts(*plot_options*)	affect rendition of the #th ROC curve; plot 1 is the "gold standard"
Reference line	
rlopts(*cline_options*)	affect rendition of the reference line
Y axis, X axis, Titles, Legend, Overall	
twoway_options	any options documented in [G] ***twoway_options***

fweights are allowed; see [U] **11.1.6 weight**.

plot_options	description
marker_options	change look of markers (color, size, etc.)
marker_label_options	add marker labels; change look or position
cline_options	change the look of the line

Description

The above commands are used to perform receiver operating characteristic (ROC) analyses with rating and discrete classification data.

The two variables *refvar* and *classvar* must be numeric. The reference variable indicates the true state of the observation, such as diseased and nondiseased or normal and abnormal, and must be coded as 0 and 1. The rating or outcome of the diagnostic test or test modality is recorded in *classvar*, which must be at least ordinal, with higher values indicating higher risk.

`roctab` performs nonparametric ROC analyses. By default, `roctab` calculates the area under the ROC curve. Optionally, `roctab` can plot the ROC curve, display the data in tabular form, and produce Lorenz-like plots.

`roccomp` tests the equality of two or more ROC areas obtained from applying two or more test modalities to the same sample or to independent samples. `roccomp` expects the data to be in wide form when comparing areas estimated from the same sample and in long form for areas estimated from independent samples.

`rocgold` independently tests the equality of the ROC area of each of several test modalities, specified by *classvar*, against a "gold standard" ROC curve, *goldvar*. For each comparison, `rocgold` reports the raw and the Bonferroni-adjusted significance probability. Optionally, Šidák's adjustment for multiple comparisons can be obtained.

See [R] **rocfit** for a command that fits maximum-likelihood ROC models.

Options for roctab

⌐ Main ⌐

`lorenz` specifies that Gini and Pietra indices be reported. Optionally, `graph` will plot the Lorenz-like curve.

`binomial` specifies that exact binomial confidence intervals be calculated.

`detail` outputs a table displaying the sensitivity, specificity, the percentage of subjects correctly classified, and two likelihood ratios for each possible cutpoint of *classvar*.

`table` outputs a $2 \times k$ contingency table displaying the raw data.

`bamber` specifies that the standard error for the area under the ROC curve be calculated using the method suggested by Bamber (1975). Otherwise, standard errors are obtained as suggested by DeLong, DeLong, and Clarke-Pearson (1988).

`hanley` specifies that the standard error for the area under the ROC curve be calculated using the method suggested by Hanley and McNeil (1982). Otherwise, standard errors are obtained as suggested by DeLong, DeLong, and Clarke-Pearson (1988).

graph produces graphical output of the ROC curve. If lorenz is specified, graphical output of a Lorenz-like curve will be produced.

norefline suppresses plotting the 45-degree reference line from the graphical output of the ROC curve.

summary reports the area under the ROC curve, its standard error, and its confidence interval. If lorenz is specified, Lorenz indices are reported. This option is needed only when also specifying graph.

specificity produces a graph of sensitivity versus specificity instead of sensitivity versus (1 − specificity). specificity implies graph.

level(#) specifies the confidence level, as a percentage, for the confidence intervals. The default is level(95) or as set by set level; see [R] **level**.

___
| Plot |
___

plotopts(*plot_options*) affects the rendition of the plotted ROC curve—the curve's plotted points connected by lines. The *plot_options* can affect the size and color of markers, whether and how the markers are labeled, and whether and how the points are connected; see [G] *marker_options*, [G] *marker_label_options*, and [G] *cline_options*.

___
| Reference line |
___

rlopts(*cline_options*) affects the rendition of the reference line; see [G] *cline_options*.

___
| Add plots |
___

addplot(*plot*) provides a way to add other plots to the generated graph; see [G] *addplot_option*.

___
| Y axis, X axis, Titles, Legend, Overall |
___

twoway_options are any of the options documented in [G] *twoway_options*, excluding by(). These include options for titling the graph (see [G] *title_options*) and for saving the graph to disk (see [G] *saving_option*).

Options for roccomp and rocgold

___
| Main |
___

by(*varname*) (roccomp only) is required when comparing independent ROC areas. The by() variable identifies the groups to be compared.

sidak (rocgold only) requests that the significance probability be adjusted for the effect of multiple comparisons by using Šidák's method. Bonferroni's adjustment is reported by default.

test(*matname*) specifies the contrast matrix to be used when comparing ROC areas. By default, the null hypothesis that all areas are equal is tested.

graph produces graphical output of the ROC curve.

norefline suppresses plotting the 45-degree reference line from the graphical output of the ROC curve.

separate is meaningful only with roccomp and specifies that each ROC curve be placed on its own graph rather than one curve on top of the other.

summary reports the area under the ROC curve, its standard error, and its confidence interval. This option is needed only when also specifying graph.

binormal specifies that the areas under the ROC curves to be compared should be estimated using the binormal distribution assumption. By default, areas to be compared are computed using the trapezoidal rule.

line#opts(*cline_options*) affects the rendition of the line representing the #th ROC curve drawn using the binormal distribution assumption; see [G] *cline_options*. These lines are drawn only if the binormal option is specified.

level(#) specifies the confidence level, as a percentage, for the confidence intervals. The default is level(95) or as set by set level; see [R] **level**.

_____| Plots |_____

plot#opts(*plot_options*) affects the rendition of the #th ROC curve—the curve's plotted points connected by lines. The *plot_options* can affect the size and color of markers, whether and how the markers are labeled, and whether and how the points are connected; see [G] *marker_options*, [G] *marker_label_options*, and [G] *cline_options*.

For rocgold, plot1opts() are applied to the ROC for the gold standard.

_____| Reference line |_____

rlopts(*cline_options*) affects the rendition of the reference line; see [G] *cline_options*.

_____| Y axis, X axis, Titles, Legend, Overall |_____

twoway_options are any of the options documented in [G] *twoway_options*. These include options for titling the graph (see [G] *title_options*), options for saving the graph to disk (see [G] *saving_option*), and the by() option (see [G] *by_option*).

Remarks

Remarks are presented under the following headings:

> *Introduction*
> *Nonparametric ROC curves*
> *Lorenz-like curves*
> *Comparing areas under the ROC curve*
> *Correlated data*
> *Independent data*
> *Comparing areas with a gold standard*

Introduction

Receiver operating characteristic (ROC) analysis quantifies the accuracy of diagnostic tests or other evaluation modalities used to discriminate between two states or conditions, which are here referred to as normal and abnormal. The discriminatory accuracy of a diagnostic test is measured by its ability to correctly classify known normal and abnormal subjects. The analysis uses the ROC curve, a graph of the sensitivity versus $1 -$ specificity of the diagnostic test. The sensitivity is the fraction of positive cases that are correctly classified by the diagnostic test, whereas the specificity is the fraction of negative cases that are correctly classified. Thus the sensitivity is the true-positive rate, and the specificity is the true-negative rate.

The global performance of a diagnostic test is commonly summarized by the area under the ROC curve. This area can be interpreted as the probability that the result of a diagnostic test of a randomly selected abnormal subject will be greater than the result of the same diagnostic test from a randomly selected normal subject. The greater the area under the ROC curve, the better the global performance of the diagnostic test.

Both nonparametric methods and parametric (semiparametric) methods have been suggested for generating the ROC curve and for calculating its area. The following sections present these approaches, and the last section presents tests for comparing areas under ROC curves.

See Pepe (2003) for a discussion of ROC analysis. Pepe has posted Stata datasets and programs used to reproduce results presented in the book (http://www.stata.com/bookstore/pepe.html).

Nonparametric ROC curves

The points on the nonparametric ROC curve are generated by using each possible outcome of the diagnostic test as a classification cutpoint and computing the corresponding sensitivity and $1 -$ specificity. These points are then connected by straight lines, and the area under the resulting ROC curve is computed using the trapezoidal rule.

▷ Example 1

Hanley and McNeil (1982) presented data from a study in which a reviewer was asked to classify, using a nine-point scale, a random sample of 109 tomographic images from patients with neurological problems. The rating scale was as follows: 1—definitely normal, 2—probably normal, 3—questionable, 4—probably abnormal, and 5—definitely abnormal. The true disease status was normal for 58 of the patients and abnormal for the remaining 51 patients.

Here we list 9 of the 109 observations.

```
. use http://www.stata-press.com/data/r10/hanley
. list disease rating in 1/9
```

	disease	rating
1.	1	5
2.	0	1
3.	1	5
4.	0	4
5.	0	1
6.	0	3
7.	1	5
8.	0	5
9.	0	1

For each observation, `disease` identifies the true disease status of the subject ($0 =$ normal, $1 =$ abnormal), and `rating` contains the classification value assigned by the reviewer.

We can use `roctab` to calculate and plot the nonparametric ROC curve by specifying both the `summary` and `graph` options. By also specifying the `table` option, we obtain a contingency table summarizing our dataset.

```
. roctab disease rating, table graph summary
```

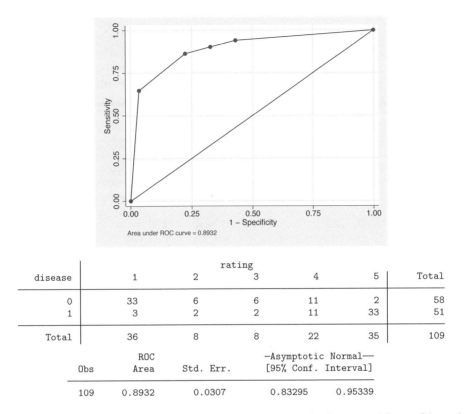

disease	rating 1	2	3	4	5	Total
0	33	6	6	11	2	58
1	3	2	2	11	33	51
Total	36	8	8	22	35	109

Obs	ROC Area	Std. Err.	—Asymptotic Normal— [95% Conf. Interval]	
109	0.8932	0.0307	0.83295	0.95339

By default, roctab reports the area under the curve, its standard error, and its confidence interval. The graph option can be used to plot the ROC curve.

The ROC curve is plotted by computing the sensitivity and specificity using each value of the rating variable as a possible cutpoint. A point is plotted on the graph for each of the cutpoints. These plotted points are joined by straight lines to form the ROC curve, and the area under the ROC curve is computed using the trapezoidal rule.

We can tabulate the computed sensitivities and specificities for each of the possible cutpoints by specifying detail.

(Continued on next page)

```
. roctab disease rating, detail
Detailed report of Sensitivity and Specificity
```

Cutpoint	Sensitivity	Specificity	Correctly Classified	LR+	LR−
(>= 1)	100.00%	0.00%	46.79%	1.0000	
(>= 2)	94.12%	56.90%	74.31%	2.1835	0.1034
(>= 3)	90.20%	67.24%	77.98%	2.7534	0.1458
(>= 4)	86.27%	77.59%	81.65%	3.8492	0.1769
(>= 5)	64.71%	96.55%	81.65%	18.7647	0.3655
(> 5)	0.00%	100.00%	53.21%		1.0000

Obs	ROC Area	Std. Err.	—Asymptotic Normal— [95% Conf. Interval]	
109	0.8932	0.0307	0.83295	0.95339

Each cutpoint in the table indicates the ratings used to classify tomographs as being from an abnormal subject. For example, the first cutpoint $(>= 1)$ indicates that all tomographs rated as 1 or greater are classified as coming from abnormal subjects. Because all tomographs have a rating of 1 or greater, all are considered abnormal. Consequently, all abnormal cases are correctly classified (sensitivity = 100%), but none of the normal patients is classified correctly (specificity = 0%). For the second cutpoint $(>= 2)$, tomographs with ratings of 1 are classified as normal, and those with ratings of 2 or greater are classified as abnormal. The resulting sensitivity and specificity are 94.12% and 56.90%, respectively. Using this cutpoint, we correctly classified 74.31% of the 109 tomographs. Similar interpretations can be used on the remaining cutpoints. As mentioned, each cutpoint corresponds to a point on the nonparametric ROC curve. The first cutpoint $(>= 1)$ corresponds to the point at (1,1), and the last cutpoint (> 5) corresponds to the point at (0,0).

detail also reports two likelihood ratios suggested by Choi (1998): the likelihood ratio for a positive test result (LR+) and the likelihood ratio for a negative test result (LR−). The likelihood ratio for a positive test result is the ratio of the probability of a positive test among the truly positive subjects to the probability of a positive test among the truly negative subjects. The likelihood ratio for a negative test result (LR−) is the ratio of the probability of a negative test among the truly positive subjects to the probability of a negative test among the truly negative subjects. Choi points out that LR+ corresponds to the slope of the line from the origin to the point on the ROC curve determined by the cutpoint. Similarly, LR− corresponds to the slope from the point (1,1) to the point on the ROC curve determined by the cutpoint.

By default, roctab calculates the standard error for the area under the curve by using an algorithm suggested by DeLong, DeLong, and Clarke-Pearson (1988) and asymptotic normal confidence intervals. Optionally, standard errors based on methods suggested by Bamber (1975) or Hanley and McNeil (1982) can be computed by specifying bamber or hanley, respectively, and an exact binomial confidence interval can be obtained by specifying binomial.

```
. roctab disease rating, bamber
```

Obs	ROC Area	Bamber Std. Err.	—Asymptotic Normal— [95% Conf. Interval]	
109	0.8932	0.0306	0.83317	0.95317

```
. roctab disease rating, hanley binomial
```

	ROC	Hanley	— Binomial Exact —	
Obs	Area	Std. Err.	[95% Conf. Interval]
109	0.8932	0.0320	0.81559	0.94180

◁

Lorenz-like curves

For applications where it is known that the risk status increases or decreases monotonically with increasing values of the diagnostic test, the ROC curve and associated indices are useful in assessing the overall performance of a diagnostic test. When the risk status does not vary monotonically with increasing values of the diagnostic test, however, the resulting ROC curve can be nonconvex and its indices can be unreliable. For these situations, Lee (1999) proposed an alternative to the ROC analysis based on Lorenz-like curves and the associated Pietra and Gini indices.

Lee (1999) mentions at least three specific situations where results from Lorenz curves are superior to those obtained from ROC curves: (1) a diagnostic test with similar means but very different standard deviations in the abnormal and normal populations, (2) a diagnostic test with bimodal distributions in either the normal or abnormal population, and (3) a diagnostic test distributed symmetrically in the normal population and skewed in the abnormal.

When the risk status increases or decreases monotonically with increasing values of the diagnostic test, the ROC and Lorenz curves yield interchangeable results.

▷ Example 2

To illustrate the use of the `lorenz` option, we constructed a fictitious dataset that yields results similar to those presented in Table III of Lee (1999). The data assume that a 12-point rating scale was used to classify 442 diseased and 442 healthy subjects. We list a few of the observations.

```
. use http://www.stata-press.com/data/r10/lorenz, clear
. list in 1/7, noobs sep(0)
```

disease	class	pop
0	5	66
1	11	17
0	6	85
0	3	19
0	10	19
0	2	7
1	4	16

The data consist of 24 observations: 12 observations from diseased individuals and 12 from nondiseased individuals. Each observation corresponds to one of the 12 classification values of the rating-scale variable, `class`. The number of subjects represented by each observation is given by the `pop` variable, making this a frequency-weighted dataset. The data were generated assuming a binormal distribution of the latent variable with similar means for the normal and abnormal populations but with the standard deviation for the abnormal population five times greater than that of the normal population.

```
. roctab disease class [fweight=pop], graph summary
```

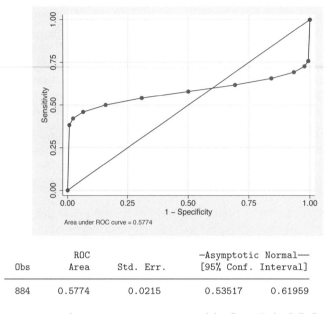

Obs	ROC Area	Std. Err.	—Asymptotic Normal— [95% Conf. Interval]	
884	0.5774	0.0215	0.53517	0.61959

The resulting ROC curve is nonconvex or, as termed by Lee, "wiggly". Lee argues that for this and similar situations, the Lorenz curve and indices are preferred.

```
. roctab disease class [fweight=pop], lorenz graph summary
```

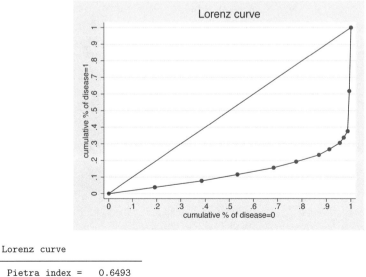

```
Lorenz curve
────────────────────────────
  Pietra index  =    0.6493
  Gini index    =    0.7441
```

Like ROC curves, a more bowed Lorenz curve suggests a better diagnostic test. This bowedness is quantified by the Pietra index, which is geometrically equivalent to twice the largest triangle that can be inscribed in the area between the curve and the diagonal line, and the Gini index, which is equivalent to twice the area between the Lorenz curve and the diagonal. Lee (1999) provides several additional interpretations for the Pietra and Gini indices.

◁

Comparing areas under the ROC curve

The area under multiple ROC curves can be compared by using `roccomp`. The command syntax is slightly different if the ROC curves are correlated (i.e., different diagnostic tests are applied to the same sample) or independent (i.e., diagnostic tests are applied to different samples).

Correlated data

▷ Example 3

Hanley and McNeil (1983) presented data from an evaluation of two computer algorithms designed to reconstruct CT images from phantoms. We will call these two algorithms' modalities 1 and 2. A sample of 112 phantoms was selected; 58 phantoms were considered normal, and the remaining 54 were abnormal. Each of the two modalities was applied to each phantom, and the resulting images were rated by a reviewer using a six-point scale: 1—definitely normal, 2—probably normal, 3—possibly normal, 4—possibly abnormal, 5—probably abnormal, and 6—definitely abnormal. Because each modality was applied to the same sample of phantoms, the two sets of outcomes are correlated.

We list the first 7 observations:

```
. use http://www.stata-press.com/data/r10/ct, clear
. list in 1/7, sep(0)
```

	mod1	mod2	status
1.	2	1	0
2.	5	5	1
3.	2	1	0
4.	2	3	0
5.	5	6	1
6.	2	2	0
7.	3	2	0

The data are in wide form, which is required when dealing with correlated data. Each observation corresponds to one phantom. The variable `mod1` identifies the rating assigned for the first modality, and `mod2` identifies the rating assigned for the second modality. The true status of the phantoms is given by `status=0` if they are normal and `status=1` if they are abnormal. The observations with at least one missing rating were dropped from the analysis.

We plot the two ROC curves and compare their areas.

(Continued on next page)

```
. roccomp status mod1 mod2, graph summary
```

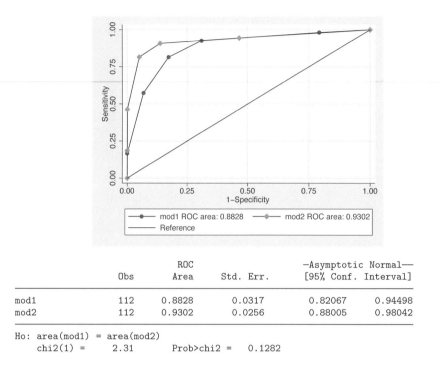

	Obs	ROC Area	Std. Err.	—Asymptotic Normal— [95% Conf. Interval]	
mod1	112	0.8828	0.0317	0.82067	0.94498
mod2	112	0.9302	0.0256	0.88005	0.98042

```
Ho: area(mod1) = area(mod2)
    chi2(1) =     2.31      Prob>chi2 =   0.1282
```

By default, roccomp, with the **graph** option specified, plots the ROC curves on the same graph. Optionally the curves can be plotted side by side, each on its own graph, by also specifying **separate**.

For each curve, roccomp reports summary statistics and provides a test for the equality of the area under the curves, using an algorithm suggested by DeLong, DeLong, and Clarke-Pearson (1988).

Although the area under the ROC curve for modality 2 is larger than that of modality 1, the chi-squared test yielded a significance probability of 0.1282, suggesting that there is no significant difference between these two areas.

The roccomp command can also be used to compare more than two ROC areas. To illustrate this, we modified the previous dataset by including a fictitious third modality.

```
. use http://www.stata-press.com/data/r10/ct2
. roccomp status mod1 mod2 mod3, graph summary
```

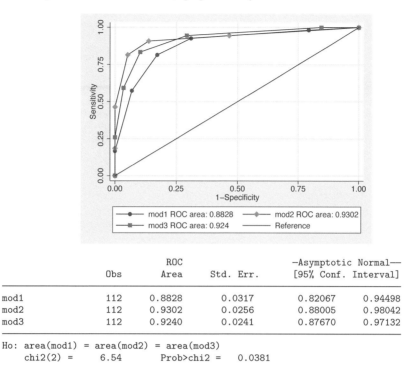

	Obs	ROC Area	Std. Err.	—Asymptotic Normal— [95% Conf. Interval]	
mod1	112	0.8828	0.0317	0.82067	0.94498
mod2	112	0.9302	0.0256	0.88005	0.98042
mod3	112	0.9240	0.0241	0.87670	0.97132

```
Ho: area(mod1) = area(mod2) = area(mod3)
    chi2(2) =    6.54      Prob>chi2 =    0.0381
```

By default, roccomp tests whether the areas under the ROC curves are all equal. Other comparisons can be tested by creating a contrast matrix and specifying test(*matname*), where *matname* is the name of the contrast matrix.

For example, assume that we are interested in testing whether the area under the ROC for mod1 is equal to that of mod3. To do this, we can first create an appropriate contrast matrix and then specify its name with the test() option.

Of course, this is a trivial example because we could have just specified

```
. roccomp status mod1 mod3
```

without including mod2 to obtain the same test results. However, for illustration, we will continue with this example.

The contrast matrix must have its number of columns equal to the number of *classvars* (i.e., the total number of ROC curves) and a number of rows less than or equal to the number of *classvars*, and the elements of each row must add to zero.

(Continued on next page)

```
. matrix C=(1,0,-1)

. roccomp status mod1 mod2 mod3, test(C)
                              ROC                     —Asymptotic Normal—
                   Obs        Area     Std. Err.     [95% Conf. Interval]

     mod1          112       0.8828     0.0317        0.82067    0.94498
     mod2          112       0.9302     0.0256        0.88005    0.98042
     mod3          112       0.9240     0.0241        0.87670    0.97132

Ho: Comparison as defined by contrast matrix: C
      chi2(1) =      5.25       Prob>chi2 =    0.0220
```

Although all three areas are reported, the comparison is made using the specified contrast matrix.

Perhaps more interesting would be a comparison of the area from mod1 and the average area of mod2 and mod3.

```
. matrix C=(1,-.5,-.5)

. roccomp status mod1 mod2 mod3, test(C)
                              ROC                     —Asymptotic Normal—
                   Obs        Area     Std. Err.     [95% Conf. Interval]

     mod1          112       0.8828     0.0317        0.82067    0.94498
     mod2          112       0.9302     0.0256        0.88005    0.98042
     mod3          112       0.9240     0.0241        0.87670    0.97132

Ho: Comparison as defined by contrast matrix: C
      chi2(1) =      3.43       Prob>chi2 =    0.0642
```

Other contrasts could be made. For example, we could test if mod3 is different from at least one of the other two by first creating the following contrast matrix:

```
. matrix C=(-1, 0, 1 \ 0, -1, 1)

. matrix list C

C[2,3]
     c1  c2  c3
r1   -1   0   1
r2    0  -1   1
```

◁

Independent data

▷ Example 4

In example 3, we noted that because each test modality was applied to the same sample of phantoms, the classification outcomes were correlated. Now assume that we have collected the same data presented by Hanley and McNeil (1983), except that we applied the first test modality to one sample of phantoms and the second test modality to a different sample of phantoms. The resulting measurements are now considered independent.

Here are a few of the observations.

```
. use http://www.stata-press.com/data/r10/ct3
. list in 1/7, sep(0)
```

	pop	status	rating	mod
1.	12	0	1	1
2.	31	0	1	2
3.	1	1	1	1
4.	3	1	1	2
5.	28	0	2	1
6.	19	0	2	2
7.	3	1	2	1

The data are in long form, which is required when dealing with independent data. The data consist of 24 observations: 6 observations corresponding to abnormal phantoms and 6 to normal phantoms evaluated using the first modality, and similarly 6 observations corresponding to abnormal phantoms and 6 to normal phantoms evaluated using the second modality. The number of phantoms corresponding to each observation is given by the pop variable. Once again we have frequency-weighted data. The variable mod identifies the modality, and rating is the assigned classification.

We can better view our data by using the table command.

```
. table status rating [fw=pop], by(mod) row col
```

mod and status			rating				
	1	2	3	4	5	6	Total
1							
0	12	28	8	6	4		58
1	1	3	6	13	22	9	54
Total	13	31	14	19	26	9	112
2							
0	31	19	5	3			58
1	3	2	5	19	15	10	54
Total	34	21	10	22	15	10	112

The status variable indicates the true status of the phantoms: status=0 if they are normal and status=1 if they are abnormal.

We now compare the areas under the two ROC curves.

(Continued on next page)

. roccomp status rating [fw=pop], by(mod) graph summary

mod	Obs	ROC Area	Std. Err.	-Asymptotic Normal— [95% Conf. Interval]	
1	112	0.8828	0.0317	0.82067	0.94498
2	112	0.9302	0.0256	0.88005	0.98042

Ho: area(1) = area(2)
    chi2(1) =   1.35   Prob>chi2 =   0.2447

◁

Comparing areas with a gold standard

The area under multiple ROC curves can be compared with a gold standard using rocgold. The command syntax is similar to that of roccomp. The tests are corrected for the effect of multiple comparisons.

▷ Example 5

We will use the same data (presented by Hanley and McNeil 1983) as in the roccomp examples. Let's assume that the first modality is considered to be the standard against which both the second and third modalities are compared.

We want to plot and compare both the areas of the ROC curves of mod2 and mod3 with mod1. Since we consider mod1 to be the gold standard, it is listed first after the reference variable in the rocgold command line.

```
. use http://www.stata-press.com/data/r10/ct2

. rocgold status mod1 mod2 mod3, graph summary
```

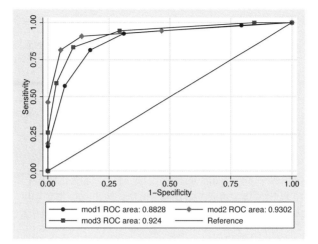

	ROC Area	Std. Err.	chi2	df	Pr>chi2	Bonferroni Pr>chi2
mod1 (standard)	0.8828	0.0317				
mod2	0.9302	0.0256	2.3146	1	0.1282	0.2563
mod3	0.9240	0.0241	5.2480	1	0.0220	0.0439

Equivalently, we could have done this in two steps by using the roccomp command.

```
. roccomp status mod1 mod2, graph summary

. roccomp status mod1 mod3, graph summary msymbol(O S)
```

◁

Saved Results

roctab saves the following in r():

Scalars

r(N)	number of observations	r(area)	area under the ROC curve
r(se)	standard error for the area under the ROC curve	r(pietra)	Pietra index
r(lb)	lower bound of CI for the area under the ROC curve	r(gini)	Gini index
r(ub)	upper bound of CI for the area under the ROC curve		

roccomp saves the following in r():

Scalars

r(N_g)	number of groups	r(df)	χ^2 degrees of freedom
r(p)	significance probability	r(chi2)	χ^2

Matrices

r(V)	variance–covariance matrix

`rocgold` saves the following in `r()`:

Scalars
    `r(N_g)`   number of groups

Matrices
    `r(V)`   variance–covariance matrix   `r(p)`   significance-probability vector

    `r(chi2)` χ^2 vector   `r(p_adj)` adjusted significance-probability vector

    `r(df)` χ^2 degrees-of-freedom vector

Methods and Formulas

`roctab`, `roccomp`, and `rocgold` are implemented as ado-files.

Assume that we applied a diagnostic test to each of N_n normal and N_a abnormal subjects. Further assume that the higher the outcome value of the diagnostic test, the higher the risk of the subject being abnormal. Let $\widehat{\theta}$ be the estimated area under the curve, and let $X_i, i = 1, 2, \ldots, N_a$ and $Y_j, j = 1, 2, \ldots, N_n$ be the values of the diagnostic test for the abnormal and normal subjects, respectively.

Nonparametric ROC

The points on the nonparametric ROC curve are generated by using each possible outcome of the diagnostic test as a classification cutpoint and computing the corresponding sensitivity and $1 -$ specificity. These points are then connected by straight lines, and the area under the resulting ROC curve is computed using the trapezoidal rule.

The default standard error for the area under the ROC curve is computed using the algorithm described by DeLong, DeLong, and Clarke-Pearson (1988). For each abnormal subject, i, define

$$V_{10}(X_i) = \frac{1}{N_n} \sum_{j=1}^{N_n} \psi(X_i, Y_j)$$

and for each normal subject, j, define

$$V_{01}(Y_j) = \frac{1}{N_a} \sum_{i=1}^{N_a} \psi(X_i, Y_j)$$

where

$$\psi(X, Y) = \begin{cases} 1 & Y < X \\ \frac{1}{2} & Y = X \\ 0 & Y > X \end{cases}$$

Define

$$S_{10} = \frac{1}{N_a - 1} \sum_{i=1}^{N_a} \{V_{10}(X_i) - \widehat{\theta}\}^2$$

and

$$S_{01} = \frac{1}{N_n - 1} \sum_{j=1}^{N_n} \{V_{01}(Y_j) - \widehat{\theta}\}^2$$

The variance of the estimated area under the ROC curve is given by

$$\text{var}(\widehat{\theta}) = \frac{1}{N_a}S_{10} + \frac{1}{N_n}S_{01}$$

The `hanley` standard error for the area under the ROC curve is computed using the algorithm described by Hanley and McNeil (1982). It requires the calculation of two quantities: Q_1 is Pr(two randomly selected abnormal subjects will both have a higher score than a randomly selected normal subject), and Q_2 is Pr(one randomly selected abnormal subject will have a higher score than any two randomly selected normal subjects). The Hanley and McNeil variance of the estimated area under the ROC curve is

$$\text{var}(\widehat{\theta}) = \frac{\widehat{\theta}(1 - \widehat{\theta}) + (N_a - 1)(Q_1 - \widehat{\theta^2}) + (N_n - 1)(Q_2 - \widehat{\theta^2})}{N_a N_n}$$

The `bamber` standard error for the area under the ROC curve is computed using the algorithm described by Bamber (1975). For any two Y values, Y_j and Y_k, and any X_i value, define

$$b_{yyx} = p(Y_j, Y_k < X_i) + p(X_i < Y_j, Y_k) - 2p(Y_j < X_i < Y_k)$$

and similarly, for any two X values, X_i and X_l, and any Y_j value, define

$$b_{xxy} = p(X_i, X_l < Y_j) + p(Y_j < X_i, X_l) - 2p(X_i < Y_j < X_l)$$

Bamber's unbiased estimate of the variance for the area under the ROC curve is

$$\text{var}(\widehat{\theta}) = \frac{1}{4}(N_a-1)(N_n-1)\{p(X \neq Y)+(N_a-1)b_{xxy}+(N_n-1)b_{yyx}-4(N_a+N_n-1)(\widehat{\theta}-0.5)^2\}$$

Asymptotic confidence intervals are constructed and reported by default, assuming a normal distribution for the area under the ROC curve.

Exact binomial confidence intervals are calculated as described in [R] **ci**, with p equal to the area under the ROC curve.

Comparing areas under the ROC curve

Areas under ROC curves are compared using an algorithm suggested by DeLong, DeLong, and Clarke-Pearson (1988). Let $\widehat{\theta} = (\widehat{\theta^1}, \widehat{\theta^2}, \ldots, \widehat{\theta^k})$ be a vector representing the areas under k ROC curves. For the rth area, define

$$V_{10}^r(X_i) = \frac{1}{N_n} \sum_{j=1}^{N_n} \psi(X_i^r, Y_j^r)$$

and for each normal subject, j, define

$$V_{01}^r(Y_j) = \frac{1}{N_a} \sum_{i=1}^{N_a} \psi(X_i^r, Y_j^r)$$

where

$$\psi(X^r, Y^r) = \begin{cases} 1 & Y^r < X^r \\ \frac{1}{2} & Y^r = X^r \\ 0 & Y^r > X^r \end{cases}$$

Define the $k \times k$ matrix $\mathbf{S_{10}}$ such that the (r, s)th element is

$$S_{10}^{r,s} = \frac{1}{N_a - 1} \sum_{i=1}^{N_a} \{V_{10}^r(X_i) - \widehat{\theta^r}\}\{V_{10}^s(X_i) - \widehat{\theta^s}\}$$

and $\mathbf{S_{01}}$ such that the (r, s)th element is

$$S_{01}^{r,s} = \frac{1}{N_n - 1} \sum_{j=1}^{N_n} \{V_{01}^r(Y_i) - \widehat{\theta^r}\}\{V_{01}^s(Y_i) - \widehat{\theta^s}\}$$

Then the covariance matrix is

$$S = \frac{1}{N_a} S_{10} + \frac{1}{N_n} S_{01}$$

Let $\mathbf{L}$ be a contrast matrix defining the comparison, so that

$$(\widehat{\theta} - \theta)' \mathbf{L}' (\mathbf{LSL}')^{-1} \mathbf{L}(\widehat{\theta} - \theta)$$

has a chi-squared distribution with degrees of freedom equal to the rank of $\mathbf{LSL}'$.

References

Bamber, D. 1975. The area above the ordinal dominance graph and the area below the receiver operating characteristic graph. *Journal of Mathematical Psychology* 12: 387–415.

Choi, B. C. K. 1998. Slopes of a receiver operating characteristic curve and likelihood ratios for a diagnostic test. *American Journal of Epidemiology* 148: 1127–1132.

Cleves, M. A. 1999. sg120: Receiver Operating Characteristic (ROC) analysis. *Stata Technical Bulletin* 52: 19–33. Reprinted in *Stata Technical Bulletin Reprints*, vol. 9, pp. 212–229.

——. 2000. sg120.2: Correction to roccomp command. *Stata Technical Bulletin* 54: 26. Reprinted in *Stata Technical Bulletin Reprints*, vol. 9, p. 231.

——. 2002a. Comparative assessment of three common algorithms for estimating the variance of the area under the nonparametric receiver operating characteristic curve. *Stata Journal* 2: 280–289.

——. 2002b. From the help desk: Comparing areas under receiver operating characteristic curves from two or more probit or logit models. *Stata Journal* 2: 301–313.

DeLong, E. R., D. M. DeLong, and D. L. Clarke-Pearson. 1988. Comparing the areas under two or more correlated receiver operating curves: A nonparametric approach. *Biometrics* 44: 837–845.

Erdreich, L. S., and E. T. Lee. 1981. Use of relative operating characteristic analysis in epidemiology: A method for dealing with subjective judgment. *American Journal of Epidemiology* 114: 649–662.

Hanley, J. A., and B. J. McNeil. 1982. The meaning and use of the area under a receiver operating characteristic (ROC) curve. *Radiology* 143: 26–36.

——. 1983. A method of comparing the areas under receiver operating characteristic curves derived from the same cases. *Radiology* 148: 839–843.

Juul, S. 2006. *An Introduction to Stata for Health Researchers*. College Station, TX: Stata Press.

Lee, W. C. 1999. Probabilistic analysis of global performances of diagnostic test: Interpreting the Lorenz curve-based summary measures. *Statistics in Medicine* 18: 455–471.

Ma, G., and W. J. Hall. 1993. Confidence bands for the receiver operating characteristic curves. *Medical Decision Making* 13: 191–197.

Pepe, M. S. 2003. *The Statistical Evaluation of Medical Tests for Classification and Prediction.* New York: Oxford University Press.

Reichenheim, M. E., and A. Ponce de Leon. 2002. Estimation of sensitivity and specificity arising from validity studies with incomplete designs. *Stata Journal* 2: 267–279.

Seed, P. T., and A. Tobias. 2001. sbe36.1: Summary statistics for diagnostic tests. *Stata Technical Bulletin* 59: 25–27. Reprinted in *Stata Technical Bulletin Reprints*, vol. 10, pp. 90–93.

Tobias, A. 2000. sbe36: Summary statistics report for diagnostic tests. *Stata Technical Bulletin* 56: 16–18. Reprinted in *Stata Technical Bulletin Reprints*, vol. 10, pp. 87–90.

Working, H., and H. Hotelling. 1929. Application of the theory of error to the interpretation of trends. *Journal of the American Statistical Association* 24: 73–85.

Also See

[R] **logistic** — Logistic regression, reporting odds ratios

[R] **rocfit** — Fit ROC models

Title

---
rocfit — Fit ROC models

Syntax

rocfit *refvar classvar* [*if*] [*in*] [*weight*] [, *rocfit_options*]

rocfit_options	description
Model	
<u>cont</u>inuous(#)	divide *classvar* into # groups of approximately equal length
<u>g</u>enerate(*newvar*)	create *newvar* containing classification groups
SE	
vce(*vcetype*)	*vcetype* may be oim or opg
Reporting	
<u>level</u>(#)	set confidence level; default is level(95)
Max options	
maximize_options	control the maximization process; seldom used

fweights are allowed; see [U] **11.1.6 weight**.
See [U] **20 Estimation and postestimation commands** for more capabilities of estimation commands.

Description

rocfit fits maximum-likelihood ROC models assuming a binormal distribution of the latent variable.

The two variables *refvar* and *classvar* must be numeric. The reference variable indicates the true state of the observation, such as diseased and nondiseased or normal and abnormal, and must be coded as 0 and 1. The rating or outcome of the diagnostic test or test modality is recorded in *classvar*, which must be at least ordinal, with higher values indicating higher risk.

See [R] **roc** for other commands designed to perform receiver operating characteristic (ROC) analyses with rating and discrete classification data.

Options

⌐ Model ⌐

<u>cont</u>inuous(#) specifies that the continuous *classvar* be divided into # groups of approximately equal length. This option is required when *classvar* takes on more than 20 distinct values.

<u>cont</u>inuous(.) may be specified to indicate that *classvar* be used as it is, even though it could have more than 20 distinct values.

<u>g</u>enerate(*newvar*) specifies the new variable that is to contain the values indicating the groups produced by continuous(#). generate() may be specified only with continuous().

SE

vce(*vcetype*) specifies the type of standard error reported. *vcetype* may be either oim or opg; see [R] *vce_option*.

Reporting

level(*#*); see [R] **estimation options**.

Max options

maximize_options: <u>diff</u>icult, <u>tech</u>nique(*algorithm_spec*), <u>iter</u>ate(*#*), [<u>no</u>]<u>log</u>, <u>tr</u>ace, <u>grad</u>ient, <u>showstep</u>, <u>hess</u>ian, <u>shownr</u>tolerance, <u>tol</u>erance(*#*), <u>ltol</u>erance(*#*), <u>gtol</u>erance(*#*), <u>nrtol</u>erance(*#*), <u>nonrtol</u>erance, from(*init_specs*); see [R] **maximize**. These options are seldom used.

Setting the optimization type to technique(bhhh) resets the default *vcetype* to vce(opg).

Remarks

Dorfman and Alf (1969) developed a generalized approach for obtaining maximum likelihood estimates of the parameters for a smooth fitting ROC curve. The most commonly used method, and the one implemented here, is based upon the binormal model; see Pepe (2003) for a discussion of different approaches to ROC analysis.

The model assumes the existence of an unobserved, continuous, latent variable that is normally distributed (perhaps after a monotonic transformation) in both the normal and abnormal populations with means μ_n and μ_a and variances σ_n^2 and σ_a^2, respectively. The model further assumes that the K categories of the rating variable result from partitioning the unobserved latent variable by $K-1$ fixed boundaries. The method fits a straight line to the empirical ROC points plotted using normal probability scales on both axes. Maximum likelihood estimates of the line's slope and intercept and the $K-1$ boundaries are obtained simultaneously. See *Methods and Formulas* for details.

The intercept from the fitted line is a measurement of $(\mu_a - \mu_n)/\sigma_a$, and the slope measures σ_n/σ_a.

Thus the intercept is the standardized difference between the two latent population means, and the slope is the ratio of the two standard deviations. The null hypothesis that there is no difference between the two population means is evaluated by testing that the intercept $= 0$, and the null hypothesis that the variances in the two populations are equal is evaluated by testing that the slope $= 1$.

▷ Example 1

We use Hanley and McNeil's (1982) dataset, described in example 1 of [R] **roc**, to fit a smooth ROC curve assuming a binormal model.

(*Continued on next page*)

```
. use http://www.stata-press.com/data/r10/hanley

. rocfit disease rating

Fitting binormal model:

Iteration 0:   log likelihood = -123.68069
Iteration 1:   log likelihood = -123.64867
Iteration 2:   log likelihood = -123.64855
Iteration 3:   log likelihood = -123.64855

Binormal model of disease on rating          Number of obs   =        109
Goodness-of-fit chi2(2) =        0.21
Prob > chi2             =        0.9006
Log likelihood         =   -123.64855
```

	Coef.	Std. Err.	z	P>|z|	[95% Conf.	Interval]
intercept	1.656782	0.310456	5.34	0.000	1.048300	2.265265
slope (*)	0.713002	0.215882	-1.33	0.092	0.289881	1.136123
/cut1	0.169768	0.165307	1.03	0.152	-0.154227	0.493764
/cut2	0.463215	0.167235	2.77	0.003	0.135441	0.790990
/cut3	0.766860	0.174808	4.39	0.000	0.424243	1.109477
/cut4	1.797938	0.299581	6.00	0.000	1.210770	2.385106

Index	Indices from binormal fit			
	Estimate	Std. Err.	[95% Conf.	Interval]
ROC area	0.911331	0.029506	0.853501	0.969161
delta(m)	2.323671	0.502370	1.339044	3.308298
d(e)	1.934361	0.257187	1.430284	2.438438
d(a)	1.907771	0.259822	1.398530	2.417012

```
(*) z test for slope==1
```

rocfit outputs the MLE for the intercept and slope of the fitted regression line along with, in this case, four boundaries (because there are five ratings) labeled /cut1 through /cut4. In addition, rocfit also computes and reports four indices based on the fitted ROC curve: the area under the curve (labeled ROC area), $\delta(m)$ (labeled delta(m)), d_e (labeled d(e)), and d_a (labeled d(a)). More information about these indices can be found in *Methods and Formulas* and in Erdreich and Lee (1981).

◁

Saved Results

rocfit saves the following in e():

Scalars

e(N)	number of observations	e(se_area)	standard error for the area under the
e(k)	number of parameters		ROC curve
e(k_eq)	number of equations	e(deltam)	delta(m)
e(k_eq_model)	number of equations in model	e(se_delm)	standard area for delta(m)
	Wald test	e(de)	d(e) index
e(k_dv)	number of dependent variables	e(se_de)	standard error for d(e) index
e(df_m)	model degrees of freedom	e(da)	d(a) index
e(ll)	log likelihood	e(se_da)	standard error for d(a) index
e(chi2_gf)	goodness-of-fit χ^2	e(rank)	rank of e(V)
e(df_gf)	goodness-of-fit degrees of freedom	e(ic)	number of iterations
e(p_gf)	χ^2 goodness-of-fit significance	e(rc)	return code
	probability	e(converged)	1 if converged, 0 otherwise
e(area)	area under the ROC curve		

Macros

e(cmd)	rocfit	e(vcetype)	title used to label Std. Err.
e(cmdline)	command as typed	e(opt)	type of optimization
e(depvar)	*refvar* and *classvar*	e(ml_method)	type of ml method
e(title)	title in estimation output	e(user)	name of likelihood-evaluator program
e(wtype)	weight type	e(technique)	maximization technique
e(wexp)	weight expression	e(crittype)	optimization criterion
e(chi2type)	GOF; type of model χ^2 test	e(properties)	b V
e(vce)	*vcetype* specified in vce()		

Matrices

e(b)	coefficient vector	e(gradient)	gradient vector
e(ilog)	iteration log (up to	e(V)	variance–covariance matrix of the
	20 iterations)		estimators

Functions

e(sample)	marks estimation sample

Methods and Formulas

rocfit is implemented as an ado-file.

Dorfman and Alf (1969) developed a general procedure for obtaining maximum likelihood estimates of the parameters of a smooth-fitting ROC curve. The most common method, and the one implemented in Stata, is based upon the binormal model.

The model assumes that there is an unobserved continuous latent variable that is normally distributed in both the normal and abnormal populations. The idea is better explained with the following illustration:

(Continued on next page)

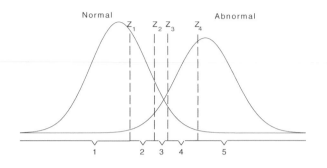

The latent variable is assumed to be normally distributed for both the normal and abnormal subjects, perhaps after a monotonic transformation, with means μ_n and μ_a and variances σ_n^2 and σ_a^2, respectively.

This latent variable is assumed to be partitioned into the k categories of the rating variable by $k-1$ fixed boundaries. In the above figure, the $k = 5$ categories of the rating variable identified on the bottom result from the partition of the four boundaries Z_1 through Z_4.

Let R_j for $j = 1, 2, \ldots, k$ indicate the categories of the rating variable, let $i = 1$ if the subject belongs to the normal group, and let $i = 2$ if the subject belongs to the abnormal group.

Then

$$p(R_j|i = 1) = F(Z_j) - F(Z_{j-1})$$

where $Z_k = (x_k - \mu_n)/\sigma_n$, F is the cumulative normal distribution, $F(Z_0) = 0$ and $F(Z_k) = 1$. Also,

$$p(R_j|i = 2) = F(bZ_j - a) - F(bZ_{j-1} - a)$$

where $b = \sigma_n/\sigma_a$ and $a = (\mu_a - \mu_n)/\sigma_a$.

The parameters a, b and the $k-1$ fixed boundaries Z_j are simultaneously estimated by maximizing the log-likelihood function

$$\log L = \sum_{i=1}^{2} \sum_{j=1}^{k} r_{ij} \log\{p(R_j|i)\}$$

where r_{ij} is the number of R_js in group i.

The area under the fitted ROC curve is computed as

$$\Phi\left(\frac{a}{\sqrt{1 + b^2}}\right)$$

where Φ is the standard normal cumulative distribution function.

Point estimates for the ROC curve indices are as follows:

$$\delta(m) = \frac{a}{b} \qquad d_e = \frac{2a}{b + 1} \qquad d_a = \frac{a\sqrt{2}}{\sqrt{1 + b^2}}$$

Variances for these indices are computed using the delta method.

The $\delta(m)$ estimates $(\mu_a - \mu_n)/\sigma_n$, d_e estimates $2(\mu_a - \mu_n)/(\sigma_a - \sigma_n)$, and d_a estimates $\sqrt{2}(\mu_a - \mu_n)/(\sigma_a^2 - \sigma_n^2)^2$.

Simultaneous confidence bands for the entire curve are obtained, as suggested by Ma and Hall (1993), by first obtaining Working–Hotelling (1929) confidence bands for the fitted straight line in normal probability coordinates and then transforming them back to ROC coordinates.

References

Bamber, D. 1975. The area above the ordinal dominance graph and the area below the receiver operating characteristic graph. *Journal of Mathematical Psychology* 12: 387–415.

Choi, B. C. K. 1998. Slopes of a receiver operating characteristic curve and likelihood ratios for a diagnostic test. *American Journal of Epidemiology* 148: 1127–1132.

Cleves, M. A. 1999. sg120: Receiver Operating Characteristic (ROC) analysis. *Stata Technical Bulletin* 52: 19–33. Reprinted in *Stata Technical Bulletin Reprints*, vol. 9, pp. 212–229.

———. 2000. sg120.1: Two new options added to rocfit command. *Stata Technical Bulletin* 53: 18–19. Reprinted in *Stata Technical Bulletin Reprints*, vol. 9, pp. 230–231.

Dorfman, D. D., and E. Alf. 1969. Maximum likelihood estimation of parameters of signal detection theory and determination of confidence intervals-rating method data. *Journal of Mathematical Psychology* 6: 487–496.

Erdreich, L. S., and E. T. Lee. 1981. Use of relative operating characteristic analysis in epidemiology: A method for dealing with subjective judgment. *American Journal of Epidemiology* 114: 649–662.

Hanley, J. A., and B. J. McNeil. 1982. The meaning and use of the area under a receiver operating characteristic (ROC) curve. *Radiology* 143: 26–36.

Ma, G., and W. J. Hall. 1993. Confidence bands for the receiver operating characteristic curves. *Medical Decision Making* 13: 191–197.

Pepe, M. S. 2003. *The Statistical Evaluation of Medical Tests for Classification and Prediction*. New York: Oxford University Press.

Working, H., and H. Hotelling. 1929. Application of the theory of error to the interpretation of trends. *Journal of the American Statistical Association* 24: 73–85.

Also See

[R] **rocfit postestimation** — Postestimation tools for rocfit

[R] **roc** — Receiver operating characteristic (ROC) analysis

[U] **20 Estimation and postestimation commands**

Title

rocfit postestimation — Postestimation tools for rocfit

Description

The following command is of special interest after `rocfit`:

command	description
rocplot	plot the fitted ROC curve and simultaneous confidence bands

For information about `rocplot`, see below.

The following standard postestimation commands are also available:

command	description
estat	AIC, BIC, VCE, and estimation sample summary
estimates	cataloging estimation results
*lincom	point estimates, standard errors, testing, and inference for linear combinations of coefficients
*test	Wald tests for simple and composite linear hypotheses

*See *Using lincom and test* below.
See the corresponding entries in the *Stata Base Reference Manual* for details.

Special-interest postestimation command

`rocplot` plots the fitted ROC curve and simultaneous confidence bands.

Syntax for rocplot

> rocplot [, *rocplot_options*]

rocplot_options	description
Main	
<u>conf</u>band	display confidence bands
<u>noref</u>line	suppress plotting the reference line
<u>l</u>evel(*#*)	set confidence level; default is level(95)
Plot	
<u>plotopts</u>(*plot_options*)	affect rendition of the ROC points
Fit line	
<u>lineopts</u>(*cline_options*)	affect rendition of the fitted ROC line
CI plot	
<u>ciopts</u>(*area_options*)	affect rendition of the confidence bands
Reference line	
<u>rlopts</u>(*cline_options*)	affect rendition of the reference line
Add plots	
addplot(*plot*)	add other plots to the generated graph
Y axis, X axis, Titles, Legend, Overall	
twoway_options	any options other than by() documented in [G] *twoway_options*

plot_options	description
marker_options	change look of markers (color, size, etc.)
marker_label_options	add marker labels; change look or position
cline_options	change the look of the line

Options for rocplot

⌐ Main ⌐

confband specifies that simultaneous confidence bands be plotted around the ROC curve.

norefline suppresses plotting the 45-degree reference line from the graphical output of the ROC curve.

level(*#*) specifies the confidence level, as a percentage, for the confidence bands. The default is level(95) or as set by set level; see [R] **level**.

⌐ Plot ⌐⎽⎽

plotopts(*plot_options*) affects the rendition of the plotted ROC points, including the size and color of markers, whether and how the markers are labeled, and whether and how the points are connected. For the full list of available *plot_options*, see [G] *marker_options*, [G] *marker_label_options*, and [G] *cline_options*.

⌐ Fit line ⌐⎽⎽

fitopts(*cline_options*) affects the rendition of the fitted ROC line; see [G] *cline_options*.

⌐ CI plot ⌐⎽⎽

ciopts(*area_options*) affects the rendition of the confidence bands; see [G] *area_options*.

⌐ Reference line ⌐⎽⎽⎽

rlopts(*cline_options*) affects the rendition of the reference line; see [G] *cline_options*.

⌐ Add plots ⌐⎽⎽

addplot(*plot*) provides a way to add other plots to the generated graph. See [G] *addplot_option*.

⌐ Y axis, X axis, Titles, Legend, Overall ⌐⎽⎽⎽⎽⎽⎽⎽⎽⎽⎽⎽⎽⎽⎽⎽⎽⎽⎽⎽⎽⎽⎽⎽⎽⎽⎽⎽⎽⎽⎽⎽⎽⎽⎽⎽⎽⎽

twoway_options are any of the options documented in [G] *twoway_options*, excluding by(). These include options for titling the graph (see [G] *title_options*) and for saving the graph to disk (see [G] *saving_option*).

Remarks

Remarks are presented under the following headings:

>   *Using lincom and test*
>   *Using rocplot*

Using lincom and test

intercept, slope, and cut#, shown in example 1 of [R] **rocfit**, are equation names and not variable names, so they need to be referenced as described in *Special syntaxes after multiple-equation estimation* in [R] **test**. For example, instead of typing

```
. test intercept
intercept not found
r(111);
```

you should type

```
. test [intercept]_cons
( 1)  [intercept]_cons = 0
         chi2(  1) =    28.48
       Prob > chi2 =    0.0000
```

Using rocplot

▷ Example 1

In [R] **rocfit**, we fitted a ROC curve by typing `rocfit disease rating`.

In the output table for our model, we are testing whether the variances of the two latent populations are equal by testing that the slope $= 1$.

We plot the fitted ROC curve.

```
. rocplot, confband
```

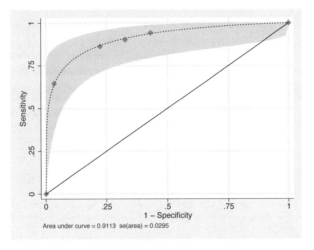

◁

Methods and Formulas

All postestimation commands listed above are implemented as ado-files.

Also See

[R] **rocfit** — Fit ROC models

[U] **20 Estimation and postestimation commands**

Title

rologit — Rank-ordered logistic regression

Syntax

rologit *depvar* *indepvars* [*if*] [*in*] [*weight*] , group(*varname*) [*options*]

options	description
Model	
*group(*varname*)	identifier variable that links the alternatives
offset(*varname*)	include *varname* in model with coefficient constrained to 1
incomplete(#)	use # to code unranked alternatives; default is incomplete(0)
reverse	reverse the preference order
notestrhs	keep RHS variables that do not vary within group
ties(*spec*)	method to handle ties: exactm, breslow, efron, or none
SE/Robust	
vce(*vcetype*)	*vcetype* may be oim, robust, cluster *clustvar*, bootstrap, or jackknife
Reporting	
level(#)	set confidence level; default is level(95)
Max options	
maximize_options	control the maximization process; seldom used

*group(*varname*) is required.
bootstrap, by, jackknife, rolling, statsby, and xi are allowed; see [U] **11.1.10 Prefix commands**.
Weights are not allowed with the bootstrap prefix.
fweights, iweights, and pweights are allowed, except with ties(efron); see [U] **11.1.6 weight**.
See [U] **20 Estimation and postestimation commands** for more capabilities of estimation commands.

Description

rologit fits the rank-ordered logistic regression model by maximum likelihood (Beggs, Cardell, and Hausman 1981). This model is also known as the Plackett–Luce model (Marden 1995), as the exploded logit model (Punj and Staelin 1978), and as the choice-based method of conjoint analysis (Hair et al. 1995).

rologit expects the data to be in long form, similar to clogit, in which each of the ranked alternatives forms an observation; all observations related to an individual are linked together by the variable that you specify in the group() option. The distinction from clogit is that *depvar* in rologit records the rankings of the alternatives, whereas for clogit, *depvar* only marks the best alternative by a value not equal to zero. rologit interprets equal scores of *depvar* as ties. The ranking information may be incomplete "at the bottom" (least preferred alternatives). That is, unranked alternatives may be coded as 0 or as a common value that may be specified with the incomplete() option.

If your data record only the unique best alternative, rologit fits the same model as clogit.

Options

group(*varname*) is required, and it specifies the identifier variable (numeric or string) that links the alternatives for an individual, which have been compared and rank ordered with respect to one another.

offset(*varname*); see [R] **estimation options**.

incomplete(#) specifies the numeric value used to code alternatives that are not ranked. It is assumed that unranked alternatives are less preferred than the ranked alternatives (i.e., the data record the ranking of the most preferred alternatives). It is not assumed that subjects are indifferent between the unranked alternatives. # defaults to 0.

reverse specifies that in the preference order, a higher number means a less attractive alternative. The default is that higher values indicate more attractive alternatives. The rank-ordered logit model is not symmetric in the sense that reversing the ordering simply leads to a change in the signs of the coefficients.

notestrhs suppresses the test that the independent variables vary within (at least some of) the groups. Effects of variables that are always constant are not identified. For instance, a rater's gender cannot directly affect his or her rankings; it could affect the rankings only via an interaction with a variable that does vary over alternatives.

ties(*spec*) specifies the method for handling ties (indifference between alternatives) (see [ST] **stcox** for details):

exactm	exact marginal likelihood (default)
breslow	Breslow's approximation (default if pweights specified)
efron	Efron's approximation (default if robust VCE)
none	no ties allowed

vce(*vcetype*) specifies the type of standard error reported, which includes types that are derived from asymptotic theory, that are robust to some kinds of misspecification, that allow for intragroup correlation, and that use bootstrap or jackknife methods; see [R] *vce_option*.

If ties(exactm) is specified, *vcetype* may be only oim, bootstrap, or jackknife.

level(#); see [R] **estimation options**.

maximize_options: iterate(#), trace, [no]log, tolerance(#), ltolerance(#); see [R] **maximize**. These options are seldom used.

Remarks

The rank-ordered logit model can be applied to analyze how decision makers combine attributes of alternatives into overall evaluations of the attractiveness of these alternatives. The model generalizes a version of McFadden's choice model without alternative-specific covariates, as fitted by the clogit command. It uses richer information about the comparison of alternatives, namely, how decision-makers rank the alternatives rather than just specifying the alternative that they like best.

Remarks are presented under the following headings:

Examples
Comparing respondents
Incomplete rankings and ties
Clustered choice data
Comparison of rologit and clogit
On reversals of rankings

Examples

A popular way to study employer preferences for characteristics of employees is the quasi-experimental "vignette method". As an example, we consider the research by De Wolf on the labor market position of social science graduates (De Wolf 2000). This study addresses how the educational portfolio (e.g., general skills versus specific knowledge) affects short term and long-term labor-market opportunities. De Wolf asked 22 human resource managers (the respondents) to rank order the six most suitable candidates of 20 fictitious applicants and to rank order these six candidates for three jobs, namely, (1) researcher, (2) management trainee, and (3) policy adviser. Applicants were described by 10 attributes, including their age, gender, details of their portfolio, and work experience. In this example, we analyze a subset of the data. Also, to simplify the output, we drop, at random, 10 nonselected applicants per case. The resulting dataset includes 29 cases, consisting of 10 applicants each. The data are in long form: observations correspond to alternatives (the applications), and alternatives that figured in one decision task are identified by the variable `caseid`. We list the observations for `caseid==7`, in which the respondent considered applicants for a social science research position.

```
. use http://www.stata-press.com/data/r10/evignet
(Vignet study employer prefs (Inge de Wolf 2000))

. list pref female age grades edufit workexp boardexp if caseid==7, noobs
```

pref	female	age	grades	edufit	workexp	boardexp
0	yes	28	A/B	no	none	no
0	no	25	C/D	yes	one year	no
0	no	25	C/D	yes	none	yes
0	yes	25	C/D	no	internship	yes
1	no	25	C/D	yes	one year	yes
2	no	25	A/B	yes	none	no
3	yes	25	A/B	yes	one year	no
4	yes	25	A/B	yes	none	yes
5	no	25	A/B	yes	internship	no
6	yes	28	A/B	yes	one year	yes

Here six applicants were selected. The rankings are stored in the variable `pref`, where a value of 6 corresponds to "best among the candidates", a value of 5 corresponds to "second-best among the candidates", etc. The applicants with a ranking of 0 were not among the best six candidates for the job. The respondent was not asked to express his preferences among these four applicants, but by the elicitation procedure, it is known that he ranks these four applicants below the six selected applicants. The best candidate was a female, 28 years old, with education fitting the job, with good grades (A/B), with 1 year of work experience, and with experience being a board member of a fraternity, a sports club, etc. The profiles of the other candidates read similarly. In this case, the respondent completed the task; that is, he selected and rank ordered the six most suitable applicants. Sometimes the respondent performed only part of the task.

```
. list pref female age grades edufit workexp boardexp if caseid==18, noobs
```

pref	female	age	grades	edufit	workexp	boardexp
0	no	25	C/D	yes	none	yes
0	no	25	C/D	no	internship	yes
0	no	28	C/D	no	internship	yes
0	yes	25	A/B	no	one year	no
2	yes	25	A/B	no	none	yes
2	no	25	A/B	no	none	yes
2	no	25	A/B	no	one year	yes
5	no	25	A/B	no	none	yes
5	no	25	A/B	no	none	yes
5	yes	25	A/B	no	none	no

The respondent selected the six best candidates and segmented these six candidates into two groups: one group with the three best candidates, and a second group of three candidates that were "still acceptable". The numbers 2 and 5, indicating these two groups, are arbitrary apart from the implied ranking of the groups. The ties between the candidates in a group indicate that the respondent was not able to rank the candidates within the group.

The purpose of the vignette experiment was to explore and test hypotheses about which of the employees' attributes are valued by employers, how these attributes are weighted depending on the type of job (described by variable job in these data), etc. In the psychometric tradition of Thurstone (1927), *value* is assumed to be linear in the attributes, with the coefficients expressing the direction and weight of the attributes. In addition, it is assumed that *valuation* is to some extent a random procedure, captured by an additive random term. For instance, if value depends only on an applicant's age and gender, we would have

$$\text{value}(\texttt{female}_i, \texttt{age}_i) = \beta_1 \texttt{female}_i + \beta_2 \texttt{age}_i + \epsilon_i$$

where the random residual ϵ_i captures all omitted attributes. Thus $\beta_1 > 0$ means that the employer assigns higher value to a woman than to a man. Given this conceptualization of value, it is straightforward to model the decision (selection) among alternatives or the ranking of alternatives: the alternative with the highest value is selected (chosen), or the alternatives are ranked according to their value. To complete the specification of a model of choice and of ranking, we assume that the random residual ϵ_i follows an "extreme value distribution of type I", introduced in this context by Luce (1959). This specific assumption is made mostly for computational convenience.

This model is known by many names. Among others, it is known as the rank-ordered logit model in economics (Beggs, Cardell, and Hausman 1981), as the exploded logit model in marketing research (Punj and Staelin 1978), as the choice-based conjoint analysis model (Hair et al. 1995), and as the Plackett–Luce model (Marden 1995). The model coefficients are estimated using the method of maximum likelihood. The implementation in rologit uses an analogy between the rank-ordered logit model and the Cox regression model observed by Allison and Christakis (1994); see *Methods and Formulas*. The rologit command implements this method for rankings, whereas clogit deals with the variant of choices, that is, only the most highly valued alternative is recorded. In the latter case, the model is also known as the Luce–McFadden choice model. In fact, when the data record the most preferred (unique) alternative and no additional ranking information about preferences is available, rologit and clogit return the same information, though formatted somewhat differently.

```
. rologit pref female age grades edufit workexp boardexp if job==1, group(caseid)
Iteration 0:   log likelihood =  -95.41087
Iteration 1:   log likelihood = -71.180903
Iteration 2:   log likelihood =  -68.47734
Iteration 3:   log likelihood = -68.345918
Iteration 4:   log likelihood = -68.345389
Refining estimates:
Iteration 0:   log likelihood = -68.345389
```

```
Rank-ordered logistic regression                Number of obs      =        80
Group variable: caseid                          Number of groups   =         8

No ties in data                                 Obs per group: min =        10
                                                               avg =     10.00
                                                               max =        10

                                                LR chi2(6)         =     54.13
Log likelihood = -68.34539                      Prob > chi2        =    0.0000
```

pref	Coef.	Std. Err.	z	P>|z|	[95% Conf. Interval]	
female	-.4487287	.3671307	-1.22	0.222	-1.168292	.2708343
age	-.0984926	.0820473	-1.20	0.230	-.2593024	.0623172
grades	3.064534	.6148245	4.98	0.000	1.8595	4.269568
edufit	.7658064	.3602366	2.13	0.034	.0597556	1.471857
workexp	1.386427	.292553	4.74	0.000	.8130341	1.959821
boardexp	.6944377	.3762596	1.85	0.065	-.0430176	1.431893

Focusing only on the variables whose coefficients are significant at the 10% level (we are analyzing 8 respondents only!), the estimated value of an applicant for a job of type 1 (research positions) can be written as

```
value = 3.06*grades + 0.77*edufit + 1.39*workexp + 0.69*boardexp
```

Thus employers prefer applicants for a research position (job==1) whose educational portfolio fits the job, who have better grades, who have more relevant work experience, and who have (extracurricular) board experience. They don't seem to care much about the sex and age of applicants, which is comforting.

Given these estimates of the valuation by employers, we consider the probabilities that each of the applications is ranked first. Under the assumption that the ϵ_i are independent and follow an extreme value type I distribution, Luce (1959) showed that the probability, π_i, that alternative i is valued higher than alternatives $2, \ldots, k$ can be written in the multinomial logit form

$$\pi_i = \Pr\{\text{value}_1 > \max(\text{value}_2, \ldots, \text{value}_m)\} = \frac{\exp(\text{value}_i)}{\sum_{j=1}^{k} \exp(\text{value}_i)}$$

The probability of observing a specific ranking can be written as the *product* of such terms, representing a sequential decision interpretation in which the rater first chooses the most preferred alternative, and then the most preferred alternative among the rest, etc.

The probabilities for alternatives to be ranked first are conveniently computed by predict.

```
. predict p if e(sample)
(option pr assumed; conditional probability that alternative is ranked first)
(210 missing values generated)
. sort caseid pref p
. list pref p grades edufit workexp boardexp if caseid==7, noobs
```

pref	p	grades	edufit	workexp	boardexp
0	.0027178	C/D	yes	none	yes
0	.0032275	C/D	no	internship	yes
0	.0064231	A/B	no	none	no
0	.0217202	C/D	yes	one year	no
1	.0434964	C/D	yes	one year	yes
2	.0290762	A/B	yes	none	no
3	.2970933	A/B	yes	one year	no
4	.0371747	A/B	yes	none	yes
5	.1163203	A/B	yes	internship	no
6	.4427504	A/B	yes	one year	yes

There clearly is a positive relation between the stated ranking and the predicted probabilities for alternatives to be ranked first, but the association is not perfect. In fact, we would not have expected a perfect association, as the model specifies a (nondegenerate) probability distribution over the possible rankings of the alternatives. These predictions for sets of 10 candidates can also be used to make predictions for subsets of the alternatives. For instance, suppose that only the last three candidates listed in this table would be available. According to parameter estimates of the rank-ordered logit model, the probability that the last of these candidates is selected equals $0.443/(0.037 + 0.116 + 0.443) = 0.743$.

Comparing respondents

The rologit model assumes that all respondents, HR managers in large public-sector organizations in The Netherlands, use the *same* valuation function; i.e., they apply the same decision weights. This is the substantive interpretation of the assumption that the β's are constant between the respondents. To probe this assumption, we could test whether the coefficients vary between different groups of respondents. For a metric characteristic of the HR manager, such as firmsize, we can consider a trend-model in the valuation weights,

$$\beta_{ij} = \alpha_{i0} + \alpha_{i1} \texttt{firmsize}_j$$

and we can test that the slopes α_{i1} of firmsize are zero.

```
. generate firmsize = employer
. generate size_edufit   = firmsize * edufit
. generate size_grades   = firmsize * grades
. generate size_workexp  = firmsize * workexp
. generate size_boardexp = firmsize * boardexp
```

(*Continued on next page*)

```
. rologit pref edufit grades workexp size* if job==1, group(caseid) nolog
Rank-ordered logistic regression              Number of obs      =         80
Group variable: caseid                        Number of groups   =          8

No ties in data                               Obs per group: min =         10
                                                             avg =      10.00
                                                             max =         10

                                              LR chi2(7)         =      57.17
Log likelihood = -66.82346                    Prob > chi2        =     0.0000
```

pref	Coef.	Std. Err.	z	P>|z|	[95% Conf. Interval]	
edufit	1.29122	1.13764	1.13	0.256	-.9385127	3.520953
grades	6.439776	2.288056	2.81	0.005	1.955267	10.92428
workexp	1.23342	.8065067	1.53	0.126	-.347304	2.814144
size_edufit	-.0173333	.0711942	-0.24	0.808	-.1568714	.1222048
size_grades	-.2099279	.1218251	-1.72	0.085	-.4487008	.028845
size_workexp	.0097508	.0525081	0.19	0.853	-.0931632	.1126649
size_board~p	.0382304	.0227545	1.68	0.093	-.0063676	.0828284

```
. testparm size*
 ( 1)   size_edufit = 0
 ( 2)   size_grades = 0
 ( 3)   size_workexp = 0
 ( 4)   size_boardexp = 0

          chi2(  4) =      7.14
        Prob > chi2 =    0.1288
```

The Wald test that the slopes size_* are jointly zero provides no evidence upon which we would reject the null hypothesis; i.e., we do not find evidence against the assumption of constant valuation weights of the attributes by firms of different size. We did not enter firmsize as a predictor variable. Characteristics of the decision-making agent do not vary between alternatives. Thus an additive effect of these characteristics on the valuation of alternatives does *not* affect the agent's ranking of alternatives and his choice. Consequently the coefficient of firmsize is not identified. rologit would in fact have diagnosed the problem and dropped firmsize from the analysis. Diagnosing this problem can slow the estimation considerably; the test may be suppressed by specifying the option notestrhs.

Incomplete rankings and ties

rologit allows incomplete rankings and ties in the rankings as proposed by Allison and Christakis (1994). rologit only permits rankings to be incomplete "at the bottom"; namely, that the ranking of the least attractive alternatives for subjects may not be known—do not confuse this with the situation that a subject is indifferent between these alternatives. This form of incompleteness occurred in the example discussed here, as the respondents were instructed to select and rank only the top 6 alternatives. It may also be that respondents refused to rank the alternatives that are very unattractive. rologit does not allow other forms of incompleteness, for instance, data in which respondents indicate which of four cars they like best, and which one they like least, but not how they rank the two intermediate cars. Another example of incompleteness that cannot be analyzed with rologit is data in which respondents select the three alternatives they like best but are not requested to express their preferences among the three selected alternatives.

rologit also permits ties in rankings. rologit assumes that if a subject expresses a tie between two or more alternatives, he or she actually holds one particular strict preference ordering, but with all possibilities of a strict ordering consistent with the expressed weak ordering being equally probable. For instance, suppose that a respondent ranks alternative 1 highest. He prefers alternatives 2 and 3 over alternative 4, and he is indifferent between alternatives 2 and 3. We assume that this respondent either has the strict preference ordering $1 > 2 > 3 > 4$ or $1 > 3 > 2 > 4$, with both possibilities being equally likely. From a psychometric perspective, it may actually be more appropriate to also assume that the alternatives 2 and 3 are close; for instance, the difference between the associated valuations (utilities) is less than some threshold or minimally discernible difference. Computationally, however, this is a more demanding model.

Clustered choice data

We have seen that applicants with work experience are in a relatively favorable position. To test whether the effects of work experience vary between the jobs, we can include interactions between the type of job and the attributes of applicants. Such interactions can be obtained via xi. To suppress the meaningless (and unidentified) main effect of job, we use xi specifications of the form i.job|*varname*.

Since some HR managers contributed data for more than one job, we cannot assume that their selection decisions for different jobs are independent. We can account for this by specifying the vce(cluster *clustvar*) option. By treating choice data as incomplete ranking data with only the most preferred alternative marked, rologit may be used to estimate the model parameters for clustered choice data.

```
. xi: rologit pref i.job|female i.job|grades i.job|edufit i.job|workexp,
> group(caseid) vce(cluster employer) nolog
i.job              _Ijob_1-3           (naturally coded; _Ijob_1 omitted)
i.job|female       _IjobXfemal_#       (coded as above)
i.job|grades       _IjobXgrade_#       (coded as above)
i.job|edufit       _IjobXedufi_#       (coded as above)
i.job|workexp      _IjobXworke_#       (coded as above)
```

```
Rank-ordered logistic regression              Number of obs      =        290
Group variable: caseid                        Number of groups   =         29

Ties handled via the efron method             Obs per group: min =         10
                                                             avg =      10.00
                                                             max =         10

                                              Wald chi2(12)      =      79.57
Log pseudolikelihood=-296.3855                Prob > chi2        =     0.0000
```

(Std. Err. adjusted for 22 clusters in employer)

pref	Coef.	Robust Std. Err.	z	P>|z|	[95% Conf. Interval]	
female	-.2286609	.2519883	-0.91	0.364	-.7225489	.2652272
_IjobXfema~2	.0293815	.4829166	0.06	0.951	-.9171177	.9758808
_IjobXfema~3	.1195538	.3688844	0.32	0.746	-.6034463	.8425538
grades	2.812555	.8517878	3.30	0.001	1.143081	4.482028
_IjobXgrad~2	-2.364247	1.005963	-2.35	0.019	-4.335898	-.3925961
_IjobXgrad~3	-1.88232	.8995277	-2.09	0.036	-3.645362	-.1192782
edufit	.7027757	.2398396	2.93	0.003	.2326987	1.172853
_IjobXeduf~2	-.267475	.4244964	-0.63	0.529	-1.099473	.5645226
_IjobXeduf~3	-.3182995	.3689972	-0.86	0.388	-1.041521	.4049217
workexp	1.224453	.3396773	3.60	0.000	.5586978	1.890208
_IjobXwork~2	-.6870077	.3692946	-1.86	0.063	-1.410812	.0367964
_IjobXwork~3	-.4656993	.4515712	-1.03	0.302	-1.350763	.4193639

The parameter estimates for the first job type are very similar to those that would have been obtained from an analysis isolated to these data. Differences are due only to an implied change in the method of handling ties. With clustered observations, `rologit` uses Efron's method. If we had specified the option `ties(efron)` with the separate analyses, then the parameter estimates would have been identical to the simultaneous results. Another difference is that `rologit` now reports robust standard errors, adjusted for clustering within respondents. These could have been obtained for the separate analyses, as well by specifying the `vce(robust)` option. In fact, this option would also have forced `rologit` to switch to Efron's approximation as well.

Given the combined results for the three types of jobs, we can test easily whether the weights for the attributes of applicants vary between the jobs, in other words, whether employers are looking for different qualifications in applicants for different jobs. A Wald test for the equality hypothesis of no difference can be obtained with the `testparm` command:

```
. testparm _I*
 ( 1)   _IjobXfemal_2 = 0
 ( 2)   _IjobXfemal_3 = 0
 ( 3)   _IjobXgrade_2 = 0
 ( 4)   _IjobXgrade_3 = 0
 ( 5)   _IjobXedufi_2 = 0
 ( 6)   _IjobXedufi_3 = 0
 ( 7)   _IjobXworke_2 = 0
 ( 8)   _IjobXworke_3 = 0

           chi2(  8) =   14.96
         Prob > chi2 =    0.0599
```

We find only mild evidence that employers look for different qualities in candidates according to the job for which they are being considered.

❑ Technical Note

Allison (1999) stressed that the comparison between groups of the coefficients of logistic regression is problematic, especially in its latent-variable interpretation. In many common latent-variable models, only the regression coefficients divided by the scale of the latent variable are identified. Thus a comparison of logit regression coefficients between, say, men and women is meaningful only if one is willing to argue that the standard deviation of the latent residual does not differ between the sexes. The rank-ordered logit model is also affected by this problem. While we formulated the model with a scale-free residual, we can actually think of the model for the value of an alternative as being scaled by the standard deviation of the random term, representing other relevant attributes of alternatives. Again comparing attribute weights between jobs is meaningful to the extent that we are willing to defend the proposition that "all omitted attributes" are equally important for different kinds of jobs.

❑

Comparison of rologit and clogit

The rank-ordered logit model also has a sequential interpretation. A subject first chooses the best among the alternatives. Next he or she selects the best alternative among the remaining alternatives, etc. The decisions at each of the subsequent stages are described by a conditional logit model, and a subject is assumed to apply the same decision weights at each stage. Some authors have expressed concern that later choices may well be made more randomly than the first few decisions. A formalization of this idea is a heteroskedastic version of the rank-ordered logit model in which the scale of the random term increases with the number of decisions made (e.g., Hausman and Ruud 1987). This extended model is currently not supported by `rologit`. However, the hypothesis that the same decision weights are applied at the first stage and at later stages can be tested by applying a Hausman test.

First, we fit the rank-ordered logit model on the full ranking data for the first type of job,

```
. rologit pref age female edufit grades workexp boardexp if job==1, group(caseid) nolog
```

```
Rank-ordered logistic regression                  Number of obs      =        80
Group variable: caseid                            Number of groups   =         8

No ties in data                                   Obs per group: min =        10
                                                                 avg =     10.00
                                                                 max =        10

                                                  LR chi2(6)         =     54.13
Log likelihood = -68.34539                        Prob > chi2        =    0.0000
```

pref	Coef.	Std. Err.	z	P>|z|	[95% Conf. Interval]	
age	-.0984926	.0820473	-1.20	0.230	-.2593024	.0623172
female	-.4487287	.3671307	-1.22	0.222	-1.168292	.2708343
edufit	.7658064	.3602366	2.13	0.034	.0597556	1.471857
grades	3.064534	.6148245	4.98	0.000	1.8595	4.269568
workexp	1.386427	.292553	4.74	0.000	.8130341	1.959821
boardexp	.6944377	.3762596	1.85	0.065	-.0430176	1.431893

and we save the estimates for later use with the `estimates` command.

```
. estimates store Ranking
```

To estimate the decision weights on the basis of the most preferred alternatives only, we create a variable `best` that is 1 for the best alternatives, and 0 otherwise. The `by` prefix is useful here.

```
. by caseid (pref), sort: gen best = pref == pref[_N] if job==1
(210 missing values generated)
```

By specifying `(pref)` with `by caseid`, we ensured that the data were sorted in increasing order on `pref` within `caseid`. Hence, the most preferred alternatives are last in the sort order. The expression `pref == pref[_N]` is true (1) for the most preferred alternatives, even if the alternative is not unique, and false (0) otherwise. If the most preferred alternatives were sometimes tied, we could still fit the model for the based-alternatives-only data via `rologit`, but `clogit` would yield different results since it deals with ties in a less appropriate way for continuous valuations. To ascertain whether there are ties in the selected data regarding applicants for research positions, we can combine `by` with `assert`:

```
. by caseid (pref), sort: assert pref[_N-1] != pref[_N] if job==1
```

There are no ties. We can now fit the model on the choice data by using either `clogit` or `rologit`.

(Continued on next page)

```
. rologit best age edufit grades workexp boardexp if job==1, group(caseid) nolog
```

Rank-ordered logistic regression   Number of obs   =   80
Group variable: caseid   Number of groups =   8

No ties in data   Obs per group: min =   10
                                             avg =   10.00
                                             max =   10

                                     LR chi2(5)   =   17.27
Log likelihood = -9.783205   Prob > chi2   =   0.0040

best	Coef.	Std. Err.	z	P>|z|	[95% Conf. Interval]	
age	-.1048959	.2017068	-0.52	0.603	-.5002339	.2904421
edufit	.4558387	.9336775	0.49	0.625	-1.374136	2.285813
grades	3.443851	1.969002	1.75	0.080	-.4153223	7.303025
workexp	2.545648	1.099513	2.32	0.021	.3906422	4.700655
boardexp	1.765176	1.112763	1.59	0.113	-.4157988	3.946152

```
. estimates store Choice
```

The same results, though with a slightly different formatted header, would have been obtained by using clogit on these data.

```
. clogit best age edufit grades workexp boardexp if job==1, group(caseid) nolog
```

Conditional (fixed-effects) logistic regression   Number of obs   =   80
                                         LR chi2(5)   =   17.27
                                         Prob > chi2   =   0.0040
Log likelihood = -9.7832046   Pseudo R2   =   0.4689

best	Coef.	Std. Err.	z	P>|z|	[95% Conf. Interval]	
age	-.1048959	.2017068	-0.52	0.603	-.5002339	.2904421
edufit	.4558387	.9336775	0.49	0.625	-1.374136	2.285813
grades	3.443851	1.969002	1.75	0.080	-.4153223	7.303025
workexp	2.545648	1.099513	2.32	0.021	.3906422	4.700655
boardexp	1.765176	1.112763	1.59	0.113	-.4157988	3.946152

The parameters of the ranking and choice models look different, but the standard errors based on the choice data are much larger. Are we estimating parameters with the ranking data that are different from those with the choice data? A Hausman test compares two estimators of a parameter. One of the estimators should be efficient under the null hypothesis, namely, that choosing the second-best alternative is determined with the same decision weights as the best, etc. In our case, the efficient estimator of the decision weights uses the ranking information. The other estimator should be consistent, even if the null hypothesis is false. In our application, this is the estimator that uses the first-choice data only.

```
. hausman Choice Ranking

                     ── Coefficients ──
                    (b)          (B)          (b-B)      sqrt(diag(V_b-V_B))
                  Choice      Ranking     Difference           S.E.

        age    -.1048959    -.0984926     -.0064033          .1842657
     edufit     .4558387     .7658064     -.3099676          .8613846
     grades     3.443851     3.064534      .3793169          1.870551
    workexp     2.545648     1.386427      1.159221          1.059878
   boardexp     1.765176      .6944377     1.070739           1.04722

                       b = consistent under Ho and Ha; obtained from rologit
            B = inconsistent under Ha, efficient under Ho; obtained from rologit

    Test:  Ho:  difference in coefficients not systematic

              chi2(5) = (b-B)'[(V_b-V_B)^(-1)](b-B)
                      =        3.05
             Prob>chi2 =       0.6918
```

We do not find evidence for misspecification. We have to be cautious, though, since Hausman-type tests are often not powerful, and the number of observations in our example is very small, which makes the quality of the approximation of the null distribution by a chi-squared test rather uncertain.

On reversals of rankings

The rank-ordered logit model has a property that you may find unexpected and even unfortunate. Compare two analyses with the rank-ordered logit model, one in which alternatives are ranked from "most attractive" to "least attractive", the other a reversed analysis in which these alternatives are ranked from "most unattractive" to "least unattractive". By unattractiveness, you probably mean just the opposite of attractiveness, and you expect that the weights of the attributes in predicting "attractiveness" to be minus the weights in predicting "unattractiveness". This is, however, *not* true for the rank-ordered logit model. The assumed distribution of the random residual takes the form $F(\epsilon) = 1 - \exp\{\exp(-\epsilon)\}$. This distribution is right-skewed. Therefore, slightly different models result from adding and subtracting the random residual, corresponding with high-to-low and low-to-high rankings. Thus the estimated coefficients will differ between the two specifications, though usually not in an important way. You may observe the difference by specifying the option `reverse` of `rologit`. Reversing the rank order makes rankings that are incomplete at the bottom become incomplete at the top. Only the first kind of incompleteness is supported by `rologit`. Thus, for this comparison, we exclude the alternatives that are not ranked, omitting the information that ranked alternatives are preferred over excluded ones.

```
. rologit pref grades edufit workexp boardexp if job==1 & pref!=0, group(caseid)
(output omitted)
. estimates store Original
. rologit pref grades edufit workexp boardexp if job==1 & pref!=0, group(caseid) reverse
(output omitted)
. estimates store Reversed
```

(*Continued on next page*)

```
. estimates table Original Reversed, stats(aic bic)
```

Variable	Original	Reversed
grades	2.0032332	-1.0955335
edufit	-.13111006	-.05710681
workexp	1.2805373	-1.2096383
boardexp	.46213212	-.27200317
aic	96.750452	99.665642
bic	104.23526	107.15045

Thus, although the weights of the attributes for reversed rankings are indeed mostly of opposite signs, the magnitudes of the weights and their standard errors differ. Which one is more appropriate? We have no advice to offer here. The specific science of the problem will determine what is appropriate, though we would be surprised indeed if this helps here. Formal testing does not help much either, as the models for the original and reversed rankings are not nested. The model-selection indices, such as the AIC and BIC, however, suggest that you stick to the rank-ordered logit model applied to the original ranking rather than to the reversed ranking.

Saved Results

rologit saves the following in e():

Scalars

e(N)	number of observations		e(N_g)	number of groups
e(ll_0)	log likelihood of the null model		e(g_min)	minimum group size
	("all rankings are equiprobable")		e(g_avg)	average group size
e(ll)	log likelihood		e(g_max)	maximum group size
e(df_m)	model degrees of freedom		e(code_inc)	value for incomplete preferences
e(chi2)	χ^2		e(N_clust)	number of clusters
e(r2_p)	pseudo-R^2			

Macros

e(cmd)	rologit		e(offset)	offset
e(cmdline)	command as typed		e(chi2type)	Wald or LR; type of model χ^2 test
e(depvar)	name of dependent variable		e(ties)	breslow, efron, exactm
e(group)	name of group() variable		e(vce)	vcetype specified in vce()
e(wtype)	weight type		e(vcetype)	title used to label Std. Err.
e(wexp)	weight expression		e(crittype)	optimization criterion
e(title)	title in estimation output		e(properties)	b V
e(clustvar)	name of cluster variable		e(predict)	program used to implement predict

Matrices

e(b)	coefficient vector		e(V)	variance–covariance matrix of the estimators

Functions

e(sample)	marks estimation sample

Methods and Formulas

`rologit` is implemented as an ado-file.

Allison and Christakis (1994) demonstrate that maximum likelihood estimates for the rank-ordered logit model can be obtained as the maximum partial-likelihood estimates of an appropriately specified Cox regression model for waiting time ([ST] **stcox**). In this analogy, a higher value for an alternative is formally equivalent to a higher hazard rate of failure. `rologit` uses `stcox` to fit the rank-ordered logit model based on such a specification of the data in Cox terms. A higher stated preference is represented by a shorter waiting time until failure. Incomplete rankings are dealt with via censoring. Moreover, decision situations (subjects) are to be treated as strata. Finally, as proposed by Allison and Christakis, ties in rankings are handled by the marginal-likelihood method, specifying that all strict preference orderings consistent with the stated weak preference ordering are equally likely. The marginal-likelihood estimator is available in `stcox` via the option `exactm`. The approximations of the marginal likelihood due to Breslow and Efron are also appropriate for the analysis of rank-ordered logit models. Since in most applications the number of ranked alternatives by one subject will be fairly small (at most, say, 20), the number of ties is small as well, and so you rarely will need to turn to approximations to restrict computer time. Since the marginal-likelihood estimator in `stcox` does not support the cluster adjustment or `pweight`s, you should use the Efron approximation in such cases.

Acknowledgment

The `rologit` command was written by Jeroen Weesie, Department of Sociology, Utrecht University, The Netherlands.

References

Allison, P. 1999. Comparing logit and probit coefficients across groups. *Sociological Methods and Research* 28: 186–208.

Allison, P., and N. Christakis. 1994. Logit models for sets of ranked items. In *Sociological Methodology*, Volume 24, ed. P. Marsden, 199–228. Oxford: Blackwell.

Beggs, S., S. Cardell, and J. Hausman. 1981. Assessing the potential demand for electric cars. *Journal of Econometrics* 16: 1–19.

Hair, J. E., Jr., R. E. Anderson, R. L. Tatham, and W. C. Black. 1995. *Multivariate Data Analysis: with Readings.* 4th ed. London: Prentice Hall.

Hausman, J., and B. Ruud. 1987. Specifying and testing econometric models for rank-ordered data. *Journal of Econometrics* 34: 83–104.

Luce, R. D. 1959. *Individual Choice Behavior: A Theoretical Analysis.* New York: Dover.

Marden, J. I. 1995. *Analyzing and Modeling Rank Data.* London: Chapman & Hall.

McCullagh, P. 1993. Permutations and regression models. In *Probability Models and Statistical Analysis for Ranking Data*, ed. M. A. Fligner and J. S. Verducci, 196–215. New York: Springer.

Punj, G. N., and R. Staelin. 1978. The choice process for graduate business schools. *Journal of Marketing Research* 15: 588–598.

Thurstone, L. L. 1927. A law of comparative judgment. *Psychological Reviews* 34: 273–286.

Wolf, I. de 2000. Opleidingsspecialisatie en arbeidsmarktsucces van sociale wetenschappers. Amsterdam: ThelaThesis.

Yellott, J. I., Jr. 1977. The relationship between Luce's choice axiom, Thurstone's theory of comparative judgment, and the double-exponential distribution. *Journal of Mathematical Psychology* 15: 109–144.

Also See

Title

rologit postestimation — Postestimation tools for rologit

Description

The following postestimation commands are available for `rologit`:

command	description
adjust	adjusted predictions of $\mathbf{x}\beta$ or $\exp(\mathbf{x}\beta)$
estat	AIC, BIC, VCE, and estimation sample summary
estimates	cataloging estimation results
hausman	Hausman's specification test
lincom	point estimates, standard errors, testing, and inference for linear combinations of coefficients
linktest	link test for model specification
lrtest	likelihood-ratio test
mfx	marginal effects or elasticities
nlcom	point estimates, standard errors, testing, and inference for nonlinear combinations of coefficients
predict	predictions, residuals, influence statistics, and other diagnostic measures
predictnl	point estimates, standard errors, testing, and inference for generalized predictions
test	Wald tests for simple and composite linear hypotheses
testnl	Wald tests of nonlinear hypotheses

See the corresponding entries in the *Stata Base Reference Manual* for details.

Syntax for predict

predict [*type*] *newvar* [*if*] [*in*] [, *statistic* <u>nooff</u>set]

statistic	description
<u>pr</u>	probability that alternatives are ranked first; the default
xb	linear prediction
stdp	standard error of the linear prediction

These statistics are available both in and out of sample; type `predict ... if esample() ...` if wanted only for the estimation sample.

Options for predict

pr, the default, calculates the probability that alternatives are ranked first.

xb calculates the linear prediction.

stdp calculates the standard error of the linear prediction.

nooffset is relevant only if you specified offset(*varname*) for rologit. It modifies the calculations made by predict so that they ignore the offset variable; the linear prediction is treated as $\mathbf{x}_j\mathbf{b}$ rather than as $\mathbf{x}_j\mathbf{b} + \text{offset}_j$.

Remarks

See *Comparing respondents* and *Clustered choice data* in [R] **rologit** for examples of the use of testparm, an alternative to the test command.

See *Comparison of rologit and clogit* and *On reversals of rankings* in [R] **rologit** for examples of the use of estimates.

See *Comparison of rologit and clogit* in [R] **rologit** for an example of the use of hausman.

Methods and Formulas

All postestimation commands listed above are implemented as ado-files.

Also See

[R] **rologit** — Rank-ordered logistic regression

[U] **20 Estimation and postestimation commands**

Title

rreg — Robust regression

Syntax

rreg *depvar* [*indepvars*] [*if*] [*in*] [, *options*]

options	description
Model	
tune(#)	use # as the biweight tuning constant; default is tune(7)
Reporting	
level(#)	set confidence level; default is level(95)
genwt(*newvar*)	create *newvar* containing the weights assigned to each observation
Opt options	
optimization_options	control the optimization process; seldom used
graph	graph weights during convergence

depvar and *indepvars* may contain time-series operators; see [U] **11.4.3 Time-series varlists**.
by, rolling, statsby, and xi are allowed; see [U] **11.1.10 Prefix commands**.
See [U] **20 Estimation and postestimation commands** for more capabilities of estimation commands.

Description

rreg performs one version of robust regression of *depvar* on *indepvars*.

Also see *Robust standard errors* in [R] **regress** for standard regression with robust variance estimates and [R] **qreg** for quantile (including median or least-absolute-residual) regression.

Options

▭ Model ▭

tune(#) is the biweight tuning constant. The default is 7, meaning seven times the median absolute deviation from the median residual (MAD); see *Methods and Formulas*. Lower tuning constants downweight outliers rapidly but may lead to unstable estimates (less than 6 is not recommended). Higher tuning constants produce milder downweighting.

▭ Reporting ▭

level(#); see [R] **estimation options**.

genwt(*newvar*) creates the new variable *newvar* containing the weights assigned to each observation.

▭ Opt options ▭

optimization_options: iterate(#), tolerance(#), [no]log. iterate() specifies the maximum number of iterations; iterations stop when the maximum change in weights drops below tolerance(); and log/nolog specifies whether or not to show the iteration log. These options are seldom used.

`graph` allows you to graphically watch the convergence of the iterative technique. The weights obtained from the most recent round of estimation are graphed against the weights obtained from the previous round.

Remarks

`rreg` first performs an initial screening based on Cook's distance > 1 to eliminate gross outliers before calculating starting values and then performs Huber iterations followed by biweight iterations, as suggested by Li (1985).

▷ Example 1

We wish to examine the relationship between mileage rating, weight, and location of manufacture for the 74 cars in our automobile data. As a point of comparison, we begin by fitting an ordinary regression:

```
. use http://www.stata-press.com/data/r10/auto
(1978 Automobile Data)

. regress mpg weight foreign
```

Source	SS	df	MS
Model	1619.2877	2	809.643849
Residual	824.171761	71	11.608053
Total	2443.45946	73	33.4720474

Number of obs =	74
F(2, 71) =	69.75
Prob > F =	0.0000
R-squared =	0.6627
Adj R-squared =	0.6532
Root MSE =	3.4071

mpg	Coef.	Std. Err.	t	P>|t|	[95% Conf. Interval]
weight	-.0065879	.0006371	-10.34	0.000	-.0078583 -.0053175
foreign	-1.650029	1.075994	-1.53	0.130	-3.7955 .4954422
_cons	41.6797	2.165547	19.25	0.000	37.36172 45.99768

We now compare this with the results from `rreg`:

```
. rreg mpg weight foreign

    Huber iteration 1:  maximum difference in weights = .80280176
    Huber iteration 2:  maximum difference in weights = .2915438
    Huber iteration 3:  maximum difference in weights = .08911171
    Huber iteration 4:  maximum difference in weights = .02697328
 Biweight iteration 5:  maximum difference in weights = .29186818
 Biweight iteration 6:  maximum difference in weights = .11988101
 Biweight iteration 7:  maximum difference in weights = .03315872
 Biweight iteration 8:  maximum difference in weights = .00721325
```

Robust regression

Number of obs =	74
F(2, 71) =	168.32
Prob > F =	0.0000

mpg	Coef.	Std. Err.	t	P>|t|	[95% Conf. Interval]
weight	-.0063976	.0003718	-17.21	0.000	-.007139 -.0056562
foreign	-3.182639	.627964	-5.07	0.000	-4.434763 -1.930514
_cons	40.64022	1.263841	32.16	0.000	38.1202 43.16025

Note the large change in the `foreign` coefficient.

◁

❑ Technical Note

It would have been better if we had fitted the previous robust regression by typing `rreg mpg weight foreign, genwt(w)`. The new variable `w` would then contain the estimated weights. Let's pretend that we did this:

```
. rreg mpg weight foreign, genwt(w)
(output omitted )
. summarize w, detail
                    Robust Regression Weight
```

	Percentiles	Smallest		
1%	0	0		
5%	.0442957	0		
10%	.4674935	0	Obs	74
25%	.8894815	.0442957	Sum of Wgt.	74
50%	.9690193		Mean	.8509966
		Largest	Std. Dev.	.2746451
75%	.9949395	.9996715		
90%	.9989245	.9996953	Variance	.0754299
95%	.9996715	.9997343	Skewness	-2.287952
99%	.9998585	.9998585	Kurtosis	6.874605

We discover that 3 observations in our data were dropped altogether (they have weight 0). We could further explore our data:

```
. sort w
. list make mpg weight w if w<.467, sep(0)
```

	make	mpg	weight	w
1.	Datsun 210	35	2,020	0
2.	Subaru	35	2,050	0
3.	VW Diesel	41	2,040	0
4.	Plym. Arrow	28	3,260	.04429567
5.	Cad. Seville	21	4,290	.08241943
6.	Toyota Corolla	31	2,200	.10443129
7.	Olds 98	21	4,060	.28141296

Being familiar with the automobile data, we immediately spotted two things: the VW is the only diesel car in our data, and the weight recorded for the Plymouth Arrow is incorrect.

❑

▷ Example 2

If we specify no explanatory variables, `rreg` produces a robust estimate of the mean:

```
. rreg mpg
    Huber iteration 1:  maximum difference in weights = .64471879
    Huber iteration 2:  maximum difference in weights = .05098336
    Huber iteration 3:  maximum difference in weights = .0099887
 Biweight iteration 4:  maximum difference in weights = .25197391
 Biweight iteration 5:  maximum difference in weights = .00358606
```

```
   Robust regression                                Number of obs =      74
                                                    F(  0,    73) =    0.00
                                                    Prob > F      =      .

─────────────┬──────────────────────────────────────────────────────────────
         mpg │    Coef.    Std. Err.      t      P>|t|    [95% Conf. Interval]
─────────────┼──────────────────────────────────────────────────────────────
       _cons │  20.68825    .641813    32.23    0.000    19.40912    21.96738
─────────────┴──────────────────────────────────────────────────────────────
```

The estimate is given by the coefficient on _cons. The mean is 20.69 with an estimated standard error of .6418. The 95% confidence interval is [19.4, 22.0]. By comparison, ci gives us the standard calculation:

```
. ci mpg
    Variable │      Obs        Mean    Std. Err.      [95% Conf. Interval]
─────────────┼─────────────────────────────────────────────────────────────
         mpg │       74     21.2973    .6725511      19.9569     22.63769
```

◁

Saved Results

rreg saves the following in e():

Scalars

e(N)	number of observations	e(r2)	R-squared
e(mss)	model sum of squares	e(r2_a)	adjusted R-squared
e(df_m)	model degrees of freedom	e(F)	F statistic
e(rss)	residual sum of squares	e(rmse)	root mean squared error
e(df_r)	residual degrees of freedom		

Macros

e(cmd)	rreg	e(title)	title in estimation output
e(cmdline)	command as typed	e(model)	ols
e(depvar)	name of dependent variable	e(properties)	b V
e(genwt)	variable containing the weights	e(predict)	program used to implement predict

Matrices

e(b)	coefficient vector	e(V)	variance–covariance matrix of the estimators

Functions

e(sample)	marks estimation sample

Methods and Formulas

rreg is implemented as an ado-file.

See Berk (1990), Goodall (1983), and Rousseeuw and Leroy (2003) for a general description of the issues and methods. Hamilton (1991, 1992) provides a more detailed description of rreg and some Monte Carlo evaluations.

rreg begins by fitting the regression (see [R] **regress**), calculating Cook's D (see [R] **predict** and [R] **regress postestimation**), and excluding any observation for which $D > 1$.

Thereafter `rreg` works iteratively: it performs a regression, calculates case weights from absolute residuals, and regresses again using those weights. Iterations stop when the maximum change in weights drops below `tolerance()`. Weights derive from one of two weight functions, Huber weights and biweights. Huber weights (Huber 1964) are used until convergence, and then, from that result, biweights are used until convergence. The biweight was proposed by Beaton and Tukey (1974, 151–152) after the Princeton robustness study (Andrews et al. 1972) had compared various estimators. Both weighting functions are used because Huber weights have problems dealing with severe outliers, whereas biweights sometimes fail to converge or have multiple solutions. The initial Huber weighting should improve the behavior of the biweight estimator.

In Huber weighting, cases with small residuals receive weights of 1; cases with larger residuals receive gradually smaller weights. Let $e_i = y_i - \mathbf{X}_i\mathbf{b}$ represent the ith-case residual. The ith scaled residual $u_i = e_i/s$ is calculated, where $s = M/.6745$ is the residual scale estimate and $M = \text{med}(|e_i - \text{med}(e_i)|)$ is the median absolute deviation from the median residual. Huber estimation obtains case weights:

$$w_i = \begin{cases} 1 & \text{if } |u_i| \le c_h \\ c_h/|u_i| & \text{otherwise} \end{cases}$$

`rreg` defines $c_h = 1.345$, so downweighting begins with cases whose absolute residual exceeds $(1.345/.6745)M \approx 2M$.

With biweights, all cases with nonzero residuals receive some downweighting, according to the smoothly decreasing biweight function

$$w_i = \begin{cases} \{1 - (u_i/c_b)^2\}^2 & \text{if } |u_i| \le c_b \\ 0 & \text{otherwise} \end{cases}$$

where $c_b = 4.685 \cdot \text{tune}()/7$. Thus when $\text{tune}() = 7$, cases with absolute residuals of $(4.685/.6745)M \approx 7M$ or more are assigned 0 weight and thus are effectively dropped. Goodall (1983, 377) suggests using a value between 6 and 9, inclusive, for `tune()` in the biweight case and states that performance is good between 6 and 12, inclusive.

The tuning constants $c_h = 1.345$ and $c_b = 4.685$ (assuming `tune()` is set at the default 7) give `rreg` about 95% of the efficiency of OLS when applied to data with normally distributed errors (Hamilton 1991). Lower tuning constants downweight outliers more drastically (but give up Gaussian efficiency); higher tuning constants make the estimator more like OLS.

Standard errors are calculated using the pseudovalues approach described in Street, Carroll, and Ruppert (1988).

Acknowledgment

The current version of `rreg` is due to the work of Lawrence Hamilton, Department of Sociology, University of New Hampshire.

References

Andrews, D. F., P. J. Bickel, F. R. Hampel, P. J. Huber, W. H. Rogers, and J. W. Tukey. 1972. *Robust Estimates of Location: Survey and Advances*. Princeton: Princeton University Press.

Beaton, A. E., and J. W. Tukey. 1974. The fitting of power series, meaning polynomials, illustrated on band-spectroscopic data. *Technometrics* 16: 146–185.

Berk, R. A. 1990. A primer on robust regression. In *Modern Methods of Data Analysis*, ed. J. Fox and J. S. Long, 292–324. Newbury Park, CA: Sage.

Goodall, C. 1983. M-estimators of location: An outline of the theory. In *Understanding Robust and Exploratory Data Analysis*, ed. D. C. Hoaglin, F. Mosteller, and J. W. Tukey, 339–431. New York: Wiley.

Gould, W. W., and W. H. Rogers. 1994. Quantile regression as an alternative to robust regression. *1994 Proceedings of the Statistical Computing Section.* Alexandria, VA: American Statistical Association.

Hamilton, L. C. 1991. srd1: How robust is robust regression? *Stata Technical Bulletin* 2: 21–26. Reprinted in *Stata Technical Bulletin Reprints*, vol. 1, pp. 169–175.

——. 1992. *Regression with Graphics: A Second Course in Applied Statistics.* Belmont, CA: Duxbury.

——. 2006. *Statistics with Stata (Update for Version 9).* Belmont, CA: Duxbury.

Huber, P. J. 1964. Robust estimation of a location parameter. *Annals of Mathematical Statistics* 35: 73–101.

Li, G. 1985. Robust regression. In *Exploring Data Tables, Trends, and Shapes*, ed. D. C. Hoaglin, F. Mosteller, and J. W. Tukey, 281–340. New York: Wiley.

Mosteller, F., and J. W. Tukey. 1977. *Data Analysis and Regression: A Second Course in Statistics.* Reading, MA: Addison–Wesley.

Relles, D. A., and W. H. Rogers. 1977. Statisticians are fairly robust estimators of location. *Journal of the American Statistical Association* 72: 107–111.

Rousseeuw, P. J., and A. M. Leroy. 2003. *Robust Regression and Outlier Detection.* New York: Wiley.

Street, J. O., R. J. Carroll, and D. Ruppert. 1988. A note on computing robust regression estimates via iteratively reweighted least squares. *American Statistician* 42: 152–154.

Also See

[R] **rreg postestimation** — Postestimation tools for rreg

[R] **qreg** — Quantile regression

[R] **regress** — Linear regression

[U] **20 Estimation and postestimation commands**

Title

rreg postestimation — Postestimation tools for rreg

Description

The following postestimation commands are available for `rreg`:

command	description
adjust[1]	adjusted predictions of $\mathbf{x}\beta$ or $\exp(\mathbf{x}\beta)$
estat	VCE and estimation sample summary
estimates	cataloging estimation results
lincom	point estimates, standard errors, testing, and inference for linear combinations of coefficients
mfx	marginal effects or elasticities
nlcom	point estimates, standard errors, testing, and inference for nonlinear combinations of coefficients
predict	predictions, residuals, influence statistics, and other diagnostic measures
predictnl	point estimates, standard errors, testing, and inference for generalized predictions
test	Wald tests for simple and composite linear hypotheses
testnl	Wald tests of nonlinear hypotheses

[1] `adjust` is not appropriate with time-series operators.

See the corresponding entries in the *Stata Base Reference Manual* for details.

Syntax for predict

predict [*type*] *newvar* [*if*] [*in*] [, *statistic*]

statistic	description
xb	linear prediction; the default
stdp	standard error of the linear prediction
<u>res</u>iduals	residuals
hat	diagonal elements of the hat matrix

These statistics are available both in and out of sample; type `predict ... if e(sample) ...` if wanted only for the estimation sample.

Options for predict

`xb`, the default, calculates the linear prediction.

`stdp` calculates the standard error of the linear prediction.

`residuals` calculates the residuals.

`hat` calculates the diagonal elements of the hat matrix. You must have run the `rreg` command with the `genwt()` option.

Methods and Formulas

All postestimation commands listed above are implemented as ado-files.

Also See

[R] **rreg** — Robust regression

[U] **20 Estimation and postestimation commands**

Title

runtest — Test for random order

Syntax

runtest *varname* $\left[in \right]$ $\left[, options \right]$

options	description
<u>c</u>ontinuity	continuity correction
<u>d</u>rop	ignore values equal to the threshold
<u>s</u>plit	randomly split values equal to the threshold as above or below the threshold; default is to count as below
<u>m</u>ean	use mean as threshold; default is median
<u>t</u>hreshold(#)	assign arbitrary threshold; default is median

Description

runtest tests whether the observations of *varname* are serially independent—that is, whether they occur in a random order—by counting how many runs there are above and below a threshold. By default, the median is used as the threshold. A small number of runs indicates positive serial correlation; a large number indicates negative serial correlation.

Options

continuity specifies a continuity correction that may be helpful in small samples. If there are fewer than 10 observations either above or below the threshold, however, the tables in Swed and Eisenhart (1943) provide more reliable critical values. By default, no continuity correction is used.

drop directs runtest to ignore any values of *varname* that are equal to the threshold value when counting runs and tabulating observations. By default, runtest counts a value as being above the threshold when it is strictly above the threshold and as being below the threshold when it is less than or equal to the threshold.

split directs runtest to randomly split values of *varname* that are equal to the threshold. In other words, when *varname* is equal to threshold, a "coin" is flipped. If it comes up heads, the value is counted as above the threshold. If it comes up tails, the value is counted as below the threshold.

mean directs runtest to tabulate runs above and below the mean rather than the median.

threshold(#) specifies an arbitrary threshold to use in counting runs. For example, if *varname* has already been coded as a 0/1 variable, the median generally will not be a meaningful separating value.

213

Remarks

`runtest` performs a nonparametric test of the hypothesis that the observations of *varname* occur in a random order by counting how many runs there are above and below a threshold. If *varname* is positively serially correlated, it will tend to remain above or below its median for several observations in a row; that is, there will be relatively few runs. If, on the other hand, *varname* is negatively serially correlated, observations above the median will tend to be followed by observations below the median and vice versa; that is, there will be relatively many runs.

By default, `runtest` uses the median for the threshold, and this is not necessarily the best choice. If `mean` is specified, the mean is used instead of the median. If `threshold(#)` is specified, `#` is used. Since `runtest` divides the data into two states—above and below the threshold—it is appropriate for data that are already binary; for example, win or lose, live or die, rich or poor, etc. Such variables are often coded as 0 for one state and 1 for the other. In this case, you should specify `threshold(0)` since, by default, `runtest` separates the observations into those that are greater than the threshold and those that are less than *or equal* to the threshold.

As with most nonparametric procedures, the treatment of ties complicates the test. Observations equal to the threshold value are ties and can be treated in one of three ways. By default, they are treated as if they were below the threshold. If `drop` is specified, they are omitted from the calculation and the total number of observations is adjusted. If `split` is specified, each is randomly assigned to the above- and below-threshold groups. The random assignment is different each time the procedure is run unless you specify the random-number seed; see [D] **generate**.

▷ Example 1

We can use `runtest` to check regression residuals for serial correlation.

```
. use http://www.stata-press.com/data/r10/run1
. scatter resid year, connect(l) yline(0) title(Regression residuals)
```

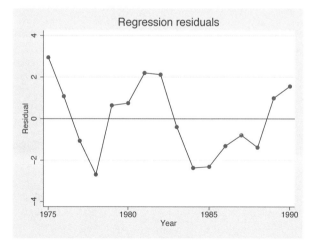

The graph gives the impression that these residuals are positively correlated. Excursions above or below zero—the natural threshold for regression residuals—tend to last for several observations. `runtest` can evaluate the statistical significance of this impression.

```
. runtest resid, thresh(0)
 N(resid <= 0) = 8
 N(resid >  0) = 8
          obs = 16
       N(runs) = 5
            z  = -2.07
     Prob>|z|  = .04
```

There are five runs in these 16 observations. Using the normal approximation to the true distribution of the number of runs, the five runs in this series are fewer than would be expected if the residuals were serially independent. The p-value is 0.03, indicating a two-sided significant result at the 5% level. If the alternative hypothesis is positive serial correlation, rather than any deviation from randomness, then the one-sided p-value is $.03/2 = .015$. With so few observations, however, the normal approximation may be inaccurate. (Tables compiled by Swed and Eisenhart list five runs as the 5% critical value for a one-sided test.)

runtest is a nonparametric test. It ignores the magnitudes of the observations and notes only whether the values are above or below the threshold. We can demonstrate this feature by reducing the information about the regression residuals in this example to a 0/1 variable that indicates only whether a residual is positive or negative.

```
. generate byte sign = resid>0
. runtest sign, thresh(0)
 N(sign <= 0) = 8
 N(sign >  0) = 8
          obs = 16
       N(runs) = 5
            z  = -2.07
     Prob>|z|  = .04
```

As expected, runtest produces the same answer as before.

◁

❑ Technical Note

The run test can also be used to test the null hypothesis that two samples are drawn from the same underlying distribution. The run test is sensitive to differences in the shapes, as well as the locations, of the empirical distributions.

Suppose, for example, that two different additives are added to the oil in 10 different cars during an oil change. The cars are run until a viscosity test determines that another oil change is needed, and the number of miles traveled between oil changes is recorded. The data are

(Continued on next page)

```
. use http://www.stata-press.com/data/r10/additive, clear
. list
```

	additive	miles
1.	1	4024
2.	1	4756
3.	1	7993
4.	1	5025
5.	1	4188
6.	2	3007
7.	2	1988
8.	2	1051
9.	2	4478
10.	2	4232

To test whether the additives generate different distributions of miles between oil changes, we sort the data by `miles` and then use `runtest` to see whether the marker for each additive occurs in random order:

```
. sort miles
. runtest additive, thresh(1)
N(additive <= 1) = 5
N(additive >  1) = 5
            obs = 10
        N(runs) = 4
              z = -1.34
        Prob>|z| = .18
```

Here the additives do not produce statistically different results.

❑

❑ Technical Note

A test that is related to the run test is the runs up-and-down test. In the latter test, the data are classified not by whether they lie above or below a threshold but by whether they are steadily increasing or decreasing. Thus an unbroken string of increases in the variable of interest is counted as one run, as is an unbroken string of decreases. According to Madansky (1988), the run test is superior to the runs up-and-down test for detecting trends in the data, but the runs up-and-down test is superior for detecting autocorrelation.

`runtest` can be used to perform a runs up-and-down test. Using the regression residuals from the example above, we can perform a `runtest` on their first differences:

```
. use http://www.stata-press.com/data/r10/run1
. generate resid_D = resid - resid[_n-1]
(1 missing value generated)
. runtest resid_D, thresh(0)
N(resid_D <= 0) = 7
N(resid_D >  0) = 8
            obs = 15
        N(runs) = 6
              z = -1.33
        Prob>|z| = .18
```

Edgington (1961) has compiled a table of the small sample distribution of the runs up-and-down statistic, and this table is reprinted in Madansky (1988). For large samples, the z statistic reported by runtest is incorrect for the runs up-and-down test. Let N be the number of observations (15 here), and let r be the number of runs (6). The expected number of runs in the runs up-and-down test is

$$\mu_r = \frac{2N - 1}{3}$$

the variance is

$$\sigma_r^2 = \frac{16N - 29}{90}$$

and the correct z statistic is

$$\widehat{z} = \frac{r - \mu_r}{\sigma_r}$$

❑

❑ Technical Note

runtest will tolerate missing values at the beginning or end of a series, as occurred in the technical note above (generating first differences resulted in a missing value for the first observation). runtest, however, will issue an error message if there are any missing observations in the interior of the series (in the portion covered by the in *range* modifier). To perform the test anyway, simply drop the missing observations before using runtest.

❑

Saved Results

runtest saves the following in r():

Scalars

r(N)	number of observations	r(p)	*p*-value of z
r(N_below)	number below the threshold	r(z)	z statistic
r(N_above)	number above the threshold	r(n_runs)	number of runs
r(mean)	expected number of runs	r(Var)	variance of the number of runs

Methods and Formulas

runtest is implemented as an ado-file.

runtest begins by calculating the number of observations below the threshold, n_0; the number of observations above the threshold, n_1; the total number of observations, $N = n_0 + n_1$; and the number of runs, r. These statistics are always reported, so the exact tables of critical values in Swed and Eisenhart (1943) may be consulted if necessary.

The expected number of runs under the null is

$$\mu_r = \frac{2n_0 n_1}{N} + 1$$

the variance is

$$\sigma_r^2 = \frac{2n_0 n_1 (2n_0 n_1 - N)}{N^2 (N - 1)}$$

and the normal approximation test statistic is

$$\hat{z} = \frac{r - \mu_r}{\sigma_r}$$

Acknowledgment

`runtest` was written by Sean Becketti, a past editor of the *Stata Technical Bulletin.*

References

Edgington, E. S. 1961. Probability table for number of runs of signs of first differences in ordered series. *Journal of the American Statistical Association* 56: 156–159.

Madansky, A. 1988. *Prescriptions for Working Statisticians.* New York: Springer.

Swed, F. S., and C. Eisenhart. 1943. Tables for testing randomness of grouping in a sequence of alternatives. *Annals of Mathematical Statistics* 14: 83–86.

Title

> **sampsi** — Sample size and power determination

Syntax

sampsi $\#_1$ $\#_2$ $\left[\, , \; options \right]$

options	description
Main	
onesample	one-sample test; default is two-sample
sd1(#)	standard deviation of sample 1
sd2(#)	standard deviation of sample 2
Options	
alpha(#)	significance level of test; default is alpha(0.05)
power(#)	power of test; default is power(0.90)
n1(#)	size of sample 1
n2(#)	size of sample 2
ratio(#)	ratio of sample sizes; default is ratio(1)
pre(#)	number of baseline measurements; default is pre(0)
post(#)	number of follow-up measurements; default is post(1)
nocontinuity	do not use continuity correction for two-sample test on proportions
r0(#)	correlation between baseline measurements; default is r0()=r1()
r1(#)	correlation between follow-up measurements
r01(#)	correlation between baseline and follow-up measurements
onesided	one-sided test; default is two-sided
method(*method*)	analysis method where *method* is post, change, ancova, or all; default is method(all)

Description

sampsi estimates required sample size or power of tests for studies comparing two groups. sampsi can be used when comparing means or proportions for simple studies where only one measurement of the outcome is planned and for comparing mean summary statistics for more complex studies where repeated measurements of the outcome on each experimental unit are planned.

If n1(#) or n2(#) is specified, sampsi computes power; otherwise, it computes sample size. For simple studies, if sd1(#) or sd2(#) is specified, sampsi assumes a comparison of means; otherwise, it assumes a comparison of proportions. For repeated measurements, sd1(#) or sd2(#) must be specified. sampsi is an immediate command; all its arguments are numbers.

(Continued on next page)

Options

onesample indicates a one-sample test. The default is two-sample.

sd1(#) and sd2(#) are the standard deviations of population 1 and population 2, respectively. One or both must be specified when doing a comparison of means. When the onesample option is used, sd1(#) is the standard deviation of the single sample (it can be abbreviated as sd(#)). If only one of sd1(#) or sd2(#) is specified, sampsi assumes that sd1 = sd2. If neither sd1(#) nor sd2(#) is specified, sampsi assumes a test of proportions. For repeated measurements, sd1(#) and/or sd2(#) must be specified.

alpha(#) is the significance level of the test. The default is alpha(0.05) unless set level has been used to reset the default significance level for confidence intervals. If a set level $\#_{lev}$ command has been issued, the default value is alpha(1 − $\#_{lev}$/100). See [R] **level**.

power(#) = $1 - \beta$ is the power of the test. The default is power(0.90).

n1(#) and n2(#) are the sizes of sample 1 and sample 2, respectively. One or both must be specified when computing power. If neither n1(#) nor n2(#) is specified, sampsi computes sample size. When the onesample option is used, n1(#) is the size of the single sample (it can be abbreviated as n(#)). If only one of n1(#) or n2(#) is specified, the unspecified one is computed using the formula ratio = n2()/n1().

ratio(#) is the ratio of sample sizes for two-sample tests: ratio = n2()/n1(). The default is ratio(1).

pre(#) specifies the number of baseline measurements (prerandomization) planned in a repeated-measure study. The default is pre(0).

post(#) specifies the number of follow-up measurements (postrandomization) planned in a repeated-measure study. The default is post(1).

nocontinuity requests power and sample size calculations without the continuity correction for the two-sample test on proportions. If not specified, the continuity correction is used.

r0(#) specifies the correlation between baseline measurements in a repeated-measure study. If r0(#) is not specified, sampsi assumes that r0() = r1().

r1(#) specifies the correlation between follow-up measurements in a repeated-measure study. For a repeated-measure study, either r1(#) or r01(#) must be specified. If r1(#) is not specified, sampsi assumes that r1() = r01().

r01(#) specifies the correlation between baseline and follow-up measurements in a repeated-measure study. For a repeated-measure study, either r01(#) or r1(#) must be specified. If r01(#) is not specified, sampsi assumes that r01() = r1().

onesided indicates a one-sided test. The default is two-sided.

method(post | change | ancova | all) specifies the analysis method to be used with repeated measures. change and ancova can be used only if baseline measurements are planned. The default is method(all), which means to use all three methods. Each method is described in *Methods and Formulas*.

Remarks

Remarks are presented under the following headings:

Studies with one measurement of the outcome
Two-sample test of equality of means
One-sample test of mean
Two-sample test of equality of proportions
One-sample test of proportion
Clinical trials with repeated measures

Studies with one measurement of the outcome

For simple studies, where only one measurement of the outcome is planned, `sampsi` computes sample size or power for four types of tests:

1. two-sample comparison of mean μ_1 of population 1 with mean μ_2 of population 2. The null hypothesis is $\mu_1 = \mu_2$, and normality is assumed. The postulated values of the means are $\mu_1 = \#_1$ and $\mu_2 = \#_2$, and the postulated standard deviations are `sd1(#)` and `sd2(#)`.

2. one-sample comparison of the mean μ of a population with a hypothesized value μ_0. The null hypothesis is $\mu = \mu_0$, and normality is assumed. The first argument, $\#_1$, to `sampsi` is μ_0. The second argument, $\#_2$, is the postulated value of μ, i.e., the alternative hypothesis is $\mu = \#_2$. The postulated standard deviation is `sd1(#)`. To get this test, the `onesample` option must be specified.

3. two-sample comparison of proportion p_1 with proportion p_2. The null hypothesis is $p_1 = p_2$, and the postulated values are $p_1 = \#_1$ and $p_2 = \#_2$.

4. one-sample comparison of a proportion p with a hypothesized value p_0. The null hypothesis is $p = p_0$, where $p_0 = \#_1$. The alternative hypothesis is $p = \#_2$. To get this test, the `onesample` option must be specified.

Examples of these follow.

Two-sample test of equality of means

▷ Example 1

We are doing a study of the relationship of oral contraceptives (OC) and blood pressure (BP) level for women ages 35–39 (Rosner 2000, 307–308). From a pilot study, it was determined that the mean and standard deviation BP of OC users were 132.86 and 15.34, respectively. The mean and standard deviation of OC nonusers in the pilot study were found to be 127.44 and 18.23. Since it is easier to find OC nonusers than users, we decide that n_2, the size of the sample of OC nonusers, should be twice n_1, the size of the sample of OC users; that is, $r = n_2/n_1 = 2$. To compute the sample sizes for $\alpha = 0.05$ (two-sided) and the power of 0.80, we issue the following command:

(Continued on next page)

```
. sampsi 132.86 127.44, p(0.8) r(2) sd1(15.34) sd2(18.23)

Estimated sample size for two-sample comparison of means

Test Ho: m1 = m2, where m1 is the mean in population 1
                  and m2 is the mean in population 2
Assumptions:

           alpha =    0.0500  (two-sided)
           power =    0.8000
              m1 =    132.86
              m2 =    127.44
             sd1 =     15.34
             sd2 =     18.23
           n2/n1 =      2.00

Estimated required sample sizes:

              n1 =        108
              n2 =        216
```

We now find out that we have only enough money to study 100 subjects from each group. We can compute the power for $n_1 = n_2 = 100$ by typing

```
. sampsi 132.86 127.44, n1(100) sd1(15.34) sd2(18.23)

Estimated power for two-sample comparison of means

Test Ho: m1 = m2, where m1 is the mean in population 1
                  and m2 is the mean in population 2
Assumptions:

           alpha =    0.0500  (two-sided)
              m1 =    132.86
              m2 =    127.44
             sd1 =     15.34
             sd2 =     18.23
   sample size n1 =       100
              n2 =        100
           n2/n1 =      1.00

Estimated power:

           power =    0.6236
```

We did not have to specify n2(#) or ratio(#) since ratio(1) is the default.

◁

One-sample test of mean

▷ Example 2

Suppose that we wish to test the effects of a low-fat diet on serum cholesterol levels. We will measure the difference in cholesterol level for each subject before and after being on the diet. Since there is only one group of subjects, all on the diet, this is a one-sample test, and we must use the onesample option with sampsi.

Our null hypothesis is that the mean of individual differences in cholesterol level will be zero; i.e., $\mu = 0$ mg/100 mL. If the effect of the diet is as large as a mean difference of -10 mg/100 mL, then we wish to have power of 0.95 for rejecting the null hypothesis. Since we expect a reduction in levels, we want to use a one-sided test with $\alpha = 0.025$. From past studies, we estimate that the standard deviation of the difference in cholesterol levels will be about 20 mg/100 mL. To compute the required sample size, we type

```
. sampsi 0 -10, sd(20) onesam a(0.025) onesided p(0.95)

Estimated sample size for one-sample comparison of mean
  to hypothesized value

Test Ho: m =       0, where m is the mean in the population

Assumptions:

           alpha =    0.0250   (one-sided)
           power =    0.9500
   alternative m =        -10
              sd =         20

Estimated required sample size:

               n =         52
```

We decide to conduct the study with $n = 60$ subjects, and we wonder what the power will be at a one-sided significance level of $\alpha = 0.01$:

```
. sampsi 0 -10, sd(20) onesam a(0.01) onesided n(60)

Estimated power for one-sample comparison of mean
  to hypothesized value

Test Ho: m =       0, where m is the mean in the population

Assumptions:

           alpha =    0.0100   (one-sided)
   alternative m =        -10
              sd =         20
   sample size n =         60

Estimated power:

           power =    0.9390
```

◁

Two-sample test of equality of proportions

▷ Example 3

We want to conduct a survey on people's opinions of the president's performance. Specifically, we want to determine whether members of the president's party have a different opinion from people with another party affiliation. Using past surveys as a guide, we estimate that only 25% of members of the president's party will say that the president is doing a poor job, whereas 40% of members of other parties will rate the president's performance as poor. We compute the required sample sizes for $\alpha = 0.05$ (two-sided) and the power of 0.90 by typing

```
. sampsi 0.25 0.4

Estimated sample size for two-sample comparison of proportions

Test Ho: p1 = p2, where p1 is the proportion in population 1
                  and p2 is the proportion in population 2

Assumptions:

           alpha =    0.0500   (two-sided)
           power =    0.9000
              p1 =    0.2500
              p2 =    0.4000
           n2/n1 =    1.00

Estimated required sample sizes:

              n1 =        216
              n2 =        216
```

To compute the power for a survey with a sample of $n_1 = 300$ members of the president's party and a sample of $n_2 = 150$ members of other parties, we type

```
. sampsi 0.25 0.4, n1(300) r(0.5)

Estimated power for two-sample comparison of proportions

Test Ho: p1 = p2, where p1 is the proportion in population 1
                  and p2 is the proportion in population 2
Assumptions:

              alpha =    0.0500  (two-sided)
                 p1 =    0.2500
                 p2 =    0.4000
     sample size n1 =       300
                 n2 =       150
              n2/n1 =      0.50

Estimated power:

              power =    0.8790
```

◁

One-sample test of proportion

▷ Example 4

Someone claims that females are more likely than males to study French. Our null hypothesis is that the proportion of female French students is 0.5. We wish to compute the sample size that will give us 80% power to reject the null hypothesis if the true proportion of female French students is 0.75:

```
. sampsi 0.5 0.75, power(0.8) onesample

Estimated sample size for one-sample comparison of proportion
   to hypothesized value

Test Ho: p = 0.5000, where p is the proportion in the population
Assumptions:

              alpha =    0.0500  (two-sided)
              power =    0.8000
        alternative p =    0.7500

Estimated required sample size:

                  n =        29
```

What is the power if the true proportion of female French students is only 0.6 and the biggest sample of French students we can survey is $n = 200$?

```
. sampsi 0.5 0.6, n(200) onesample

Estimated power for one-sample comparison of proportion
   to hypothesized value

Test Ho: p = 0.5000, where p is the proportion in the population
Assumptions:

              alpha =    0.0500  (two-sided)
        alternative p =    0.6000
        sample size n =       200

Estimated power:

              power =    0.8123
```

◁

❑ Technical Note

r(warning) is saved only for power calculations for one- and two-sample tests on proportions. If sample sizes are not large enough (Tamhane and Dunlop 2000, 300, 307) for sample proportions to be approximately normally distributed, r(warning) is set to 1. Otherwise, a note is displayed in the output, and r(warning) is set to 0.

❑

Clinical trials with repeated measures

In randomized controlled trials (RCTs), when comparing a standard treatment with an experimental therapy, it is not unusual for the study design to allow for repeated measurements of the outcome. Typically, one or more measurements are taken at baseline immediately before randomization, and additional measurements are taken at regular intervals during follow-up. Depending on the analysis method planned and the correlations between measurements at different time points, there can be a great increase in efficiency (variance reduction) from such designs over a simple study with only one measurement of the outcome.

Frison and Pocock (1992) discuss three methods used in RCTs to compare two treatments using a continuous outcome measured at different times on each patient.

Posttreatment means (POST) uses the mean of each patient's follow-up measurements as the summary measurement. It compares the two groups by using a simple t test. This method ignores any baseline measurements.

Mean changes (CHANGE) uses each patient's difference between the mean of the follow-up measurements and the mean of baseline measurements as the summary measurement. It compares the two groups by using a simple t test.

Analysis of covariance (ANCOVA) uses the mean baseline measurement for each patient as a covariate in a linear model for treatment comparisons of follow-up means.

method() specifies which of these three analyses is planned to be used. sampsi will calculate the decrease in variance of the estimate of treatment effect from the number of measurements at baseline, the number of measurements during follow-up, and the correlations between measurements at different times, and use the calculation to estimate power and/or sample size.

▷ Example 5

We are designing a clinical trial comparing a new medication for the treatment of angina to a placebo. We are planning on performing an exercise stress test on each patient four times during the study: once at time of treatment randomization and three more times at 4, 6, and 8 weeks after randomization. From each test, we will measure the time in seconds from the beginning of the test until the patient is unable to continue because of angina pain. From a previous pilot study, we estimated the means (sds) for the new drug and the placebo group to be 498 seconds (20.2) and 485 seconds (19.5), respectively, and an overall correlation at follow-up of 0.7. We will analyze these data by comparing each patient's difference between the mean of posttreatment measurements and the mean of baseline measurements, i.e., the change method. To compute the number of subjects needed for allocation to each treatment group for $\alpha = 0.05$ (two-sided) and power of 90%, we issue the following command:

```
. sampsi 498 485, sd1(20.2) sd2(19.5) method(change) pre(1) post(3) r1(.7)

Estimated sample size for two samples with repeated measures

Assumptions:
        alpha =    0.0500  (two-sided)
                                        power =    0.9000
                                           m1 =       498
                                           m2 =       485
                                          sd1 =      20.2
                                          sd2 =      19.5
                                        n2/n1 =      1.00
              number of follow-up measurements =         3
   correlation between follow-up measurements =     0.700
             number of baseline measurements =         1
   correlation between baseline & follow-up =     0.700

Method: CHANGE
   relative efficiency =     2.500
      adjustment to sd =     0.632
           adjusted sd1 =    12.776
           adjusted sd2 =    12.333

   Estimated required sample sizes:
                    n1 =        20
                    n2 =        20
```

The output from sampsi for repeated measurements includes the specified parameters used to estimate the sample sizes or power, the relative efficiency of the design, and the adjustment to the standard deviation. These last two are the inverse and the square root of the calculated improvement in the variance compared to a similar study where only one measurement is planned.

We see that we need to allocate 20 subjects to each treatment group. Assume that we have funds to enroll only 30 patients into our study. If we randomly assigned 15 patients to each treatment group, what would be the expected power of our study, assuming that all other parameters remain the same?

```
. sampsi 498 485, sd1(20.2) sd2(19.5) meth(change) pre(1) post(3) r1(.7) n1(15) n2(15)

Estimated power for two samples with repeated measures

Assumptions:
                                        alpha =    0.0500  (two-sided)
                                           m1 =       498
                                           m2 =       485
                                          sd1 =      20.2
                                          sd2 =      19.5
                             sample size n1 =        15
                                           n2 =        15
                                        n2/n1 =      1.00
              number of follow-up measurements =         3
   correlation between follow-up measurements =     0.700
             number of baseline measurements =         1
   correlation between baseline & follow-up =     0.700

Method: CHANGE
   relative efficiency =     2.500
      adjustment to sd =     0.632
           adjusted sd1 =    12.776
           adjusted sd2 =    12.333

   Estimated power:
                 power =     0.809
```

If we enroll 30 patients into our study instead of the recommended 40, the power of the study decreases from 90% to approximately 81%.

◁

Saved Results

sampsi saves the following in r():

Scalars

r(N_1)	sample size n_1
r(N_2)	sample size n_2
r(power)	power
r(adj)	adjustment to the SE
r(warning)	0 if assumptions are satisfied and 1 otherwise

Methods and Formulas

sampsi is implemented as an ado-file.

In the following formulas, α is the significance level, $1 - \beta$ is the power, $z_{1-\alpha/2}$ is the $(1 - \alpha/2)$ quantile of the normal distribution, and $r = n_2/n_1$ is the ratio of sample sizes. The formulas below are for two-sided tests. The formulas for one-sided tests can be obtained by replacing $z_{1-\alpha/2}$ with $z_{1-\alpha}$.

1. The required sample sizes for a two-sample test of equality of means (assuming normality) are

$$n_1 = \frac{(\sigma_1^2 + \sigma_2^2/r)(z_{1-\alpha/2} + z_{1-\beta})^2}{(\mu_1 - \mu_2)^2}$$

and $n_2 = rn_1$ (Rosner 2000, 308).

2. For a one-sample test of a mean where the null hypothesis is $\mu = \mu_0$ and the alternative hypothesis is $\mu = \mu_A$, the required sample size (assuming normality) is

$$n = \left\{ \frac{(z_{1-\alpha/2} + z_{1-\beta})\sigma}{\mu_A - \mu_0} \right\}^2$$

(Pagano and Gauvreau 2000, 247–248).

3. The required sample sizes for a two-sample test of equality of proportions (using a normal approximation with a continuity correction) are

$$n_1 = \frac{n'}{4} \left[1 + \left\{ 1 + \frac{2(r+1)}{n'r \, |p_1 - p_2|} \right\}^{1/2} \right]^2$$

$$n_2 = rn_1$$

where

$$n' = \frac{\left[z_{1-\alpha/2}\left\{(r+1)\overline{pq}\right\}^{1/2} + z_{1-\beta}\left(rp_1q_1 + p_2q_2\right)^{1/2} \right]^2}{r(p_1 - p_2)^2}$$

and $\overline{p} = (p_1 + rp_2)/(r+1)$ and $\overline{q} = 1 - \overline{p}$ (Fleiss, Levin, and Paik 2003, 76).

Without a continuity correction, the sample sizes are

$$n_1 = n'$$

$$n_2 = rn_1$$

where n' is defined above.

4. For a one-sample test of proportion where the null hypothesis is $p = p_0$ and the alternative hypothesis is $p = p_A$, the required sample size (using a normal approximation) is

$$n = \left[\frac{z_{1-\alpha/2}\{p_0(1-p_0)\}^{1/2} + z_{1-\beta}\{p_A(1-p_A)\}^{1/2}}{p_A - p_0} \right]^2$$

(Pagano and Gauvreau 2000, 332).

5. For repeated measurements, Frison and Pocock (1992) discuss three methods for use in randomized clinical trials to compare two treatments using a continuous outcome measured at different times on each patient. Each uses the average of baseline measurements, $\bar{x}_0$, and follow-up measurements, $\bar{x}_1$:

POST outcome is $\bar{x}_1$, where the analysis is by simple t test.

CHANGE outcome is $\bar{x}_1 - \bar{x}_0$, where the analysis is by simple t test.

ANCOVA outcome is $\bar{x}_1 - \beta\bar{x}_0$, where the β is estimated by analysis of covariance, correcting for the average at baseline.

ANCOVA will always be the most efficient of the three approaches. β is set so that $\beta\bar{x}_0$ accounts for the largest possible variation of $\bar{x}_1$.

For a study with one measurement each at baseline and follow-up, CHANGE will be more efficient than POST, provided that the correlation between measurements at baseline and measurements at follow-up is more than 0.5. POST ignores all baseline measurements, which tends to make it unpopular. CHANGE is the method most commonly used. With more than one baseline measurement, CHANGE and ANCOVA tend to produce similar sample sizes and power.

The improvements in variance of the estimate of treatment effect over a study with only one measurement depend on the number of measurements p at baseline; the number of measurements during follow-up; and the correlations between measurements at baseline $\bar{\rho}_{\mathrm{pre}}$, between measurements at follow-up $\bar{\rho}_{\mathrm{post}}$, and between measurements at baseline and measurements at follow-up $\bar{\rho}_{\mathrm{mix}}$. The improvements in variance for the POST method are given by

$$\frac{1 + (r-1)\bar{\rho}_{\mathrm{post}}}{r}$$

for the CHANGE method by

$$\frac{1 + (r-1)\bar{\rho}_{\mathrm{post}}}{r} + \frac{1 + (p-1)\bar{\rho}_{\mathrm{pre}}}{p} - 2\bar{\rho}_{\mathrm{mix}}$$

and for the ANCOVA method by

$$\frac{1 + (r-1)\bar{\rho}_{\mathrm{post}}}{r} - \frac{\bar{\rho}_{\mathrm{mix}}^2 p}{1 + (p-1)\bar{\rho}_{\mathrm{pre}}}$$

Often the three correlations are assumed equal. In data from several trials, Frison and Pocock found that $\bar{\rho}_{\mathrm{pre}}$ and $\bar{\rho}_{\mathrm{post}}$ typically had values around 0.7, whereas $\bar{\rho}_{\mathrm{mix}}$ was nearer 0.5. This finding is consistent with the common finding that measurements closer in time are more strongly correlated.

Power calculations are based on estimates of one variance at all time points.

Acknowledgments

`sampsi` is based on the `sampsiz` command written by Joseph Hilbe of Arizona State University (Hilbe 1993). Paul Seed of Maternal and Fetal Research Unit, St. Thomas' Hospital (Seed 1997, 1998), expanded the command to allow for repeated measurements.

References

Fleiss, J. L., B. Levin, and M. C. Paik. 2003. *Statistical Methods for Rates and Proportions.* 3rd ed. New York: Wiley.

Frison, L., and S. Pocock. 1992. Repeated measures in clinical trials: Analysis using mean summary statistics and its implications for design. *Statistics in Medicine* 11: 1685-1704.

Hilbe, J. M. 1993. sg15: Sample size determination for means and proportions. *Stata Technical Bulletin* 11: 17–20. Reprinted in *Stata Technical Bulletin Reprints*, vol. 2, pp. 145–149.

Pagano, M., and K. Gauvreau. 2000. *Principles of Biostatistics.* 2nd ed. Belmont, CA: Duxbury.

Rosner, B. 2000. *Fundamentals of Biostatistics.* 5th ed. Belmont, CA: Duxbury.

Royston, J. P., and A. Babiker. 2002. A menu-drive facility for complex sample size calculation in randomized controlled trials with a survival or a binary outcome. *Stata Journal* 2: 151–163.

Seed, P. T. 1997. sbe18: Sample size calculations for clinical trials with repeated measures data. *Stata Technical Bulletin* 40: 16–18. Reprinted in *Stata Technical Bulletin Reprints*, vol. 7, pp. 121–125.

———. 1998. sbe18.1: Update of sampsi. *Stata Technical Bulletin* 45: 21. Reprinted in *Stata Technical Bulletin Reprints*, vol. 8, p. 84.

Tamhane, A., and D. Dunlop. 2000. *Statistics and Data Analysis: From Elementary to Intermediate.* Upper Saddle River, NJ: Prentice Hall.

Also See

[ST] **stpower** — Sample-size, power, and effect-size determination for survival analysis

Title

saved results — Saved results

Syntax

List results stored in r()

> return list

List results stored in e()

> ereturn list

List results stored in s()

> sreturn list

Description

Results of calculations are saved by many Stata commands so that they can be easily accessed and substituted into later commands.

return list lists results stored in r().

ereturn list lists results stored in e().

sreturn list lists results stored in s().

This entry discusses using saved results. Programmers wishing to save results should see [P] **return** and [P] **ereturn**.

Remarks

Stata commands are classified as being

r-class	general commands that save results in r()
e-class	estimation commands that save results in e()
s-class	parsing commands that save results in s()
n-class	commands that do not save in r(), e(), or s()

There is also a c-class, c(), containing the values of system parameters and settings, along with certain constants, such as the value of pi; see [P] **creturn**. A program, however, cannot be c-class.

You can look at the *Saved Results* section of the manual entry of a command to determine whether it is r-, e-, s-, or n-class, but it is easy enough to guess.

Commands producing statistical results are either r-class or e-class. They are e-class if they present estimation results and r-class otherwise. s-class is a class used by programmers and is primarily used in subprograms performing parsing. n-class commands explicitly state where the result is to go. For instance, generate and replace are n-class because their syntax is generate *varname* = ... and replace *varname* = ...

After executing a command, you can type return list, ereturn list, or sreturn list to see what has been saved:

230

```
. use http://www.stata-press.com/data/r10/auto
(1978 Automobile Data)

. summarize mpg
    Variable |       Obs        Mean   Std. Dev.       Min        Max
-------------+--------------------------------------------------------
         mpg |        74     21.2973    5.785503        12         41

. return list

scalars:
                  r(N) =  74
              r(sum_w) =  74
               r(mean) =  21.2972972972973
                r(Var) =  33.47204738985561
                 r(sd) =  5.785503209735141
                r(min) =  12
                r(max) =  41
                r(sum) =  1576
```

After summarize, you can use r(N), r(mean), r(sd), etc., in expressions:

```
. gen double mpgstd = (mpg-r(mean))/r(sd)

. summarize mpgstd
    Variable |       Obs        Mean   Std. Dev.       Min        Max
-------------+--------------------------------------------------------
      mpgstd |        74   -1.64e-16           1  -1.606999    3.40553
```

Be careful to use results stored in r() soon because they will be replaced the next time you execute another r-class command. For instance, although r(mean) was 21.3 (approximately) after summarize mpg, it is $-1e{-}16$ now because you just ran summarize again.

e-class is really no different from r-class, except for where results are stored and that, when an estimation command stores results, it tends to store a lot of them:

```
. regress mpg weight displ
(output omitted )

. ereturn list

scalars:
                  e(N) =  74
               e(df_m) =  2
               e(df_r) =  71
                  e(F) =  66.78504752026517
                 e(r2) =  .6529306984682528
               e(rmse) =  3.45606176570828
                e(mss) =  1595.409691543724
                e(rss) =  848.0497679157352
               e(r2_a) =  .643154098425105
                 e(ll) =  -195.2397979466294
               e(ll_0) =  -234.3943376482347

macros:
            e(cmdline) : "regress mpg weight displ"
              e(title) : "Linear regression"
                e(vce) : "ols"
             e(depvar) : "mpg"
                e(cmd) : "regress"
         e(properties) : "b V"
            e(predict) : "regres_p"
              e(model) : "ols"
          e(estat_cmd) : "regress_estat"
```

```
matrices:
                    e(b) :  1 x 3
                    e(V) :  3 x 3
functions:
              e(sample)
```

These e-class results will stick around until you run another estimation command. Typing `return list` and `ereturn list` is the easy way to find out what a command stores.

Both r- and e-class results come in four flavors: scalars, macros, matrices, and functions. (s-class results come in only one flavor—macros—and as earlier noted, s class is used solely by programmers, so ignore it.)

Scalars are just that—numbers by any other name. You can subsequently refer to `r(mean)` or `e(rmse)` in numeric expressions and obtain the result to full precision.

Macros are strings. For instance, `e(depvar)` contains "mpg". You can refer to it, too, in subsequent expressions, but really that would be of most use to programmers, who will refer to it using constructs like `"'e(depvar)'"`. In any case, macros are macros, and you obtain their contents just as you would a local macro, by enclosing their name in single quotes. The name here is the full name, so `'e(depvar)'` is mpg.

Matrices are matrices, and all estimation commands store `e(b)` and `e(V)` containing the coefficient vector and variance–covariance matrix of the estimates (VCE).

Functions are saved by e-class commands only, and the only function existing is `e(sample)`. `e(sample)` evaluates to 1 (meaning true) if the observation was used in the previous estimation and to 0 (meaning false) otherwise.

❑ Technical Note

Say that some command set `r(scalar)` and `r(macro)`, the first being stored as a scalar and the second as a macro. In theory, in subsequent use you are supposed to refer to `r(scalar)` and `'r(macro)'`. In fact, however, you can refer to either one with or without quotes, so you could refer to `'r(scalar)'` and `r(macro)`. Programmers sometimes do this.

When you refer to `r(scalar)`, you are referring to the full double-precision saved result. Think of `r(scalar)` without quotes as a function returning the value of the saved result `scalar`. When you refer to `r(scalar)` in quotes, Stata understands `'r(scalar)'` to mean "substitute the printed result of evaluating `r(scalar)`". Pretend that `r(scalar)` equals the number 23. Then `'r(scalar)'` is 23, the character 2 followed by 3.

Referring to `r(scalar)` in quotes is sometimes useful. Say that you want to use the immediate command `ci` with `r(scalar)`. The immediate command `ci` requires its arguments to be numbers— numeric literals in programmer's jargon—and it will not take an expression. Thus you could not type `'ci r(scalar) ...'`. You could, however, type `'ci 'r(scalar)' ...'` because `'r(scalar)'` is just a numeric literal.

For `r(macro)`, you are supposed to refer to it in quotes: `'r(macro)'`. If, however, you omit the quotes in an expression context, Stata evaluates the macro and then pretends that it is the result of function-returning-string. There are side effects of this, the most important being that the result is trimmed to 80 characters.

Referring to `r(macro)` without quotes is never a good idea; the feature was included merely for completeness.

You can even refer to r(matrix) in quotes (assume that r(matrix) is a matrix). 'r(matrix)' does not result in the matrix being substituted; it returns the word matrix. Programmers sometimes find that useful.

❑

Reference

Jann, B. 2005. Making regression tables from stored estimates. *Stata Journal* 5: 288–308.

Also See

[P] **ereturn** — Post the estimation results

[P] **return** — Return saved results

[U] **18.8 Accessing results calculated by other programs**

[U] **18.9 Accessing results calculated by estimation commands**

Title

scobit — Skewed logistic regression

Syntax

scobit *depvar* [*indepvars*] [*if*] [*in*] [*weight*] [, *options*]

options	description
Model	
<u>nocon</u>stant	suppress constant term
<u>off</u>set(*varname*)	include *varname* in model with coefficient constrained to 1
asis	retain perfect predictor variables
<u>constr</u>aints(*constraints*)	apply specified linear constraints
SE/Robust	
vce(*vcetype*)	*vcetype* may be oim, <u>r</u>obust, <u>c</u>luster *clustvar*, opg, <u>boot</u>strap, or <u>jack</u>knife
Reporting	
<u>level</u>(#)	set confidence level; default is level(95)
or	report odds ratios
Max options	
maximize_options	control the maximization process

bootstrap, by, jackknife, nestreg, rolling, statsby, stepwise, svy, and xi are allowed; see
   [U] **11.1.10 Prefix commands**.
Weights are not allowed with the bootstrap prefix.
vce() and weights are not allowed with the svy prefix.
fweights, iweights, and pweights are allowed; see [U] **11.1.6 weight**.
See [U] **20 Estimation and postestimation commands** for more capabilities of estimation commands.

Description

scobit fits a maximum-likelihood skewed logit model.

See [R] **logistic** for a list of related estimation commands.

Options

_____| Model |_____

noconstant, offset(*varname*), constraints(*constraints*); see [R] **estimation options**.

asis forces retention of perfect predictor variables and their associated perfectly predicted observations
   and may produce instabilities in maximization; see [R] **probit**.

_____| SE/Robust |_____

vce(*vcetype*) specifies the type of standard error reported, which includes types that are derived
   from asymptotic theory, that are robust to some kinds of misspecification, that allow for intragroup
   correlation, and that use bootstrap or jackknife methods; see [R] *vce_option*.

Reporting

level(#); see [R] **estimation options**.

or reports the estimated coefficients transformed to odds ratios, i.e., e^b rather than b. Standard errors and confidence intervals are similarly transformed. This option affects how results are displayed, not how they are estimated. or may be specified at estimation or when replaying previously estimated results.

Max options

maximize_options: <u>diff</u>icult, <u>tech</u>nique(*algorithm_spec*), <u>iter</u>ate(#), [<u>no</u>]log, <u>tra</u>ce, <u>grad</u>ient, showstep, <u>hess</u>ian, <u>shownr</u>tolerance, <u>tol</u>erance(#), <u>ltol</u>erance(#), <u>gtol</u>erance(#), <u>nrtol</u>erance(#), <u>nonrtol</u>erance, from(*init_specs*); see [R] **maximize**.

Setting the optimization type to technique(bhhh) resets the default *vcetype* to vce(opg).

Remarks

Remarks are presented under the following headings:

> *Skewed logistic model*
> *Robust standard errors*

Skewed logistic model

scobit fits maximum likelihood models with dichotomous dependent variables coded as 0/1 (or, more precisely, coded as 0 and not 0).

▷ Example 1

We have data on the make, weight, and mileage rating of 22 foreign and 52 domestic automobiles. We wish to fit a model explaining whether a car is foreign based on its mileage. Here is an overview of our data:

```
. use http://www.stata-press.com/data/r10/auto
(1978 Automobile Data)

. keep make mpg weight foreign

. describe

Contains data from http://www.stata-press.com/data/r10/auto.dta
  obs:            74                          1978 Automobile Data
  vars:            4                          13 Apr 2007 17:45
  size:         1,998 (99.9% of memory free)  (_dta has notes)
```

variable name	storage type	display format	value label	variable label
make	str18	%-18s		Make and Model
mpg	int	%8.0g		Mileage (mpg)
weight	int	%8.0gc		Weight (lbs.)
foreign	byte	%8.0g	origin	Car type

```
Sorted by:  foreign
     Note:  dataset has changed since last saved
```

```
. inspect foreign
foreign:  Car type                          Number of Observations

                                        Total   Integers   Nonintegers
    |                                       -          -             -
    #                         Negative
    #                         Zero          52         52            -
    #                         Positive      22         22            -
    #
    #     #                   Total         74         74            -
    #     #                   Missing        -
    |
    +-----------------------------                     ----------
    0                     1                   74
    (2 unique values)

        foreign is labeled and all values are documented in the label.
```

The variable `foreign` takes on two unique values, 0 and 1. The value 0 denotes a domestic car, and 1 denotes a foreign car.

The model that we wish to fit is

$$\Pr(\texttt{foreign} = 1) = F(\beta_0 + \beta_1 \texttt{mpg})$$

where $F(z) = 1 - 1/\{1 + \exp(z)\}^{\alpha}$.

To fit this model, we type

```
. scobit foreign mpg

Fitting logistic model:

Iteration 0:   log likelihood =  -45.03321
Iteration 1:   log likelihood = -39.380959
Iteration 2:   log likelihood = -39.288802
Iteration 3:   log likelihood =  -39.28864

Fitting full model:

Iteration 0:   log likelihood =  -39.28864
Iteration 1:   log likelihood = -39.286393
Iteration 2:   log likelihood = -39.284415
Iteration 3:   log likelihood = -39.284234
Iteration 4:   log likelihood = -39.284197
Iteration 5:   log likelihood = -39.284196
```

```
Skewed logistic regression                  Number of obs    =        74
                                            Zero outcomes    =        52
Log likelihood =  -39.2842                  Nonzero outcomes =        22
```

foreign	Coef.	Std. Err.	z	P>|z|	[95% Conf. Interval]	
mpg	.1813879	.2407362	0.75	0.451	-.2904463	.6532222
_cons	-4.274883	1.399305	-3.06	0.002	-7.017471	-1.532295
/lnalpha	-.4450405	3.879885	-0.11	0.909	-8.049476	7.159395
alpha	.6407983	2.486224			.0003193	1286.133

```
Likelihood-ratio test of alpha=1:    chi2(1) =    0.01     Prob > chi2 = 0.9249
note: likelihood-ratio tests are recommended for inference with scobit models.
```

We find that cars yielding better gas mileage are less likely to be foreign. The likelihood-ratio test at the bottom of the output indicates that the model is not significantly different from a logit model. Therefore, we should use the more parsimonious model.

◁

❑ Technical Note

Stata interprets a value of 0 as a negative outcome (failure) and treats all other values (except missing) as positive outcomes (successes). Thus if the dependent variable takes on the values 0 and 1, 0 is interpreted as failure and 1 as success. If the dependent variable takes on the values 0, 1, and 2, 0 is still interpreted as failure, but both 1 and 2 are treated as successes.

Formally, when we type scobit y x, Stata fits the model

$$\Pr(y_j \neq 0 \mid \mathbf{x}_j) = 1 - 1 \Big/ \Big\{ 1 + \exp(\mathbf{x}_j \boldsymbol{\beta}) \Big\}^{\alpha}$$

❑

Robust standard errors

If you specify the vce(robust) option, scobit reports robust standard errors as described in [U] **20.15 Obtaining robust variance estimates**. For the model of foreign on mpg, the robust calculation increases the standard error of the coefficient on mpg by around 25%:

```
. scobit foreign mpg, vce(robust) nolog
Skewed logistic regression                        Number of obs   =       74
                                                  Zero outcomes   =       52
Log pseudolikelihood =  -39.2842                  Nonzero outcomes =      22
```

foreign	Coef.	Robust Std. Err.	z	P>|z|	[95% Conf. Interval]	
mpg	.1813879	.3028487	0.60	0.549	-.4121847	.7749606
_cons	-4.274883	1.335521	-3.20	0.001	-6.892455	-1.657311
/lnalpha	-.4450405	4.71561	-0.09	0.925	-9.687466	8.797385
alpha	.6407983	3.021755			.0000621	6616.919

```
Likelihood-ratio test of alpha=1:    chi2(1) =     0.01    Prob > chi2 = 0.9249
note: likelihood-ratio tests are recommended for inference with scobit models.
```

Without vce(robust), the standard error for the coefficient on mpg was reported to be .241, with a resulting confidence interval of $[-.29, .65]$.

Specifying the vce(cluster *clustvar*) option relaxes the independence assumption required by the skewed logit estimator to being just independence between clusters. To demonstrate this, we will switch to a different dataset.

▷ Example 2

We are studying the unionization of women in the United States and have a dataset with 26,200 observations on 4,434 women between 1970 and 1988. For our purposes, we will use the variables age (the women were 14–26 in 1968 and the data thus span the age range of 16–46), grade (years of schooling completed, ranging from 0 to 18), not_smsa (28% of the person-time was spent living outside an SMSA—standard metropolitan statistical area), south (41% of the person-time was in the South), and southXt (south interacted with year, treating 1970 as year 0). We also have variable union. Overall, 22% of the person-time is marked as time under union membership and 44% of these women have belonged to a union.

We fit the following model, ignoring that women are observed an average of 5.9 times each in these data:

```
. use http://www.stata-press.com/data/r10/union, clear
(NLS Women 14-24 in 1968)

. scobit union age grade not_smsa south southXt, nolog nonrtol
```

Skewed logistic regression
```
                                        Number of obs    =      26200
                                        Zero outcomes    =      20389
Log likelihood = -13544.2               Nonzero outcomes =       5811
```

union	Coef.	Std. Err.	z	P>|z|	[95% Conf. Interval]	
age	.0085889	.0023036	3.73	0.000	.004074	.0131039
grade	.0447166	.0057073	7.83	0.000	.0335304	.0559027
not_smsa	-.1906374	.0317694	-6.00	0.000	-.2529042	-.1283707
south	-.6446248	.0557704	-11.56	0.000	-.7539328	-.5353169
southXt	.0068271	.0047299	1.44	0.149	-.0024433	.0160976
_cons	-10.82928	63.79149	-0.17	0.865	-135.8583	114.1997
/lnalpha	8.862483	63.79072	0.14	0.890	-116.165	133.89
alpha	7061.993	450489.6			3.55e-51	1.41e+58

```
Likelihood-ratio test of alpha=1:    chi2(1) =      3.07    Prob > chi2 = 0.0799
note: likelihood-ratio tests are recommended for inference with scobit models.
```

The reported standard errors in this model are probably meaningless. Women are observed repeatedly, so the observations are not independent. Looking at the coefficients, we find a large southern effect against unionization and little time trend. The vce(cluster *clustvar*) option provides a way to fit this model and obtains correct standard errors:

```
. scobit union age grade not_smsa south southXt, vce(cluster id) nolog nonrtol
```

Skewed logistic regression
```
                                        Number of obs    =      26200
                                        Zero outcomes    =      20389
Log pseudolikelihood = -13544.2         Nonzero outcomes =       5811
```
```
                        (Std. Err. adjusted for 4434 clusters in idcode)
```

union	Coef.	Robust Std. Err.	z	P>|z|	[95% Conf. Interval]	
age	.0085889	.0033835	2.54	0.011	.0019575	.0152204
grade	.0447166	.0125938	3.55	0.000	.0200332	.0693999
not_smsa	-.1906374	.0641961	-2.97	0.003	-.3164594	-.0648155
south	-.6446248	.0833872	-7.73	0.000	-.8080608	-.4811889
southXt	.0068271	.0063044	1.08	0.279	-.0055292	.0191834
_cons	-10.82928	.9169089	-11.81	0.000	-12.62639	-9.032174
/lnalpha	8.862483	.7787163	11.38	0.000	7.336227	10.38874
alpha	7061.993	5499.289			1534.909	32491.65

```
Likelihood-ratio test of alpha=1:    chi2(1) =      3.07    Prob > chi2 = 0.0799
note: likelihood-ratio tests are recommended for inference with scobit models.
```

scobit, vce(cluster *clustvar*) is robust to assumptions about within-cluster correlation. That is, it inefficiently sums within cluster for the standard error calculation rather than attempting to exploit what might be assumed about the within-cluster correlation (as do the xtgee population-averaged models; see [XT] **xtgee**).

◁

❑ Technical Note

The scobit model can be difficult to fit because of the functional form. Often it requires many iterations, or the optimizer prints out warning and informative messages during the optimization. For example, without the `nrtolerance` option, the model using the `union` dataset will not converge. See [R] **maximize** for details about the optimizer.

❑

❑ Technical Note

The main reason for using `scobit` rather that `logit` is that the effects of the regressors on the probability of success are not constrained to be the largest when the probability is 0.5. Rather, the independent variables might show their largest impact when the probability of success is 0.3 or 0.6. This added flexibility results because the scobit function, unlike the logit function, can be skewed and is not constrained to be mirror symmetric about the 0.5 probability of success.

As Nagler (1994) pointed out, the point of maximum impact is constrained under the scobit model to fall within the interval $(0, 1 - e^{(-1)})$ or approximately (0, 0.63). Achen (2002) notes that if we believe the maximum impact to be outside that range, we can instead estimate the "power logit" model by simply reversing the 0s and 1s of our outcome variable and estimating a scobit model on failure, rather than success. We would need to reverse the signs of the coefficients if we wanted to interpret them in terms of impact on success, or we could leave them as they are and interpret them in terms of impact on failure. The important thing to remember is that the scobit model, unlike the logit model, is not invariant to the choice of which result is assigned to success.

❑

(Continued on next page)

Saved Results

scobit saves the following in e():

Scalars

e(N)	number of observations	e(N_s)	number of successes (nonzero
e(k)	number of parameters		outcomes)
e(k_eq)	number of equations	e(alpha)	alpha
e(k_aux)	number of auxiliary parameters	e(N_clust)	number of clusters
e(k_dv)	number of dependent variables	e(chi2)	χ^2
e(df_m)	model degrees of freedom	e(chi2_c)	χ^2 for comparison test
e(ll)	log likelihood	e(p)	significance
e(ll_0)	log likelihood, constant-only model	e(rank)	rank of e(V)
e(ll_c)	log likelihood, comparison model	e(ic)	number of iterations
e(N_f)	number of failures (zero outcomes)	e(rc)	return code
		e(converged)	1 if converged, 0 otherwise

Macros

e(cmd)	scobit	e(vcetype)	title used to label Std. Err.
e(cmdline)	command as typed	e(diparm#)	display transformed parameter #
e(depvar)	name of dependent variable	e(opt)	type of optimization
e(wtype)	weight type	e(ml_method)	type of ml method
e(wexp)	weight expression	e(user)	name of likelihood-evaluator program
e(title)	title in estimation output	e(crittype)	optimization criterion
e(clustvar)	name of cluster variable	e(technique)	maximization technique
e(offset)	offset	e(properties)	b V
e(chi2type)	Wald or LR; type of model χ^2 test	e(footnote)	program used to implement the
e(chi2_ct)	Wald or LR; type of model χ^2 test		footnote display
	corresponding to e(chi2_c)	e(predict)	program used to implement predict
e(vce)	*vcetype* specified in vce()		

Matrices

e(b)	coefficient vector	e(V)	variance–covariance matrix of the
e(ilog)	iteration log (up to 20 iterations)		estimators
e(gradient)	gradient vector		

Functions

e(sample)	marks estimation sample

Methods and Formulas

scobit is implemented as an ado-file.

Skewed logit analysis is an alternative to logit that relaxes the assumption that individuals with initial probability of .5 are most sensitive to changes in independent variables.

The log-likelihood function for skewed logit is

$$\ln L = \sum_{j \in S} w_j \ln F(\mathbf{x}_j \mathbf{b}) + \sum_{j \notin S} w_j \ln\left\{1 - F(\mathbf{x}_j \mathbf{b})\right\}$$

where S is the set of all observations j such that $y_j \neq 0$, $F(z) = 1 - 1/\left\{1 + \exp(z)\right\}^{\alpha}$, and w_j denotes the optional weights. $\ln L$ is maximized as described in [R] **maximize**.

If robust standard errors are requested, the calculation described in *Methods and Formulas* of [R] **regress** is carried forward with

$$\mathbf{u}_j^1 = \mathbf{x}_j \frac{\alpha \exp(\mathbf{x}_j \mathbf{b})}{\{1 + \exp(\mathbf{x}_j \mathbf{b})\}[\{1 + \exp(\mathbf{x}_j \mathbf{b})\}^\alpha - 1]}$$

$$\mathbf{u}_j^2 = \frac{\alpha \ln\{1 + \exp(\mathbf{x}_j \mathbf{b})\}}{\{1 + \exp(\mathbf{x}_j \mathbf{b})\}[\{1 + \exp(\mathbf{x}_j \mathbf{b})\}^\alpha - 1]}$$

for the positive outcomes and

$$\mathbf{u}_j^1 = -\mathbf{x}_j \frac{\alpha \exp(\mathbf{x}_j \mathbf{b})}{1 + \exp(\mathbf{x}_j \mathbf{b})}$$

$$\mathbf{u}_j^2 = \alpha \ln\{1 + \exp(\mathbf{x}_j \mathbf{b})\}$$

for the negative outcomes.

References

Achen, C. H. 2002. Toward a new political methodology: Microfoundations and ART. *Annual Review of Political Science* 5: 423–450.

Nagler, J. 1994. Scobit: An alternative estimator to logit and probit. *American Journal of Political Science* 38: 230–255.

Also See

[R] **scobit postestimation** — Postestimation tools for scobit

[R] **cloglog** — Complementary log-log regression

[R] **glm** — Generalized linear models

[R] **logistic** — Logistic regression, reporting odds ratios

[U] **20 Estimation and postestimation commands**

Title

scobit postestimation — Postestimation tools for scobit

Description

The following postestimation commands are available for `scobit`:

command	description
adjust	adjusted predictions of $x\beta$
estat	AIC, BIC, VCE, and estimation sample summary
estat (svy)	postestimation statistics for survey data
estimates	cataloging estimation results
lincom	point estimates, standard errors, testing, and inference for linear combinations of coefficients
lrtest[1]	likelihood-ratio test
mfx	marginal effects or elasticities
nlcom	point estimates, standard errors, testing, and inference for nonlinear combinations of coefficients
predict	predictions, residuals, influence statistics, and other diagnostic measures
predictnl	point estimates, standard errors, testing, and inference for generalized predictions
suest	seemingly unrelated estimation
test	Wald tests for simple and composite linear hypotheses
testnl	Wald tests of nonlinear hypotheses

[1] `lrtest` is not appropriate with `svy` estimation results.

See the corresponding entries in the *Stata Base Reference Manual* for details, but see
   [SVY] **estat** for details about `estat` (svy).

Syntax for predict

predict [*type*] *newvar* [*if*] [*in*] [, *statistic* <u>nooff</u>set]

predict [*type*] { *stub** | *newvar*_{reg} *newvar*_{lnalpha} } [*if*] [*in*] , <u>sc</u>ores

statistic	description
<u>pr</u>	probability of a positive outcome; the default
xb	$x_j b$, linear prediction
stdp	standard error of the linear prediction

These statistics are available both in and out of sample; type `predict ... if e(sample) ...` if wanted only for
   the estimation sample.

Options for predict

pr, the default, calculates the probability of a positive outcome.

xb calculates the linear prediction.

stdp calculates the standard error of the linear prediction.

nooffset is relevant only if you specified offset(*varname*) for scobit. It modifies the calculations made by predict so that they ignore the offset variable; the linear prediction is treated as $x_j b$ rather than as $x_j b + \text{offset}_j$.

scores calculates equation-level score variables.

The first new variable will contain $\partial \ln L / \partial (x_j \beta)$.

The second new variable will contain $\partial \ln L / \partial \ln \alpha$.

Remarks

Once you have fitted a model, you can obtain the predicted probabilities by using the predict command for both the estimation sample and other samples; see [U] **20 Estimation and postestimation commands** and [R] **predict**. Here we will make only a few additional comments.

predict without arguments calculates the predicted probability of a positive outcome. With the xb option, it calculates the linear combination $x_j b$, where x_j are the independent variables in the jth observation and b is the estimated parameter vector.

With the stdp option, predict calculates the standard error of the prediction, which is *not* adjusted for replicated covariate patterns in the data.

▷ Example 1

In example 1 of [R] **scobit**, we fitted the model scobit foreign mpg. To obtain predicted probabilities, we type

```
. use http://www.stata-press.com/data/r10/auto
(1978 Automobile Data)

. keep make mpg weight foreign

. scobit foreign mpg
 (output omitted )
. predict p
(option pr assumed; Pr(foreign))

. summarize foreign p
```

Variable	Obs	Mean	Std. Dev.	Min	Max
foreign	74	.2972973	.4601885	0	1
p	74	.2974049	.182352	.0714664	.871624

◁

Methods and Formulas

All postestimation commands listed above are implemented as ado-files.

Also See

[R] **scobit** — Skewed logistic regression

[U] **20 Estimation and postestimation commands**

Title

sdtest — Variance-comparison tests

Syntax

One-sample variance-comparison test

sdtest *varname* == # $\left[\,if\,\right]$ $\left[\,in\,\right]$ $\left[\,,\,\underline{l}evel(\#)\,\right]$

Two-sample variance-comparison test

sdtest *varname*$_1$ == *varname*$_2$ $\left[\,if\,\right]$ $\left[\,in\,\right]$ $\left[\,,\,\underline{l}evel(\#)\,\right]$

Two-group variance-comparison test

sdtest *varname* $\left[\,if\,\right]$ $\left[\,in\,\right]$, by(*groupvar*) $\left[\,\underline{l}evel(\#)\,\right]$

Immediate form of one-sample variance-comparison test

sdtesti #$_{obs}$ $\left\{\,\#_{mean}\,|\,.\,\right\}$ #$_{sd}$ #$_{val}$ $\left[\,,\,\underline{l}evel(\#)\,\right]$

Immediate form of two-sample variance-comparison test

sdtesti #$_{obs,1}$ $\left\{\,\#_{mean,1}\,|\,.\,\right\}$ #$_{sd,1}$ #$_{obs,2}$ $\left\{\,\#_{mean,2}\,|\,.\,\right\}$ #$_{sd,2}$ $\left[\,,\,\underline{l}evel(\#)\,\right]$

Robust tests for equality of variance

robvar *varname* $\left[\,if\,\right]$ $\left[\,in\,\right]$, by(*groupvar*)

by is allowed with sdtest and robvar; see [D] **by**.

Description

sdtest performs tests on the equality of standard deviations (variances). In the first form, sdtest tests that the standard deviation of *varname* is #. In the second form, sdtest tests that *varname*$_1$ and *varname*$_2$ have the same standard deviation. In the third form, sdtest performs the same test, using the standard deviations of the two groups defined by *groupvar*.

sdtesti is the immediate form of sdtest; see [U] **19 Immediate commands**.

Both the traditional F test for the homogeneity of variances and Bartlett's generalization of this test to K samples are sensitive to the assumption that the data are drawn from an underlying Gaussian distribution. See, for example, the cautionary results discussed by Markowski and Markowski (1990). Levene (1960) proposed a test statistic for equality of variance that was found to be robust under nonnormality. Then Brown and Forsythe (1974) proposed alternative formulations of Levene's test statistic that use more robust estimators of central tendency in place of the mean. These reformulations were demonstrated to be more robust than Levene's test when dealing with skewed populations.

robvar reports Levene's robust test statistic (W_0) for the equality of variances between the groups defined by *groupvar* and the two statistics proposed by Brown and Forsythe that replace the mean in Levene's formula with alternative location estimators. The first alternative (W_{50}) replaces the mean with the median. The second alternative replaces the mean with the 10% trimmed mean (W_{10}).

Options

level(*#*) specifies the confidence level, as a percentage, for confidence intervals of the means. The default is level(95) or as set by set level; see [U] **20.7 Specifying the width of confidence intervals**.

by(*groupvar*) specifies the *groupvar* that defines the groups to be compared. For sdtest, there should be two groups, but for robvar there may be more than two groups. Do not confuse the by() option with the by prefix; both may be specified.

Remarks

Remarks are presented under the following headings:

> *Basic form*
> *Immediate form*
> *Robust test*

Basic form

sdtest performs two different statistical tests: one testing equality of variances and the other testing that the standard deviation is equal to a known constant. Which test it performs is determined by whether you type a variable name or a number to the right of the equal sign.

▷ Example 1: One-sample test of variance

We have a sample of 74 automobiles. For each automobile, we know the mileage rating. We wish to test whether the overall standard deviation is 5 mpg:

```
. use http://www.stata-press.com/data/r10/auto
(1978 Automobile Data)

. sdtest mpg == 5

One-sample test of variance
```

Variable	Obs	Mean	Std. Err.	Std. Dev.	[95% Conf. Interval]	
mpg	74	21.2973	.6725511	5.785503	19.9569	22.63769

```
      sd = sd(mpg)                                        c = chi2 =  97.7384
Ho: sd = 5                                     degrees of freedom =       73

      Ha: sd < 5              Ha: sd != 5                      Ha: sd > 5
   Pr(C < c) = 0.9717      2*Pr(C > c) = 0.0565           Pr(C > c) = 0.0283
```

◁

▷ Example 2: Variance ratio test

We are testing the effectiveness of a new fuel additive. We run an experiment on 12 cars, running each without and with the additive. The data can be found in [R] **ttest**. The results for each car are stored in the variables mpg1 and mpg2:

```
. use http://www.stata-press.com/data/r10/fuel
. sdtest mpg1==mpg2
Variance ratio test
```

Variable	Obs	Mean	Std. Err.	Std. Dev.	[95% Conf. Interval]
mpg1	12	21	.7881701	2.730301	19.26525 22.73475
mpg2	12	22.75	.9384465	3.250874	20.68449 24.81551
combined	24	21.875	.6264476	3.068954	20.57909 23.17091

```
        ratio = sd(mpg1) / sd(mpg2)                              f =   0.7054
Ho: ratio = 1                              degrees of freedom =    11, 11

    Ha: ratio < 1              Ha: ratio != 1              Ha: ratio > 1
 Pr(F < f) = 0.2862       2*Pr(F < f) = 0.5725         Pr(F > f) = 0.7138
```

We cannot reject the hypothesis that the standard deviations are the same.

In [R] **ttest**, we draw an important distinction between paired and unpaired data, which, in this example, means whether there are 12 cars in a before-and-after experiment or 24 different cars. For sdtest, on the other hand, there is no distinction. If the data had been unpaired and stored as described in [R] **ttest**, we could have typed sdtest mpg, by(treated), and the results would have been the same.

◁

Immediate form

▷ Example 3: sdtesti

Immediate commands are used not with data, but with reported summary statistics. For instance, to test whether a variable on which we have 75 observations and a reported standard deviation of 6.5 comes from a population with underlying standard deviation 6, we would type

```
. sdtesti 75 . 6.5 6
One-sample test of variance
```

	Obs	Mean	Std. Err.	Std. Dev.	[95% Conf. Interval]
x	75	.	.7505553	6.5	. .

```
       sd = sd(x)                                  c = chi2 =  86.8472
Ho: sd = 6                              degrees of freedom =        74

    Ha: sd < 6                Ha: sd != 6                 Ha: sd > 6
 Pr(C < c) = 0.8542       2*Pr(C > c) = 0.2916        Pr(C > c) = 0.1458
```

The mean plays no role in the calculation, so it may be omitted.

To test whether the variable comes from a population with the same standard deviation as another for which we have a calculated standard deviation of 7.5 over 65 observations, we would type

```
. sdtesti 75 . 6.5  65 . 7.5
```

Variance ratio test

	Obs	Mean	Std. Err.	Std. Dev.	[95% Conf. Interval]	
x	75	.	.7505553	6.5	.	.
y	65	.	.9302605	7.5	.	.
combined	140	.	.	.	.	.

ratio = sd(x) / sd(y)		f = 0.7511
Ho: ratio = 1		degrees of freedom = 74, 64
Ha: ratio < 1	Ha: ratio != 1	Ha: ratio > 1
Pr(F < f) = 0.1172	2*Pr(F < f) = 0.2344	Pr(F > f) = 0.8828

◁

Robust test

▷ Example 4: robvar

We wish to test whether the standard deviation of the length of stay for patients hospitalized for a given medical procedure differs by gender. Our data consist of observations on the length of hospital stay for 1778 patients: 884 males and 894 females. Length of stay, lengthstay, is highly skewed (skewness coefficient = 4.912591) and thus violates Bartlett's normality assumption. Therefore, we use robvar to compare the variances.

```
. use http://www.stata-press.com/data/r10/stay
. robvar lengthstay, by(sex)
```

	Summary of Length of stay in days		
sex	Mean	Std. Dev.	Freq.
male	9.0874434	9.7884747	884
female	8.800671	9.1081478	894
Total	8.9432508	9.4509466	1778

```
W0  = .55505315   df(1, 1776)    Pr > F = .45635888
W50 = .42714734   df(1, 1776)    Pr > F = .51347664
W10 = .44577674   df(1, 1776)    Pr > F = .50443411
```

For these data, we cannot reject the null hypothesis that the variances are equal. However, Bartlett's test yields a significance probability of 0.0319 because of the pronounced skewness of the data.

◁

❑ Technical Note

robvar implements both the conventional Levene's test centered at the mean and a median-centered test. In a simulation study, Conover, Johnson, and Johnson (1981) compare the properties of the two tests and recommend using the median test for asymmetric data, although for small sample sizes the test is somewhat conservative. See Carroll and Schneider (1985) for an explanation of why both mean- and median-centered tests have approximately the same level for symmetric distributions, but for asymmetric distributions the median test is closer to the correct level.

❑

Saved Results

sdtest and sdtesti save the following in r():

Scalars

r(N)	number of observations	r(sd_1)	standard deviation for first variable
r(p_l)	lower one-sided p-value	r(sd_2)	standard deviation for second variable
r(p_u)	upper one-sided p-value	r(df)	degrees of freedom
r(p)	two-sided p-value	r(df_1)	numerator degrees of freedom
r(F)	F statistic	r(df_2)	denominator degrees of freedom
r(sd)	standard deviation	r(chi2)	χ^2

robvar saves the following in r():

Scalars

r(N)	number of observations	r(w10)	Brown and Forsythe's F statistic
r(w50)	Brown and Forsythe's F statistic (median)		(trimmed mean)
r(p_w50)	Brown and Forsythe's p-value	r(p_w10)	Brown and Forsythe's p-value
r(w0)	Levene's F statistic		(trimmed mean)
r(p_w0)	Levene's p-value	r(df_1)	numerator degrees of freedom
		r(df_2)	denominator degrees of freedom

Methods and Formulas

sdtest, sdtesti, and robvar are implemented as ado-files.

See Hoel (1984; 279–284, 298–300) or Pagano and Gauvreau (2000, chap. 11) for an introduction and explanation of the calculation of these tests.

The test for $\sigma = \sigma_0$ is given by

$$\chi^2 = \frac{(n-1)s^2}{\sigma_0^2}$$

which is distributed as χ^2 with $n-1$ degrees of freedom.

The test for $\sigma_x^2 = \sigma_y^2$ is given by

$$F = \frac{s_x^2}{s_y^2}$$

which is distributed as F with $n_x - 1$ and $n_y - 1$ degrees of freedom.

robvar is also implemented as an ado-file.

Let X_{ij} be the jth observation of X for the ith group. Let $Z_{ij} = |X_{ij} - \overline{X}_i|$, where $\overline{X}_i$ is the mean of X in the ith group. Levene's test statistic is

$$W_0 = \frac{\sum_i n_i (\overline{Z}_i - \overline{Z})^2 / (g-1)}{\sum_i \sum_j (Z_{ij} - \overline{Z}_i)^2 / \sum_i (n_i - 1)}$$

where n_i is the number of observations in group i and g is the number of groups. W_{50} is obtained by replacing $\overline{X}_i$ with the ith group median of X_{ij}, whereas W_{10} is obtained by replacing $\overline{X}_i$ with the 10% trimmed mean for group i.

References

Brown, M. B., and A. B. Forsythe. 1974. Robust test for the equality of variances. *Journal of the American Statistical Association* 69: 364–367.

Carroll, R. J., and H. Schneider. 1985. A note on Levene's test for equality of variances. *Statistics & Probability Letters* 3: 191–194.

Cleves, M. A. 1995. sg35: Robust tests for the equality of variances. *Stata Technical Bulletin* 25: 13–15. Reprinted in *Stata Technical Bulletin Reprints*, vol. 5, pp. 91–93.

——. 2000. sg35.2: Robust tests for the equality of variances update to Stata 6. *Stata Technical Bulletin* 53: 17–18. Reprinted in *Stata Technical Bulletin Reprints*, vol. 9, pp. 158–159.

Conover, W. J., M. E. Johnson, and M. M. Johnson. 1981. A comparative study of tests for homogeneity of variances, with applications to the outer continental shelf bidding data. *Technometrics* 23: 351–361.

Hoel, P. G. 1984. *Introduction to Mathematical Statistics.* 5th ed. New York: Wiley.

Levene, H. 1960. Robust tests for equality of variances. In *Contributions to Probability and Statistics: Essays in Honor of Harold Hotelling.* ed. I. Olkin, S. G. Ghurye, W. Hoeffding, W. G. Madow, and H. B. Mann, 278–292. Menlo Park, CA: Stanford University Press.

Markowski, C. A., and E. P. Markowski. 1990. Conditions for the effectiveness of a preliminary test of variance. *American Statistician* 44: 322–326.

Pagano, M., and K. Gauvreau. 2000. *Principles of Biostatistics.* 2nd ed. Belmont, CA: Duxbury.

Seed, P. 2000. sbe33: Comparing several methods of measuring the same quantity. *Stata Technical Bulletin* 55: 2–9. Reprinted in *Stata Technical Bulletin Reprints*, vol. 10, pp. 73–82.

Tobias, A. 1998. gr28: A graphical procedure to test equality of variances. *Stata Technical Bulletin* 42: 4–6. Reprinted in *Stata Technical Bulletin Reprints*, vol. 7, pp. 68–70.

Also See

[R] **ttest** — Mean-comparison tests

Title

┌───┐
│ **search** — Search Stata documentation   │
└───┘

Syntax

search *word* [*word* ...] [, *search_options*]

<u>se</u>t searchdefault { local | net | all } [, <u>perma</u>nently]

findit *word* [*word* ...]

search_options	description
local	search using Stata's keyword database; the default
net	search across materials available via Stata's net command
all	search across both the local keyword database and the net material
<u>au</u>thor	search by author's name
<u>en</u>try	search by entry ID
<u>ex</u>act	search across both the local keyword database and the net materials; prevents matching on abbreviations
faq	search the FAQs posted to the Stata web site
<u>h</u>istorical	search entries that are of historical interest only
or	list an entry if *any* of the words typed after search are associated with the entry
<u>man</u>ual	search the entries in the *Stata Documentation*
sj	search the entries in the *Stata Journal* and the STB

Description

search searches a keyword database and the Internet.

Capitalization of the words following search is irrelevant, as is the inclusion or exclusion of special characters such as commas, hyphens, and the like.

set searchdefault affects the default behavior of the search command. local is the default.

findit is equivalent to search *word* [*word* ...], all. findit results are displayed in the Viewer. findit is the best way to search for information on a topic across all sources, including the online help, the FAQs at the Stata web site, the *Stata Journal*, and all Stata-related Internet sources including user-written additions. From findit, you can click to go to a source or to install additions.

See [R] **hsearch** for a command that searches help files.

Options for search

local, the default (unless changed by set searchdefault), specifies that the search be performed using only Stata's keyword database.

net specifies that the search be performed across the materials available via Stata's net command. Using search *word* [*word* ...], net is equivalent to typing net search *word* [*word* ...] (without options); see [R] **net search**.

251

all specifies that the search be performed across both the local keyword database and the `net` materials.

author specifies that the search be performed on the basis of author's name rather than keywords. A search with the `author` option is performed on the local keyword database only.

entry specifies that the search be performed on the basis of entry IDs rather than keywords. A search with the `entry` option is performed on the local keyword database only.

exact prevents matching on abbreviations. A search with the `exact` option is performed across both the local keyword database and the `net` materials.

faq limits the search to the FAQs posted on the Stata web site: http://www.stata.com. A search with the `faq` option is performed on the local keyword database only.

historical adds to the search entries that are of historical interest only. By default, such entries are not listed. Past entries are classified as historical if they discuss a feature that later became an official part of Stata. Updates to historical entries will always be found, even if `historical` is not specified. A search with the `historical` option is performed on the local keyword database only.

or specifies that an entry be listed if any of the words typed after `search` are associated with the entry. The default is to list the entry only if all the words specified are associated with the entry. A search with the `or` option is performed on the local keyword database only.

manual limits the search to entries in the *Stata Documentation*; that is, the search is limited to the *Stata User's Guide* and all the *Stata Reference* manuals. A search with the `manual` option is performed on the local keyword database only.

sj limits the search to entries in the *Stata Journal* and its predecessor, the *Stata Technical Bulletin*. A search with the `sj` option is performed on the local keyword database only.

Option for set searchdefault

permanently specifies that, in addition to making the change right now, the `searchdefault` setting be remembered and become the default setting when you invoke Stata.

Remarks

Remarks are presented under the following headings:

> *Introduction*
> *Internet searches*
> *Author searches*
> *Entry ID searches*
> *Return codes*

Introduction

See [U] **4 Stata's online help and search facilities** for a tutorial introduction to `search`. `search` is one of Stata's most useful commands. To understand the advanced features of `search`, you need to know how it works.

search has a database—files—containing the titles, etc., of every entry in the *Stata User's Guide*, the *Stata Base Reference Manual*, the *Stata Graphics Reference Manual*, the *Stata Longitudinal/Panel-Data Reference Manual*, the *Stata Multivariate Statistics Reference Manual*, the *Stata Programming Reference Manual*, the *Stata Survey Data Reference Manual*, the *Stata Survival Analysis and Epidemiological Tables Reference Manual*, the *Stata Time-Series Reference Manual*, the *Mata Reference Manual*, the articles in the *Stata Journal* and the *Stata Technical Bulletin*, and the FAQs posted on the Stata web site. In these files is a list of words, called keywords, associated with each entry.

When you type search *xyz*, search reads the database and compares the list of keywords with *xyz*. If it finds *xyz* in the list or a keyword that allows an abbreviation of *xyz*, it displays the entry.

When you type search *xyz abc*, search does the same thing but displays an entry only if it contains both keywords. The order does not matter, so you can search linear regression or search regression linear.

Obviously, how many entries search finds depends on how the search database was constructed. We have included a plethora of keywords under the theory that, for a given request, it is better to list too much rather than risk listing nothing at all. Still, you are in the position of guessing the keywords. Do you look up normality test, normality tests, or tests of normality? Answer: normality test would be best, but all would work. In general, use the singular, and strike the unnecessary words. For guidelines for specifying keywords, see [U] **4.5 More on search**.

set searchdefault allows you to specify where search searches. set searchdefault local, the default, restricts search to using only Stata's keyword database. set searchdefault net restricts search to searching only the Internet. set searchdefault all indicates that both the keyword database and the Internet are to be searched.

Internet searches

search with the net option searches the Internet for user-written additions to Stata, including, but not limited to, user-written additions published in the *Stata Journal* (SJ) and the *Stata Technical Bulletin* (STB). search *keywords*, net performs the same search as the command net search (with no options); see [R] **net search**.

```
. search random effect, net

Keyword search

        Keywords:   random effect
          Search:   (1) Web resources from Stata and from other users

Web resources from Stata and other users

(contacting http://www.stata.com)

92 packages found (Stata Journal and STB listed first)
--------------------------------------------------

st0044_1 from http://www.stata-journal.com/software/sj6-4
    SJ6-4 st0044_1.  Update:  Do-it-yourself shuffling and the... / Update:
    Do-it-yourself shuffling and the number of runs / under randomness / by N.
    Smeeton, King's College, London, UK / N. J. Cox, Durham University /
    Support:  nigel.smeeton@kcl.ac.uk, n.j.cox@durham.ac.uk / After

st0106 from http://www.stata-journal.com/software/sj6-2
    SJ6-2 st0106.  Maximum simulated likelihood estimation of... / Maximum
    simulated likelihood estimation of random effects / dynamic probit models
    with autocorrelated errors / by Mark B. Stewart, University of Warwick,
    Coventry, UK / Support:  mark.stewart@warwick.ac.uk / After installation,
```

  (*output omitted*)

    permutation from http://www-gene.cimr.cam.ac.uk/clayton/software/stata
        Routines for generalized permutation testing / David Clayton / / Diabetes
        and Inflammation Laboratory Tel: 44 (0)1223 762669 / Cambridge Institute
        for Medical Research Fax: 44 (0)1223 762102 / Wellcome Trust/MRC Building
        david.claytoncimr.cam.ac.uk / Addenbrooke's Hospital, Cambridge, CB2 2XY
    hammersley from http://www.rpi.edu/ simonk/stata
        hammersley. Hammersley sequence (non-random sampling). / Program by
        Kenneth L. Simons. / hammersley creates hammersley sequence points, which
        can be used as an / alternative to random numbers (quasi-random numbers)
        for simulation, or as / an alternative to grid search to explore a
    (*output omitted*)

    33 references found in tables of contents
    ----

    http://www.stata-journal.com/software/sj6-4/
        Update: Generalized Lorenz curves and related graphs / Update: Quantile
        plots, generalized / Update: Confidence intervals for rank statistics: /
        Somers' D and extensions / Update: Do-it-yourself shuffling and the
        number of runs / under randomness / Update: From the help desk: Local
    http://www.stata-journal.com/software/sj6-2/
        Update: Maximum R-squared and pure error / lack-of-fit test / Update:
        Concordance correlation coefficient / and associated measures, tests, and
        graphs / Update: Multivariate probit regression using / simulated maximum
        likelihood / Calculation of multivariate normal probabilities / by
    (*output omitted*)
    http://fmwww.bc.edu/RePEc/bocode/x/
        module to transform the logit scores into probabilities / module to make
        data set of summary statistics on disk or in memory / module to make a
        data set of frequencies and percents on disk or in memory / module to
        calculate and graph cross-correlation function / module to create

    (end of search)

Author searches

search ordinarily compares the words following search with the keywords for the entry. If you specify the author option, however, it compares the words with the author's name. In the search database, we have filled in author names for all SJ and STB inserts.

For instance, in [R] **kdensity** in this manual you will discover that Isaías H. Salgado-Ugarte wrote the first version of Stata's kdensity command and published it in the STB. Assume that you read his original insert and found the discussion useful. You might now wonder what else he has written in the SJ or STB. To find out, you type

    . search Salgado-Ugarte, author
      (*output omitted*)

Names like Salgado-Ugarte are confusing to many people. search does not require you to specify the entire name; what you type is compared with each "word" of the name and, if any part matches, the entry is listed. The hyphen is a special character, and you can omit it. Thus you can obtain the same list by looking up Salgado, Ugarte, or Salgado Ugarte without the hyphen.

Actually, to find all entries written by Salgado-Ugarte, you need to type

    . search Salgado-Ugarte, author historical
      (*output omitted*)

Prior inserts in the SJ or STB that provide a feature that later was superseded by a built-in feature of Stata are marked as historical in the search database and, by default, are not listed. The historical option ensures that all entries are listed.

Entry ID searches

If you specify the `entry` option, `search` compares what you have typed with the entry ID. The entry ID is not the title—it is the reference listed to the left of the title that tells you where to look. For instance, in

```
[R]     regress  . . . . . . . . . . . . . . . . . . . . . Linear regression
        (help regress)
```

[R] **regress** is the entry ID. This is a reference, of course, to this manual. In

```
FAQ        . . . . . . . . . . Analysis of multiple failure-time survival data
           . . . . . . . . . . . . . . . . . . . . . . . . . . . . . . M. Cleves
           11/99   How do I analyze multiple failure-time data using Stata?
                   http://www.stata.com/support/faqs/stat/stmfail.html
```

"FAQ" is the entry ID. In

```
SJ-7-1  st0118  . . A survey on survey stat.: What is and can be done in Stata
                . . . . . . . . . . . . . . . . . . . . . . . F. Kreuter and R. Valliant
                Q1/07  SJ7(1):1--21                        (no commands)
                discusses survey issues in analyzing complex survey
                data and describes some of Stata's capabilities for
                such analyses
```

"SJ-7-1" is the entry ID.

`search` with the `entry` option searches these entry ids.

Thus you could generate a table of contents for the *User's Guide* by typing

```
. search [U], entry
  (output omitted )
```

You could generate a table of contents for *Stata Journal*, Volume 1, Issue 1, by typing

```
. search sj-1-1, entry
  (output omitted )
```

To generate a table of contents for the 26th issue of the STB, you would type

```
. search STB-26, entry historical
  (output omitted )
```

The `historical` option here is possibly important. STB-26 was published in July 1995, and perhaps some of its inserts have already been marked historical.

You could obtain a list of all inserts associated with sg53 by typing

```
. search sg53, entry historical
  (output omitted )
```

Again we include the `historical` option in case any of the relevant inserts have been marked historical.

Return codes

In addition to indexing the entries in the *User's Guide* and all the *Reference* manuals, `search` also can be used to search return codes.

To see information on return code 131, type

```
. search rc 131
[R]      . . . . . . . . . . . . . . . . . . . . . . . . . . . Return code 131
         not possible with test;
         You requested a test of a hypothesis that is nonlinear in the
         variables.  test tests only linear hypotheses.  Use testnl.
```

If you want a list of all Stata return codes, type

```
. search error, entry
  (output omitted)
```

Methods and Formulas

findit is implemented as an ado-file.

Acknowledgment

findit grew from a suggestion by Nicholas J. Cox, Durham University.

Also See

[R] **hsearch** — Search help files

[R] **help** — Display online help

[R] **net search** — Search the Internet for installable packages

[U] **4 Stata's online help and search facilities**

Title

serrbar — Graph standard error bar chart

Syntax

serrbar *mvar svar xvar* $\big[$ *if* $\big]$ $\big[$ *in* $\big]$ $\big[$, *options* $\big]$

options	description
Main	
<u>sc</u>ale(#)	scale length of graph bars; default is scale(1)
Error bars	
rcap_options	affect rendition of capped spikes
Plotted points	
<u>m</u>vopts(*scatter_options*)	affect rendition of plotted points
Add plots	
addplot(*plot*)	add other plots to generated graph
Y axis, X axis, Titles, Legend, Overall	
twoway_options	any options other than by() documented in [G] ***twoway_options***

Description

serrbar graphs *mvar* $\pm$ scale() $\times$ *svar* against *xvar*. Usually, but not necessarily, *mvar* and *svar* will contain means and standard errors or standard deviations of some variable so that a standard error bar chart is produced.

Options

⌐ **Main** ⌐

scale(#) controls the length of the bars. The upper and lower limits of the bars will be *mvar* + scale() $\times$ *svar* and *mvar* − scale() $\times$ *svar*. The default is scale(1).

⌐ **Error bars** ⌐

rcap_options affect the rendition of the plotted error bars (the capped spikes). See [G] **graph twoway rcap**.

⌐ **Plotted points** ⌐

mvopts(*scatter_options*) affects the rendition of the plotted points (*mvar* versus *xvar*). See [G] **graph twoway scatter**.

⌐ **Add plots** ⌐

addplot(*plot*) provides a way to add other plots to the generated graph; see [G] ***addplot_option***.

┌ Y axis, X axis, Titles, Legend, Overall ┐

twoway_options are any of the options documented in [G] *twoway_options*, excluding by(). These include options for titling the graph (see [G] *title_options*) and for saving the graph to disk (see [G] *saving_option*).

Remarks

▷ Example 1

In quality-control applications, the three most commonly used variables with this command are the process mean, process standard deviation, and time. For instance, we have data on the average weights and standard deviations from an assembly line in San Francisco for the period January 8 to January 16. Our data are

```
. use http://www.stata-press.com/data/r10/assembly
. list, sep(0) divider
```

	date	mean	std
1.	108	192.22	3.94
2.	109	192.64	2.83
3.	110	192.37	4.58
4.	113	194.76	3.25
5.	114	192.69	2.89
6.	115	195.02	1.73
7.	116	193.40	2.62

We type serrbar mean std date, scale(2) but, after seeing the result, decide to make it fancier:

```
. serrbar mean std date, scale(2) title("Observed Weight Variation")
> sub("San Francisco plant, 1/8 to 1/16") yline(195) yaxis(1 2)
> ylab(195, axis(2)) ytitle("", axis(2))
```

◁

Methods and Formulas

serrbar is implemented as an ado-file.

Acknowledgment

serrbar was written by Nicholas J. Cox of Durham University.

Also See

[R] **qc** — Quality control charts

Title

set — Overview of system parameters

Syntax

set [*setcommand* ...]

set typed without arguments is equivalent to query typed without arguments.

Description

This entry provides a reference to Stata's set commands. For many entries, more thorough information is provided elsewhere; see the Reference field in each entry below for the location of this information.

To reset system parameters to factory defaults, see [R] **set_defaults**.

Remarks

set adosize

Syntax: set adosize # [, permanently]

Default: 500

Description: sets the maximum amount of memory that automatically loaded do-files may consume. $10 \leq \# \leq 1000$.

Reference: [P] **sysdir**

set autotabgraphs (Windows only)

Syntax: set autotabgraphs {on | off} [, permanently]

Default: off

Description: determines whether graphs are created as tabs within one window or as separate windows.

set checksum

Syntax: set checksum {on | off} [, permanently]

Default: off

Description: determines whether files should be prevented from being downloaded from the Internet if checksums do not match.

Reference: [D] **checksum**

set conren (Unix console only)
   Syntax 1:   set conren
   Syntax 2:   set conren clear
   Syntax 3:   set conren $[$sf $\mid$ bf $\mid$ it$]$
                              $\{$<u>res</u>ult $\mid$ $[$txt $\mid$ text$]$ $\mid$ input $\mid$ <u>err</u>or $\mid$ <u>li</u>nk $\mid$ <u>hi</u>lite$\}$
                              $[$*char*$[$*char*...$]]$
   Syntax 4:   set conren $\{$ulon $\mid$ <u>ul</u>off$\}$ $[$*char* $[$*char* ...$]]$
   Syntax 5:   set conren reset $[$*char* $[$*char* ...$]]$
   Description: this can possibly make the output on your screen appear prettier.
                set conren displays a list of the currently defined display codes.
                set conren clear clears all codes.
                set conren followed by a font type (bf, sf, or it) and display context (result, error, link or hilite) and then followed by a series of space-separated characters sets the code for the specified font type and display context. If the font type is omitted, the code is set to the same specified code for all three font types. set conren ulon and set conren uloff set the codes for turning on and off underlining.
                set conren reset sets the code that will turn off all display and underlining codes.
   Reference: [GSU] conren

set copycolor (Macintosh and Windows only)
   Syntax: set copycolor $\{$<u>auto</u>matic $\mid$ asis $\mid$ gs1 $\mid$ gs2 $\mid$ gs3$\}$ $[$, <u>perm</u>anently$]$
   Default: automatic
   Description: determines how colors are handled when graphs are copied to the clipboard.
   Reference: [G] **set printcolor**

set dockable (Windows only)
   Syntax: set dockable $\{$on $\mid$ off$\}$ $[$, <u>perm</u>anently$]$
   Default: off
   Description: determines whether to enable the use of dockable window characteristics, including the ability to dock or tab a window into another window.

set dockingguides (Windows only)
   Syntax: set <u>dockingg</u>uides $\{$on $\mid$ off$\}$ $[$, <u>perm</u>anently$]$
   Default: on
   Description: determines whether to enable the use of dockable guides when repositioning a dockable window.

set doublebuffer (Windows only)
   Syntax: set doublebuffer $\{$on $\mid$ off$\}$ $[$, <u>perm</u>anently$]$
   Default: on
   Description: enables or disables double buffering of the Results, Viewer, and Data Editor windows. Double buffering prevents the windows from flickering when redrawn or resized. Users who encounter performance problems such as the Results window outputting very slowly should disable double buffering.

set dp

Syntax: set dp {comma|period} [, permanently]

Default: period

Description: determines whether a period or a comma is to be used as the decimal point.

Reference: [D] **format**

set eolchar (Macintosh only)

Syntax: set eolchar {mac|unix} [, permanently]

Default: mac

Description: sets the default end-of-line delimiter for text files created in Stata.

set fastscroll

Syntax: set fastscroll {on|off} [, permanently]

Default: on

Description: sets the scrolling method for new output in the Results window. Setting fastscroll to on is faster but can be jumpy. Setting fastscroll to off is slower but smoother.

set floatresults (Windows only)

Syntax: set floatresults {on|off}

Default: off

Description: determines whether to enable floating window behavior for the Results window. The term "float" in this context means the Results window will always float over the main Stata window; the Results window can never be place behind the main Stata window. There is no permanently option because permanently is implied.

set floatwindows (Windows only)

Syntax: set floatwindows {on|off}

Default: off

Description: determines whether to enable floating window behavior for all Stata windows except the main Stata window and the Results window. The term "float" in this context means a window will always float over the main Stata window; these windows can never be place behind the main Stata window. There is no permanently option because permanently is implied.

set graphics

Syntax: set graphics {on|off}

Default: on; default is off for console Stata

Description: determines whether graphs are displayed on your monitor.

Reference: [G] **set graphics**

set httpproxy

Syntax: set httpproxy {on|off} [, init]

Default: off

Description: turns on/off the use of a proxy server. There is no permanently option because permanently is implied.

Reference: [R] **netio**

set httpproxyauth
 Syntax: set <u>httpproxyauth</u> {on | off}
 Default: off
 Description: determines whether authorization is required for the proxy server.
      There is no permanently option because permanently is implied.
 Reference: [R] **netio**

set httpproxyhost
 Syntax: set httpproxyhost ["]*name*["]
 Description: sets the name of a host to be used as a proxy server. There is no permanently
      option because permanently is implied.
 Reference: [R] **netio**

set httpproxyport
 Syntax: set httpproxyport #
 Default: 8080 if Stata cannot autodetect the proper setting for your computer.
 Description: sets the port number for a proxy server. There is no permanently option
      because permanently is implied.
 Reference: [R] **netio**

set httpproxypw
 Syntax: set httpproxypw ["]*password*["]
 Description: sets the appropriate password. There is no permanently option because
      permanently is implied.
 Reference: [R] **netio**

set httpproxyuser
 Syntax: set httpproxyuser ["]*name*["]
 Description: sets the appropriate user ID. There is no permanently option because
      permanently is implied.
 Reference: [R] **netio**

set level
 Syntax: set <u>l</u>evel # [, permanently]
 Default: 95
 Description: sets the default significance level for confidence intervals for all commands
      that report confidence intervals. $10.00 \leq \# \leq 99.99$, and # can have at
      most two digits after the decimal point.
 Reference: [R] **level**

set linegap
 Syntax: set <u>linegap</u> #
 Default: 1
 Description: sets the space between lines, in pixels, in the Results window. There is no
      permanently option because permanently is implied.

set linesize
    Syntax: set <u>line</u>size #
    Default: 1 less than the full width of the screen
    Description: sets the line width, in characters, for both the screen and the log file.
    Reference: [R] **log**

set locksplitters (Windows only)
    Syntax: set <u>lock</u>splitters {on | off} [, <u>perman</u>ently]
    Default: off
    Description: determines whether splitters should be locked so that docked windows
                   cannot be resized.

set logtype
    Syntax: set <u>log</u>type {<u>text</u> | <u>smcl</u>} [, <u>perman</u>ently]
    Default: smcl
    Description: sets the default log filetype.
    Reference: [R] **log**

set macgphengine (Macintosh only)
    Syntax: set <u>macgph</u>engine {quartz | quickdraw} [, <u>perman</u>ently]
    Default: quartz
    Description: determines whether Quartz or QuickDraw is used to draw graphs. Quartz is the
                   primary graphics engine for OS X and supports smooth lines, curves, and text.
                   QuickDraw was the graphics engine for Mac OS prior to OS X and lacks many
                   of the advanced features of Quartz. Graphs drawn by Quartz are visibly
                   superior to those drawn by QuickDraw and closely resemble printed output.

set matacache, set matafavor, set matalibs, set matalnum, set matamofirst,
    set mataoptimize, and set matastrict; see [M-3] **mata set**.

set matsize
    Syntax: set <u>mat</u>size # [, <u>perman</u>ently]
    Default: 400 for Stata/MP and Stata/SE; 200 for Stata/IC; 40 for Small Stata
    Description: sets the maximum number of variables that can be included in any estimation
                   command. This setting cannot be changed in Small Stata.
    Reference: [R] **matsize**

set maxdb
    Syntax: set maxdb # [, <u>perman</u>ently]
    Default: 50
    Description: sets the maximum number of dialog boxes whose contents are remembered
                   from one invocation to the next during a session.
    Reference: [R] **db**

set maxiter
    Syntax: set maxiter # [, <u>perman</u>ently]
    Default: 16000
    Description: sets the default maximum number of iterations for estimation commands.
                   $0 \leq \# \leq 16000$.
    Reference: [R] **maximize**

set maxvar
    Syntax: set maxvar # $\left[\text{, permanently}\right]$
    Default: 5000 for Stata/MP and Stata/SE; 2048 for Stata/IC; 99 for Small Stata
    Description: sets the maximum number of variables. This can be changed only in Stata/MP and Stata/SE.
    Reference: [D] **memory**

set memory
    Syntax: set memory #[b|k|m|g] $\left[\text{, permanently}\right]$
    Default: 10m for Stata/MP and Stata/SE; 1m for Stata/IC; fixed data area of 99 variables on roughly 1,000 observations for Small Stata
    Description: set memory allocated to Stata's data areas. In Small Stata this cannot be changed.
    Reference: [D] **memory**

set more
    Syntax: set more {on|off} $\left[\text{, permanently}\right]$
    Default: on
    Description: pause when —more— is displayed, continuing only when the user presses a key.
    Reference: [R] **more**

set obs
    Syntax: set obs #
    Default: current number of observations
    Description: changes the number of observations in the current dataset. # must be at least as large as the current number of observations. If there are variables in memory, the values of all new observations are set to *missing*.
    Reference: [D] **obs**

set odbcmgr (Unix only)
    Syntax: set odbcmgr {iodbc|unixodbc} $\left[\text{, permanently}\right]$
    Default: iodbc
    Description: determines whether iODBC or unixODBC is your ODBC driver manager.
    Reference: [D] **odbc**

set output
    Syntax: set output {proc|inform|error}
    Default: proc
    Description: specifies the output to be displayed. proc means display all output; inform suppresses procedure output but displays informative messages and error messages; error suppresses all output except error messages. set output is seldom used.
    Reference: [P] **quietly**

set pagesize
    Syntax: set pagesize #
    Default: 2 less than the physical number of lines on the screen
    Description: sets the number of lines between —more— messages.
    Reference: [R] **more**

set persistfv (Windows only)
    Syntax: set persistfv {on | off}
    Default: off
    Description: determines whether floating viewers are to persist between Stata sessions or when saving and loading a windowing preference.

set persistvtopic (Windows only)
    Syntax: set persistvtopic {on | off}
    Default: off
    Description: determines whether viewer topics are to persist between Stata sessions or when saving and loading a windowing preference.

set piccomments (Macintosh only)
    Syntax: set piccomments {on | off} [, permanently]
    Default: on
    Description: determines whether PICT files use PICCOMMENTS for drawing true rotated text.

set pinnable (Windows only)
    Syntax: set pinnable {on | off} [, permanently]
    Default: off
    Description: determines whether to enable the use of pinnable window characteristics for certain windows in Stata.

set printcolor
    Syntax: set printcolor { automatic | asis | gs1 | gs2 | gs3 } [, permanently]
    Default: automatic
    Description: determines how colors are handled when graphs are printed.
    Reference: [G] **set printcolor**

set processors
    Syntax: set processors #
    Description: sets the number of processors or cores that Stata/MP will use. The default is the number of processors available on the computer, or the number of processors allowed by Stata/MP's license, whichever is less.

set reventries (Windows only)
    Syntax: set reventries # [, permanently]
    Default: 5000
    Description: sets the number of scrollback lines available in the Review window.

set revwindow (Macintosh only)
    Syntax: set revwindow {float | nofloat} , permanently
    Default: float
    Description: specifies whether the Review window should be placed behind other windows. float indicates that the window should stay in front. Stata must be restarted for a new setting to take effect. permanently is required.

set rmsg
   Syntax: set rmsg {on | off} [, permanently]
   Default: off
   Description: indicates whether a return message telling the execution time is to be displayed at
               the completion of each command.
   Reference: [P] **rmsg**

set scheme
   Syntax: set scheme *schemename* [, permanently]
   Default: s2color
   Description: determines the overall look for graphs.
   Reference: [G] **set scheme**

set scrollbufsize
   Syntax: set scrollbufsize #
   Default: 32000
   Description: sets the scrollback buffer size, in bytes, for the Results window;
               may be set between 10k and 500k.

set searchdefault
   Syntax: set searchdefault {local | net | all} [, permanently]
   Default: local
   Description: sets the default behavior of the search command. set searchdefault local
               restricts search to use only Stata's keyword database. set searchdefault net
               restricts search to searching only the Internet. set searchdefault all
               indicates that both the keyword database and the Internet are to be searched.
   Reference: [R] **search**

set seed
   Syntax: set seed {# | *code*}
   Default: 123456789
   Description: specifies initial value of the random-number seed used by the uniform() function.
   Reference: [D] **generate**

set smoothfonts (Macintosh only)
   Syntax: set smoothfonts {on | off} [, permanently]
   Default: on
   Description: determines whether to use the Quartz graphics engine to draw smooth text edges
               in all Stata windows except the Graph windows, which use the setting in
               macgphengine.

set smoothsize (Macintosh only)
   Syntax: set smoothsize # [, permanently]
   Default: 10
   Description: sets the minimum font size for font smoothing; see set smoothfonts.

set timeout1
- Syntax: set timeout1 *#seconds* [, permanently]
- Default: 120
- Description: sets the number of seconds Stata will wait for a remote host to respond to an initial contact before giving up. In general, users should not modify this value unless instructed to do so by Stata Technical Services.
- Reference: [R] **netio**

set timeout2
- Syntax: set timeout2 *#seconds* [, permanently]
- Default: 300
- Description: sets the number of seconds Stata will keep trying to get information from a remote host after initial contact before giving up. In general, users should not modify this value unless instructed to do so by Stata Technical Services.
- Reference: [R] **netio**

set trace
- Syntax: set trace {on | off}
- Default: off
- Description: determines whether to trace the execution of programs for debugging.
- Reference: [P] **trace**

set tracedepth
- Syntax: set tracedepth #
- Default: 32000 (equivalent to ∞)
- Description: if trace is set on, traces execution of programs and nested programs up to tracedepth. For example, if tracedepth is 2, the current program and any subroutine called would be traced, but subroutines of subroutines would not be traced.
- Reference: [P] **trace**

set traceexpand
- Syntax: set traceexpand {on | off} [, permanently]
- Default: on
- Description: if trace is set on, shows lines both before and after macro expansion. If traceexpand is set off, only the line before macro expansion is shown.
- Reference: [P] **trace**

set tracehilite
- Syntax: set tracehilite "*pattern*" [, word]
- Default: ""
- Description: highlights *pattern* in the trace output.
- Reference: [P] **trace**

set traceindent
    Syntax: set traceindent {on|off} [, permanently]
    Default: on
    Description: if trace is set on, indents displayed lines according to their nesting level. The
                  lines of the main program are not indented. Two spaces of indentation are used for
                  each level of nested subroutine.
    Reference: [P] **trace**

set tracenumber
    Syntax: set tracenumber {on|off} [, permanently]
    Default: off
    Description: if trace is set on, shows the nesting level numerically in front of the line.
                  Lines of the main program are preceded by 01, lines of subroutines called by the
                  main program are preceded by 02, etc.
    Reference: [P] **trace**

set tracesep
    Syntax: set tracesep {on|off} [, permanently]
    Default: on
    Description: if trace is set on, displays a horizontal separator line that displays the name
                  of the subroutine whenever a subroutine is called or exits.
    Reference: [P] **trace**

set type
    Syntax: set type {float|double} [, permanently]
    Default: float
    Description: specifies the default storage type assigned to new variables.
    Reference: [D] **generate**

set update_interval (Macintosh and Windows only)
    Syntax: set update_interval #
    Default: 7
    Description: sets the number of days to elapse before performing the next automatic
                  update query.
    Reference: [R] **update**

set update_prompt (Macintosh and Windows only)
    Syntax: set update_prompt {on|off}
    Default: on
    Description: determines wheter a dialog is to be displayed before performing an automatic
                  update query. There is no permanently option because permanently is implied.
    Reference: [R] **update**

set update_query (Macintosh and Windows only)
    Syntax: set update_query {on|off}
    Default: on
    Description: determines whether update query is to be automatically performed when Stata
                  is launched. There is no permanently option because permanently is implied.
    Reference: [R] **update**

set use_atsui_graph (Macintosh only)
    Syntax: set use_atsui_graph {on | off} [, permanently]
    Default: off
    Description: indicates whether ATSUI (Apple Type Services for Unicode Imaging) should be used
               for rendering text in graphs.

set use_qd_text (Macintosh only)
    Syntax: set use_qd_text {on | off} [, permanently]
    Default: on
    Description: uses the QuickDraw graphics engine to draw text while still allowing Quartz to
               smooth the edges of the text. This is faster than drawing text with Quartz but is
               potentially less accurate at rendering certain characters due to differences in
               font metrics between QuickDraw and Quartz. In most cases, you will notice little,
               if any, difference between text drawn by QuickDraw and Quartz.

set varabbrev
    Syntax: set varabbrev {on | off} [, permanently]
    Default: on
    Description: indicates whether Stata should allow variable abbreviations.

set varlabelpos (Unix only)
    Syntax: set varlabelpos #
    Default: 32
    Description: sets the maximum number of characters to display for variable names in the
               Variables window.

set varwindow (Macintosh only)
    Syntax: set varwindow {float | nofloat} , permanently
    Default: float
    Description: specifies whether the Variables window should be placed behind other windows.
               float indicates that the window should stay in front. Stata must be restarted
               for a new setting to take effect. permanently is required.

set virtual
    Syntax: set virtual {on | off}
    Default: off
    Description: indicates whether Stata should work to arrange its memory to keep objects
               close together.
    Reference: [D] **memory**

Also See

[R] **query** — Display system parameters

[R] **set_defaults** — Reset system parameters to original Stata defaults

[M-3] **mata set** — Set and display Mata system parameters

Title

set_defaults — Reset system parameters to original Stata defaults

Syntax

set_defaults { *category* | _all } [, permanently]

where *category* is one of memory | output | interface | graphics | efficiency |

network | update | trace | mata | other

Description

set_defaults resets settings made by set to the original default settings that were shipped with Stata.

set_defaults _all resets all the categories, whereas set defaults *category* resets only the settings for the specified category.

Option

permanently specifies that, in addition to making the change right now, the settings be remembered and become the default settings when you invoke Stata.

Remarks

▷ Example 1

To assist us in debugging a new command, we modified some of the trace settings. To return them to their original values, we type

```
. set_defaults trace
-> set trace off
-> set tracedepth 32000
-> set traceexpand on
-> set tracesep on
-> set traceindent on
-> set tracenumber off
-> set tracehilite ""
(preferences reset)
```

◁

Methods and Formulas

set_defaults is implemented as an ado-file.

271

Also See

[R] **query** — Display system parameters

[R] **set** — Overview of system parameters

[M-3] **mata set** — Set and display Mata system parameters

Title

signrank — Equality tests on matched data

Syntax

Wilcoxon matched-pairs signed-ranks test

signrank *varname* = *exp* [*if*] [*in*]

Sign test of matched pairs

signtest *varname* = *exp* [*if*] [*in*]

by is allowed with signrank and signtest; see [D] **by**.

Description

signrank tests the equality of matched pairs of observations by using the Wilcoxon matched-pairs signed-ranks test (Wilcoxon 1945). The null hypothesis is that both distributions are the same.

signtest also tests the equality of matched pairs of observations (Arbuthnott 1712, but better explained by Snedecor and Cochran 1989) by calculating the differences between *varname* and the expression. The null hypothesis is that the median of the differences is zero; no further assumptions are made about the distributions. This, in turn, is equivalent to the hypothesis that the true proportion of positive (negative) signs is one-half.

For equality tests on unmatched data, see [R] **ranksum**.

Remarks

▷ Example 1: signrank

We are testing the effectiveness of a new fuel additive. We run an experiment with 12 cars. We first run each car without the fuel treatment and measure the mileage. We then add the fuel treatment and repeat the experiment. The results of the experiment are

Without treatment	With treatment	Without treatment	With treatment
20	24	18	17
23	25	24	28
21	21	20	24
25	22	24	27
18	23	23	21
17	18	19	23

We create two variables called mpg1 and mpg2, representing mileage without and with the treatment, respectively. We can test the null hypothesis that the treatment had no effect by typing

```
. use http://www.stata-press.com/data/r10/fuel

. signrank mpg1=mpg2
```

Wilcoxon signed-rank test

sign	obs	sum ranks	expected
positive	3	13.5	38.5
negative	8	63.5	38.5
zero	1	1	1
all	12	78	78

```
unadjusted variance      162.50
adjustment for ties       -1.62
adjustment for zeros      -0.25
                        _____
adjusted variance        160.62

Ho: mpg1 = mpg2
            z =  -1.973
   Prob > |z| =   0.0485
```

The output indicates that we can reject the null hypothesis at any level above 4.85%.   ◁

▷ Example 2: signtest

signtest tests that the median of the differences is zero, making no further assumptions, whereas signrank assumed that the distributions are equal as well. Using the data above, we type

```
. signtest mpg1=mpg2
```

Sign test

sign	observed	expected
positive	3	5.5
negative	8	5.5
zero	1	1
all	12	12

```
One-sided tests:
  Ho: median of mpg1 - mpg2 = 0 vs.
  Ha: median of mpg1 - mpg2 > 0
      Pr(#positive >= 3) =
        Binomial(n = 11, x >= 3, p = 0.5) =  0.9673

  Ho: median of mpg1 - mpg2 = 0 vs.
  Ha: median of mpg1 - mpg2 < 0
      Pr(#negative >= 8) =
        Binomial(n = 11, x >= 8, p = 0.5) =  0.1133

Two-sided test:
  Ho: median of mpg1 - mpg2 = 0 vs.
  Ha: median of mpg1 - mpg2 != 0
      Pr(#positive >= 8 or #negative >= 8) =
        min(1, 2*Binomial(n = 11, x >= 8, p = 0.5)) =  0.2266
```

The summary table indicates that there were three comparisons for which mpg1 exceeded mpg2, eight comparisons for which mpg2 exceeded mpg1, and one comparison for which they were the same.

The output below the summary table is based on the binomial distribution. The significance of the one-sided test, where the alternative hypothesis is that the median of mpg2 − mpg1 is greater than zero, is 0.1133. The significance of the two-sided test, where the alternative hypothesis is simply that the median of the differences is different from zero, is $0.2266 = 2 \times 0.1133$.   ◁

Saved Results

signrank saves the following in r():

Scalars

r(N_neg)	number of negative comparisons	r(sum_neg)	sum of the negative ranks
r(N_pos)	number of positive comparisons	r(z)	z statistic
r(N_tie)	number of tied comparisons	r(Var_a)	adjusted variance
r(sum_pos)	sum of the positive ranks		

signtest saves the following in r():

Scalars

r(N_neg)	number of negative comparisons	r(p_2)	two-sided probability
r(N_pos)	number of positive comparisons	r(p_neg)	one-sided probability of negative comparison
r(N_tie)	number of tied comparisons	r(p_pos)	one-sided probability of positive comparison

Methods and Formulas

signrank and signtest are implemented as ado-files.

For a practical introduction to these techniques with an emphasis on examples rather than theory, see Bland (2000) or Sprent and Smeeton (2007). For a summary of these tests, see Snedecor and Cochran (1989).

signrank

Both the sign test and Wilcoxon signed-rank tests test the null hypothesis that the distribution of a random variable $D = varname - exp$ has median zero. The sign test makes no additional assumptions, but the Wilcoxon signed-rank test makes the additional assumption that the distribution of D is symmetric. If $D = X_1 - X_2$, where X_1 and X_2 have the same distribution, then it follows that the distribution of D is symmetric about zero. Thus the Wilcoxon signed-rank test is often described as a test of the hypothesis that two distributions are the same, i.e., $X_1 \sim X_2$.

Let d_j denote the difference for any matched pair of observations,

$$d_j = x_{1j} - x_{2j} = varname - exp$$

for $j = 1, 2, \ldots, n$.

Rank the absolute values of the differences $|d_j|$, and assign any tied values the average rank. Consider the signs of d_j, and let

$$r_j = \text{sign}(d_j) \, \text{rank}(|d_j|)$$

be the signed ranks. The test statistic is

$$T_{\text{obs}} = \sum_{j=1}^{n} r_j = (\text{sum of ranks for + signs}) - (\text{sum of ranks for - signs})$$

The null hypothesis is that the distribution of d_j is symmetric about 0. Hence the likelihood is unchanged if we flip signs on the d_j, and thus the randomization datasets are the 2^n possible sign changes for the d_j. Thus the randomization distribution of our test statistic T can be computed by considering all the 2^n possible values of

$$T = \sum_{j=1}^{n} S_j r_j$$

where the r_j are the observed signed ranks (considered fixed) and S_j is either $+1$ or -1.

With this distribution, the mean and variance of T are given by

$$E(T) = 0 \quad \text{and} \quad \text{Var}_{\text{adj}}(T) = \sum_{j=1}^{n} r_j^2$$

The test statistic for the Wilcoxon signed-rank test is often expressed (equivalently) as the sum of the positive signed-ranks, T_+, where

$$E(T_+) = \frac{n(n+1)}{4} \quad \text{and} \quad \text{Var}_{\text{adj}}(T_+) = \frac{1}{4} \sum_{j=1}^{n} r_j^2$$

Zeros and ties do not affect the theory above, and the exact variance is still given by the above formula for $\text{Var}_{\text{adj}}(T_+)$. When $d_j = 0$ is observed, d_j will always be zero in each of the randomization datasets (using $\text{sign}(0) = 0$). When there are ties, you can assign averaged ranks for each group of ties and then treat them the same as the other ranks.

The "unadjusted variance" reported by `signrank` is the variance that the randomization distribution would have had if there had been no ties or zeros:

$$\text{Var}_{\text{unadj}}(T_+) = \frac{1}{4} \sum_{j=1}^{n} j^2 = \frac{n(n+1)(2n+1)}{24}$$

The adjustment for zeros is the change in the variance when the ranks for the zeros are signed to make $r_j = 0$,

$$\Delta \text{Var}_{\text{zero adj}}(T_+) = -\frac{1}{4} \sum_{j=1}^{n_0} j^2 = -\frac{n_0(n_0+1)(2n_0+1)}{24}$$

where n_0 is the number of zeros. The adjustment for ties is the change in the variance when the ranks (for nonzero observations) are replaced by averaged ranks:

$$\Delta \text{Var}_{\text{ties adj}}(T_+) = \text{Var}_{\text{adj}}(T_+) - \text{Var}_{\text{unadj}}(T_+) - \Delta \text{Var}_{\text{zero adj}}(T_+)$$

A normal approximation is used to calculate

$$z = \frac{T_+ - E(T_+)}{\sqrt{\text{Var}_{\text{adj}}(T_+)}}$$

signtest

The test statistic for the sign test is the number n_+ of differences

$$d_j = x_{1j} - x_{2j} = varname - exp$$

greater than zero. Assuming that the probability of a difference being equal to zero is exactly zero, then, under the null hypothesis, $n_+ \sim \text{binomial}(n, p = 1/2)$, where n is the total number of observations.

But what if some differences are zero? This question has a ready answer if you view the test from the perspective of Fisher's Principle of Randomization (Fisher 1935). Fisher's idea (stated in a modern way) was to look at a family of transformations of the observed data such that the a priori likelihood (under the null hypothesis) of the transformed data is the same as the likelihood of the observed data. The distribution of the test statistic is then produced by calculating its value for each of the transformed "randomization" datasets, assuming that each dataset is equally likely.

For the sign test, the "data" are simply the set of signs of the differences. Under the null hypothesis of the sign test, the probability that d_j is less than zero is equal to the probability that d_j is greater than zero. Thus you can transform the observed signs by flipping any number of them, and the set of signs will have the same likelihood. The 2^n possible sign changes form the family of randomization datasets. If you have no zeros, this procedure again leads to $n_+ \sim \text{binomial}(n, p = 1/2)$.

If you do have zeros, changing their signs leaves them as zeros. So, if you observe n_0 zeros, each of the 2^n sign-change datasets will also have n_0 zeros. Hence, the values of n_+ calculated over the sign-change datasets range from 0 to $n - n_0$, and the "randomization" distribution of n_+ is $\text{binomial}(n - n_0, p = 1/2)$.

The work of Arbuthnott (1712) and later 18th-century contributions is discussed by Hald (2003, chap. 17).

Frank Wilcoxon (1892–1965) was born in Ireland to American parents. After working in various occupations (including merchant seaman, oil-well pump attendant, and tree surgeon), he settled in chemistry, gaining degrees from Rutgers and Cornell and employment from various companies. Working mainly on the development of fungicides and insecticides, Wilcoxon became interested in statistics in 1925 and made several key contributions to nonparametric methods. After retiring from industry, he taught statistics at Florida State until his death.

References

Arbuthnott, J. 1712. An argument for divine providence, taken from the constant regularity observ'd in the births of both sexes. *Philosophical Transactions* 27: 186–190.

Bland, M. 2000. *An Introduction to Medical Statistics*. 3rd ed. Oxford: Oxford University Press.

Bradley, R. A. 2001. Frank Wilcoxon. In *Statisticians of the Centuries*, ed. C. C. Heyde and E. Seneta, 420–424. New York: Springer.

Fisher, R. A. 1935. *Design of Experiments*. Edinburgh: Oliver & Boyd.

Hald, A. 2003. *A History of Probability and Statistics and Their Applications before 1750*. New York: Wiley.

Newson, R. 2006. Confidence intervals for rank statistics: Somers' D and extensions. *Stata Journal* 6: 309–334.

Snedecor, G. W., and W. G. Cochran. 1989. *Statistical Methods*. 8th ed. Ames, IA: Iowa State University Press.

Sprent, P., and N. C. Smeeton. 2007. *Applied Nonparametric Statistical Methods*. 4th ed. Boca Raton, FL: Chapman & Hall/CRC.

Sribney, W. M. 1995. crc40: Correcting for ties and zeros in sign and rank tests. *Stata Technical Bulletin* 26: 2–4. Reprinted in *Stata Technical Bulletin Reprints*, vol. 5, pp. 5–8.

Wilcoxon, F. 1945. Individual comparisons by ranking methods. *Biometrics* 1: 80–83.

Also See

[R] **ranksum** — Equality tests on unmatched data

[R] **ttest** — Mean-comparison tests

Title

> **simulate** — Monte Carlo simulations

Syntax

> simulate [*exp_list*] , <u>r</u>eps(#) [*options*] : *command*

options	description
nodots	suppress the replication dots
<u>no</u>isily	display any output from *command*
<u>trace</u>	trace the *command*
<u>sa</u>ving(*filename*, ...)	save results to *filename*
<u>nol</u>egend	suppress table legend
<u>v</u>erbose	display the full table legend
seed(#)	set random-number seed to #

All weight types supported by *command* are allowed; see [U] **11.1.6 weight**.

exp_list contains	(*name*: *elist*)
	elist
	eexp
elist contains	*newvar* = (*exp*)
	(*exp*)
eexp is	*specname*
	[*eqno*]*specname*
specname is	_b
	_b[]
	_se
	_se[]
eqno is	# #
	name

exp is a standard Stata expression; see [U] **13 Functions and expressions**.

Distinguish between [ ], which are to be typed, and [ ], which indicate optional arguments.

Description

simulate eases the programming task of performing Monte Carlo–type simulations. Typing

> . simulate *exp_list*, reps(#): *command*

runs *command* for # replications and collects the results in *exp_list*.

command defines the command that performs one simulation. Most Stata commands and user-written programs can be used with simulate, as long as they follow standard Stata syntax; see [U] **11 Language syntax**. The by prefix may not be part of *command*.

exp_list specifies the expression to be calculated from the execution of *command*. If no expressions are given, *exp_list* assumes a default, depending upon whether *command* changes results in `e()` or `r()`. If *command* changes results in `e()`, the default is `_b`. If *command* changes results in `r()` (but not `e()`), the default is all the scalars posted to `r()`. It is an error not to specify an expression in *exp_list* otherwise.

Options

reps(#) is required—it specifies the number of replications to be performed.

nodots suppresses display of the replication dots. By default, one dot character is displayed for each successful replication. A red 'x' is displayed if *command* returns an error or if one of the values in *exp_list* is missing.

noisily requests that any output from *command* be displayed. This option implies the nodots option.

trace causes a trace of the execution of *command* to be displayed. This option implies the noisily option.

saving(*filename*[, *suboptions*]) creates a Stata data file (.dta file) consisting of, for each statistic in *exp_list*, a variable containing the simulated values.

double specifies that the results for each replication be stored as doubles, meaning 8-byte reals. By default, they are stored as floats, meaning 4-byte reals.

every(#) specifies that results be written to disk every #th replication. every() should be specified only in conjunction with saving() when *command* takes a long time for each replication. This will allow recovery of partial results should some other software crash your computer. See [P] **postfile**.

replace specifies that *filename* be overwritten if it exists.

nolegend suppresses display of the table legend. The table legend identifies the rows of the table with the expressions they represent.

verbose requests that the full table legend be displayed. By default, coefficients and standard errors are not displayed.

seed(#) sets the random-number seed. Specifying this option is equivalent to typing the following command before calling simulate:

. set seed #

Remarks

For an introduction to Monte Carlo methods, see Johnston and DiNardo (1997, 348–358).

▷ Example 1

We have a dataset containing means and variances of 100-observation samples from a lognormal distribution (as a first step in evaluating, say, the coverage of a 95%, *t*-based confidence interval). Then we perform the experiment 1,000 times.

The following command definition will generate 100 independent observations from a lognormal distribution and compute the summary statistics for this sample.

```
program lnsim, rclass
        version 10
        drop _all
        set obs 100
        gen z = exp(invnormal(uniform()))
        summarize z
        return scalar mean = r(mean)
        return scalar Var  = r(Var)
end
```

We can save 1,000 simulated means and variances from lnsim by typing

```
. set seed 1234
. simulate mean=r(mean) var=r(Var), reps(1000) nodots: lnsim
    command:  lnsim
       mean:  r(mean)
        var:  r(Var)
. describe *
```

variable name	storage type	display format	value label	variable label
mean	float	%9.0g		r(mean)
var	float	%9.0g		r(Var)

```
. summarize
```

Variable	Obs	Mean	Std. Dev.	Min	Max
mean	1000	1.641109	.2230346	1.053379	2.47622
var	1000	4.794263	4.597844	.8202238	50.07779

◁

❑ Technical Note

Before executing our lnsim simulator, we can verify that it works by executing it interactively.

```
. set seed 1234
. lnsim
obs was 0, now 100
```

Variable	Obs	Mean	Std. Dev.	Min	Max
z	100	1.581671	1.908433	.0338387	11.18932

```
. return list
scalars:
            r(Var) =  3.642116429498976
           r(mean) =  1.581670569628477
```

❑

▷ Example 2

Consider a more complicated problem. Let's experiment with fitting $y_j = a + bx_j + u_j$ when the true model has $a = 1$, $b = 2$, $u_j = z_j + cx_j$, and when z_j is $N(0,1)$. We will save the parameter estimates and standard errors and experiment with varying c. x_j will be fixed across experiments but will originally be generated as $N(0,1)$. We begin by interactively making the true data:

```
. drop _all
. set obs 100
obs was 0, now 100
```

```
. set seed 54321
. gen x = invnormal(uniform())
. gen true_y = 1+2*x
. save truth
file truth.dta saved
```

Our program is

```
program hetero1
        version 10
        args c
        use truth, clear
        gen y = true_y + (invnormal(uniform()) + `c'*x)
        regress y x
end
```

Note the use of `c` in our statement for generating y. c is a local macro generated from `args c` and thus refers to the first argument supplied to `hetero1`. If we want $c = 3$ for our experiment, we type

```
. simulate _b _se, reps(10000): hetero1 3
  (output omitted )
```

Our program `hetero1` could, however, be more efficient because it rereads the file `truth` once every replication. It would be better if we could read the data just once. In fact, if we read in the data right before running `simulate`, we really shouldn't have to reread for each subsequent replication. A faster version reads

```
program hetero2
        version 10
        args c
        capture drop y
        gen y = true_y + (invnormal(uniform()) + `c'*x)
        regress y x
end
```

Requiring that the current dataset has the variables `true_y` and x may become inconvenient. Another improvement would be to require that the user supply variable names, such as in

```
program hetero3
        version 10
        args truey x c
        capture drop y
        gen y = `truey' + (invnormal(uniform()) + `c'*`x')
        regress y x
end
```

Thus we can type

```
. simulate _b _se, reps(10000): hetero3 true_y x 3
  (output omitted )
```

◁

▷ Example 3

Now let's consider the problem of simulating the ratio of two medians. Suppose that each sample of size n_i comes from a normal population with a mean μ_i and standard deviation σ_i, where $i = 1, 2$. We write the program below and save it as a text file called `myratio.ado` (see [U] **17 Ado-files**). Our program is an `rclass` command that requires six arguments as input, identified by the local macros n1, mu1, sigma1, n2, mu2, and sigma2, which correspond to n_1, μ_1, σ_1, n_2, μ_2, and σ_2, respectively. With these arguments, `myratio` will generate the data for the two samples, use `summarize` to compute the two medians and save the ratio of the medians in `r(ratio)`.

```
program myratio, rclass
        version 10
        args n1 mu1 sigma1 n2 mu2 sigma2
        // generate the data
        drop _all
        local N = 'n1'+'n2'
        set obs 'N'
        tempvar y
        generate 'y' = invnormal(uniform())
        replace 'y' = cond(_n<='n1','mu1'+'y'*'sigma1','mu2'+'y'*'sigma2')
        // calculate the medians
        tempname m1
        summarize 'y' if _n<='n1', detail
        scalar 'm1' = r(p50)
        summarize 'y' if _n>'n1', detail
        // save the results
        return scalar ratio = 'm1' / r(p50)
end
```

The result of running our simulation is

```
. set seed 19192
. simulate ratio=r(ratio), reps(1000) nodots: myratio 5 3 1 10 3 2
      command:  myratio 5 3 1 10 3 2
        ratio:  r(ratio)

. summarize
```

Variable	Obs	Mean	Std. Dev.	Min	Max
ratio	1000	1.085591	.4203416	.3174837	5.936007

◁

❑ Technical Note

Stata lets us do simulations of simulations and simulations of bootstraps. Stata's bootstrap command (see [R] **bootstrap**) works much like simulate, except that it feeds the user-written program a bootstrap sample. Say that we want to evaluate the bootstrap estimator of the standard error of the median when applied to lognormally distributed data. We want to perform a simulation, resulting in a dataset of medians and bootstrap estimated standard errors.

As background, summarize calculates summary statistics, leaving the mean in r(mean) and the variance in r(Var). summarize with the detail option also calculates summary statistics, but more of them, and leaves the median in r(p50).

Thus our plan is to perform simulations by randomly drawing a dataset: we calculate the median of our random sample, we use bootstrap to obtain a dataset of medians calculated from bootstrap samples of our random sample, the standard deviation of those medians is our estimate of the standard error, and the summary statistics are saved in the results of summarize.

Our simulator is

(Continued on next page)

```
program define bsse, rclass
        version 10
        drop _all
        set obs 100
        gen x = invnormal(uniform())
        tempfile bsfile
        bootstrap midp=r(p50), rep(100) saving('bsfile'): summarize x, detail
        use 'bsfile', clear
        summarize midp
        return scalar mean = r(mean)
        return scalar sd   = r(sd)
end
```

We can obtain final results, running our simulation 1,000 times, by typing

```
. set seed 48901

. simulate med=r(mean) bs_se=r(sd), reps(1000): bsse

      command:  bsse
          med:  r(mean)
        bs_se:  r(sd)

Simulations (1000)
———+— 1 —+— 2 —+— 3 —+— 4 —+— 5
.................................................    50
.................................................   100
.................................................   150
.................................................   200
  (output omitted )
.................................................   850
.................................................   900
.................................................   950
.................................................  1000

. summarize
    Variable │       Obs        Mean    Std. Dev.        Min        Max
─────────────┼──────────────────────────────────────────────────────────
         med │      1000    .0002238    .1175051   -.3625843   .3817343
       bs_se │      1000    .1252704    .0294073    .0552211   .2785387
```

This is a case where the simulation dots (drawn by default, unless the `nodots` option is specified) will give us an idea of how long this simulation will take to finish as it runs.   ❏

Methods and Formulas

`simulate` is implemented as an ado-file.

References

Gould, W. W. 1994. ssi6.1: Simplified Monte Carlo simulations. *Stata Technical Bulletin* 20: 22–24. Reprinted in *Stata Technical Bulletin Reprints*, vol. 4, pp. 207–210.

Hamilton, L. C. 2006. *Statistics with Stata (Updated for Version 9)*. Belmont, CA: Duxbury.

Johnston, J., and J. DiNardo. 1997. *Econometric Methods*. 4th ed. New York: McGraw–Hill.

Weesie, J. 1998. ip25: Parameterized Monte Carlo simulations: An enhancement to the simulation command. *Stata Technical Bulletin* 43: 13–15. Reprinted in *Stata Technical Bulletin Reprints*, vol. 8, pp. 75–77.

Also See

Title

sj — Stata Journal and STB installation instructions

Description

The *Stata Journal* (SJ) is a quarterly journal containing articles about statistics, data analysis, teaching methods, and effective use of Stata's language. The SJ publishes reviewed papers together with shorter notes and comments, regular columns, tips, book reviews, and other material of interest to researchers applying statistics in a variety of disciplines. You can read all about the *Stata Journal* at http://www.stata-journal.com.

The *Stata Journal* is a printed and electronic journal with corresponding software. If you want the journal, you must subscribe, but the software is available for free from our web site at http://www.stata-journal.com.

The predecessor to the *Stata Journal* was the *Stata Technical Bulletin* (STB). The STB was also a printed and electronic journal with corresponding software. Individual STB journals may still be purchased. The STB software is available for free from our web site at http://www.stata.com.

Below are instructions for installing the *Stata Journal* and the *Stata Technical Bulletin* software from our web site.

Remarks

Remarks are presented under the following headings:

> *Installing the Stata Journal software*
> > *Obtaining from the Internet by pointing and clicking*
> > *Obtaining from the Internet via command mode*
> *Installing the STB software*
> > *Obtaining from the Internet by pointing and clicking*
> > *Obtaining from the Internet via command mode*

Installing the Stata Journal software

Each issue of the *Stata Journal* is labeled Volume #, Number #. Volume 1 refers to the first year of publication, Volume 2 to the second, and so on. Issues are numbered 1, 2, 3, and 4 within each year. The first issue of the *Journal* was published fourth quarter, 2001, and that issue is numbered Volume 1, Number 1. For installation purposes, we refer to this issue as `sj1-1`.

The articles, columns, notes, and comments that make up the *Stata Journal* are assigned a letter-and-number code, called an insert tag, such as st0001, an0034, or ds0011. The letters represent a category: st is the statistics category, an is the announcements category, etc. The numbers are assigned sequentially, so st0001 is the first article in the statistics category.

Sometimes inserts are subsequently updated, either to fix bugs or to add new features. A number such as st0001_1 indicates that this article, column, note, or comment is an update to the original st0001 article. Updates are complete; that is, installing st0001_1 provides all the features of the original article and more.

The *Stata Journal* software may be obtained by pointing and clicking or by using command mode.

The sections below detail how to install an insert. In all cases, pretend that you wish to install insert st0001_1 from sj2-2.

Obtaining from the Internet by pointing and clicking

1. Select **Help > SJ and User-written Programs**.

2. Click on *Stata Journal*.

3. Click on *sj2-2*.

4. Click on *st0001_1*.

5. Click on *(click here to install)*.

Obtaining from the Internet via command mode

Type the following:

```
. net from http://www.stata-journal.com/software
. net cd sj2-2
. net describe st0001_1
. net install st0001_1
```

The above could be shortened to

```
. net from http://www.stata-journal.com/software/sj2-2
. net describe st0001_1
. net install st0001_1
```

Alternatively, you could type

```
. net sj 2-2
. net describe st0001_1
. net install st0001_1
```

but going about it the long way around is more entertaining, at least the first time.

Installing the STB software

Each issue of the STB is numbered. STB-1 refers to the first issue (published May 1991), STB-2 to the second (published July 1991), and so on.

An issue of the STB consists of inserts—articles—and these are assigned letter-and-number combinations, such as sg84, dm80, sbe26.1, etc. The letters represent a category: sg is the general statistics category and dm the data-management category. The numbers are assigned sequentially, so sbe39 is the 39th insert in the biostatistics and epidemiology series.

Insert sbe39, it turns out, provides a method of accounting for publication bias in meta-analysis; it adds a new command called `metatrim` to Stata. If you installed sbe39, you would have that command and its online help. Insert sbe39 was published in STB-57 (September 2000). Obtaining `metatrim` simply requires going to STB-57 and getting sbe39.

Sometimes inserts were subsequently updated, either to fix bugs or to add new features. sbe39 was updated: the first update is sbe39.1 and the second is sbe39.2. You could install insert sbe39.2, and it would not matter whether you had previously installed sbe39.1. Updates are complete: installing sbe39.2 provides all the features of the original insert and more.

For computer naming purposes, insert sbe39.2 is referred to as `sbe39_2`. When referred to in normal text, however, the insert is still called sbe39.2 because that looks nicer.

Inserts are easily available from the Internet. Inserts may be obtained by pointing and clicking or by using command mode.

The sections below detail how to install an insert. In all cases, pretend that you wish to install insert sbe39.2 from STB-61.

Obtaining from the Internet by pointing and clicking

1. Select **Help > SJ and User-written Programs**.

2. Click on *STB*.

3. Click on *stb61*.

4. Click on *sbe39_2*.

5. Click on *(click here to install)*.

Obtaining from the Internet via command mode

Type the following:

```
. net from http://www.stata.com
. net cd stb
. net cd stb61
. net describe sbe39_2
. net install sbe39_2
```

The above could be shortened to

```
. net from http://www.stata.com/stb/stb61
. net describe sbe39_2
. net install sbe39_2
```

but going about it the long way around is more entertaining, at least the first time.

Also See

[R] **search** — Search Stata documentation

[R] **net** — Install and manage user-written additions from the Internet

[R] **net search** — Search the Internet for installable packages

[R] **update** — Update Stata

[U] **3.5 The Stata Journal and the Stata Technical Bulletin**

[U] **28 Using the Internet to keep up to date**

[GSM] **20 Updating and extending Stata—Internet functionality**

[GSU] **20 Updating and extending Stata—Internet functionality**

[GSW] **20 Updating and extending Stata—Internet functionality**

Title

> **sktest** — Skewness and kurtosis test for normality

Syntax

> sktest *varlist* [*if*] [*in*] [*weight*] [, <u>noa</u>djust]
>
> aweights and fweights are allowed; see [U] **11.1.6 weight**.

Description

For each variable in *varlist*, sktest presents a test for normality based on skewness and another based on kurtosis and then combines the two tests into an overall test statistic. sktest requires a minimum of 8 observations to make its calculations.

Option

⌐ Main ⌐

<u>noa</u>djust suppresses the empirical adjustment made by Royston (1991c) to the overall χ^2 and its significance level and presents the unaltered test as described by D'Agostino, Balanger, and D'Agostino Jr. (1990).

Remarks

Also see [R] **swilk** for the Shapiro–Wilk and Shapiro–Francia tests for normality. Those tests are, in general, preferred for nonaggregated data (Gould and Rogers 1991, Gould 1992, Royston 1991c). Moreover, a normal quantile plot should be used with any test for normality; see [R] **diagnostic plots** for more information.

▷ Example 1

Using our automobile dataset, we will test whether the variables mpg and trunk are normally distributed:

```
. use http://www.stata-press.com/data/r10/auto
(1978 Automobile Data)

. sktest mpg trunk
```

Skewness/Kurtosis tests for Normality

Variable	Pr(Skewness)	Pr(Kurtosis)	adj chi2(2)	joint Prob>chi2
mpg	0.002	0.080	10.95	0.0042
trunk	0.912	0.044	4.19	0.1228

We can reject the hypothesis that mpg is normally distributed, but we cannot reject the hypothesis that trunk is normally distributed, at least at the 12% level. The kurtosis for trunk is 2.19, as can be verified by issuing the command

```
. summarize trunk, detail
```
   (*output omitted*)

and the *p*-value of 0.044 shown in the table above indicates that it is significantly different from the kurtosis of a normal distribution at the 5% significance level. However, on the basis of skewness alone, we cannot reject the hypothesis that trunk is normally distributed.

◁

❑ Technical Note

sktest implements the test as described by D'Agostino, Balanger, and D'Agostino, Jr. (1990) but with the adjustment made by Royston (1991c). In the above example, if we had specified the noadjust option, the χ^2 values would have been 13.13 for mpg and 4.05 for trunk. With the adjustment, the χ^2 value might show as '.'. This result should be interpreted as an absurdly large number; the data are most certainly not normal.

❑

Saved Results

sktest saves the following in r():

Scalars

r(chi2)	χ^2
r(P_skew)	Pr(skewness)
r(P_kurt)	Pr(kurtosis)
r(P_chi2)	Prob > chi2

Methods and Formulas

sktest is implemented as an ado-file.

sktest implements the test described by D'Agostino, Balanger, and D'Agostino, Jr. (1990) with the empirical correction developed by Royston (1991c).

Let g_1 denote the coefficient of skewness and b_2 denote the coefficient of kurtosis as calculated by summarize, and let n denote the sample size. If weights are specified, then g_1, b_2, and n denote the weighted coefficients of skewness and kurtosis and weighted sample size, respectively. See [R] **summarize** for the formulas for skewness and kurtosis.

To perform the test of skewness, we compute

$$Y = g_1 \left\{ \frac{(n+1)(n+3)}{6(n-2)} \right\}^{1/2}$$

$$\beta_2(g_1) = \frac{3(n^2 + 27n - 70)(n+1)(n+3)}{(n-2)(n+5)(n+7)(n+9)}$$

$$W^2 = -1 + \left[2\left\{ \beta_2(g_1) - 1 \right\} \right]^{1/2}$$

and
$$\alpha = \left\{ 2/(W^2 - 1) \right\}^{1/2}$$

Then the distribution of the test statistic

$$Z_1 = \frac{1}{\sqrt{\ln W}} \ln \left[Y/\alpha + \left\{ (Y/\alpha)^2 + 1 \right\}^{1/2} \right]$$

is approximately standard normal under the null hypothesis that the data are distributed normally.

To perform the test of kurtosis, we compute

$$E(b_2) = \frac{3(n-1)}{n+1}$$

$$\text{var}(b_2) = \frac{24n(n-2)(n-3)}{(n+1)^2(n+3)(n+5)}$$

$$X = \left\{ b_2 - E(b_2) \right\} / \sqrt{\text{var}(b_2)}$$

$$\sqrt{\beta_1(b_2)} = \frac{6(n^2 - 5n + 2)}{(n+7)(n+9)} \left\{ \frac{6(n+3)(n+5)}{n(n-2)(n-3)} \right\}^{1/2}$$

and

$$A = 6 + \frac{8}{\sqrt{\beta_1(b_2)}} \left[\frac{2}{\sqrt{\beta_1(b_2)}} + \left\{ 1 + \frac{4}{\beta_1(b_2)} \right\}^{1/2} \right]$$

Then the distribution of the test statistic

$$Z_2 = \frac{1}{\sqrt{2/(9A)}} \left[\left(1 - \frac{2}{9A} \right) - \left\{ \frac{1 - 2/A}{1 + X\sqrt{2/(A-4)}} \right\}^{1/3} \right]$$

is approximately standard normal under the null hypothesis that the data are distributed normally.

D'Agostino, Balanger, and D'Agostino, Jr.'s omnibus test of normality uses the statistic

$$K^2 = Z_1^2 + Z_2^2$$

which has approximately a χ^2 distribution with 2 degrees of freedom under the null of normality.

Royston (1991c) proposed the following adjustment to the test of normality, which `sktest` uses by default. Let $\Phi(x)$ denote the cumulative standard normal distribution function for x, and let $\Phi^{-1}(p)$ denote the inverse cumulative standard normal function [i.e., $x = \Phi^{-1}\{\Phi(x)\}$]. Define the following terms:

$$Z_c = -\Phi^{-1} \left\{ \exp\left(-\frac{1}{2} K^2 \right) \right\}$$

$$Z_t = 0.55 n^{0.2} - 0.21$$

$$a_1 = (-5 + 3.46 \ln n) \exp(-1.37 \ln n)$$

$$b_1 = 1 + (0.854 - 0.148 \ln n) \exp(-0.55 \ln n)$$

$$a_2 = a_1 - \left\{ 2.13/(1 - 2.37 \ln n) \right\} Z_t$$

and

$$b_2 = 2.13/(1 - 2.37 \ln n) + b1$$

If $Z_c < -1$ set $Z = Z_c$; else if $Z_c < Z_t$ set $Z = a_1 + b_1 Z_c$; else set $Z = a_2 + b_2 Z_c$. Define $P = 1 - \Phi(Z)$. Then $K^2 = -2 \ln P$ is approximately distributed χ^2 with 2 degrees of freedom.

The relative merits of the skewness and kurtosis test versus the Shapiro–Wilk and Shapiro–Francia tests have been a subject of debate. The interested reader is directed to the articles in the *Stata Technical Bulletin*. Our recommendation is to use the Shapiro–Francia test whenever possible, that is, whenever dealing with nonaggregated or ungrouped data (Gould and Rogers 1991, Gould 1992); see [R] **swilk**. If normality is rejected, use `sktest` to determine the source of the problem.

As both D'Agostino, Balanger, and D'Agostino, Jr. (1990) and Royston (1991d) mention, researchers should also examine the normal quantile plot to determine normality rather than blindly relying on a few test statistics. See the `qnorm` command documented in [R] **diagnostic plots** for more information on normal quantile plots.

`sktest` is similar in spirit to the Jarque–Bera (1987) test of normality. The Jarque–Bera test statistic is also calculated from the sample skewness and kurtosis, though it is based on asymptotic standard errors with no corrections for sample size. In effect, `sktest` offers two adjustments for sample size, that of Royston (1991c) and that of D'Agostino, Balanger, and D'Agostino, Jr. (1990).

Acknowledgments

`sktest` has benefited greatly by the comments and work of Patrick Royston of the MRC Clinical Trials Unit, London; at this point, the program should be viewed as due as much to Royston as to us, except, of course, for any errors. We are also indebted to Nicholas J. Cox, Durham University, for helpful comments.

References

D'Agostino, R. B., A. Balanger, and R. B. D'Agostino, Jr. 1990. A suggestion for using powerful and informative tests of normality. *American Statistician* 44: 316–321.

———. 1991. sg3.3: Comment on tests of normality. *Stata Technical Bulletin* 3: 20. Reprinted in *Stata Technical Bulletin Reprints*, vol. 1, pp. 105–106.

Gould, W. W. 1991. sg3: Skewness and kurtosis tests of normality. *Stata Technical Bulletin* 1: 20–21. Reprinted in *Stata Technical Bulletin Reprints*, vol. 1, pp. 99–101.

———. 1992. sg3.7: Final summary of tests of normality. *Stata Technical Bulletin* 5: 10–11. Reprinted in *Stata Technical Bulletin Reprints*, vol. 1, pp. 114–115.

Gould, W. W., and W. H. Rogers. 1991. sg3.4: Summary of tests of normality. *Stata Technical Bulletin* 3: 20–23. Reprinted in *Stata Technical Bulletin Reprints*, vol. 1, pp. 106–110.

Jarque, C. M., and A. K. Bera. 1987. A test for normality of observations and regression residuals. *International Statistical Review* 2: 163–172.

Royston, P. 1991a. sg3.1: Tests for departure from normality. *Stata Technical Bulletin* 2: 16–17. Reprinted in *Stata Technical Bulletin Reprints*, vol. 1, pp. 101–104.

———. 1991b. sg3.2: Shapiro–Wilk and Shapiro–Francia tests. *Stata Technical Bulletin* 3: 19. Reprinted in *Stata Technical Bulletin Reprints*, vol. 1, p. 105.

———. 1991c. sg3.5: Comment on sg3.4 and an improved D'Agostino test. *Stata Technical Bulletin* 3: 23–24. Reprinted in *Stata Technical Bulletin Reprints*, vol. 1, pp. 110–112.

———. 1991d. sg3.6: A response to sg3.3: Comment on tests of normality. *Stata Technical Bulletin* 4: 8–9. Reprinted in *Stata Technical Bulletin Reprints*, vol. 1, pp. 112–114.

Also See

[R] **diagnostic plots** — Distributional diagnostic plots

[R] **ladder** — Ladder of powers

[R] **lv** — Letter-value displays

[R] **swilk** — Shapiro–Wilk and Shapiro–Francia tests for normality

Title

> **slogit** — Stereotype logistic regression

Syntax

slogit *depvar* $\left[\,indepvars\,\right]$ $\left[\,if\,\right]$ $\left[\,in\,\right]$ $\left[\,weight\,\right]$ $\left[\,,\ options\,\right]$

options	description
Model	
<u>dim</u>ension(#)	dimension of the model; default is dimension(1)
<u>base</u>outcome(# | *lbl*)	set the base outcome to level # or *lbl*; default is the last outcome
<u>constr</u>aints(*numlist*)	apply specified linear constraints
<u>col</u>linear	keep collinear variables
<u>nocor</u>ner	do not generate the corner constraints
SE/Robust	
vce(*vcetype*)	*vcetype* may be oim, <u>r</u>obust, <u>cl</u>uster *clustvar*, opg, <u>boot</u>strap, or <u>jack</u>knife
Reporting	
<u>l</u>evel(#)	set confidence level; default is level(95)
Max options	
maximize_options	control the maximization process; seldom used
<u>init</u>ialize(*initype*)	method of initializing scale parameters; *initype* can be constant, random, or svd; see *Options* for details
<u>nonorm</u>alize	do not normalize the numeric variables

bootstrap, by, jackknife, rolling, statsby, svy, and xi are allowed; see [U] **11.1.10 Prefix commands**.
Weights are not allowed with the bootstrap prefix.
vce() and weights are not allowed with the svy prefix.
fweights, iweights, and pweights are allowed; see [U] **11.1.6 weight**.
See [U] **20 Estimation and postestimation commands** for more capabilities of estimation commands.

Description

slogit fits maximum-likelihood stereotype logistic regression models as developed by Anderson (1984). Like multinomial logistic and ordered logistic models, stereotype logistic models are for use with categorical dependent variables. In a multinomial logistic model, the categories cannot be ranked, whereas in an ordered logistic model the categories follow a natural ranking scheme. You can view stereotype logistic models as a compromise between those two models. You can use them when you are unsure of the relevance of the ordering, as is often the case when subjects are asked to assess or judge something. You can also use them in place of multinomial logistic models when you suspect that some of the alternatives are similar. Unlike ordered logistic models, stereotype logistic models do not impose the proportional-odds assumption.

Options

dimension(*#*) specifies the dimension of the model, which is the number of equations required to describe the relationship between the dependent variable and the independent variables. The maximum dimension is $\min(m-1, p)$, where m is the number of categories of the dependent variable and p is the number of independent variables in the model. The stereotype model with maximum dimension is a reparameterization of the multinomial logistic model.

baseoutcome(*# | lbl*) specifies the outcome level whose scale parameters and intercept are constrained to be zero. The base outcome may be specified as a number of a label. By default, slogit assumes that the outcome levels are ordered and uses the largest level of the dependent variable as the base outcome.

constraints(*numlist*), collinear; see [R] **estimation options**.

By default, the linear equality constraints suggested by Anderson (1984), termed the corner constraints, are generated for you. You can add constraints to these as needed, or you can turn off the corner constraints by specifying nocorner. These constraints are in addition to the constraints placed on the ϕ parameters corresponding to baseoutcome(*#*).

nocorner specifies that slogit not generate the corner constraints. If you specify nocorner, you must specify at least dimension() $\times$ dimension() constraints for the model to be identified.

vce(*vcetype*) specifies the type of standard error reported, which includes types that are derived from asymptotic theory, that are robust to some kinds of misspecification, that allow for intragroup correlation, and that use bootstrap or jackknife methods; see [R] *vce_option*.

If specifying vce(bootstrap) or vce(jackknife), you must also specify baseoutcome().

level(*#*); see [R] **estimation options**.

maximize_options: difficult, technique(*algorithm_spec*), iterate(*#*), [no]log, trace, gradient, showstep, hessian, shownrtolerance, tolerance(*#*), ltolerance(*#*), gtolerance(*#*), nrtolerance(*#*), nonrtolerance, from(*init_specs*); see [R] **maximize**. These options are seldom used.

Setting the optimization type to technique(bhhh) resets the default *vcetype* to vce(opg).

initialize(constant | random | svd) specifies how initial estimates are computed. The default, initialize(constant), is to set the scale parameters to the constant $\min(1/2, 1/d)$, where d is the dimension specified in dimension().

initialize(random) requests that uniformly distributed random numbers between 0 and 1 be used as initial values for the scale parameters. If you specify this option, you should also use set seed to ensure that you can replicate your results (see [D] **generate**).

initialize(svd) requests that a singular value decomposition be performed on the matrix of regression estimates from mlogit to reduce its rank to the dimension specified in dimension(). slogit uses the reduced-rank components of the SVD as initial estimates for the scale and regression coefficients. For details, see *Methods and Formulas*.

`nonormalize` specifies that the numeric variables not be normalized. Normalization of the numeric variables improves numerical stability but consumes more memory in generating temporary double-precision variables. Variables that are of type `byte` are not normalized, and if initial estimates are specified using the `from()` option, normalization of variables is not performed. See *Methods and Formulas* for more information.

Remarks

Remarks are presented under the following headings:

> *Introduction*
> *One-dimensional model*
> *Higher-dimension models*

Introduction

Stereotype logistic models are often used when subjects are requested to assess or judge something. For example, consider a survey in which consumers may be asked to rate the quality of a product on a scale from 1 to 5, 1 indicating poor quality and 5 indicating excellent quality. If the categories are monotonically related to an underlying latent variable, the ordered logistic model is appropriate. However, suppose that consumers assess quality not just along one dimension, but rather weigh two or three latent factors. Stereotype logistic regression allows you to specify multiple equations to capture the effects of those latent variables, which you then parameterize in terms of observable characteristics. Unlike with multinomial logit, the number of equations you specify could be less than $m - 1$, where m is the number of categories of the dependent variable.

Stereotype logistic models are also used when categories may be indistinguishable. Suppose that a consumer must choose among A, B, C, or D. Multinomial logistic modeling assumes that the four choices are distinct in the sense that a consumer choosing one of the goods can distinguish its characteristics from the others. If goods A and B are in fact similar, consumers may be randomly picking between the two. One alternative is to combine the two categories and fit a three-category multinomial logistic model. A more flexible alternative is to use a stereotype logistic model.

In the multinomial logistic model, you estimate $m - 1$ parameter vectors $\widetilde{\boldsymbol{\beta}}_k$, $k = 1, \ldots, m - 1$, where m is the number of categories of the dependent variable. The stereotype logistic model is a restriction on the multinomial model in the sense that there are d parameter vectors, where d is between one and $\min(m - 1, p)$, and p is the number of regressors. The relationship between the stereotype model's coefficients $\boldsymbol{\beta}_j$, $j = 1, \ldots, d$, and the multinomial model's coefficients is $\widetilde{\boldsymbol{\beta}}_k = -\sum_{j=1}^{d} \phi_{jk} \boldsymbol{\beta}_j$. The ϕs are scale parameters to be estimated along with the $\boldsymbol{\beta}_j$s.

Given a row vector of covariates $\mathbf{x}$, let $\eta_k = \theta_k - \sum_{j=1}^{d} \phi_{jk} \mathbf{x} \boldsymbol{\beta}_j$. The probability of observing outcome k is

$$
\Pr(Y_i = k) = \begin{cases} \dfrac{\exp(\eta_k)}{1 + \sum_{l=1}^{m-1} \exp(\eta_l)} & k < m \\[2ex] \dfrac{1}{1 + \sum_{l=1}^{m-1} \exp(\eta_l)} & k = m \end{cases}
$$

This model includes a set of θ parameters so that each equation has an unrestricted constant term. If $d = m - 1$, the stereotype model is just a reparameterization of the multinomial logistic model. To identify the ϕs and the $\boldsymbol{\beta}$s, you must place at least d^2 restrictions on the parameters. By default, `slogit` uses the "corner constraints" $\phi_{jj} = 1$ and $\phi_{jk} = 0$ for $j \neq k$, $k \leq d$, and $j \leq d$.

One-dimensional model

▷ Example 1

We have 2 years of repair rating data on the make, price, mileage rating, and gear ratio of 104 foreign and 44 domestic automobiles (with 13 missing values on repair rating). We wish to fit a stereotype logistic model to discriminate between the levels of repair rating using mileage, price, gear ratio, and origin of the manufacturer. Here is an overview of our data:

```
. use http://www.stata-press.com/data/r10/auto2yr
(Automobile Models)
. tabulate repair
     repair |      Freq.     Percent        Cum.
------------+-----------------------------------
       Poor |          5        3.70        3.70
       Fair |         19       14.07       17.78
    Average |         57       42.22       60.00
       Good |         38       28.15       88.15
  Excellent |         16       11.85      100.00
------------+-----------------------------------
      Total |        135      100.00
```

The variable `repair` can take five values, 1, ..., 5, which represent the subjective rating of the car model's repair record as *Poor*, *Fair*, *Average*, *Good*, and *Excellent*.

We wish to fit the following one-dimensional stereotype logistic model

$$\eta_k = \theta_k - \phi_k \left(\beta_1 \texttt{foreign} + \beta_2 \texttt{mpg} + \beta_3 \texttt{price} + \beta_4 \texttt{gratio}\right)$$

for $k < 5$ and $\eta_5 = 0$. To fit this model, we type

```
. slogit repair foreign mpg price gratio
Iteration 0:    log likelihood =  -173.7818  (not concave)
Iteration 1:    log likelihood = -164.77318
Iteration 2:    log likelihood =  -161.7069
Iteration 3:    log likelihood = -159.76139
Iteration 4:    log likelihood = -159.34328
Iteration 5:    log likelihood = -159.25914
Iteration 6:    log likelihood = -159.25691
Iteration 7:    log likelihood = -159.25691

Stereotype logistic regression                Number of obs   =        135
                                              Wald chi2(4)    =       9.33
Log likelihood = -159.25691                   Prob > chi2     =     0.0535

 ( 1)  [phi1_1]_cons = 1
```

repair	Coef.	Std. Err.	z	P>|z|	[95% Conf. Interval]	
foreign	5.947382	2.094126	2.84	0.005	1.84297	10.05179
mpg	.1911968	.08554	2.24	0.025	.0235414	.3588521
price	-.0000576	.0001357	-0.42	0.671	-.0003236	.0002083
gratio	-4.307571	1.884713	-2.29	0.022	-8.00154	-.6136017
/phi1_1	1	.	.	.	.	.
/phi1_2	1.262268	.3530565	3.58	0.000	.5702904	1.954247
/phi1_3	1.17593	.3169397	3.71	0.000	.5547394	1.79712
/phi1_4	.8657195	.2411228	3.59	0.000	.3931275	1.338311
/phi1_5	0	(base outcome)				
/theta1	-6.864749	4.21252	-1.63	0.103	-15.12114	1.391639
/theta2	-7.613977	4.861803	-1.57	0.117	-17.14294	1.914981
/theta3	-5.80655	4.987508	-1.16	0.244	-15.58189	3.968786
/theta4	-3.85724	3.824132	-1.01	0.313	-11.3524	3.637922
/theta5	0	(base outcome)				

```
(repair=Excellent is the base outcome)
```

The coefficient associated with the first scale parameter, ϕ_{11}, is 1, and its standard error and other statistics are missing. This is the corner constraint applied to the one-dimensional model; in the header, this constraint is listed as [phi1_1]_cons = 1. Also the ϕ and θ parameters that are associated with the base outcome are identified. Keep in mind, though, that there are no coefficient estimates for [phi1_5]_cons or [theta5]_cons in the ereturn matrix e(b). The Wald statistic is for a test of the joint significance of the regression coefficients on foreign, mpg, price, and gratio.

The one-dimensional stereotype model restricts the multinomial logistic regression coefficients $\widetilde{\beta}_k$, $k = 1, \ldots, m - 1$ to be parallel; that is, $\widetilde{\beta}_k = -\phi_k \beta$. As Lunt (2001) discusses, in the one-dimensional stereotype model, one linear combination $x_i \beta$ best discriminates the outcomes of the dependent variable, and the scale parameters ϕ_k measure the distance between the outcome levels and the linear predictor. If $\phi_1 \geq \phi_2 \geq \cdots \phi_{m-1} \geq \phi_m \equiv 0$, the model suggests that the subjective assessment of the dependent variable is indeed ordered. Here the maximum likelihood estimates of the ϕs are not monotonic, as would be assumed in an ordered logit model.

We test that $\phi_1 = \phi_2$ by typing

```
. test [phi1_2]_cons = [phi1_1]_cons
 ( 1)  - [phi1_1]_cons + [phi1_2]_cons = 0
        chi2( 1) =     0.55
      Prob > chi2 =    0.4576
```

Since the two parameters are not statistically different, we decide to add a constraint to force $\phi_1 = \phi_2$:

```
. constraint 1 [phi1_2]_cons = [phi1_1]_cons

. slogit repair foreign mpg price gratio, constraint(1) nolog
```

```
Stereotype logistic regression              Number of obs   =        135
                                            Wald chi2(4)    =      21.28
Log likelihood = -159.65769                 Prob > chi2     =     0.0003

 ( 1)  [phi1_1]_cons = 1
 ( 2)  - [phi1_1]_cons + [phi1_2]_cons = 0
```

repair	Coef.	Std. Err.	z	P>|z|	[95% Conf. Interval]	
foreign	7.166515	1.690177	4.24	0.000	3.853829	10.4792
mpg	.2340043	.0807042	2.90	0.004	.0758271	.3921816
price	-.000041	.0001618	-0.25	0.800	-.0003581	.000276
gratio	-5.218107	1.798717	-2.90	0.004	-8.743528	-1.692686
/phi1_1	1	.	.	.	.	.
/phi1_2	1	.	.	.	.	.
/phi1_3	.9751096	.1286563	7.58	0.000	.7229478	1.227271
/phi1_4	.7209343	.1220353	5.91	0.000	.4817494	.9601191
/phi1_5	0	(base outcome)				
/theta1	-8.293452	4.645182	-1.79	0.074	-17.39784	.8109368
/theta2	-6.958451	4.629292	-1.50	0.133	-16.0317	2.114795
/theta3	-5.620232	4.953981	-1.13	0.257	-15.32986	4.089392
/theta4	-3.745624	3.809189	-0.98	0.325	-11.2115	3.720249
/theta5	0	(base outcome)				

```
(repair=Excellent is the base outcome)
```

The ϕ estimates are now monotonically decreasing and the standard errors of the ϕs are small relative to the size of the estimates, so we conclude that, with the exception of outcomes *Poor* and *Fair*, the groups are distinguishable for the one-dimensional model and that the quality assessment can be ordered.

◁

Higher-dimension models

The stereotype logistic model is not limited to ordered categorical dependent variables; you can use it on nominal data to reduce the dimension of the regressions. Recall that a multinomial model fitted to a categorical dependent variable with m levels will have $m - 1$ sets of regression coefficients. However, a model with fewer dimensions may fit the data equally well, suggesting that some of the categories are indistinguishable.

▷ Example 2

As discussed in [R] **mlogit**, we have data on the type of health insurance available to 616 psychologically depressed subjects in the United States (Tarlov et al. 1989; Wells et al. 1989). Patients may have either an indemnity (fee-for-service) plan or a prepaid plan, such as an HMO, or may be uninsured. Demographic variables include age, gender, race, and site.

First, we fit the saturated, two-dimensional model that is equivalent to a multinomial logistic model. We choose the base outcome to be 1 (indemnity insurance) since that is the default for `mlogit`.

```
. use http://www.stata-press.com/data/r10/sysdsn3
(Health insurance data)

. slogit insure age male nonwhite site2 site3, dim(2) base(1)

Iteration 0:   log likelihood = -534.36165
Iteration 1:   log likelihood = -534.36165

Stereotype logistic regression                  Number of obs   =        615
                                                Wald chi2(10)   =      38.17
Log likelihood = -534.36165                     Prob > chi2     =     0.0000

 ( 1)  [phi1_2]_cons = 1
 ( 2)  [phi1_3]_cons = 0
 ( 3)  [phi2_2]_cons = 0
 ( 4)  [phi2_3]_cons = 1
```

insure	Coef.	Std. Err.	z	P>|z|	[95% Conf. Interval]	
dim1						
age	.011745	.0061946	1.90	0.058	-.0003962	.0238862
male	-.5616934	.2027465	-2.77	0.006	-.9590693	-.1643175
nonwhite	-.9747768	.2363213	-4.12	0.000	-1.437958	-.5115955
site2	-.1130359	.2101903	-0.54	0.591	-.5250013	.2989296
site3	.5879879	.2279351	2.58	0.010	.1412433	1.034733
dim2						
age	.0077961	.0114418	0.68	0.496	-.0146294	.0302217
male	-.4518496	.3674867	-1.23	0.219	-1.17211	.268411
nonwhite	-.2170589	.4256361	-0.51	0.610	-1.05129	.6171725
site2	1.211563	.4705127	2.57	0.010	.2893747	2.133751
site3	.2078123	.3662926	0.57	0.570	-.510108	.9257327
/phi1_1	0	(base outcome)				
/phi1_2	1	.	.	.	.	.
/phi1_3	0	.	.	.	.	.
/phi2_1	0	(base outcome)				
/phi2_2	0	.	.	.	.	.
/phi2_3	1	.	.	.	.	.
/theta1	0	(base outcome)				
/theta2	.2697127	.3284422	0.82	0.412	-.3740222	.9134476
/theta3	-1.286943	.5923219	-2.17	0.030	-2.447872	-.1260134

```
(insure=Indemnity is the base outcome)
```

For comparison, we also fit the model by using `mlogit`:

(Continued on next page)

```
. mlogit insure age male nonwhite site2 site3, nolog
```

Multinomial logistic regression

Number of obs	=	615
LR chi2(10)	=	42.99
Prob > chi2	=	0.0000

Log likelihood = -534.36165

Pseudo R2	=	0.0387

insure	Coef.	Std. Err.	z	P>|z|	[95% Conf. Interval]	
Prepaid						
age	-.011745	.0061946	-1.90	0.058	-.0238862	.0003962
male	.5616934	.2027465	2.77	0.006	.1643175	.9590693
nonwhite	.9747768	.2363213	4.12	0.000	.5115955	1.437958
site2	.1130359	.2101903	0.54	0.591	-.2989296	.5250013
site3	-.5879879	.2279351	-2.58	0.010	-1.034733	-.1412433
_cons	.2697127	.3284422	0.82	0.412	-.3740222	.9134476
Uninsure						
age	-.0077961	.0114418	-0.68	0.496	-.0302217	.0146294
male	.4518496	.3674867	1.23	0.219	-.268411	1.17211
nonwhite	.2170589	.4256361	0.51	0.610	-.6171725	1.05129
site2	-1.211563	.4705127	-2.57	0.010	-2.133751	-.2893747
site3	-.2078123	.3662926	-0.57	0.570	-.9257327	.510108
_cons	-1.286943	.5923219	-2.17	0.030	-2.447872	-.1260135

(insure==Indemnity is the base outcome)

Apart from having opposite signs, the coefficients from the stereotype logistic model are identical to those from the multinomial logit model. Recall the definition of η_k given in the *Remarks*, particularly the minus sign in front of the summation. One other difference in the output is that the constant estimates labeled /theta in the slogit output are the constants labeled _cons in the mlogit output.

Next we examine the one-dimensional model.

```
. slogit insure age male nonwhite site2 site3, dim(1) base(1) nolog
```

Stereotype logistic regression

Number of obs	=	615
Wald chi2(5)	=	28.20
Prob > chi2	=	0.0000

Log likelihood = -539.75205

(1)   [phi1_2]_cons = 1

insure	Coef.	Std. Err.	z	P>|z|	[95% Conf. Interval]	
age	.0108366	.0061918	1.75	0.080	-.0012992	.0229723
male	-.5032537	.2078171	-2.42	0.015	-.9105678	-.0959396
nonwhite	-.9480351	.2340604	-4.05	0.000	-1.406785	-.489285
site2	-.2444316	.2246366	-1.09	0.277	-.6847113	.1958481
site3	.556665	.2243799	2.48	0.013	.1168886	.9964415
/phi1_1	0	(base outcome)				
/phi1_2	1	.	.	.	.	.
/phi1_3	.0383539	.4079705	0.09	0.925	-.7612535	.8379613
/theta1	0	(base outcome)				
/theta2	.187542	.3303847	0.57	0.570	-.4600001	.835084
/theta3	-1.860134	.2158898	-8.62	0.000	-2.28327	-1.436997

(insure=Indemnity is the base outcome)

We have reduced a two-dimensional multinomial model to one dimension, reducing the number of estimated parameters by four and decreasing the model likelihood by ≈ 5.4.

slogit does not report a model likelihood-ratio test. The test of $d = 1$ (a one-dimensional model) versus $d = 0$ (the null model) does not have an asymptotic χ^2 distribution because the unconstrained ϕ parameters (/phi1_3 in the previous example) cannot be identified if $\boldsymbol{\beta} = 0$. More generally, this problem precludes testing any hierarchical model of dimension d versus $d - 1$. Of course, the likelihood-ratio test of a full-dimension model versus $d = 0$ is valid because the full model is just multinomial logistic, and all the ϕ parameters are fixed at 0 or 1.

◁

❑ Technical Note

The stereotype model is a special case of the reduced-rank vector generalized linear model discussed by Yee and Hastie (2003). If we define $\eta_{ik} = \theta_k - \sum_{j=1}^{d} \phi_{jk} \mathbf{x}_i \boldsymbol{\beta}_j$, for $k = 1, \ldots, m - 1$, we can write the expression in matrix notation as

$$\boldsymbol{\eta}_i = \boldsymbol{\theta} + \boldsymbol{\Phi} \left(\mathbf{x}_i \mathbf{B} \right)'$$

where $\boldsymbol{\Phi}$ is a $(m - 1) \times d$ matrix containing the ϕ_{jk} parameters and $\mathbf{B}$ is a $p \times d$ matrix with columns containing the $\boldsymbol{\beta}_j$ parameters, $j = 1, \ldots, d$. The factorization $\boldsymbol{\Phi} \mathbf{B}'$ is not unique since $\boldsymbol{\Phi} \mathbf{B}' = \boldsymbol{\Phi} \mathbf{M} \mathbf{M}^{-1} \mathbf{B}'$ for any nonsingular $d \times d$ matrix $\mathbf{M}$. To avoid this identifiability problem, we choose $\mathbf{M} = \boldsymbol{\Phi}_1^{-1}$, where

$$\boldsymbol{\Phi} = \begin{pmatrix} \boldsymbol{\Phi}_1 \\ \boldsymbol{\Phi}_2 \end{pmatrix}$$

and $\boldsymbol{\Phi}_1$ is $d \times d$ of rank d so that

$$\boldsymbol{\Phi} \mathbf{M} = \begin{pmatrix} \mathbf{I}_d \\ \boldsymbol{\Phi}_2 \boldsymbol{\Phi}_1^{-1} \end{pmatrix}$$

and $\mathbf{I}_d$ is a $d \times d$ identity matrix. Thus the corner constraints used by slogit are $\phi_{jj} \equiv 1$ and $\phi_{jk} \equiv 0$ for $j \neq k$ and $k, j \leq d$.

❑

(Continued on next page)

Saved Results

slogit saves the following in e():

Scalars

e(N)	number of observations	e(i_base)	base outcome index
e(k)	number of parameters	e(ll)	log likelihood
e(k_indvars)	number of independent variables	e(ll_0)	null model log likelihood
e(k_out)	number of outcomes	e(N_clust)	number of clusters
e(k_eq)	number of equations in e(b)	e(chi2)	χ^2
e(k_eq_model)	number of equations in model Wald test	e(p)	significance
		e(ic)	number of iterations
e(df_m)	Wald test degrees of freedom	e(rank)	rank of e(V)
e(df_0)	null model degrees of freedom	e(rc)	return code
e(k_dim)	model dimension	e(converged)	1 if converged, 0 otherwise

Macros

e(cmd)	slogit	e(vcetype)	title used to label Std. Err.
e(cmdline)	command as typed	e(diparm#)	display transformed parameter #
e(depvar)	name of dependent variable	e(opt)	type of optimization
e(indvars)	independent variables	e(ml_method)	type of ml method
e(wtype)	weight type	e(user)	name of likelihood-evaluator program
e(wexp)	weight expression	e(technique)	maximization technique
e(title)	title in estimation output	e(crittype)	optimization criterion
e(clustvar)	name of cluster variable	e(properties)	b V
e(out#)	outcome labels, $\# = 1 \ldots$ e(k_out)	e(footnote)	program implementing the footnote display
e(chi2type)	Wald		
e(labels)	outcome labels or numeric levels	e(predict)	program implementing predict
e(vce)	*vcetype* specified in vce()		

Matrices

e(b)	coefficient vector	e(V)	variance–covariance matrix of the estimators
e(outcomes)	outcome values		
e(ilog)	iteration log (up to 20 iterations)	e(gradient)	gradient vector

Functions

e(sample)	marks estimation sample

Methods and Formulas

slogit is implemented as an ado-file.

slogit obtains the maximum likelihood estimates for the stereotype logistic model using ml; see [R] **ml**. Each set of regression estimates, one set of β_js for each dimension, constitutes one ml model equation. The $d \times (m-1)$ ϕs and the $(m-1)$ θs are ml ancillary parameters.

Without loss of generality, let the base outcome level be the mth level of the dependent variable. Define the row vector $\phi_k = (\phi_{1k}, \ldots, \phi_{dk})$ for $k = 1, \ldots, m-1$, and define the $p \times d$ matrix $\mathbf{B} = (\beta_1, \ldots, \beta_d)$. For observation i, the log odds of outcome level k relative to level m, $k = 1, \ldots, m-1$ is the index

$$\ln \left\{ \frac{\Pr(Y_i = k)}{\Pr(Y_i = m)} \right\} = \eta_{ik} = \theta_k - \phi_k \left(\mathbf{x}_i \mathbf{B} \right)'$$
$$= \theta_k - \phi_k \nu_i'$$

The row vector $\boldsymbol{\nu}_i$ can be interpreted as a latent variable reducing the p-dimensional vector of covariates to a more interpretable $d < p$ dimension.

The probability of the ith observation having outcome level k is then

$$
\Pr(Y_i = k) = p_{ik} =
\begin{cases}
\dfrac{e^{\eta_{ik}}}{1 + \sum_{j=1}^{m-1} e^{\eta_{ij}}}, & \text{if } \ k < m \\[2ex]
\dfrac{1}{1 + \sum_{j=1}^{m-1} e^{\eta_{ij}}}, & \text{if } \ k = m
\end{cases}
$$

from which the log-likelihood function is computed as

$$
L = \sum_{i=1}^{n} w_i \sum_{k=1}^{m} I_k(y_i) \ln(p_{ik}) \tag{1}
$$

Here w_i is the weight for observation i and

$$
I_k(y_i) =
\begin{cases}
1, & \text{if observation } y_i \text{ has outcome } k \\
0, & \text{otherwise}
\end{cases}
$$

Numeric variables are normalized for numerical stability during optimization where a new double-precision variable $\widetilde{x}_j$ is created from variable x_j, $j = 1, \ldots, p$, such that $\widetilde{x}_j = (x_j - \bar{x}_j)/s_j$. This feature is turned off if you specify $\texttt{nonormalize}$, or if you use the $\texttt{from()}$ option for initial estimates. Normalization is not performed on byte variables, including the indicator variables generated by [R] $\textbf{xi}$. The linear equality constraints for regression parameters, if specified, must be scaled also. Assume that a constraint is applied to the regression parameter associated with variable j and dimension i, β_{ji}, and the corresponding element of the constraint matrix (see [P] $\textbf{makecns}$) is divided by s_j.

After convergence, the parameter estimates for variable j and dimension i—$\widetilde{\beta}_{ji}$, say—are transformed back to their original scale, $\beta_{ji} = \widetilde{\beta}_{ji}/s_j$. For the intercepts, you compute

$$
\theta_k = \widetilde{\theta}_k + \sum_{i=1}^{d} \phi_{ik} \sum_{j=1}^{p} \frac{\widetilde{\beta}_{ji} \bar{x}_j}{s_j}
$$

Initial values are computed using estimates obtained by using $\texttt{mlogit}$ to fit a multinomial logistic model. Let the $p \times (m-1)$ matrix $\widetilde{\mathbf{B}}$ contain the multinomial logistic regression parameters less the $m-1$ intercepts. Each ϕ is initialized with constant values $\min(1/2, 1/d)$, option $\texttt{initialize(constant)}$ (the default), or, with uniform random numbers, option $\texttt{initialize(random)}$. Constraints are then applied to the starting values so that the structure of the $(m-1) \times d$ matrix $\boldsymbol{\Phi}$ is

$$
\boldsymbol{\Phi} =
\begin{pmatrix}
\phi_1 \\
\phi_2 \\
\vdots \\
\phi_{m-1}
\end{pmatrix}
=
\begin{pmatrix}
\mathbf{I}_d \\
\widetilde{\boldsymbol{\Phi}}
\end{pmatrix}
$$

where $\mathbf{I}_d$ is a $d \times d$ identity matrix. Assume that only the corner constraints are used, but any constraints you place on the scale parameters are also applied to the initial scale estimates, so the structure of $\boldsymbol{\Phi}$ will change accordingly. The ϕ parameters are invariant to the scale of the covariates, so initial estimates in $[0, 1]$ are reasonable. The constraints guarantee that the rank of $\boldsymbol{\Phi}$ is at least d, so the initial estimates for the stereotype regression parameters are obtained from $\mathbf{B} = \widetilde{\mathbf{B}}\boldsymbol{\Phi}(\boldsymbol{\Phi}'\boldsymbol{\Phi})^{-1}$.

One other approach for initial estimates is provided: `initialize(svd)`. It starts with the `mlogit` estimates and computes $\widetilde{\mathbf{B}}' = \mathbf{UDV}'$, where $\mathbf{U}_{m-1 \times p}$ and $\mathbf{V}_{p \times p}$ are orthonormal matrices and $\mathbf{D}_{p \times p}$ is a diagonal matrix containing the singular values of $\widetilde{\mathbf{B}}$. The estimates for $\boldsymbol{\Phi}$ and $\mathbf{B}$ are the first d columns of $\mathbf{U}$ and $\mathbf{VD}$, respectively (Yee and Hastie 2003).

The score for regression coefficients is

$$\mathbf{u}_i(\boldsymbol{\beta}_j) = \frac{\partial L_{ik}}{\partial \boldsymbol{\beta}_j} = \mathbf{x}_i \left(\sum_{l=1}^{m-1} \phi_{jl} p_{il} - \phi_{jk} \right)$$

the score for the scale parameters is

$$u_i(\phi_{jl}) = \frac{\partial L_{ik}}{\partial \phi_{jl}} = \begin{cases} \mathbf{x}_i \boldsymbol{\beta}_j (p_{ik} - 1), & \text{if } l = k \\ \mathbf{x}_i \boldsymbol{\beta}_j p_{il}, & \text{if } l \neq k \end{cases}$$

for $l = 1, \ldots, m - 1$; and the score for the intercepts is

$$u_i(\theta_l) = \frac{\partial L_{ik}}{\partial \theta_l} = \begin{cases} 1 - p_{ik}, & \text{if } l = k \\ -p_{il}, & \text{if } l \neq k \end{cases}$$

References

Anderson, J. A. 1984. Regression and ordered categorical variables. *Journal of the Royal Statistical Society, Series B* 46: 1–30.

Lunt, M. 2001. sg163: Stereotype ordinal regression. *Stata Technical Bulletin* 61: 12–18. Reprinted in *Stata Technical Bulletin Reprints*, vol. 10, pp. 298–307.

Tarlov, A. R., J. E. Ware, Jr., S. Greenfield, E. C. Nelson, E. Perrin, and M. Zubkoff. 1989. The medical outcomes study. *Journal of the American Medical Association* 262: 925–930.

Wells, K. E., R. D. Hays, M. A. Burnam, W. H. Rogers, S. Greenfield, and J. E. Ware, Jr. 1989. Detection of depressive disorder for patients receiving prepaid or fee-for-service care. *Journal of the American Medical Association* 262: 3298–3302.

Yee, T. W., and T. J. Hastie. 2003. Reduced-rank vector generalized linear models. *Statistical Modelling* 3: 1–41.

Also See

[R] **slogit postestimation** — Postestimation tools for slogit

[R] **roc** — Receiver operating characteristic (ROC) analysis

[R] **logistic** — Logistic regression, reporting odds ratios

[R] **mlogit** — Multinomial (polytomous) logistic regression

[R] **ologit** — Ordered logistic regression

[R] **oprobit** — Ordered probit regression

[U] **20 Estimation and postestimation commands**

Title

slogit postestimation — Postestimation tools for slogit

Description

The following postestimation commands are available for `slogit`:

command	description
adjust	adjusted predictions of $\mathbf{x}\beta$ or $\exp(\mathbf{x}\beta)$
estat	AIC, BIC, VCE, and estimation sample summary
estat (svy)	postestimation statistics for survey data
estimates	cataloging estimation results
lincom	point estimates, standard errors, testing, and inference for linear combinations of coefficients
lrtest[1]	likelihood-ratio test
mfx	marginal effects or elasticities
nlcom	point estimates, standard errors, testing, and inference for nonlinear combinations of coefficients
predict	predicted probabilities, estimated index and its approximate standard error
predictnl	point estimates, standard errors, testing, and inference for generalized predictions
suest	seemingly unrelated estimation
test	Wald tests for simple and composite linear hypotheses
testnl	Wald tests of nonlinear hypotheses

[1] `lrtest` is not appropriate with `svy` estimation results.

See the corresponding entries in the *Stata Base Reference Manual* for details, but see [SVY] **estat** for details about `estat` (svy).

Syntax for predict

predict [*type*] { *stub* | *newvar* | *newvarlist* } [*if*] [*in*] [, *statistic* $\underline{\text{o}}$utcome(*outcome*)]

predict [*type*] { *stub* | *newvarlist* } [*if*] [*in*], $\underline{\text{sc}}$ores

statistic	description
Main	
p̲r	probability of a positive outcome; the default
xb	linear prediction
stdp	standard error of the linear prediction

You specify one new variable with `pr` and you specify `outcome()`, or you specify *stub* or k new variables with `pr`, where k is the number of outcomes.

You specify one new variable with `xb` and `stdp` and you specify `outcome()`.

These statistics are available both in and out of sample; type `predict ... if e(sample) ...` if wanted only for the estimation sample.

Options for predict

⌐‾‾ Main ⌐‾‾

pr, the default, calculates the probability of each of the categories of the dependent variable or the probability of the level specified in outcome(*outcome*). If you specify the outcome(*outcome*) option, you need to specify only one new variable; otherwise, you must specify a new variable for each category of the dependent variable.

xb calculates the index, $\theta_k - \sum_{j=1}^{d} \phi_{jk} \mathbf{x}_i \boldsymbol{\beta}_j$, for outcome level $k \neq$ e(i_base) and dimension $d =$ e(k_dim). It returns a vector of zeros if $k =$ e(i_base). A synonym for xb is index. This option requires the outcome(*outcome*) option.

stdp calculates the standard error of the index. A synonym for stdp is seindex. This option requires the outcome(*outcome*) option.

outcome(*outcome*) specifies the outcome for which the statistic is to be calculated. equation() is a synonym for outcome(): it does not matter which you use. outcome() or equation() can be specified using

> #1, #2, ..., where #1 means the first category of the dependent variable, #2 the second category, etc.;

> the values of the dependent variable; or

> the value labels of the dependent variable if they exist.

scores calculates the equation-level score variables. For models with d dimensions and m levels, $d + (d + 1)(m - 1)$ new variables are created. Assume $j = 1, \ldots, d$ and $k = 1, \ldots, m$ in the following.

The first d new variables will contain $\partial \ln L / \partial(\mathbf{x}\boldsymbol{\beta}_j)$.

The next $d(m - 1)$ new variables will contain $\partial \ln L / \partial \phi_{jk}$.

The last $m - 1$ new variables will contain $\partial \ln L / \partial \theta_k$.

Remarks

Once you have fitted a stereotype logistic model, you can obtain the predicted probabilities by using the predict command for both the estimation sample and other samples; see [U] **20 Estimation and postestimation commands** and [R] **predict**.

predict without arguments (or with the pr option) calculates the predicted probability of each outcome of the dependent variable. You must therefore give a new variable name for each of the outcomes. To compute the estimated probability of one outcome, you use the outcome(*outcome*) option where *outcome* is the level encoding the outcome. If the dependent variable's levels are labeled, the outcomes can also be identified by the label values (see [D] **label**).

The option xb in conjunction with outcome(*outcome*) specifies that the index be computed for the outcome encoded by level *outcome*. Its approximate standard error is computed if the stdp option is specified. Only one of the options pr, xb, or stdp can be specified with a call to predict.

▷ Example 1

In example 2 of [R] **slogit**, we fitted the one-dimensional stereotype model, where the *depvar* is insure with levels $k = 1$ for outcome *Indemnity*, $k = 2$ for *Prepaid*, and $k = 3$ for *Uninsure*. The base outcome for the model is *Indemnity*, so for $k \neq 1$ the vector of indices for the kth level is

$$\boldsymbol{\eta}_k = \theta_k - \phi_k \left(\beta_1 \text{age} + \beta_2 \text{male} + \beta_3 \text{nonwhite} + \beta_4 \text{site2} + \beta_5 \text{site3}\right)$$

We estimate the group probabilities by calling `predict` after `slogit`.

```
. predict pIndemnity pPrepaid pUninsure, p
. list pIndemnity pPrepaid pUninsure insure in 1/10
```

	pIndemn~y	pPrepaid	pUninsure	insure
1.	.54193437	.37548746	.08257817	Indemnity
2.	.43596386	.496328	.06770813	Prepaid
3.	.51115833	.41051074	.07833093	Indemnity
4.	.39411326	.54422342	.06166332	Prepaid
5.	.46556509	.46250642	.07192849	.
6.	.44017793	.49151025	.06831183	Prepaid
7.	.46321218	.46519304	.07159478	Prepaid
8.	.37723017	.5635696	.05920023	.
9.	.48677573	.43830181	.07492246	Uninsure
10.	.58236678	.32958016	.08805306	Prepaid

Observations 5 and 8 are not used to fit the model since `insure` is missing at these points, but `predict` estimates the probabilities for these observations since none of the independent variables is missing. You can use `if e(sample)` in the call to `predict` to use only those observations that are used to fit the model.

◁

Methods and Formulas

All postestimation commands listed above are implemented as ado-files.

predict

Let level b be the base outcome that is used to fit the stereotype logistic regression model of dimension d. The index for observation i and level $k \neq b$ is $\eta_{ik} = \theta_k - \sum_{j=1}^{d} \phi_{jk}\mathbf{x}_i\boldsymbol{\beta}_j$. This is the log odds of outcome encoded as level k relative to that of b so that we define $\eta_{ib} \equiv 0$. The outcome probabilities for this model are defined as $\Pr(Y_i = k) = e^{\eta_{ik}} / \sum_{j=1}^{m} e^{\eta_{ij}}$. Unlike in `mlogit`, `ologit`, and `oprobit`, the index is no longer a linear function of the parameters. The standard error of index η_{ik} is thus computed using the delta method (see also [R] **predictnl**).

The equation-level score for regression coefficients is

$$\frac{\partial \ln L_{ik}}{\partial \mathbf{x}_i\boldsymbol{\beta}_j} = \left(\sum_{l=1}^{m-1} \phi_{jl}p_{il} - \phi_{jk} \right)$$

the equation-level score for the scale parameters is

$$\frac{\partial \ln L_{ik}}{\partial \phi_{jl}} = \begin{cases} \mathbf{x}_i\boldsymbol{\beta}_j(p_{ik} - 1), & \text{if } l = k \\ \mathbf{x}_i\boldsymbol{\beta}_j p_{il}, & \text{if } l \neq k \end{cases}$$

for $l = 1, \ldots, m-1$; and the equation-level score for the intercepts is

$$\frac{\partial \ln L_{ik}}{\partial \theta_l} = \begin{cases} 1 - p_{ik}, & \text{if } l = k \\ -p_{il}, & \text{if } l \neq k \end{cases}$$

Also See

[R] **slogit** — Stereotype logistic regression

[U] **20 Estimation and postestimation commands**

Title

smooth — Robust nonlinear smoother

Syntax

smooth *smoother* [,twice] *varname* [*if*] [*in*], generate(*newvar*)

where *smoother* is specified as $Sm[Sm[\ldots]]$ and *Sm* is one of

$$\{\,1\,|\,2\,|\,3\,|\,4\,|\,5\,|\,6\,|\,7\,|\,8\,|\,9\,\}[R]$$
$$3[R]S[S|R][S|R]\ldots$$
$$E$$
$$H$$

Letters may be specified in lowercase if preferred. Examples of *smoother* [,twice] include

3RSSH	3RSSH,twice	4253H	4253H,twice	43RSR2H,twice
3rssh	3rssh,twice	4253h	4253h,twice	43rsr2h,twice

Description

smooth applies the specified resistant, nonlinear smoother to *varname* and stores the smoothed series in *newvar*.

Option

generate(*newvar*) is required; it specifies the name of the new variable that will contain the smoothed values.

Remarks

Smoothing is an exploratory data-analysis technique for making the general shape of a series apparent. In this approach (Tukey 1977), the observed data series is assumed to be the sum of an underlying process that evolves smoothly (the smooth) and of an unsystematic noise component (the rough); that is,

$$\text{data} = \text{smooth} + \text{rough}$$

Smoothed values z_t are obtained by taking medians (or some other location estimate) of each point in the original data y_t and a few of the points around it. The number of points used is called the span of the smoother. Thus a span-3 smoother produces z_t by taking the median of y_{t-1}, y_t, and y_{t+1}. smooth provides running median smoothers of spans 1 to 9—indicated by the digit that specifies their span. Median smoothers are resistant to isolated outliers, so they provide robustness to spikes in the data. Since the median is also a nonlinear operator, such smoothers are known as robust (or resistant) nonlinear smoothers.

smooth also provides the Hanning linear, nonrobust smoother, indicated by the letter H. Hanning is a span-3 smoother with binomial weights. Repeated applications of H—for example, HH, HHH, etc.— provide binomial smoothers of span 5, 7, etc. See Cox (1997, 2004) for a graphical application of this fact.

Because one smoother usually cannot adequately separate the smooth from the rough, compound smoothers—multiple smoothers applied in sequence—are used. The smoother 35H, for instance, smooths the data with a span-3 median smoother, smooths the result with a span-5 median smoother, and finally smooths that result with the Hanning smoother. smooth allows you to specify any number of smoothers in any sequence.

Three refinements can be combined with the running median and Hanning smoothers. First, the endpoints of a smooth can be given special treatment. This is specified by the E operator. Second, smoothing by 3, the span-3 running median, tends to produce flat-topped hills and valleys. The splitting operator, S, "splits" these repeated values, applies the endpoint operator to them, and then "rejoins" the series. Finally, it is sometimes useful to repeat an odd-span median smoother or the splitting operator until the smooth no longer changes. Following a digit or an S with an R specifies this type of repetition.

Even the best smoother may fail to separate the smooth from the rough adequately. To guard against losing any systematic components of the data series, after smoothing, the smoother can be reapplied to the resulting rough, and any recovered signal can be added back to the original smooth. The twice operator specifies this procedure. More generally, an arbitrary smoother can be applied to the rough (using a second smooth command), and the recovered signal can be added back to the smooth. This more general procedure is called reroughing (Tukey 1977).

The details of each of the smoothers and operators are explained in *Methods and Formulas* below.

▷ Example 1

smooth is designed to recover the general features of a series that has been contaminated with noise. To demonstrate this, we construct a series, add noise to it, and then smooth the noisy version to recover an estimate of the original data. First, we construct and display the data:

```
. drop _all
. set obs 10
. set seed 123456789
. generate time = _n
. label variable time "Time"
. generate x = _n^3 - 10*_n^2 + 5*_n
. label variable x "Signal"
. generate z = x + 50*invnormal(uniform())
. label variable z "Observed series"
```

(Continued on next page)

```
. scatter x z time, c(l .) m(i o) ytitle("")
```

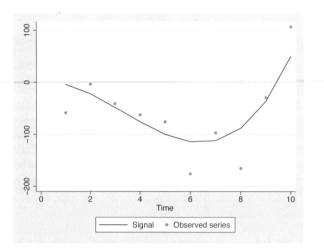

Now we smooth the noisy series, z, assumed to be the only data we would observe:

```
. smooth 4253eh,twice z, gen(sz)
. label variable sz "Smoothed series"
. scatter x z sz time, c(l . l) m(i o i) ytitle("") || scatter sz time,
> c(l . l) m(i o i) ytitle("") clpatter(dash_dot)
```

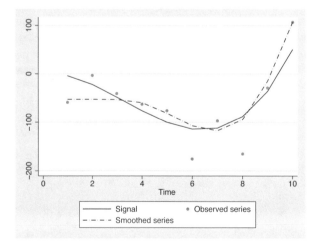

◁

▷ Example 2

Salgado-Ugarte and Curts-García (1993) provide data on the frequencies of observed fish lengths. In this example, the series to be smoothed—the frequencies—is ordered by fish length rather than by time.

```
. use http://www.stata-press.com/data/r10/fishdata, clear
. smooth 4253eh,twice freq, gen(sfreq)
. label var sfreq "4253EH,twice of frequencies"
```

```
. scatter sfreq freq length, c(l .) m(i o)
> title("Smoothed frequencies of fish lengths") ytitle("") xlabel(#4)
```

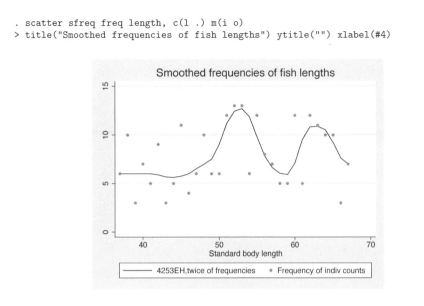

◁

❑ Technical Note

smooth allows missing values at the beginning and end of the series, but missing values in the middle are not allowed. Leading and trailing missing values are ignored. If you wish to ignore missing values in the middle of the series, you must drop the missing observations before using smooth. Doing so, of course, would violate smooth's assumption that observations are equally spaced—each observation represents a year, a quarter, or a month (or a 1-year birth-rate category). In practice, smooth produces good results as long as the spaces between adjacent observations do not vary too much.

Smoothing is usually applied to time series, but any variable with a natural order can be smoothed. For example, a smoother might be applied to the birth rate recorded by the age of the mothers (birth rate for 17-year-olds, birth rate for 18-year-olds, and so on).

❑

Methods and Formulas

smooth is implemented as an ado-file.

Running median smoothers of odd span

The smoother 3 defines

$$z_t = \text{median}(y_{t-1}, y_t, y_{t+1})$$

The smoother 5 defines

$$z_t = \text{median}(y_{t-2}, y_{t-1}, y_t, y_{t+1}, y_{t+2})$$

and so on. The smoother 1 defines $z_t = \text{median}(y_t)$, so it does nothing.

Endpoints are handled by using smoothers of shorter, odd span. Thus for 3,

$$z_1 = y_1$$
$$z_2 = \text{median}(y_1, y_2, y_3)$$
$$\vdots$$
$$z_{N-1} = \text{median}(y_{N-2}, y_{N-1}, y_N)$$
$$Z_N = y_N$$

For 5,

$$z_1 = y_1$$
$$z_2 = \text{median}(y_1, y_2, y_3)$$
$$z_3 = \text{median}(y_1, y_2, y_3, y_4, y_5)$$
$$z_4 = \text{median}(y_2, y_3, y_4, y_5, y_6)$$
$$\vdots$$
$$z_{N-2} = \text{median}(y_{N-4}, y_{N-3}, y_{N-2}, y_{N-1}, y_N)$$
$$z_{N-1} = \text{median}(y_{N-2}, y_{N-1}, y_N)$$
$$Z_N = y_N$$

and so on.

Running median smoothers of even span

Define the median() function as returning the linearly interpolated value when given an even number of arguments. Thus the smoother 2 defines

$$z_{t+.5} = (y_t + y_{t+1})/2$$

The smoother 4 defines $z_{t+.5}$ as the linearly interpolated median of $(y_{t-1}, y_t, y_{t+1}, y_{t+2})$, and so on. In all cases, endpoints are handled by using smoothers of shorter, even span. Thus for 4,

$$z_{.5} = y_1$$
$$z_{1.5} = \text{median}(y_1, y_2) = (y_1 + y_2)/2$$
$$z_{2.5} = \text{median}(y_1, y_2, y_3, y_4)$$
$$\vdots$$
$$z_{N-2.5} = \text{median}(y_{N-4}, y_{N-3}, y_{N-2}, y_N)$$
$$z_{N-1.5} = \text{median}(y_{N-2}, y_{N-1})$$
$$z_{N-.5} = \text{median}(y_{N-1}, y_N)$$
$$z_{N+.5} = y_N$$

As defined above, an even-span smoother increases the length of the series by 1 observation. However, the series can be recentered on the original observation numbers, and the "extra" observation can be eliminated by smoothing the series again with another even-span smoother. For instance, the smooth of 4 illustrated above could be followed by a smooth of 2 to obtain

$$z_1^* = (z_{.5} + z_{1.5})/2$$
$$z_2^* = (z_{1.5} + z_{2.5})/2$$
$$z_3^* = (z_{2.5} + z_{3.5})/2$$
$$\vdots$$
$$z_{N-2}^* = (z_{N-2.5} + z_{N-1.5})/2$$
$$z_{N-1}^* = (z_{N-1.5} + z_{N-.5})/2$$
$$z_N^* = (z_{N-.5} + z_{N+.5})/2$$

smooth keeps track of the number of even smoothers applied to the data and expands and shrinks the length of the series accordingly. To ensure that the final smooth has the same number of observations as *varname*, smooth requires you to specify an even number of even-span smoothers. However, the pairs of even-span smoothers need not be contiguous; for instance, 4253 and 4523 are both allowed.

Repeat operator

R indicates that a smoother is to be repeated until convergence, that is, until repeated applications of the smoother produce the same series. Thus 3 applies the smoother of running medians of span 3. 33 applies the smoother twice. 3R produces the result of repeating 3 an infinite number of times. R should be used only with odd-span smoothers since even-span smoothers are not guaranteed to converge.

The smoother 453R2 applies a span-4 smoother, followed by a span-5 smoother, followed by repeated applications of a span-3 smoother, followed by a span-2 smoother.

Endpoint rule

The endpoint rule E modifies the values z_1 and z_N according to the following formulas:

$$z_1 = \text{median}(3z_2 - 2z_3, z_1, z_2)$$
$$z_N = \text{median}(3z_{N-2} - 2z_{N-1}, z_N, z_{N-1})$$

When the endpoint rule is not applied, endpoints are typically "copied in"; that is, $z_1 = y_1$ and $z_N = y_N$.

Splitting operator

The smoothers 3 and 3R can produce flat-topped hills and valleys. The split operator attempts to eliminate such hills and valleys by splitting the sequence, applying the endpoint rule E, rejoining the series, and then resmoothing by 3R.

The S operator may be applied only after 3, 3R, or S.

We recommend that the S operator be repeated once (SS) or until no further changes take place (SR).

Hanning smoother

H is the Hanning linear smoother:

$$z_t = (y_{t-1} + 2y_t + y_{t+1})/4$$

Endpoints are copied in: $z_1 = y_1$ and $z_N = y_N$. H should be applied only after all nonlinear smoothers.

Twicing

A smoother divides the data into a smooth and a rough:

$$\text{data} = \text{smooth} + \text{rough}$$

If the smoothing is successful, the rough should exhibit no pattern. Twicing refers to applying the smoother to the observed, calculating the rough, and then applying the smoother to the rough. The resulting "smoothed rough" is then added back to the smooth from the first step.

Acknowledgments

smooth was originally written by William Gould (1992)—at which time it was named nlsm—and was inspired by Salgado-Ugarte and Curts-García (1992). Salgado-Ugarte and Curts-García (1993) subsequently reported anomalies in nlsm's treatment of even-span median smoothers. smooth corrects these problems and incorporates other improvements but otherwise is essentially the same as originally published.

References

Cox, N. J. 1997. gr22: Binomial smoothing plot. *Stata Technical Bulletin* 35: 7–9. Reprinted in *Stata Technical Bulletin Reprints*, vol. 6, pp. 36–38.

———. 2004. Software update for bsmplot. *Stata Journal* 4: 490.

———. 2005. Speaking Stata: Smoothing in various directions. *Stata Journal* 5: 574–593.

Gould, W. W. 1992. sed7.1: Resistant nonlinear smoothing using Stata. *Stata Technical Bulletin* 8: 9–12. Reprinted in *Stata Technical Bulletin Reprints*, vol. 2, pp. 104–107.

Royston, P., and N. J. Cox. 2005. A multivariable scatterplot smoother. *Stata Journal* 5: 405–412.

Salgado-Ugarte, I., and J. Curts-García. 1992. sed7: Resistant smoothing using Stata. *Stata Technical Bulletin* 7: 8–11. Reprinted in *Stata Technical Bulletin Reprints*, vol. 2, pp. 99–103.

———. 1993. sed7.2: Twice reroughing procedure for resistant nonlinear smoothing. *Stata Technical Bulletin* 11: 14–16. Reprinted in *Stata Technical Bulletin Reprints*, vol. 2, pp. 108–111.

Sasieni, P. 1998. gr27: An adaptive variable span running line smoother. *Stata Technical Bulletin* 41: 4–7. Reprinted in *Stata Technical Bulletin Reprints*, vol. 7, pp. 63–68.

Tukey, J. W. 1977. *Exploratory Data Analysis*. Reading, MA: Addison–Wesley.

Velleman, P. F. 1977. Robust nonlinear data smoothers: Definitions and recommendations. *Proceedings of the National Academy of Sciences USA* 74: 434–436.

———. 1980. Definition and comparison of robust nonlinear data smoothing algorithms. *Journal of the American Statistical Association* 75: 609–615.

Velleman, P. F., and D. C. Hoaglin. 1981. *Applications, Basics, and Computing of Exploratory Data Analysis*. Boston: Duxbury.

Also See

[R] **lowess** — Lowess smoothing

[TS] **tssmooth** — Smooth and forecast univariate time-series data

Title

> **spearman** — Spearman's and Kendall's correlations

Syntax

Spearman's rank correlation coefficients

> spearman [*varlist*] [*if*] [*in*] [, *spearman_options*]

Kendall's rank correlation coefficients

> ktau [*varlist*] [*if*] [*in*] [, *ktau_options*]

spearman_options	description
Main	
stats(*spearman_list*)	list of statistics; select up to three statistics; default is stats(rho)
print(#)	significance level for displaying coefficients
star(#)	significance level for displaying with a star
bonferroni	use Bonferroni-adjusted significance level
sidak	use Šidák-adjusted significance level
pw	calculate all pairwise correlation coefficients by using all available data
matrix	display output in matrix form

ktau_options	description
Main	
stats(*ktau_list*)	list of statistics; select up to six statistics; default is stats(taua)
print(#)	significance level for displaying coefficients
star(#)	significance level for displaying with a star
bonferroni	use Bonferroni-adjusted significance level
sidak	use Šidák-adjusted significance level
pw	calculate all pairwise correlation coefficients by using all available data
matrix	display output in matrix form

by is allowed with spearman and ktau; see [D] **by**.

where the elements of *spearman_list* may be

rho	correlation coefficient
obs	number of observations
p	significance level

and the elements of *ktau_list* may be

taua	correlation coefficient τ_a
taub	correlation coefficient τ_b
score	score
se	standard error of score
obs	number of observations
p	significance level

Description

spearman displays Spearman's rank correlation coefficients for all pairs of variables in *varlist* or, if *varlist* is not specified, all the variables in the dataset.

ktau displays Kendall's rank correlation coefficients between the variables in *varlist* or, if *varlist* is not specified, all the variables in the dataset. ktau is intended for use on small- and moderate-sized datasets; it requires considerable computation time for larger datasets.

Options for spearman

⌐ Main ⌐

stats(*spearman_list*) specifies the statistics to be displayed in the matrix of output. stats(rho) is the default. Up to three statistics may be specified. stats(rho obs p) would display the correlation coefficient, number of observations, and significance level. If *varlist* contains only two variables, all statistics are shown in tabular form, and stats(), print(), and star() have no effect unless the matrix option is specified.

print(#) specifies the significance level of correlation coefficients to be printed. Correlation coefficients with larger significance levels are left blank in the matrix. Typing spearman, print(.10) would list only those correlation coefficients that are significant at the 10% level or lower.

star(#) specifies the significance level of correlation coefficients to be marked with a star. Typing spearman, star(.05) would "star" all correlation coefficients significant at the 5% level or lower.

bonferroni makes the Bonferroni adjustment to calculated significance levels. This adjustment affects printed significance levels and the print() and star() options. Thus spearman, print(.05) bonferroni prints coefficients with Bonferroni-adjusted significance levels of .05 or less.

sidak makes the Šidák adjustment to calculated significance levels. This adjustment affects printed significance levels and the print() and star() options. Thus spearman, print(.05) sidak prints coefficients with Šidák-adjusted significance levels of .05 or less.

pw specifies that correlations be calculated by using pairwise deletion of observations with missing values. By default, spearman uses casewise deletion, where observations are ignored if any of the variables in *varlist* are missing.

matrix forces spearman to display the statistics as a matrix, even if *varlist* contains only two variables. matrix is implied if more than two variables are specified.

Options for ktau

⌐ Main ⌐

stats(*ktau_list*) specifies the statistics to be displayed in the matrix of output. stats(taua) is the default. Up to six statistics may be specified; stats(taua taub score se obs p) would display the correlation coefficients τ_a, τ_b, score, standard error of score, number of observations, and significance level. If *varlist* contains only two variables, all statistics are shown in tabular form and stats(), print(), and star() have no effect unless the matrix option is specified.

print(#) specifies the significance level of correlation coefficients to be printed. Correlation coefficients with larger significance levels are left blank in the matrix. Typing ktau, print(.10) would list only those correlation coefficients that are significant at the 10% level or lower.

star(#) specifies the significance level of correlation coefficients to be marked with a star. Typing ktau, star(.05) would "star" all correlation coefficients significant at the 5% level or lower.

bonferroni makes the Bonferroni adjustment to calculated significance levels. This adjustment affects printed significance levels and the print() and star() options. Thus ktau, print(.05) bonferroni prints coefficients with Bonferroni-adjusted significance levels of .05 or less.

sidak makes the Šidák adjustment to calculated significance levels. This adjustment affects printed significance levels and the print() and star() options. Thus ktau, print(.05) sidak prints coefficients with Šidák-adjusted significance levels of .05 or less.

pw specifies that correlations be calculated by using pairwise deletion of observations with missing values. By default, ktau uses casewise deletion, where observations are ignored if any of the variables in *varlist* are missing.

matrix forces ktau to display the statistics as a matrix, even if *varlist* contains only two variables. matrix is implied if more than two variables are specified.

Remarks

▷ Example 1

We wish to calculate the correlation coefficients among marriage rate (mrgrate), divorce rate (divorce_rate), and median age (medage) in state data. We can calculate the standard Pearson correlation coefficients and significance by typing

```
.use http://www.stata-press.com/data/r10/states2
(State data)
. pwcorr mrgrate divorce_rate medage, sig
```

	mrgrate	divorc~e	medage
mrgrate	1.0000		
divorce_rate	0.7895	1.0000	
	0.0000		
medage	0.0011	-0.1526	1.0000
	0.9941	0.2900	

(*Continued on next page*)

We can calculate Spearman's rank correlation coefficients by typing

```
. spearman mrgrate divorce_rate medage, stats(rho p)
(obs=50)
```

```
┌─────────────┐
│ Key         │
├─────────────┤
│ rho         │
│ Sig. level  │
└─────────────┘
```

	mrgrate	divorc~e	medage
mrgrate	1.0000		
divorce_rate	0.6933	1.0000	
	0.0000		
medage	-0.4869	-0.2455	1.0000
	0.0003	0.0857	

The large difference in the results is caused by one observation. Nevada's marriage rate is almost 10 times higher than the state with the next-highest marriage rate. An important feature of the Spearman rank correlation coefficient is its reduced sensitivity to extreme values compared with the Pearson correlation coefficient.

We can calculate Kendall's rank correlations by typing

```
. ktau mrgrate divorce_rate medage, stats(taua taub p)
(obs=50)
```

```
┌─────────────┐
│ Key         │
├─────────────┤
│ tau_a       │
│ tau_b       │
│ Sig. level  │
└─────────────┘
```

	mrgrate	divorc~e	medage
mrgrate	0.9829		
	1.0000		
divorce_rate	0.5110	0.9804	
	0.5206	1.0000	
	0.0000		
medage	-0.3486	-0.1698	0.9845
	-0.3544	-0.1728	1.0000
	0.0004	0.0828	

There are tied values for variables `mrgrate`, `divorce_rate`, and `medage`, so tied ranks are used. As a result, $\tau_a < 1$ on the diagonal (see *Methods and Formulas* for the definition of τ_a).

◁

❑ Technical Note

According to Conover (1999, 323), "Spearman's ρ tends to be larger than Kendall's τ in absolute value. However, as a test of significance, there is no strong reason to prefer one over the other because both will produce nearly identical results in most cases."

❑

▷ Example 2

We illustrate `spearman` and `ktau` with the auto data, which contains some missing values.

```
.use http://www.stata-press.com/data/r10/auto
(1978 Automobile Data)
. spearman mpg rep78
 Number of obs =        69
Spearman's rho =        0.3098

Test of Ho: mpg and rep78 are independent
     Prob > |t| =       0.0096
```

Because we specified two variables, `spearman` displayed the sample size, correlation, and p-value in tabular form. To obtain just the correlation coefficient displayed in matrix form, we type

```
. spearman mpg rep78, stats(rho) matrix
(obs=69)
```

	mpg	rep78
mpg	1.0000	
rep78	0.3098	1.0000

The `pw` option instructs `spearman` and `ktau` to use all nonmissing observations between a pair of variables when calculating their correlation coefficient. In the output below, some correlations are based on 74 observations, whereas others are based on 69 because 5 observations contain a missing value for `rep78`.

```
. spearman mpg price rep78, pw stats(rho obs p)  star(0.01)
```

Key
rho
Number of obs
Sig. level

	mpg	price	rep78
mpg	1.0000		
	74		
price	-0.5419*	1.0000	
	74	74	
	0.0000		
rep78	0.3143*	0.1025	1.0000
	69	69	69
	0.0085	0.4022	

Finally, the `bonferroni` and `sidak` options provide adjusted significance levels:

```
. ktau mpg price rep78, stats(taua taub score se p) bonferroni
(obs=69)
```

Key
tau_a
tau_b
score
se of score
Sig. level

	mpg	price	rep78
mpg	0.9471		
	1.0000		
	2222.0000		
	191.8600		
price	-0.3973	1.0000	
	-0.4082	1.0000	
	-932.0000	2346.0000	
	192.4561	193.0682	
	0.0000		
rep78	0.2076	0.0648	0.7136
	0.2525	0.0767	1.0000
	487.0000	152.0000	1674.0000
	181.7024	182.2233	172.2161
	0.0224	1.0000	

◁

Charles Edward Spearman (1863–1945) was a British psychologist who made contributions to correlation, factor analysis, test reliability, and psychometrics. After several years' military service, he obtained a Ph.D. in experimental psychology at Leipzig and became a professor at University College London, where he sustained a long program of work on the interpretation of intelligence tests. Ironically, the rank correlation version bearing his name is not the formula he advocated.

Maurice George Kendall (1907–1983) was a British statistician who contributed to rank correlation, time series, multivariate analysis, and other topics, and wrote many statistical texts. Most notably, perhaps, his advanced survey of the theory of statistics went through several editions, later ones with Alan Stuart; the baton has since passed to others. Kendall was employed in turn as a government and business statistician, as professor at the London School of Economics, as a consultant, and as director of the World Fertility Survey. He was knighted in 1974.

Saved Results

spearman saves the following in r():

Scalars

r(N)	number of observations (last variable pair)	r(p)	two-sided p-value (last variable pair)
r(rho)	ρ (last variable pair)		

Matrices

r(Nobs)	number of observations	r(P)	two-sided p-value
r(Rho)	ρ		

ktau saves the following in r():

Scalars

r(N)	number of observations (last variable pair)	r(score)	Kendall's score (last variable pair)
r(tau_a)	τ_a (last variable pair)	r(se_score)	s.e. of score (last variable pair)
r(tau_b)	τ_b (last variable pair)	r(p)	two-sided p-value (last variable pair)

Matrices

r(Nobs)	number of observations	r(Score)	Kendall's score
r(Tau_a)	τ_a	r(Se_Score)	standard error of score
r(Tau_b)	τ_b	r(P)	two-sided p-value

Methods and Formulas

spearman and ktau are implemented as ado-files.

Spearman's (1904) rank correlation is calculated as Pearson's correlation computed on the ranks and average ranks (Conover 1999, 314–315). Ranks are as calculated by egen; see [D] **egen**. The significance is calculated using the approximation

$$p = 2 \times \texttt{ttail}(n - 2, |\widehat{\rho}| \sqrt{n - 2} / \sqrt{1 - \widehat{\rho}^2})$$

For any two pairs of ranks (x_i, y_i) and (x_j, y_j) of one variable pair (*varname*$_1$, *varname*$_2$), $1 \le i, j \le n$, where n is the number of observations, define them as concordant if

$$(x_i - x_j)(y_i - y_j) > 0$$

and discordant if this product is less than zero.

Kendall's (1938; also see Kendall and Gibbons 1990) score S is defined as $C - D$, where C (D) is the number of concordant (discordant) pairs. Let $N = n(n - 1)/2$ be the total number of pairs, so τ_a is given by

$$\tau_a = S/N$$

and τ_b is given by

$$\tau_b = \frac{S}{\sqrt{N - U}\sqrt{N - V}}$$

where

$$U = \sum_{i=1}^{N_1} u_i(u_i - 1)/2$$

$$V = \sum_{j=1}^{N_2} v_j(v_j - 1)/2$$

and where N_1 is the number of sets of tied x values, u_i is the number of tied x values in the ith set, N_2 is the number of sets of tied y values, and v_j is the number of tied y values in the jth set. Under the null hypothesis of independence between $varname_1$ and $varname_2$, the variance of S is exactly (Kendall and Gibbons 1990, 66)

$$\text{Var}(S) = \frac{1}{18}\left\{ n(n-1)(2n+5) - \sum_{i=1}^{N_1} u_i(u_i-1)(2u_i+5) - \sum_{j=1}^{N_2} v_j(v_j-1)(2v_j+5) \right\}$$

$$+ \frac{1}{9n(n-1)(n-2)}\left\{\sum_{i=1}^{N_1} u_i(u_i-1)(u_i-2)\right\}\left\{\sum_{j=1}^{N_2} v_j(v_j-1)(v_j-2)\right\}$$

$$+ \frac{1}{2n(n-1)}\left\{\sum_{i=1}^{N_1} u_i(u_i-1)\right\}\left\{\sum_{j=1}^{N_2} v_j(v_j-1)\right\}$$

Using a normal approximation with a continuity correction,

$$z = \frac{|S| - 1}{\sqrt{\text{Var}(S)}}$$

For the hypothesis of independence, the statistics S, τ_a, and τ_b produce equivalent tests and give the same significance.

For Kendall's τ, the normal approximation is surprisingly accurate for sample sizes as small as 8, at least for calculating p-values under the null hypothesis for continuous variables. (See Kendall and Gibbons, 1990, chap. 4, who also present some tables for calculating exact p-values for $n < 10$.) For Spearman's ρ, the normal approximation requires larger samples to be valid.

Let v be the number of variables specified so that $k = v(v-1)/2$ correlation coefficients are to be estimated. If `bonferroni` is specified, the adjusted significance level is $p' = \min(1, kp)$. If `sidak` is specified, $p' = \min\left\{1, 1-(1-p)^n\right\}$. See *Methods and Formulas* in [R] **oneway** for a more complete description of the logic behind these adjustments.

Early work on rank correlation is surveyed by Kruskal (1958).

Acknowledgment

The original version of `ktau` was written by Sean Becketti, a past editor of the *Stata Technical Bulletin*.

References

Barnard, G. A. 1997. Kendall, Maurice George. In *Leading Personalities in Statistical Sciences from the Seventeenth Century to the Present*, ed. N. L. Johnson and S. Kotz, 130–132. New York: Wiley.

Conover, W. J. 1999. *Practical Nonparametric Statistics*. 3rd ed. New York: Wiley.

David, H. A., and W. A. Fuller. 2007. Sir Maurice Kendall (1907–1983): A centenary appreciation. *American Statistician* 61: 41–46.

Jeffreys, H. 1961. *Theory of Probability*. 3rd ed. Oxford: Oxford University Press.

Kendall, M. G. 1938. A new measure of rank correlation. *Biometrika* 30: 81–93.

Kendall, M. G., and J. D. Gibbons. 1990. *Rank Correlation Methods*. 5th ed. New York: Oxford University Press.

Kruskal, W. H. 1958. Ordinal measures of association. *Journal of the American Statistical Association* 53: 814–861.

Lovie, P., and A. D. Lovie. 1996. Charles Edward Spearman, F.R.S. (1863–1945). *Notes and Records of the Royal Society of London* 50: 1–14.

Newson, R. 2000a. snp15: somersd—Confidence intervals for nonparametric statistics and their differences. *Stata Technical Bulletin* 55: 47–55. Reprinted in *Stata Technical Bulletin Reprints*, vol. 10, pp. 312–322.

——. 2000b. snp15.1: Update to somersd. *Stata Technical Bulletin* 57: 35. Reprinted in *Stata Technical Bulletin Reprints*, vol. 10, pp. 322–323.

——. 2000c. snp15.2: Update to somersd. *Stata Technical Bulletin* 58: 30. Reprinted in *Stata Technical Bulletin Reprints*, vol. 10, p. 323.

——. 2001. snp15.3: Update to somersd. *Stata Technical Bulletin* 61: 22. Reprinted in *Stata Technical Bulletin Reprints*, vol. 10, p. 324.

——. 2003. snp15_4: Software updates for somersd. *Stata Journal* 3: 325.

——. 2005. snp15_5: Software updates for somersd. *Stata Journal* 5: 470.

——. 2006. snp15_6: Confidence intervals for rank statistics: Somers' D and extensions. *Stata Journal* 6: 309–334.

Seed, P. T. 2001. sg159: Confidence intervals for correlations. *Stata Technical Bulletin* 59: 27–28. Reprinted in *Stata Technical Bulletin Reprints*, vol. 10, pp. 267–269.

Spearman, C. 1904. The proof and measurement of association between two things. *American Journal of Psychology* 15: 72–101.

Wolfe, F. 1997. sg64: pwcorrs: An enhanced correlation display. *Stata Technical Bulletin* 35: 22–25. Reprinted in *Stata Technical Bulletin Reprints*, vol. 6, pp. 163–167.

——. 1999. sg64.1: Update to pwcorrs. *Stata Technical Bulletin* 49: 17. Reprinted in *Stata Technical Bulletin Reprints*, vol. 9, p. 159.

Also See

Title

spikeplot — Spike plots and rootograms

Syntax

spikeplot *varname* $\left[\textit{if}\right]$ $\left[\textit{in}\right]$ $\left[\textit{weight}\right]$ $\left[\textit{, options}\right]$

options	description
Main	
<u>r</u>ound(*#*)	round *varname* to nearest multiple of *#* (bin width)
<u>frac</u>tion	make vertical scale the proportion of total values; default is frequencies
root	make vertical scale show square roots of frequencies
Plot	
spike_options	affect rendition of plotted spikes
Add plots	
addplot(*plot*)	add other plots to generated graph
Y axis, X axis, Titles, Legend, Overall, By	
twoway_options	any options documented in [G] ***twoway_options***

fweights, aweights, and iweights are allowed; see [U] **11.1.6 weight**

Description

spikeplot produces a frequency plot for a variable in which the frequencies are depicted as vertical lines from zero. The frequency may be a count, a fraction, or the square root of the count (Tukey's rootogram, circa 1965). The vertical lines may also originate from a baseline other than zero at the user's option.

Options

⌐ Main ⌐

round(*#*) rounds the values of *varname* to the nearest multiple of *#*. This action effectively specifies the bin width.

fraction specifies that the vertical scale be the proportion of total values (percentage) rather than the count.

root specifies that the vertical scale show square roots. This option may not be specified if fraction is specified.

⌐ Plot ⌐

spike_options affect the rendition of the plotted spikes; see [G] **graph twoway spike**.

_____ Add plots _____

addplot(*plot*) provides a way to add other plots to the generated graph. See [G] ***addplot_option***.

_____ Y axis, X axis, Titles, Legend, Overall, By _____

twoway_options are any of the options documented in [G] ***twoway_options***. These include options for titling the graph (see [G] ***title_options***), options for saving the graph to disk (see [G] ***saving_option***), and the by() option (see [G] ***by_option***).

Remarks

▷ Example 1

Cox and Brady (1997a) present an illustrative example using the age structure of the population of Ghana from the 1960 census (rounded to the nearest 1,000). The dataset has ages from 0 (less than 1 year) to 90. To view the distribution of ages, we would like to use each integer from 0 to 90 as the bins for the dataset.

```
. use http://www.stata-press.com/data/r10/ghanaage
. spikeplot age [fw=pop], ytitle("Population in 1000s") xlab(0(10)90) xmtick(5(10)85)
```

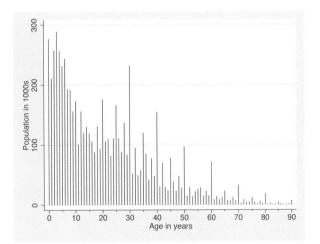

The resulting graph shows a "heaping" of ages at the multiples of 5. In addition, ages ending in even numbers are more frequent than ages ending in odd numbers (except for 5). This preference for reporting ages is well known in demography and other social sciences.

Note also that we used the ytitle() option to override the default title of "Frequency" and that we used the xlab() and xmtick() options with *numlist*s to further customize the resulting graph. See [U] **11.1.8 numlist** for details on specifying *numlist*s.

◁

▷ Example 2

The rootogram is a plot of the square-root transformation of the frequency counts. The square root of a normal distribution is a multiple of another normal distribution:

```
. clear
. set seed 1234567
. set obs 5000
obs was 0, now 5000
. generate normal = invnormal(uniform())
. label variable normal "Gaussian(0,1) random numbers"
. spikeplot normal, round(.10) xlab(-4(1)4)
```

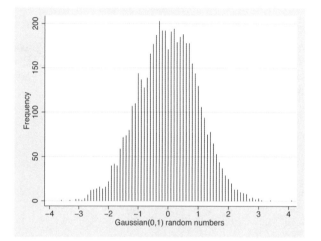

```
. spikeplot normal, round(.10) xlab(-4(1)4) root
```

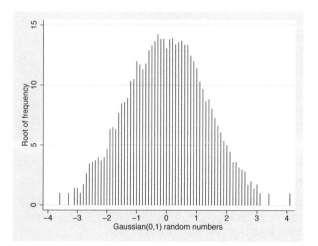

Interpreting a histogram in terms of normality is thus similar to interpreting the rootogram for normality.

This example also shows how the round() option is used to bin the values for a spike plot of a continuous variable.

◁

▷ Example 3

spikeplot can also be used to produce time-series plots. *varname* should be the time variable, and weights should be specified as the values for those times. To get a plot of daily rainfalls, we type

```
. spikeplot day [w=rain] if rain, ytitle("Daily rainfall in mm")
```

The base() option of graph twoway spike may be used to set a different baseline, such as when we want to show variations relative to an average or to some other measure of level. ◁

Methods and Formulas

spikeplot is implemented as an ado-file.

Acknowledgments

The original version of spikeplot was written by Nicholas J. Cox of Durham University and Anthony R. Brady of the Imperial College School of Medicine (1997a,b).

References

Cox, N. J., and A. R. Brady. 1997a. gr25: Spike plots for histograms, rootograms, and time-series plots. *Stata Technical Bulletin* 36: 8–11. Reprinted in *Stata Technical Bulletin Reprints*, vol. 6, pp. 50–54.

——. 1997b. gr25.1: Spike plots for histograms, rootograms, and time series plots: Update. *Stata Technical Bulletin* 40: 12. Reprinted in *Stata Technical Bulletin Reprints*, vol. 7, p. 58.

Tukey, J. W. 1965. The future of processes of data analysis. Reprinted in *The Collected Works of John W. Tukey, Volume IV: Philosophy and Principles of Data Analysis: 1965–1986*, ed. L. V. Jones, 517–547 (1986). Monterey, CA: Wadsworth & Brooks/Cole.

Also See

[R] **histogram** — Histograms for continuous and categorical variables

Title

ssc — Install and uninstall packages from SSC

Syntax

Summary of packages most recently added or updated at SSC

> ssc <u>whats</u>new [, <u>saving</u>(*filename*[, replace]) type]

Describe a specified package at SSC

> ssc <u>d</u>escribe { *pkgname* | *letter* } [, saving(*filename*[, replace])]

Install a specified package from SSC

> ssc <u>inst</u>all *pkgname* [, all replace]

Uninstall from your computer a previously installed package from SSC

> ssc uninstall *pkgname*

Type a specific file stored at SSC

> ssc type *filename*

Copy a specific file from SSC to your computer

> ssc copy *filename* [, <u>p</u>lus <u>p</u>ersonal replace <u>pub</u>lic <u>bin</u>ary]

where *letter* in ssc describe is a–z or _.

Description

ssc works with packages (and files) from the Statistical Software Components (SSC) archive, which is often called the Boston College Archive and is provided by http://repec.org/.

The SSC has become the premier Stata download site for user-written software on the web. ssc provides a convenient command-line interface to the resources available there. For example, on Statalist (see http://www.stata.com/statalist/), users will often write

> The program can be found by typing ssc install newprogramname.

Typing that would load everything associated with newprogramname, including the help files, if a package of that name existed. Since there is no package named newprogramname, typing the above literally would just produce an error message, but no harm would be done.

If you are searching for what is available, see [R] **search**. search searches and provides a GUI interface from which programs can be loaded, including the programs at the SSC archive.

After installing user-written materials, see [R] **adoupdate** for keeping those materials up to date.

Command overview

ssc whatsnew summarizes the packages made available or updated recently. Output is presented in the Stata Viewer, and from there you may click to find out more or install.

ssc describe *pkgname* describes, but does not install, the specified package. Use search to find packages; see [R] **search**. If you know the package name but do not know the exact spelling, type ssc describe followed by one letter, a–z or _ (underscore), to list all the packages starting with that letter.

ssc install *pkgname* installs the specified package. You do not have to describe a package before installing it. You may also install a package by using net; see [R] **net**.

ssc uninstall *pkgname* removes the previously installed package from your computer. It does not matter whether the package was installed using ssc install or using some other method, such as net. (ssc uninstall is a synonym for ado uninstall, so it can uninstall any installed package, not just packages obtained from SSC.)

ssc type *filename* types a specific file stored at SSC. ssc cat is a synonym for ssc type, which may appeal to those familiar with Unix.

ssc copy *filename* copies a specific file stored at SSC to your computer. By default, the file is copied to the current directory, but you can use options to change this. ssc copy is a rarely used alternative to ssc install ... , all. ssc cp is a synonym for ssc copy.

Options for use with ssc whatsnew

saving(*filename*[, replace]) specifies that the "what's new" summary be saved in *filename*. If *filename* is specified without a suffix, *filename*.smcl is assumed. If saving() is not specified, saving(ssc_results.smcl) is assumed.

type specifies that the "what's new" results be displayed in the Results window rather than in the Viewer.

Option for use with ssc describe

saving(*filename*[, replace]) specifies that, in addition to the descriptions being displayed on your screen, it be saved in the specified file.

If *filename* is specified without an extension, .smcl will be assumed, and the file will be saved as a SMCL file.

If *filename* is specified with an extension, no default extension is added. If the extension is .log, the file will be stored as an ASCII text file.

If replace is specified, *filename* is replaced if it already exists.

Options for use with ssc install

all specifies that any ancillary files associated with the package be downloaded to your current directory, in addition to the program and help files being installed. Ancillary files are files that do not end in .ado or .sthlp and typically contain datasets or examples of the use of the new command.

You can find out which files are associated with the package by typing ssc describe *pkgname* before or after installing. If you install without using the all option and then want the ancillary files, you can ssc install again.

replace specifies that any files being downloaded that already exist on your computer be replaced by the downloaded files. If replace is not specified and any files already exist, none of the files from the package is downloaded or installed.

It is better not to specify the replace option and wait to see if there is a problem. If there is a problem, it is usually better to uninstall the old package by using ssc uninstall or ado uninstall (which are, in fact, the same command).

Options for use with ssc copy

plus specifies that the file be copied to the PLUS directory, the directory where user-written additions are installed. Typing sysdir will display the identity of the PLUS directory on your computer.

personal specifies that the file be copied to your PERSONAL directory as reported by sysdir.

If neither plus nor personal is specified, the default is to copy the file to the current directory.

replace specifies that, if the file already exists on your computer, the new file replace it.

public specifies that the new file be made readable by everyone; otherwise, the file will be created according to the default permission you have set with your operating system.

binary specifies that the file being copied is a binary file and that it is to be copied as is. The default is to assume that the file is a text file and change the end-of-line characters to those appropriate for your computer/operating system.

Remarks

A new command to Stata is implemented as one or more ado-files, and together with their corresponding help files and any other associated files, they form a package.

Many packages are available at the SSC. Packages have names, such as oaxaca. When you type

```
. ssc install oaxaca
```

all the files associated with the package named oaxaca are downloaded and installed on your computer. Usually, names correspond to the name of the command being added to Stata, and so you would expect that installing the package oaxaca would add the new command oaxaca to Stata on your computer and that typing help oaxaca would provide the online help. That is indeed the situation in this case, but that is not always so. Before or after installing a package, it is a good idea to type ssc describe *pkgname* to find out the details.

▷ Example 1

ssc whatsnew summarizes the packages most recently made available or updated. Output is presented in the Viewer, from which you may click on a package name to find out more or install it. For example,

```
. ssc whatsnew
(contacting http://repec.org)
```
 (*output omitted*)

```
TABOUT
 module to export publication quality cross-tabulations
 Authors: Ian Watson
 Req: Stata version 9.2
 Revised: 2007-03-20
ICE
 module for multiple imputation of missing values
 Authors: Patrick Royston
 Req: Stata version 8
 Revised: 2007-03-18
```
 (*output omitted*)

```
End of recent additions and updates
```

ssc describe *pkgname* describes, but does not install, the specified package. You must already know the name of the package. See [R] **search** for assistance in searching for packages. Sometimes you know the package name, but you do not know the exact spelling. Then you can type ssc describe followed by one letter, a–z or _, to list all the packages starting with that letter; even so, using search is better.

```
. ssc describe ice
```

```
package ice from http://fmwww.bc.edu/repec/bocode/i
```

TITLE
        'ICE': module for multiple imputation of missing values
DESCRIPTION/AUTHOR(S)
            This is a version, for Stata 8 up, of the package originally
            publicised in Stata Journal 4(3):227-241 (2004) implementing
        the MICE (multivariate imputation by chained equations)
        method of multiple   multivariate data imputation. The most
        recent published update is   in the Stata Journal 5(4):
        527-536 (2005). Report problems to the author   (who is _not_
        on Statalist). It is intended that the present version   of
        ice will be described via another update in the Stata Journal.
          This package was previously circulated as mice.

        KW: multivariate imputation
        KW: missing values
        KW: chained equations

        Requires: Stata version 8

        Distribution-Date: 20070318

        Author: Patrick Royston, UK Medical Research Council
        Support: email pr@ctu.mrc.ac.uk

(Continued on next page)

```
INSTALLATION FILES                              (type net install ice)
     ice.ado
     ice.hlp
     ../c/cmdchk.ado
     ../m/micombine.ado
     ../m/micombine.hlp
     ../m/mijoin.ado
     ../m/mijoin.hlp
     ../m/misplit.ado
     ../m/misplit.hlp
     ../u/uvis.ado
     ../u/uvis.hlp
```

```
(type -ssc install ice- to install)
```

The default setting for the saving() option is for the output to be saved with the .smcl extension. You could also save the file with a log extension, and in this case, the file would be stored as an ASCII text file.

```
. ssc describe i, saving(i.index)
(output omitted )
. ssc describe ice, saving(ice.log)
(output omitted )
```

ssc install *pkgname* installs the specified package. You do not have to describe a package before installing it. There are ways of installing packages other than ssc install, such as net. It does not matter how a package is installed. For instance, a package can be installed using net and still be uninstalled using ssc.

```
. ssc install ice
checking ice consistency and verifying not already installed...
installing into C:\ado\plus\...
installation complete.
```

ssc uninstall *pkgname* removes the specified, previously installed package from your computer. You can uninstall immediately after installation or at any time in the future. (Technical note: ssc uninstall is a synonym for ado uninstall, so it can uninstall any installed package, not just packages obtained from the SSC.)

```
. ssc uninstall ice
package ice from http://fmwww.bc.edu/repec/bocode/i
     'ICE': module for multiple imputation of missing values
(package uninstalled)
```

ssc type *filename* types a specific file stored at the SSC. Although not shown in the syntax diagram, ssc cat is a synonym for ssc type, which may appeal to those familiar with Unix. To view only the ice help file for the ice package, you would type

```
. ssc type ice.hlp
```

```
help for ice, uvis                                        Patrick Royston
```

```
Multiple imputation by the MICE system of chained equations
```

```
(output omitted )
```

ssc copy *filename* copies a specific file stored at the SSC to your computer. By default, the file is copied to the current directory, but you can use options to change this. ssc copy is a rarely used alternative to ssc install ... , all. ssc cp is a synonym for ssc copy.

```
. ssc copy ice.ado
(file ice.ado copied to current directory)
```

◁

Methods and Formulas

`ssc` is implemented as an ado-file.

Acknowledgments

`ssc` is based on `archutil` by Nicholas J. Cox, Durham University, and Christopher Baum, Boston College. The reworking of the original was done with their blessing and their participation.

References

Baum, C. F., and N. J. Cox. 1999. ip29: Metadata for user-written contributions to the Stata programming language. *Stata Technical Bulletin* 52: 10–12. Reprinted in *Stata Technical Bulletin Reprints*, vol. 9, pp. 121–124.

Cox, N. J., and C. F. Baum. 2000. ip29.1: Metadata for user-written contributions to the Stata programming language. *Stata Technical Bulletin* 54: 21–22. Reprinted in *Stata Technical Bulletin Reprints*, vol. 9, pp. 124–126.

Also See

[R] **adoupdate** — Update user-written ado-files

[R] **net** — Install and manage user-written additions from the Internet

[R] **search** — Search Stata documentation

[R] **sj** — Stata Journal and STB installation instructions

[P] **sysdir** — Query and set system directories

Title

> **stem** — Stem-and-leaf displays

Syntax

$$\texttt{stem } \textit{varname } \left[\textit{if}\right] \left[\textit{in}\right] \left[\text{, }\textit{options}\right]$$

options	description
Main	
prune	do not print stems that have no leaves
round(#)	round data to this value; default is round(1)
digits(#)	digits per leaf; default is digits(1)
lines(#)	number of stems per interval of $10^{\texttt{digits}}$
width(#)	stem width; equal to $10^{\texttt{digits}}/\texttt{width}$

by is allowed; see [D] **by**.

Description

stem displays stem-and-leaf plots.

Options

Main

prune prevents printing any stems that have no leaves.

round(#) rounds the data to this value and displays the plot in these units. If round() is not specified, noninteger data will be rounded automatically.

digits(#) sets the number of digits per leaf. The default is 1.

lines(#) sets the number of stems per every data interval of $10^{\texttt{digits}}$. The value of lines() must divide $10^{\texttt{digits}}$; that is, if digits(1) is specified, then lines() must divide 10. If digits(2) is specified, then lines() must divide 100, etc. Only one of lines() or width() may be specified. If neither is specified, an appropriate value will be set automatically.

width(#) sets the width of a stem. lines() is equal to $10^{\texttt{digits}}/\texttt{width}$, and this option is merely an alternative way of setting lines(). The value of width() must divide $10^{\texttt{digits}}$. Only one of width() or lines() may be specified. If neither is specified, an appropriate value will be set automatically.

Note: if lines() or width() is not specified, digits() may be decreased in some circumstances to make a better-looking plot. If lines() or width() is set, the user-specified value of digits() will not be altered.

Remarks

▷ Example 1

Stem-and-leaf displays are a compact way to present considerable information about a batch of data. For instance, using our automobile data (described in [U] **1.2.1 Sample datasets**):

```
. use http://www.stata-press.com/data/r10/auto
(1978 Automobile Data)
. stem mpg
Stem-and-leaf plot for mpg (Mileage (mpg))
   1t | 22
   1f | 44444455
   1s | 66667777
   1. | 88888888899999999
   2* | 00011111
   2t | 22222333
   2f | 444455555
   2s | 666
   2. | 8889
   3* | 001
   3t |
   3f | 455
   3s |
   3. |
   4* | 1
```

The stem-and-leaf display provides a way to list our data. The expression to the left of the vertical bar is called the stem; the digits to the right are called the leaves. All the stems that begin with the same digit and the corresponding leaves, written beside each other, reconstruct an observation of the data. Thus, if we look at the four stems that begin with the digit 1 and their corresponding leaves, we see that we have two cars rated at 12 mpg, 6 cars at 14, 2 at 15, and so on. The car with the highest mileage rating in our data is rated at 41 mpg.

The above plot is a five-line plot with `lines()` equal to 5 (five lines per interval of 10) and `width()` equal to 2 (two leaves per stem).

Instead, we could specify `lines(2)`:

```
. stem mpg, lines(2)
Stem-and-leaf plot for mpg (Mileage (mpg))
   1* | 22444444
   1. | 556666777788888888899999999
   2* | 00011111222223334444
   2. | 555556668889
   3* | 0014
   3. | 55
   4* | 1
```

`stem mpg, width(5)` would produce the same plot as above.

The stem-and-leaf display provides a crude histogram of our data, one not so pretty as that produced by `histogram` (see [R] **histogram**), but one that is nonetheless informative.

◁

▷ Example 2

The miles per gallon rating fits easily into a stem-and-leaf display because, in our data, it has two digits. However, stem does not require two digits.

```
. stem price, lines(1) digits(3)

Stem-and-leaf plot for price (Price)
    3*** │ 291,299,667,748,798,799,829,895,955,984,995
    4*** │ 010,060,082,099,172,181,187,195,296,389,424,425,453,482,499, ... (26)
    5*** │ 079,104,172,189,222,379,397,705,719,788,798,799,886,899
    6*** │ 165,229,295,303,342,486,850
    7*** │ 140,827
    8*** │ 129,814
    9*** │ 690,735
   10*** │ 371,372
   11*** │ 385,497,995
   12*** │ 990
   13*** │ 466,594
   14*** │ 500
   15*** │ 906
```

The (26) at the right of the second stem shows that there were 26 leaves on this stem—too many to display on one line.

We can make a more compact stem-and-leaf plot by rounding. To display stem in units of 100, we could type

```
. stem price, round(100)

Stem-and-leaf plot for price (Price)

price rounded to nearest multiple of 100
plot in units of 100
    3* │ 33778889
    4* │ 000011122223444555555667777899
    5* │ 11222447788899
    6* │ 2233359
    7* │ 18
    8* │ 18
    9* │ 77
   10* │ 44
   11* │ 45
   12* │ 0
   13* │ 056
   14* │ 5
   15* │ 9
```

price, in our data, has four or five digits. stem presented the display in terms of units of 100, so a car that cost $3,291 was treated for display purposes as $3,300.

◁

❑ Technical Note

Stem-and-leaf diagrams have been used in Japanese railway timetables, as shown in Tufte (1990, 46–47).

❑

Saved Results

stem saves the following in r():

Scalars
r(width) width of a stem
r(digits) number of digits per leaf; default is 1

Methods and Formulas

stem is implemented as an ado-file.

References

Emerson, J. D., and D. C. Hoaglin. 1983. Stem-and-leaf displays. In *Understanding Robust and Exploratory Data Analysis*, ed. D. C. Hoaglin, F. Mosteller, and J. W. Tukey, 7–30. New York: Wiley.

Tufte, E. R. 1990. *Envisioning Information.* Cheshire, CT: Graphics Press.

Tukey, J. W. 1972. Some graphic and semigraphic displays. In *Statistical Papers in Honor of George W. Snedecor*, ed. T. A. Bancroft and S. A. Brown, 293–316. Ames, IA: Iowa State University Press.

——. 1977. *Exploratory Data Analysis.* Reading, MA: Addison–Wesley.

Also See

[R] **histogram** — Histograms for continuous and categorical variables

[R] **lv** — Letter-value displays

Title

> **stepwise** — Stepwise estimation

Syntax

stepwise [, *options*] : *command*

options	description
Model	
*pr(#)	significance level for removal from the model
*pe(#)	significance level for addition to the model
Model2	
forward	perform forward-stepwise selection
hierarchical	perform hierarchical selection
lockterm1	keep the first term
lr	perform likelihood-ratio test instead of Wald test

* At least one of pr(#) or pe(#) must be specified.

by and xi are allowed; see [U] **11.1.10 Prefix commands**.

Weights are allowed if *command* allows them; see [U] **11.1.6 weight**.

All postestimation commands behave as they would after *command* without the stepwise prefix; see the postestimation manual entry for *command*.

See [U] **20 Estimation and postestimation commands** for more capabilities of estimation commands.

Description

stepwise performs stepwise estimation. Typing

    . stepwise, pr(#): *command*

performs backward-selection estimation for *command*. The stepwise selection method is determined by the following option combinations:

options	description
pr(#)	backward selection
pr(#) hierarchical	backward hierarchical selection
pr(#) pe(#)	backward stepwise
pe(#)	forward selection
pe(#) hierarchical	forward hierarchical selection
pr(#) pe(#) forward	forward stepwise

command defines the estimation command to be executed. The following Stata commands are supported by stepwise:

clogit cloglog cnreg glm intreg logistic logit nbreg ologit oprobit
poisson probit qreg regress scobit stcox streg tobit

stepwise expects *command* to have the following form:

command_name [*depvar*] *term* [*term* ...] [*if*] [*in*] [*weight*] [, *command_options*]

where *term* is either *varname* or (*varlist*) (a *varlist* in parentheses indicates that this group of variables is to be included or excluded together). *depvar* is not present when *command_name* is `streg` or `stcox`; otherwise, *depvar* is assumed to be present. For `intreg`, *depvar* is actually two dependent variable names (*depvar₁* and *depvar₂*).

`sw` is a synonym for `stepwise`.

Options

___| Model |_____

`pr(#)` specifies the significance level for removal from the model; terms with $p \geq$ `pr()` are eligible for removal.

`pe(#)` specifies the significance level for addition to the model; terms with $p <$ `pe()` are eligible for addition.

___| Model 2 |_____

`forward` specifies the forward-stepwise method and may be specified only when both `pr()` and `pe()` are also specified. Specifying both `pr()` and `pe()` without `forward` results in backward-stepwise selection. Specifying only `pr()` results in backward selection, and specifying only `pe()` results in forward selection.

`hierarchical` specifies hierarchical selection.

`lockterm1` specifies that the first term be included in the model and not be subjected to the selection criteria.

`lr` specifies that the test of term significance be the likelihood-ratio test. The default is the less computationally expensive Wald test; that is, the test is based on the estimated variance–covariance matrix of the estimators.

Remarks

Remarks are presented under the following headings:

> *Introduction*
> *Search logic for a step*
> *Full search logic*
> *Examples*
> *Estimation sample considerations*
> *Messages*
> *Programming for stepwise*

(*Continued on next page*)

Introduction

Typing

```
. stepwise, pr(.10): regress y1 x1 x2 d1 d2 d3 x4 x5
```

performs a backward-selection search for the regression model y1 on x1, x2, d1, d2, d3, x4, and x5. In this search, each explanatory variable is said to be a term. Typing

```
. stepwise, pr(.10): regress y1 x1 x2 (d1 d2 d3) (x4 x5)
```

performs a similar backward-selection search, but the variables d1, d2, and d3 are treated as one term, as are x4 and x5. That is, d1, d2, and d3 may or may not appear in the final model, but they appear or do not appear together.

▷ Example 1

Using the automobile dataset, we fit a backward-selection model of mpg:

```
. use http://www.stata-press.com/data/r10/auto
. gen weight2 = weight*weight
. stepwise, pr(.2): regress mpg weight weight2 displ gear turn headroom
> foreign price
                    begin with full model
p = 0.7116 >= 0.2000  removing headroom
p = 0.6138 >= 0.2000  removing displacement
p = 0.3278 >= 0.2000  removing price
```

Source	SS	df	MS		Number of obs =	74
					F(5, 68) =	33.39
Model	1736.31455	5	347.262911		Prob > F =	0.0000
Residual	707.144906	68	10.3991898		R-squared =	0.7106
					Adj R-squared =	0.6893
Total	2443.45946	73	33.4720474		Root MSE =	3.2248

mpg	Coef.	Std. Err.	t	P>|t|	[95% Conf. Interval]	
weight	-.0158002	.0039169	-4.03	0.000	-.0236162	-.0079842
weight2	1.77e-06	6.20e-07	2.86	0.006	5.37e-07	3.01e-06
foreign	-3.615107	1.260844	-2.87	0.006	-6.131082	-1.099131
gear_ratio	2.011674	1.468831	1.37	0.175	-.9193321	4.94268
turn	-.3087038	.1763099	-1.75	0.084	-.6605248	.0431172
_cons	59.02133	9.3903	6.29	0.000	40.28327	77.75938

This estimation treated each variable as its own term and thus considered each one separately. The engine displacement and gear ratio should really be considered together:

```
. stepwise, pr(.2): regress mpg weight weight2 (displ gear) turn headroom
> foreign price
                        begin with full model
p = 0.7116 >= 0.2000   removing headroom
p = 0.3944 >= 0.2000   removing displacement gear_ratio
p = 0.2798 >= 0.2000   removing price
```

Source	SS	df	MS			
Model	1716.80842	4	429.202105			
Residual	726.651041	69	10.5311745			
Total	2443.45946	73	33.4720474			

Number of obs =   74
F(4,   69) =   40.76
Prob > F   = 0.0000
R-squared   = 0.7026
Adj R-squared = 0.6854
Root MSE   = 3.2452

mpg	Coef.	Std. Err.	t	P>|t|	[95% Conf. Interval]	
weight	-.0160341	.0039379	-4.07	0.000	-.0238901	-.0081782
weight2	1.70e-06	6.21e-07	2.73	0.008	4.58e-07	2.94e-06
foreign	-2.758668	1.101772	-2.50	0.015	-4.956643	-.5606925
turn	-.2862724	.176658	-1.62	0.110	-.6386955	.0661508
_cons	65.39216	8.208778	7.97	0.000	49.0161	81.76823

◁

Search logic for a step

Before discussing the complete search logic, consider the logic for a step—the first step—in detail. The other steps follow the same logic. If you type

```
. stepwise, pr(.20): regress y1 x1 x2 (d1 d2 d3) (x4 x5)
```

the logic is

1. Fit the model y on x1 x2 d1 d2 d3 x4 x5.
2. Consider dropping x1.
3. Consider dropping x2.
4. Consider dropping d1 d2 d3.
5. Consider dropping x4 x5.
6. Find the term above that is least significant. If its significance level is $\geq .20$, remove that term.

If you type

```
. stepwise, pr(.20) hierarchical: regress y1 x1 x2 (d1 d2 d3) (x4 x5)
```

the logic would be different because the hierarchical option states that the terms are ordered. The initial logic would become

1. Fit the model y on x1 x2 d1 d2 d3 x4 x5.
2. Consider dropping x4 x5—the last term.
3. If the significance of this last term is $\geq .20$, remove the term.

The process would then stop or continue. It would stop if x4 x5 were not dropped, and otherwise, stepwise would continue to consider the significance of the next-to-last term, d1 d2 d3.

Specifying pe() rather than pr() switches to forward estimation. If you type

```
. stepwise, pe(.20): regress y1 x1 x2 (d1 d2 d3) (x4 x5)
```

stepwise performs forward-selection search. The logic for the first step is

1. Fit a model of y on nothing (meaning a constant).
2. Consider adding x1.
3. Consider adding x2.
4. Consider adding d1 d2 d3.
5. Consider adding x4 x5.
6. Find the term above that is most significant. If its significance level is < .20, add that term.

As with backward estimation, if you specify hierarchical,

```
. stepwise, pe(.20) hierarchical: regress y1 x1 x2 (d1 d2 d3) (x4 x5)
```

the search for the most significant term is restricted to the next term:

1. Fit a model of y on nothing (meaning a constant).
2. Consider adding x1—the first term.
3. If the significance is < .20, add the term.

If x1 were added, stepwise would next consider x2; otherwise, the search process would stop.

stepwise can also use a stepwise selection logic that alternates between adding and removing terms. The full logic for all the possibilities is given below.

Full search logic

Option	Logic
pr() (backward selection)	Fit the full model on all explanatory variables. While the least-significant term is "insignificant", remove it and reestimate.
pr() hierarchical (backward hierarchical selection)	Fit full model on all explanatory variables. While the last term is "insignificant", remove it and reestimate.

pr() pe()
(backward stepwise)
	Fit full model on all explanatory variables.
	If the least-significant term is "insignificant", remove it and
	    reestimate; otherwise, stop.
	Do that again: if the least-significant term is "insignificant",
	    remove it and reestimate; otherwise, stop.
	Repeatedly,
	    if the most-significant excluded term is "significant", add
	        it and reestimate;
	    if the least-significant included term is "insignificant",
	        remove it and reestimate;
	until neither is possible.

pe() (forward selection)	Fit "empty" model. While the most-significant excluded term is "significant", add it and reestimate.
pe() hierarchical (forward hierarchical selection)	Fit "empty" model. While the next term is "significant", add it and reestimate.

pr() pe() forward
(forward stepwise)
	Fit "empty" model.
	If the most-significant excluded term is "significant",
	    add it and reestimate; otherwise, stop.
	If the most-significant excluded term is "significant",
	    add it and reestimate; otherwise, stop.
	Repeatedly,
	    if the least-significant included term is "insignificant",
	        remove it and reestimate;
	    if the most-significant excluded term is "significant",
	        add it and reestimate;
	until neither is possible.

Examples

The following two statements are equivalent; both include solely single-variable terms:

```
. stepwise, pr(.2): reg price mpg weight displ
. stepwise, pr(.2): reg price (mpg) (weight) (displ)
```

The following two statements are equivalent; the last term in each is r1, ..., r4:

```
. stepwise, pr(.2) hierarchical: reg price mpg weight displ (r1-r4)
. stepwise, pr(.2) hierarchical: reg price (mpg) (weight) (displ) (r1-r4)
```

To group variables `weight` and `displ` into one term, type

```
. stepwise, pr(.2) hierarchical: reg price mpg (weight displ) (r1-r4)
```

`stepwise` can be used with commands other than `regress`; for instance,

```
. stepwise, pr(.2): logit outcome (sex weight) treated1 treated2
. stepwise, pr(.2): logistic outcome (sex weight) treated1 treated2
```

Either statement would fit the same model because `logistic` and `logit` both perform logistic regression; they differ only in how they report results; see [R] **logit** and [R] **logistic**.

We use the `lockterm1` option to force the first term to be included in the model. To keep `treated1` and `treated2` in the model no matter what, we type

```
. stepwise, pr(.2) lockterm1: logistic outcome (treated1 treated2) ...
```

After `stepwise` estimation, we can type `stepwise` without arguments to redisplay results,

```
. stepwise
(output from logistic appears )
```

or type the underlying estimation command:

```
. logistic
(output from logistic appears )
```

At estimation time, we can specify options unique to the command being stepped:

```
. stepwise, pr(.2): logit outcome (sex weight) treated1 treated2, or
```

`or` is `logit`'s option to report odds ratios rather than coefficients; see [R] **logit**.

Estimation sample considerations

Whether you use backward or forward estimation, `stepwise` forms an estimation sample by taking observations with nonmissing values of all the variables specified (except for *depvar*$_1$ and *depvar*$_2$ for `intreg`). The estimation sample is held constant throughout the stepping. Thus if you type

```
. stepwise, pr(.2) hierarchical: regress amount sk edul sval
```

and variable `sval` is missing in half the data, that half of the data will not be used in the reported model, even if `sval` is not included in the final model.

The function `e(sample)` identifies the sample that was used. `e(sample)` contains 1 for observations used and 0 otherwise. For instance, if you type

```
. stepwise, pr(.2) pe(.10): logistic outcome x1 x2 (x3 x4) (x5 x6 x7)
```

and the final model is outcome on x1, x5, x6, and x7, you could recreate the final regression by typing

>   . logistic outcome x1 x5 x6 x7 if e(sample)

You could obtain summary statistics within the estimation sample of the independent variables by typing

>   . summarize x1 x5 x6 x7 if e(sample)

If you fit another model, e(sample) will automatically be redefined. Typing

>   . stepwise, lock pr(.2): logistic outcome (x1 x2) (x3 x4) (x5 x6 x7)

would automatically drop e(sample) and recreate it.

Messages

note: _____ dropped because of collinearity

Each term is checked for collinearity, and variables within the term are dropped if collinearity is found. For instance, say that you type

>   . stepwise, pr(.2): regress y x1 x2 (r1-r4) (x3 x4)

and assume that variables r1 through r4 are mutually exclusive and exhaustive dummy variables—perhaps r1, ..., r4 indicate in which of four regions the subject resides. One of the r1, ..., r4 variables will be automatically dropped to identify the model.

This message should cause you no concern.

Error message: between-term collinearity, variable _____

After removing any within-term collinearity, if stepwise still finds collinearity between terms, it refuses to continue. For instance, assume that you type

>   . stepwise, pr(.2): regress y1 x1 x2 (d1-d8) (r1-r4)

Assume that r1, ..., r4 identify in which of four regions the subject resides, and that d1, ..., d8 identify the same sort of information, but more finely. r1, say, amounts to d1 and d2; r2 to d3, d4, and d5; r3 to d6 and d7; and r4 to d8. You can estimate the d* variables or the r* variables, but not both.

It is your responsibility to specify noncollinear terms.

note: _____ dropped because of estimability
note: _____ obs. dropped because of estimability

You probably received this message in fitting a logistic or probit model. Regardless of estimation strategy, stepwise checks that the full model can be fitted. The indicated variable had a 0 or infinite standard error.

For logistic, logit, and probit, this message is typically caused by one-way causation. Assume that you type

>   . stepwise, pr(.2): logistic outcome (x1 x2 x3) d1

and assume that variable d1 is an indicator (dummy) variable. Further assume that whenever d1 = 1, outcome = 1 in the data. Then the coefficient on d1 is infinite. One (conservative) solution to this problem is to drop the d1 variable and the d1==1 observations. The underlying estimation commands probit, logit, and logistic report the details of the difficulty and solution; stepwise simply accumulates such problems and reports the above summary messages. Thus if you see this message, you could type

```
. logistic outcome x1 x2 x3 d1
```

to see the details. Although you should think carefully about such situations, Stata's solution of dropping the offending variables and observations is, in general, appropriate.

Programming for stepwise

stepwise requires that *command_name* follow standard Stata syntax and allow the if qualifier; see [U] **11 Language syntax**. Furthermore, *command_name* must have sw or swml as a program property; see [P] **program properties**. If *command_name* has swml as a property, *command_name* must save the log-likelihood value in e(ll) and model degrees of freedom in e(df_m).

Saved Results

stepwise saves whatever is saved by the underlying estimation command.

Also, stepwise saves stepwise in e(stepwise).

Methods and Formulas

stepwise is implemented as an ado-file.

Some statisticians do not recommend stepwise procedures; see Sribney (1998) for a summary.

References

Beale, E. M. L. 1970. Note on procedures for variable selection in multiple regression. *Technometrics* 12: 909–914.

Bendel, R. B., and A. A. Afifi. 1977. Comparison of stopping rules in forward "stepwise" regression. *Journal of the American Statistical Association* 72: 46–53.

Berk, K. N. 1978. Comparing subset regression procedures. *Technometrics* 20: 1–6.

Draper, N., and H. Smith. 1998. *Applied Regression Analysis*. 3rd ed. New York: Wiley.

Efroymson, M. A. 1960. Multiple regression analysis. In *Mathematical Methods for Digital Computers*, ed. A. Ralston and H. S. Wilf, 191–203. New York: Wiley.

Gorman, J. W., and R. J. Toman. 1966. Selection of variables for fitting equations to data. *Technometrics* 8: 27–51.

Hocking, R. R. 1976. The analysis and selection of variables in linear regression. *Biometrics* 32: 1–50.

Hosmer, D. W., Jr., and S. Lemeshow. 2000. *Applied Logistic Regression*. 2nd ed. New York: Wiley.

Kennedy, W. J., Jr., and T. A. Bancroft. 1971. Model-building for prediction in regression based on repeated significance tests. *Annals of Mathematical Statistics* 42: 1273–1284.

Mantel, N. 1970. Why stepdown in variable selection. *Technometrics* 12: 621–625.

———. 1971. More on variable selection and an alternative approach (letter to the editor). *Technometrics* 13: 455–457.

Sribney, W. M. 1998. FAQ: What are some problems with stepwise regression? http://www.stata.com/support/faqs/stat/.

Wang, Z. 2000. sg134: Model selection using the Akaike information criterion. *Stata Technical Bulletin* 54: 47–49. Reprinted in *Stata Technical Bulletin Reprints*, vol. 9, pp. 335–337.

Title

> **suest** — Seemingly unrelated estimation

Syntax

suest *namelist* [, *options*]

options	description
SE/Robust	
svy	survey data estimation
vce(*vcetype*)	*vcetype* may be <u>r</u>obust or <u>cl</u>uster *clustvar*
Reporting	
<u>l</u>evel(*#*)	set confidence level; default is level(95)
<u>dir</u>	display a table describing the models
<u>ef</u>orm(*string*)	report exponentiated coefficients and label as *string*

where *namelist* is a list of one or more names under which estimation results were saved via estimates store. Wildcards may be used. * and _all refer to all stored results. A period (.) may be used to refer to the last estimation results, even if they have not (yet) been stored.

Description

suest is a postestimation command; see [U] **20 Estimation and postestimation commands**.

suest combines the estimation results—parameter estimates and associated (co)variance matrices—stored under *namelist* into one parameter vector and *simultaneous* (co)variance matrix of the sandwich/robust type. This (co)variance matrix is appropriate even if the estimates were obtained on the same or on overlapping data.

Typical applications of suest are tests for intramodel and cross-model hypotheses using test or testnl, for example, a generalized Hausman specification test. lincom and nlcom may be used after suest to estimate linear combinations and nonlinear functions of coefficients. suest may also be used to adjust a standard VCE for clustering or survey design effects.

Different estimators are allowed, for example, a regress model and a probit model. The only requirement is that predict produce equation-level scores with the score option after an estimation command. The models may be estimated on different samples, due either to explicit if or in selection or to missing values. If weights are applied, the same weights (type and values) should be applied to all models in *namelist*. The estimators should be estimated without vce(robust) or vce(cluster *clustvar*) options. suest returns the robust VCE, allows the vce(cluster *clustvar*) option, and automatically works with results from the svy prefix command (only for vce(linearized)). See [SVY] **svy postestimation** for an example using suest with svy: ologit.

Since suest posts its results like a proper estimation command, its results can be stored via estimates store. Moreover, like other estimation commands, suest typed without arguments replays the results.

Options

SE/Robust

svy specifies that estimation results should be modified to reflect the survey design effects according to the svyset specifications, see [SVY] **svyset**.

The svy option is implied when suest encounters survey estimation results from the svy prefix. Poststratification is allowed only with survey estimation results from the svy prefix.

vce(*vcetype*) specifies the type of standard error reported, which includes types that are robust to some kinds of misspecification and that allow for intragroup correlation; see [R] *vce_option*.

The vce() option may not be combined with the svy option or estimation results from the svy prefix.

Reporting

level(*#*) specifies the confidence level, as a percentage, for confidence intervals of the coefficients; see [R] **level**.

dir displays a table describing the models in *namelist* just like estimates dir *namelist*.

eform(*str*) displays the coefficient table in exponentiated form: for each coefficient, $\exp(b)$ rather than b is displayed, and standard errors and confidence intervals are transformed. Display of the intercept, if any, is suppressed. *str* is the table header that will be displayed above the transformed coefficients and must be 11 characters or fewer, for example, eform("Odds ratio").

Remarks

Remarks are presented under the following headings:

> *Using suest*
> *Remarks on regress*
> *Testing the assumption of the independence of irrelevant alternatives (IIA)*
> *Testing proportionality*
> *Testing cross-model hypotheses*

Using suest

If you plan to use suest, you must take precautions when fitting the original models. These restrictions are relaxed when using svy commands; see [SVY] **svy postestimation**.

1. suest works with estimation commands that allow predict to generate equation-level score variables when supplied with the score (or scores) option. For example, equation-level score variables are generated after running mlogit by typing

   ```
   . predict sc*, scores
   ```

2. Estimation should take place *without* the options vce(robust) or vce(cluster *clustvar*). suest always computes the robust estimator of the (co)variance, and suest has a vce(cluster *clustvar*) option.

   The within-model covariance matrices computed by suest are identical to those obtained by specifying a vce(robust) or vce(cluster *clustvar*) option during estimation. suest, however, also estimates the between-model covariances of parameter estimates.

3. Finally, the estimation results to be combined should be stored by `estimates store`; see [R] **estimates store**.

After estimating and storing a series of estimation results, you are ready to combine the estimation results with `suest`,

. suest *name1* [*name2* ...] [, vce(cluster *clustvar*)]

and you can subsequently use postestimation commands, such as `test`, to test hypotheses. Here an important issue is how `suest` assigns names to the equations. If you specify one model *name*, the original equation names are left unchanged; otherwise, `suest` constructs new equation names. The coefficients of a single-equation model (such as `logit` and `poisson`) that was `estimate` stored under name *X* are collected under equation *X*. With a multiequation model stored under name *X*, `suest` prefixes *X_* to an original equation name *eq*, forming equation name, *X_eq*.

❑ Technical Note

Earlier we said that standard errors from `suest` are identical to those obtained by specifying the `vce(robust)` option with each command individually. Thus if you fit a logistic model using `logit` with the `vce(robust)` option, you will get the same standard errors when you type

. suest .

directly after `logit` using the same data without the `vce(robust)` option.

This is not true for multiple estimation results when the estimation samples are not all the same. The standard errors from `suest` will be slightly smaller than those from individual model fits using the `vce(robust)` option since `suest` uses a larger number of observations to estimate the simultaneous (co)variance matrix.

❑

❑ Technical Note

In rare circumstances, `suest` may have to truncate equation names to 32 characters. When equation names are not unique because of truncation, `suest` numbers the equations within models, using equations named *X_#*.

❑

Remarks on regress

`regress` does not include its ancillary parameter, the residual variance, in its coefficient vector and (co)variance matrix. Moreover, while the `score` option is allowed with `predict` after `regress`, a score variable is generated for the mean but not for the variance parameter. `suest` contains special code that assigns the equation name `mean` to the coefficients for the mean, adds the equation `lnvar` for the log variance, and computes the appropriate two score variables itself.

Testing the assumption of the independence of irrelevant alternatives (IIA)

The multinomial logit model and the closely related conditional logit model satisfy a probabilistic version of the assumption of the independence of irrelevant alternatives, implying that the ratio of the probabilities for two alternatives does not depend on what other alternatives are available. Hausman and McFadden (1984) proposed a test for this assumption that is implemented in the hausman command. The standard Hausman test has several limitations. First, the test statistic may be undefined because the estimated VCE does not satisfy the required asymptotic properties of the test. Second, the classic Hausman test applies only to the test of the equality of two estimators. Third, the test requires access to a fully efficient estimator; such an estimator may not be available, for example, if you are analyzing complex survey data. Using suest can overcome these three limitations.

▷ Example 1

In our first example, we follow the analysis of the type of health insurance reported in [R] **mlogit** and demonstrate the hausman command with the suest/test combination. We fit the full multinomial logit model for all three alternatives and two restricted multinomial models in which one alternative is excluded. After fitting each of these models, we store the results using the store subcommand of estimates. title() simply documents the models.

```
. use http://www.stata-press.com/data/r10/sysdsn4
(Health insurance data)

. mlogit insure age male
Iteration 0:   log likelihood = -555.85446
Iteration 1:   log likelihood = -551.32973
Iteration 2:   log likelihood = -551.32802
Multinomial logistic regression                 Number of obs   =        615
                                                LR chi2(4)      =       9.05
                                                Prob > chi2     =     0.0598
Log likelihood = -551.32802                     Pseudo R2       =     0.0081
```

insure	Coef.	Std. Err.	z	P>|z|	[95% Conf. Interval]	
Prepaid						
age	-.0100251	.0060181	-1.67	0.096	-.0218204	.0017702
male	.5095747	.1977893	2.58	0.010	.1219148	.8972345
_cons	.2633838	.2787574	0.94	0.345	-.2829708	.8097383
Uninsure						
age	-.0051925	.0113821	-0.46	0.648	-.027501	.017116
male	.4748547	.3618446	1.31	0.189	-.2343477	1.184057
_cons	-1.756843	.5309591	-3.31	0.001	-2.797504	-.7161824

```
(insure==Indemnity is the base outcome)
. estimates store m1, title(all three insurance forms)
. quietly mlogit insure age male if insure != "Uninsure":insure
. estimates store m2, title(insure != "Uninsure":insure)
. quietly mlogit insure age male if insure != "Prepaid":insure
. estimates store m3, title(insure != "Prepaid":insure)
```

Having performed the three estimations, we inspect the results. estimates dir provides short descriptions of the models that were stored using estimates store. Typing estimates table lists the coefficients, displaying blanks for a coefficient not contained in a model.

```
. est dir
```

model		command	depvar	npar	title
m1		mlogit	insure	6	all three insurance forms
m2		mlogit	insure	3	insure != Uninsure :insure
m3		mlogit	insure	3	insure != Prepaid :insure

```
. est table m1 m2 m3, star stats(N ll)
```

Variable		m1	m2	m3
Prepaid				
age		-.01002511	-.01015205	
male		.50957468**	.51440033**	
_cons		.26338378	.26780432	
Uninsure				
age		-.00519249		-.00410547
male		.47485472		.45910724
_cons		-1.7568431***		-1.8017743***
Statistics				
N		615	570	338
ll		-551.32802	-390.48643	-131.76807

legend: * p<0.05; ** p<0.01; *** p<0.001

Comparing the coefficients between models does not suggest substantial differences. We can formally test that coefficients are the same for the full model m1 and the restricted models m2 and m3 by using the hausman command. hausman expects the models to be specified in the order "always consistent" first and "efficient under H_0" second.

```
. hausman m2 m1, alleqs constant
```

		\u2014\u2014 Coefficients \u2014\u2014			
		(b)	(B)	(b-B)	sqrt(diag(V_b-V_B))
		m2	m1	Difference	S.E.
age		-.0101521	-.0100251	-.0001269	.
male		.5144003	.5095747	.0048256	.012334
_cons		.2678043	.2633838	.0044205	.

b = consistent under Ho and Ha; obtained from mlogit
B = inconsistent under Ha, efficient under Ho; obtained from mlogit

Test: Ho: difference in coefficients not systematic

$$\text{chi2(3)} = (b-B)'[(V\_b-V\_B)^{(-1)}](b-B)$$
$$= \qquad 0.08$$
$$\text{Prob>chi2} = \qquad 0.9944$$
(V_b-V_B is not positive definite)

(Continued on next page)

```
. hausman m3 m1, alleqs constant
```

	── Coefficients ──			
	(b)	(B)	(b-B)	sqrt(diag(V_b-V_B))
	m3	m1	Difference	S.E.
age	-.0041055	-.0051925	.001087	.0021357
male	.4591072	.4748547	-.0157475	.
_cons	-1.801774	-1.756843	-.0449311	.1333464

```
                    b = consistent under Ho and Ha; obtained from mlogit
        B = inconsistent under Ha, efficient under Ho; obtained from mlogit
   Test:  Ho:  difference in coefficients not systematic

          chi2(3) = (b-B)'[(V_b-V_B)^(-1)](b-B)
                  =    -0.18    chi2<0 ==> model fitted on these
                               data fails to meet the asymptotic
                               assumptions of the Hausman test;
                               see suest for a generalized test
```

According to the test of m1 against m2, we cannot reject the hypothesis that the coefficients of m1 and m2 are the same. The second Hausman test is not well defined—something that happens fairly often. The problem is due to the estimator of the variance V(b-B) as V(b)-V(B), which is a feasible estimator only asymptotically. Here it simply is not a proper variance matrix, and the Hausman test becomes undefined.

suest m1 m2 estimates the simultaneous (co)variance of the coefficients of models m1 and m2. Although suest is technically a postestimation command, it acts like an estimation command in that it stores the simultaneous coefficients in e(b) and the full (co)variance matrix in e(V). We use the command estat vce to display this full (co)variance matrix to show that the cross-model covariances have indeed been estimated. Typically, we do not expect to have a direct interest in e(V).

```
. suest m1 m2
Simultaneous results for m1, m2
```

Number of obs   =   615

	Coef.	Robust Std. Err.	z	P>|z|	[95% Conf. Interval]	
m1_Prepaid						
age	-.0100251	.0059403	-1.69	0.091	-.0216679	.0016176
male	.5095747	.1988158	2.56	0.010	.1199029	.8992465
_cons	.2633838	.277307	0.95	0.342	-.280128	.8068955
m1_Uninsure						
age	-.0051925	.0109004	-0.48	0.634	-.0265569	.0161719
male	.4748547	.3677294	1.29	0.197	-.2458817	1.195591
_cons	-1.756843	.4971363	-3.53	0.000	-2.731212	-.782474
m2_Prepaid						
age	-.0101521	.0058988	-1.72	0.085	-.0217135	.0014094
male	.5144003	.1996132	2.58	0.010	.1231656	.9056351
_cons	.2678043	.2744018	0.98	0.329	-.2700133	.805622

```
. estat vce
```

Covariance matrix of coefficients of suest model

e(V)	m1_Prepaid age	male	_cons	m1_Unins~e age	male
m1_Prepaid					
age	.00003529				
male	-.00011951	.03952773			
_cons	-.00154346	-.00415209	.07689918		
m1_Uninsure					
age	.00001529	-.00010823	-.00067479	.00011882	
male	-.00008815	.0216828	-.00037879	-.00102264	.13522494
_cons	-.00067516	.00041156	.03409656	-.00501568	.00833699
m2_Prepaid					
age	.00003501	-.00013472	-.00152737	.00001604	-.00011163
male	-.00013269	.03967268	-.00359027	-.00011897	.02204034
_cons	-.00152755	-.0035127	.07603858	-.00070546	.00058334

e(V)	m1_Unins~e _cons	m2_Prepaid age	male	_cons
m1_Uninsure				
_cons	.24714447			
m2_Prepaid				
age	-.00070509	.0000348		
male	.00091027	-.00014889	.03984544	
_cons	.0353047	-.00151433	-.00291391	.07529635

suest created equation names by combining the name under which we stored the results using estimates store with the original equation names. Thus, in the simultaneous estimation result, equation Prepaid originating in model m1 is named m1_Prepaid. According to the McFadden–Hausman specification of a test for IIA, the coefficients of the equations m1_PrePaid and m2_PrePaid should be equal. This equality can be tested easily with the test command. The option cons specifies that the intercept _cons be included in the test.

```
. test [m1_Prepaid = m2_Prepaid], cons

 ( 1)  [m1_Prepaid]age - [m2_Prepaid]age = 0
 ( 2)  [m1_Prepaid]male - [m2_Prepaid]male = 0
 ( 3)  [m1_Prepaid]_cons - [m2_Prepaid]_cons = 0

          chi2(  3) =      0.89
        Prob > chi2 =    0.8267
```

The Hausman test via suest is comparable to that computed by hausman, but they use different estimators of the variance of the difference of the estimates. The hausman command estimates $V(b-B)$ by $V(b) - V(B)$, whereas suest estimates $V(b-B)$ by $V(b) - \mathrm{cov}(b, B) - \mathrm{cov}(B, b) + V(B)$. One advantage of the second estimator is that it is always admissible, so the resulting test is always well defined. This quality is illustrated in the Hausman-type test of IIA comparing models m1 and m3.

(Continued on next page)

```
. suest m1 m3

Simultaneous results for m1, m3
```

```
                                                          Number of obs   =        615
```

	Coef.	Robust Std. Err.	z	P>|z|	[95% Conf. Interval]	
m1_Prepaid						
age	-.0100251	.0059403	-1.69	0.091	-.0216679	.0016176
male	.5095747	.1988158	2.56	0.010	.1199029	.8992465
_cons	.2633838	.277307	0.95	0.342	-.280128	.8068955
m1_Uninsure						
age	-.0051925	.0109004	-0.48	0.634	-.0265569	.0161719
male	.4748547	.3677294	1.29	0.197	-.2458817	1.195591
_cons	-1.756843	.4971363	-3.53	0.000	-2.731212	-.782474
m3_Uninsure						
age	-.0041055	.0111185	-0.37	0.712	-.0258974	.0176865
male	.4591072	.3601307	1.27	0.202	-.2467359	1.16495
_cons	-1.801774	.5226351	-3.45	0.001	-2.82612	-.7774283

```
. test [m1_Uninsure = m3_Uninsure], cons

 ( 1)  [m1_Uninsure]age - [m3_Uninsure]age = 0
 ( 2)  [m1_Uninsure]male - [m3_Uninsure]male = 0
 ( 3)  [m1_Uninsure]_cons - [m3_Uninsure]_cons = 0

           chi2(  3) =     1.49
         Prob > chi2 =    0.6843
```

Although the classic Hausman test computed by hausman is not defined in this case, the suest-based test is just fine. We cannot reject the equality of the common coefficients across m1 and m3.

A second advantage of the suest approach is that we can estimate the (co)variance matrix of the multivariate normal distribution of the estimators of the three models m1, m2, and m3 and test that the common coefficients are equal.

```
. suest m*

Simultaneous results for m1, m2, m3
```

	Number of obs	=	615

	Coef.	Robust Std. Err.	z	P>|z|	[95% Conf. Interval]	
m1_Prepaid						
age	-.0100251	.0059403	-1.69	0.091	-.0216679	.0016176
male	.5095747	.1988158	2.56	0.010	.1199029	.8992465
_cons	.2633838	.277307	0.95	0.342	-.280128	.8068955
m1_Uninsure						
age	-.0051925	.0109004	-0.48	0.634	-.0265569	.0161719
male	.4748547	.3677294	1.29	0.197	-.2458817	1.195591
_cons	-1.756843	.4971363	-3.53	0.000	-2.731212	-.782474
m2_Prepaid						
age	-.0101521	.0058988	-1.72	0.085	-.0217135	.0014094
male	.5144003	.1996132	2.58	0.010	.1231656	.9056351
_cons	.2678043	.2744018	0.98	0.329	-.2700133	.805622
m3_Uninsure						
age	-.0041055	.0111185	-0.37	0.712	-.0258974	.0176865
male	.4591072	.3601307	1.27	0.202	-.2467359	1.16495
_cons	-1.801774	.5226351	-3.45	0.001	-2.82612	-.7774283

```
. test [m1_Prepaid = m2_Prepaid], cons notest

 ( 1)  [m1_Prepaid]age - [m2_Prepaid]age = 0
 ( 2)  [m1_Prepaid]male - [m2_Prepaid]male = 0
 ( 3)  [m1_Prepaid]_cons - [m2_Prepaid]_cons = 0

. test [m1_Uninsure = m3_Uninsure], cons acc

 ( 1)  [m1_Prepaid]age - [m2_Prepaid]age = 0
 ( 2)  [m1_Prepaid]male - [m2_Prepaid]male = 0
 ( 3)  [m1_Prepaid]_cons - [m2_Prepaid]_cons = 0
 ( 4)  [m1_Uninsure]age - [m3_Uninsure]age = 0
 ( 5)  [m1_Uninsure]male - [m3_Uninsure]male = 0
 ( 6)  [m1_Uninsure]_cons - [m3_Uninsure]_cons = 0

          chi2(  6) =    1.95
        Prob > chi2 =    0.9239
```

Again we do not find evidence against the correct specification of the multinomial logit for type of insurance. The classic Hausman test assumes that one of the estimators (named B in hausman) is efficient, that is, it has minimal (asymptotic) variance. This assumption ensures that $V(b) - V(B)$ is an admissible, viable estimator for $V(b - B)$. The assumption that we have an efficient estimator is a restrictive one. It is violated, for instance, if our data are clustered. We want to adjust for clustering via a vce(cluster *clustvar*) option by requesting the cluster-adjusted sandwich estimator of variance. Consequently, in such a case, hausman cannot be used. This problem does not exist with the suest version of the Hausman test. To illustrate this feature, we suppose that the data are clustered by city—we constructed an imaginary variable cityid for this illustration. If we plan to apply suest, we would not specify the vce(cluster *clustvar*) option at the time of estimation. suest has a vce(cluster *clustvar*) option. Thus we do not need to refit the models; we can call suest and test right away.

```
. suest m1 m2, vce(cluster cityid)
```

Simultaneous results for m1, m2

```
                                          Number of obs   =        615
                            (Std. Err. adjusted for 260 clusters in cityid)
```

	Coef.	Robust Std. Err.	z	P>|z|	[95% Conf. Interval]	
m1_Prepaid						
age	-.0100251	.005729	-1.75	0.080	-.0212538	.0012035
male	.5095747	.1910495	2.67	0.008	.1351246	.8840248
_cons	.2633838	.2698797	0.98	0.329	-.2655708	.7923383
m1_Uninsure						
age	-.0051925	.0104374	-0.50	0.619	-.0256494	.0152644
male	.4748547	.3773989	1.26	0.208	-.2648334	1.214543
_cons	-1.756843	.4916593	-3.57	0.000	-2.720478	-.7932087
m2_Prepaid						
age	-.0101521	.0057164	-1.78	0.076	-.0213559	.0010518
male	.5144003	.1921384	2.68	0.007	.137816	.8909847
_cons	.2678043	.2682193	1.00	0.318	-.2578958	.7935044

```
. test [m1_Prepaid = m2_Prepaid], cons

 ( 1)  [m1_Prepaid]age - [m2_Prepaid]age = 0
 ( 2)  [m1_Prepaid]male - [m2_Prepaid]male = 0
 ( 3)  [m1_Prepaid]_cons - [m2_Prepaid]_cons = 0

          chi2(  3) =      0.79
        Prob > chi2 =      0.8529
```

suest provides some descriptive information about the clustering on cityid. Like any other estimation command, suest informs us that the standard errors are adjusted for clustering. The Hausman-type test obtained from the test command uses a simultaneous (co)variance of m1 and m2 appropriately adjusted for clustering. In this example, we still do not have reason to conclude that the multinomial logit model in this application is misspecified, that is, that IIA is violated.

◁

The multinomial logistic regression model is a special case of the conditional logistic regression model; see [R] **clogit**. Like the multinomial logistic regression model, the conditional logistic regression model also makes the IIA assumption. [R] **clogit** discusses an example in which the demand for American, Japanese, and European cars is modeled in terms of the number of local dealers of the respective brands and of some individual attributes incorporated in interaction with the nationality of cars. We want to perform a Hausman-type test for IIA comparing the decision between all nationalities with the decision between non-American cars. The following code fragment demonstrates how to conduct a Hausman test for IIA via suest in this case.

```
. clogit choice japan europe maleJap maleEur incJap incEur dealer, group(id)
. estimates store allcars
. clogit choice japan maleJap incJap dealer if car!=1 , group(id)
. estimates store foreign
. suest allcars foreign
. test [allcars=foreign], common
```

Testing proportionality

The applications of `suest` that we have discussed so far concern Hausman-type tests for mis-specification. To test such a hypothesis, we compared two estimators that have the same probability limit if the hypothesis holds true, but otherwise have different limits. We may also want to compare the coefficients of models (estimators) for other substantive reasons. Although we most often want to test whether coefficients differ between models or estimators, we may occasionally want to test other constraints (see Ruud and Hausman 1987).

▷ Example 2

In this example, using simulated labor market data for siblings, we consider two dependent variables, income (`inc`) and whether or not a person was promoted in the last year (`promo`). We apply familiar economic arguments regarding human capital, according to which employees have a higher income and a higher probability of being promoted, by having more human capital. Human capital is acquired through formal education (`edu`) and on-the-job training experience (`exp`). We study whether income and promotion are "two sides of the same coin", that is, whether they reflect a common latent variable, "human capital". Accordingly, we want to compare the effects of different aspects of human capital on different outcome variables.

We estimate fairly simple labor market equations. The income model is estimated with `regress`, and the estimation results are stored under the name `Inc`.

```
. use http://www.stata-press.com/data/r10/income

. regress inc edu exp male
```

Source	SS	df	MS		Number of obs =	277
					F(3, 273) =	42.34
Model	2058.44672	3	686.148908		Prob > F =	0.0000
Residual	4424.05183	273	16.2053181		R-squared =	0.3175
					Adj R-squared =	0.3100
Total	6482.49855	276	23.4873136		Root MSE =	4.0256

inc	Coef.	Std. Err.	t	P>|t|	[95% Conf. Interval]	
edu	2.213707	.243247	9.10	0.000	1.734828	2.692585
exp	1.47293	.231044	6.38	0.000	1.018076	1.927785
male	.5381153	.4949466	1.09	0.278	-.436282	1.512513
_cons	1.255497	.3115808	4.03	0.000	.642091	1.868904

```
. est store Inc
```

Being sibling data, the observations are clustered on family of origin, `famid`. In the estimation of the regression parameters, we did not specify a `vce(cluster famid)` option to adjust standard errors for clustering on family (`famid`). Thus the standard errors reported by `regress` are potentially flawed. This problem will, however, be corrected by specifying a `vce(cluster clustvar)` option with `suest`.

Next we estimate the promotion equation with `probit` and again store the results under an appropriate name.

(Continued on next page)

```
. probit promo edu exp male, nolog
```

```
Probit regression                              Number of obs   =        277
                                               LR chi2(3)      =      49.76
                                               Prob > chi2     =     0.0000
Log likelihood = -158.43888                    Pseudo R2       =     0.1357
```

promo	Coef.	Std. Err.	z	P>|z|	[95% Conf. Interval]	
edu	.4593002	.0898535	5.11	0.000	.2831906	.6354097
exp	.3593023	.0805773	4.46	0.000	.2013737	.5172309
male	.2079983	.1656412	1.26	0.209	-.1166524	.532649
_cons	-.464622	.1088164	-4.27	0.000	-.6778982	-.2513458

```
. est store Promo
```

The coefficients in the income and promotion equations definitely seem to be different. However, since the scales of the two variables are different, we would not expect the coefficients to be equal. The correct hypothesis here is that the proportionality of the coefficients of the two models, apart from the constant, are equal. This formulation would still reflect that the relative effects of the different aspects of human capital do not differ between the dependent variables. We can obtain a nonlinear Wald test for the hypothesis of proportionality by using the `testnl` command on the combined estimation results of the two estimators. Thus we first have to form the combined estimation results. At this point we specify the `vce(cluster famid)` option to adjust for the clustering of observations on `famid`.

```
. suest Inc Promo, vce(cluster famid)
```

```
Simultaneous results for Inc, Promo
                                               Number of obs   =        277
```

```
                              (Std. Err. adjusted for 135 clusters in famid)
```

	Coef.	Robust Std. Err.	z	P>|z|	[95% Conf. Interval]	
Inc_mean						
edu	2.213707	.2483907	8.91	0.000	1.72687	2.700543
exp	1.47293	.1890583	7.79	0.000	1.102383	1.843478
male	.5381153	.4979227	1.08	0.280	-.4377952	1.514026
_cons	1.255497	.3374977	3.72	0.000	.594014	1.916981
Inc_lnvar						
_cons	2.785339	.079597	34.99	0.000	2.629332	2.941347
Promo						
edu	.4593002	.0886977	5.18	0.000	.2854558	.6331445
exp	.3593023	.0797717	4.50	0.000	.2029526	.5156521
male	.2079983	.1691049	1.23	0.219	-.1234412	.5394379
_cons	-.464622	.1042165	-4.46	0.000	-.6688826	-.2603614

The standard errors reported by suest are identical to those reported by the respective estimation commands when invoked with the option vce(cluster famid). We are now ready to test for proportionality:

$$H_0 : \frac{\beta_{\text{edu}}^{\text{Income}}}{\beta_{\text{edu}}^{\text{Promotion}}} = \frac{\beta_{\text{exp}}^{\text{Income}}}{\beta_{\text{exp}}^{\text{Promotion}}} = \frac{\beta_{\text{male}}^{\text{Income}}}{\beta_{\text{male}}^{\text{Promotion}}}$$

It is straightforward to translate this into syntax suitable for testnl, recalling that the coefficient of variable v in equation eq is denoted by $[eq]v$.

```
. testnl [Inc_mean]edu/[Promo]edu =
>          [Inc_mean]exp/[Promo]exp =
>          [Inc_mean]male/[Promo]male

  (1)  [Inc_mean]edu/[Promo]edu = [Inc_mean]exp/[Promo]exp
  (2)  [Inc_mean]edu/[Promo]edu = [Inc_mean]male/[Promo]male

              chi2(2) =        0.61
          Prob > chi2 =        0.7385
```

From the evidence, we fail to reject the hypotheses that the coefficients of the income and promotion equations are proportional. Thus it is not unreasonable to assume that income and promotion can be explained by the same latent variable, "labor market success".

A disadvantage of the nonlinear Wald test is that it is not invariant with respect to representation: a Wald test for a mathematically equivalent formulation of the nonlinear constraint usually leads to a different test result. An equivalent formulation of the proportionality hypothesis is

$$H_0: \quad \beta_{\text{edu}}^{\text{Income}} \beta_{\text{exp}}^{\text{Promotion}} = \beta_{\text{edu}}^{\text{Promotion}} \beta_{\text{exp}}^{\text{Income}} \quad \text{and}$$
$$\beta_{\text{edu}}^{\text{Income}} \beta_{\text{male}}^{\text{Promotion}} = \beta_{\text{edu}}^{\text{Promotion}} \beta_{\text{male}}^{\text{Income}}$$

This formulation is "more linear" in the coefficients. The asymptotic χ^2 distribution of the nonlinear Wald statistic can be expected to be more accurate for this representation.

```
. testnl ([Inc_mean]edu*[Promo]exp = [Inc_mean]exp*[Promo]edu)
>         ([Inc_mean]edu*[Promo]male = [Inc_mean]male*[Promo]edu)

  (1)  [Inc_mean]edu*[Promo]exp = [Inc_mean]exp*[Promo]edu
  (2)  [Inc_mean]edu*[Promo]male = [Inc_mean]male*[Promo]edu

              chi2(2) =        0.46
          Prob > chi2 =        0.7936
```

Here the two representations lead to similar test statistics and p-values. As before, we fail to reject the hypothesis of proportionality of the coefficients of the two models.

◁

Testing cross-model hypotheses

▷ Example 3

In this example, we demonstrate how some cross-model hypotheses can be tested using the facilities already available in most estimation commands. This demonstration will explain the intricate relationship between the cluster adjustment of the robust estimator of variance and the suest command. It will also be made clear that a new facility is required to perform more general cross-model testing.

We want to test whether the effect of x_1 on the binary variable y_1 is the same as the effect of x_2 on the binary y_2; see Clogg, Petkova, and Haritou 1995. In this setting, x_1 may equal x_2, and y_1 may equal y_2. We assume that logistic regression models can be used to model the responses, and for simplicity, we ignore further predictor variables in these models. If the two logit models are fitted on independent samples so that the estimators are (stochastically) independent, a Wald test for `_b[x1] = _b[x2]` rejects the null hypothesis if

$$\frac{\widehat{b}(x_1) - \widehat{b}(x_2)}{\left[\widehat{\sigma}^2\left\{\widehat{b}(x_1)\right\} + \widehat{\sigma}^2\left\{\widehat{b}(x_2)\right\}\right]^{1/2}}$$

is larger than the appropriate χ_1^2 threshold. If the models are fitted on the *same* sample (or on dependent samples), so that the estimators are stochastically dependent, the above test that ignores the covariance between the estimators is not appropriate.

It is instructive to see how this problem can be tackled by "stacking" data. In the stacked format, we doubled the number of observations. The dependent variable is y_1 in the first half of the data and is y_2 in the second half of the data. The predictor variable z_1 is set to x_1 in the first half of the expanded data and to 0 in the rest. Similarly, z_2 is 0 in the first half and x_2 in the second half. The following diagram illustrates the transformation, in the terminology of the `reshape` command, from wide to long format.

$$
\begin{pmatrix}
\text{id} & y_1 & y_2 & x_1 & x_2 \\
\hline
1 & y_{11} & y_{21} & x_{11} & x_{21} \\
2 & y_{12} & y_{22} & x_{12} & x_{22} \\
3 & y_{13} & y_{23} & x_{13} & x_{23}
\end{pmatrix}
\implies
\begin{pmatrix}
\text{id} & y & z_1 & z_2 & \text{model} \\
\hline
1 & y_{11} & x_{11} & 0 & 1 \\
2 & y_{12} & x_{12} & 0 & 1 \\
3 & y_{13} & x_{13} & 0 & 1 \\
1 & y_{21} & 0 & x_{21} & 2 \\
2 & y_{22} & 0 & x_{22} & 2 \\
3 & y_{23} & 0 & x_{23} & 2
\end{pmatrix}
$$

The observations in the long format data organization are *clustered* on the original subjects and are identified with the identifier `id`. The clustering on `id` has to be accounted for when fitting a simultaneous model. The simplest way to deal with clustering is to use the cluster adjustment of the robust or sandwich estimator; see [P] **_robust**. The data manipulation can be accomplished easily with the `stack` command; see [D] **stack**. Subsequently, we fit a simultaneous logit model and perform a Wald test for the hypothesis that the coefficients of `z1` and `z2` are the same. A full setup to obtain the cross-model Wald test could then be as follows:

```
. generate zero = 0          // a variable that is always 0
. generate one  = 1          // a variable that is always 1
. generate two  = 2          // a variable that is always 2
. stack  id y1 x1 zero one   id y2 zero x2 two, into(id y z1 z2 model)
. generate model2 = (model==2)
. logit y model2 z1 z2, vce(cluster id)
. test _b[z1] = _b[z2]
```

The coefficient of `z1` represents the effect of `x1` on `y1`, and similarly, `z2` for the effect of `x2` on `y2`. The variable `model2` is a dummy for the "second model", which is included to allow the intercept in the second model to differ from that in the first model. The estimates of the coefficient of `z1` and its standard error in the combined model are the same as the estimates of the coefficient of `z1` and its standard error if we fit the model on the unstacked data.

```
. logit y1 x1, vce(robust)
```

The vce(cluster *clustvar*) option specified with the logit command for the stacked data ensures that the covariances of _b[z1] and _b[z2] are indeed estimated. This estimation ensures that the Wald test for the equality of the coefficients is correct. If we had not specified the vce(cluster *clustvar*) option, the (co)variance matrix of the coefficients would have been block-diagonal; i.e., the covariances of _b[z1] and _b[z2] would have been 0. Then test would have effectively used the invalid formula for the Wald test for two independent samples.

In this example, the two logit models were fitted on the same data. The same setup would apply, without modification, when the two logit models were fitted on overlapping data that resulted, for instance, if the y or x variables were missing in some observations.

The suest command allows us to obtain the above Wald test more efficiently by avoiding the data manipulation, obviating the need to fit a model with twice the number of coefficients. The test statistic produced by the above code fragment is *identical* to that obtained via suest on the original (unstacked) data:

```
. logit y1 x1
. estimates store M1
. logit y2 x2
. estimates store M2
. suest M1 M2
. test [M1]x1=[M2]x2
```

The stacking method can be applied not only to the testing of cross-model hypotheses for logit models but also to any estimation command that supports the vce(cluster *clustvar*) option. The stacking approach clearly generalizes to stacking more than two logit or other models, testing more general linear hypotheses, and testing nonlinear cross-model hypotheses (see [R] **testnl**). In all these cases, suest would yield identical statistical results but at smaller costs in terms of data management, computer storage, and computer time.

Is suest nothing but a convenience command? No, there are two disadvantages to the stacking method, both of which are resolved via suest. First, if the models include ancillary parameters (in a regression model, the residual variance; in an ordinal response model, the cutpoints; and in lognormal survival-time regression, the time scale parameter), these parameters are constrained to be equal between the stacked models. In suest, this constraint is relaxed. Second, the stacking method does not generalize to compare different statistical models, such as a probit model and a regression model. As demonstrated in the previous section, suest can deal with this situation.

◁

(*Continued on next page*)

Saved Results

suest saves the following in e():

Scalars
e(N)	number of observations
e(N_clust)	number of clusters
e(rank)	rank of e(V)

Macros
e(cmd)	suest
e(eqnames#)	original names of equations of model #
e(names)	list of model names
e(wtype)	weight type
e(wexp)	weight expression
e(clustvar)	name of cluster variable
e(vce)	*vcetype* specified in vce()
e(vcetype)	title used to label Std. Err.

Matrices
e(b)	stacked coefficient vector of the models
e(V)	(co)variance matrix

Functions
e(sample)	marks estimation sample

Methods and Formulas

suest is implemented as an ado-file.

The estimation of the simultaneous (co)variance of a series of k estimators is a nonstandard application of the sandwich estimator, as implemented by the command [P] **_robust**. You may want to read this manual entry before reading further.

The starting point is that we have fitted k different models on the *same* data—partially overlapping or nonoverlapping data are special cases. We want to derive the *simultaneous* distribution of these k estimators, for instance, to test a cross-estimator hypothesis H_0. As in the framework of Hausman testing, H_0 will often be of the form that different estimators have the same probability limit under some hypothesis, while the estimators have different limits if the hypothesis is violated.

We consider (vector) estimators $\widehat{\beta}_i$ to be defined as "the" solution of the estimation equations $\mathbf{G}_i$,

$$\mathbf{G}_i(\mathbf{b}_i) = \sum_j w_{ij}\mathbf{u}_{ij}(\mathbf{b}_i) = \mathbf{0}, \qquad i = 1, \ldots, k$$

We refer to the $\mathbf{u}_{ij}$ as the "scores". Specifying some weights $w_{ij} = 0$ trivially accommodates for partially overlapping or even disjointed data. Under "suitable regularity conditions" (see White 1982 and 1996 for details), the $\widehat{\beta}_i$ are asymptotically normally distributed, with the variance estimated consistently by the sandwich estimator

$$V_i = \mathrm{Var}(\widehat{\beta}_i) = \mathbf{D}_i^{-1} \sum_j w_{ij}\mathbf{u}_{ij}\mathbf{u}_{ij}' \, \mathbf{D}_i^{-1}$$

where $\mathbf{D}_i$ is the Jacobian of $\mathbf{G}_i$ evaluated at $\widehat{\beta}_i$. In the context of maximum likelihood estimation, $\mathbf{D}_i$ can be estimated consistently by (minus) the Hessian of the log likelihood or by the Fisher information matrix. If the model is also well specified, the sandwiched term $(\sum_j w_{ij}\mathbf{u}_{ij}\mathbf{u}'_{ij})$ converges in probability to $\mathbf{D}_i$, so V_i may be consistently estimated by $\mathbf{D}_i^{-1}$.

To derive the simultaneous distribution of the estimators, we consider the "stacked" estimation equation,

$$\mathbf{G}(\widehat{\beta}) = \left[\mathbf{G}_1(\widehat{\beta}_1)' \quad \mathbf{G}_1(\widehat{\beta}_2)' \quad \dots \quad \mathbf{G}_k(\widehat{\beta}_k)'\right]' = \mathbf{0}$$

Under "suitable regularity conditions" (see White 1996 for details), $\widehat{\beta}$ is asymptotically *jointly* normally distributed. The Jacobian and scores of the simultaneous equation are easily expressed in the Jacobian and scores of the separate equations. The Jacobian of $\mathbf{G}$,

$$\mathbf{D}(\widehat{\beta}) = \left.\frac{d\mathbf{G}(\beta)}{d\beta}\right|_{\beta=\widehat{\beta}}$$

is block diagonal with blocks $\mathbf{D}_1, \dots, \mathbf{D}_k$. The inverse of $\mathbf{D}(\widehat{\beta})$ is again block diagonal, with the inverses of $\mathbf{D}_i$ on the diagonal. The scores $\mathbf{u}$ of $\mathbf{G}$ are simply obtained as the *concatenated* scores of the separate equations:

$$\mathbf{u}_j = (\mathbf{u}'_{1j} \quad \mathbf{u}'_{2j} \quad \dots \quad \mathbf{u}'_{kj})'$$

Out-of-sample (i.e., where $w_{ij} = 0$) values of the score variables are defined as 0 (thus we drop the i subscript from the common weight variable). The sandwich estimator for the asymptotic variance of $\widehat{\beta}$ reads

$$V = \mathrm{Var}(\widehat{\beta}) = \mathbf{D}(\widehat{\beta})^{-1}\left(\sum_j w_j \mathbf{u}_j \mathbf{u}'_j\right)\mathbf{D}(\widehat{\beta})^{-1}$$

Taking a "partitioned" look at this expression, we see that $V(\widehat{\beta}_i)$ is estimated by

$$\mathbf{D}_i^{-1}\left(\sum_j w_j \mathbf{u}_{ij} \mathbf{u}'_{ij}\right)\mathbf{D}_i^{-1}$$

which is, yet again, the familiar sandwich estimator for $\widehat{\beta}_i$ based on the separate estimation equation $\mathbf{G}_i$. Thus considering several estimators simultaneously in this way does not affect the estimators of the asymptotic variances of these estimators. However, as a bonus of stacking, we obtained a sandwich-type estimate of the *covariance* V_{ih} of estimators $\widehat{\beta}_i$ and $\widehat{\beta}_h$,

$$V_{ih} = \mathrm{Cov}(\widehat{\beta}_i, \widehat{\beta}_h) = \mathbf{D}_i^{-1}\left(\sum_j w_j \mathbf{u}_{ij} \mathbf{u}'_{ih}\right)\mathbf{D}_h^{-1}$$

which is also obtained by White (1982).

This estimator for the covariance of estimators is an application of the cluster modification of the sandwich estimator proposed by Rogers (1993). Consider the stacked data format as discussed in the logit example, and assume that Stata would be able to estimate a "stacked model" in which different models apply to different observations, for example, a probit model for the first half, a regression model for the second half, and a one-to-one cluster relation between the first and second half. If there are no common parameters to both models, the score statistics of parameters for the stacked models are zero in the half of the data in which they do not occur. In Rogers' method, we have to sum the score statistics over the observations within a cluster. This step boils down to concatenating the score statistics at the level of the cluster.

We compare the sandwich estimator of the (co)variance V_{12} of two estimators with the estimator of variance $\widetilde{V}_{12}$ applied in the classic Hausman test. Hausman (1978) showed that if $\widehat{\beta}_1$ is consistent under H_0 and $\widehat{\beta}_2$ is efficient under H_0, then asymptotically

$$\text{Cov}(\widehat{\beta}_1, \widehat{\beta}_2) = \text{Var}(\widehat{\beta}_2)$$

and so $\text{var}(\widehat{\beta}_1 - \widehat{\beta}_2)$ is consistently estimated by $V_1 - V_2$.

Acknowledgment

suest was written by Jeroen Weesie, Department of Sociology, Utrecht University, The Netherlands. This research is supported by grant PGS 50-370 by The Netherlands Organization for Scientific Research.

An earlier version of suest was published in the *Stata Technical Bulletin* (1999). The current version of suest is not backward compatible with the STB version because of the introduction of new ways to manage estimation results via the estimates command.

References

Arminger, G. 1990. Testing against misspecification in parametric rate models. In *Event History Analysis in Life Course Research*, ed. K. U. Mayer and N. B. Tuma, 146–158. Madison: University of Wisconsin Press.

Clogg, C. C., E. Petkova, and A. Haritou. 1995. Statistical methods for comparing regression coefficients between models. *American Journal of Sociology* 100: 1261–1293. With comments by P. D. Allison and a reply by the authors.

Gourieroux, C., and A. Monfort. 1989. *Statistics and Econometric Models. Vol. 2*. New York: Cambridge University Press.

Hausman, J. 1978. Specification tests in econometrics. *Econometrica* 46: 1251–1271.

Hausman, J., and D. McFadden. 1984. Specification tests in econometrics. *Econometrica* 52: 1219–1240.

Huber, P. J. 1967. The behavior of maximum likelihood estimates under nonstandard conditions. In *Proceedings of the Fifth Symposium on Mathematical Statistics and Probability*, vol. 1, 221–233. Berkeley: University of California Press.

Rogers, W. H. 1993. sg17: Regression standard errors in clustered samples. *Stata Technical Bulletin* 13: 19–23. Reprinted in *Stata Technical Bulletin Reprints*, vol. 3, pp. 88–94.

Ruud, B., and J. Hausman. 1987. Specifying and testing econometric models for rank-ordered data. *Journal of Econometrics* 34: 83–104.

Weesie, J. 1999. sg121: Seemingly unrelated estimation and the cluster-adjusted sandwich estimator. *Stata Technical Bulletin* 52: 34–47. Reprinted in *Stata Technical Bulletin Reprints*, vol. 9, pp. 231–248.

White, H. 1982. Maximum-likelihood estimation of misspecified models. *Econometrica* 50: 1–25.

——. 1996. *Estimation, Inference and Specification Analysis*. Cambridge: Cambridge University Press.

Also See

[R] **estimates** — Save and manipulate estimation results

[R] **hausman** — Hausman specification test

[R] **lincom** — Linear combinations of estimators

[R] **nlcom** — Nonlinear combinations of estimators

[R] **test** — Test linear hypotheses after estimation

[R] **testnl** — Test nonlinear hypotheses after estimation

[P] **_robust** — Robust variance estimates

Title

summarize — Summary statistics

Syntax

$\underline{\text{summarize}}$ $\left[\textit{varlist}\right]$ $\left[\textit{if}\right]$ $\left[\textit{in}\right]$ $\left[\textit{weight}\right]$ $\left[,\ \textit{options}\right]$

options	description
Main	
$\underline{\text{d}}$etail	display additional statistics
$\underline{\text{mea}}$nonly	suppress the display; calculate only the mean; programmer's option
$\underline{\text{f}}$ormat	use variable's display format
$\underline{\text{sep}}$arator(#)	draw separator line after every # variables; default is separator(5)

varlist may contain time-series operators; see [U] **11.4.3 Time-series varlists**.

by is allowed; see [D] **by**.

aweights, fweights, and iweights are allowed. However, iweights may not be used with the detail
  option; see [U] **11.1.6 weight**.

Description

summarize calculates and displays a variety of univariate summary statistics. If no *varlist* is
specified, summary statistics are calculated for all the variables in the dataset.

Also see [R] **ci** for calculating the standard error and confidence intervals of the mean.

Options

┌─ Main ┐

detail produces additional statistics, including skewness, kurtosis, the four smallest and largest
  values, and various percentiles.

meanonly, which is allowed only when detail is not specified, suppresses the display of results
  and calculation of the variance. Ado-file writers will find this useful for fast calls.

format requests that the summary statistics be displayed using the display formats associated with
  the variables rather than the default g display format; see [U] **12.5 Formats: controlling how data
  are displayed**.

separator(#) specifies how often to insert separation lines into the output. The default is sepa-
  rator(5), meaning that a line is drawn after every five variables. separator(10) would draw
  a line after every 10 variables. separator(0) suppresses the separation line.

Remarks

summarize can produce two different sets of summary statistics. Without the detail option,
the number of nonmissing observations, the mean and standard deviation, and the minimum and
maximum values are presented. With detail, the same information is presented along with the
variance, skewness, and kurtosis; the four smallest and four largest values; and the 1st, 5th, 10th,
25th, 50th (median), 75th, 90th, 95th, and 99th percentiles.

▷ Example 1

We have data containing information on various automobiles, among which is the variable mpg, the mileage rating. We can obtain a quick summary of the mpg variable by typing

```
. use http://www.stata-press.com/data/r10/auto
(1978 Automobile Data)
. summarize mpg
```

Variable	Obs	Mean	Std. Dev.	Min	Max
mpg	74	21.2973	5.785503	12	41

We see that we have 74 observations. The mean of mpg is 21.3 miles per gallon, and the standard deviation is 5.79. The minimum is 12, and the maximum is 41.

If we had not specified the variable (or variables) we wanted to summarize, we would have obtained summary statistics on all the variables in the dataset:

```
. summarize, separator(4)
```

Variable	Obs	Mean	Std. Dev.	Min	Max
make	0				
price	74	6165.257	2949.496	3291	15906
mpg	74	21.2973	5.785503	12	41
rep78	69	3.405797	.9899323	1	5
headroom	74	2.993243	.8459948	1.5	5
trunk	74	13.75676	4.277404	5	23
weight	74	3019.459	777.1936	1760	4840
length	74	187.9324	22.26634	142	233
turn	74	39.64865	4.399354	31	51
displacement	74	197.2973	91.83722	79	425
gear_ratio	74	3.014865	.4562871	2.19	3.89
foreign	74	.2972973	.4601885	0	1

There are only 69 observations on rep78, so some of the observations are missing. There are no observations on make since it is a string variable.

◁

▷ Example 2

The detail option provides all the information of a normal summarize and more. The format of the output also differs, as shown here:

```
. summarize mpg, detail
```

Mileage (mpg)

	Percentiles	Smallest		
1%	12	12		
5%	14	12		
10%	14	14	Obs	74
25%	18	14	Sum of Wgt.	74
50%	20		Mean	21.2973
		Largest	Std. Dev.	5.785503
75%	25	34		
90%	29	35	Variance	33.47205
95%	34	35	Skewness	.9487176
99%	41	41	Kurtosis	3.975005

As in the previous example, we see that the mean of mpg is 21.3 miles per gallon and that the standard deviation is 5.79. We also see the various percentiles. The median of mpg (the 50th percentile) is 20 miles per gallon. The 25th percentile is 18, and the 75th percentile is 25.

When we performed summarize, we learned that the minimum and maximum were 12 and 41, respectively. We now see that the four smallest values in our dataset are 12, 12, 14, and 14. The four largest values are 34, 35, 35, and 41. The skewness of the distribution is 0.95, and the kurtosis is 3.98. (A normal distribution would have a skewness of 0 and a kurtosis of 3.)

Skewness is a measure of the lack of symmetry of a distribution. If the distribution is symmetric, the coefficient of skewness is 0. If the coefficient is negative, the median is usually greater than the mean and the distribution is said to be skewed left. If the coefficient is positive, the median is usually less than the mean and the distribution is said to be skewed right. *Kurtosis* (from the Greek *kyrtosis*, meaning curvature) is a measure of peakedness of a distribution. The smaller the coefficient of kurtosis, the flatter the distribution. The normal distribution has a coefficient of kurtosis of 3 and provides a convenient benchmark.

◁

The idea of the mean is quite old (Plackett 1958), but its extension to a scheme of moment-based measures was not done until the end of the 19th century. Between 1893 and 1905, Pearson discussed and named the standard deviation, skewness, and kurtosis, but he was not the first to use any of these. Thiele (1889), in contrast, had earlier grasped firmly the notion that the m_r provide a systematic basis for discussing distributions. However, even earlier anticipations can also be found. For example, Euler in 1778 used m_2 and m_3 in passing in a treatment of estimation (Hald 1998, 87), but seemingly did not build on that.

Similarly, the idea of the median is quite old. The history of the interquartile range is tangled up with that of the probable error, a long-popular measure. Extending this in various ways to a more general approach based on quantiles (to use a later term) occurred to several people in the nineteenth century. Galton (1875) is a nice example, particularly because he seems so close to the key idea of the quantiles as a function, which took another century to reemerge strongly.

Thorvald Nicolai Thiele (1838–1910) was a Danish scientist who worked in astronomy, mathematics, actuarial science, and statistics. He made many pioneering contributions to statistics, several of which were overlooked until recently. Thiele advocated graphical analysis of residuals checking for trends, symmetry of distributions, and changes of sign, and he even warned against overinterpreting such graphs.

▷ Example 3

summarize can usefully be combined with the by *varlist*: prefix. In our dataset, we have a variable, foreign, that distinguishes foreign and domestic cars. We can obtain summaries of mpg and weight within each subgroup by typing

(Continued on next page)

```
. by foreign: summarize mpg weight
```

```
-> foreign = Domestic
       Variable |       Obs        Mean    Std. Dev.       Min        Max
----------------+--------------------------------------------------------
            mpg |        52    19.82692    4.743297         12         34
         weight |        52    3317.115    695.3637       1800       4840
```

```
-> foreign = Foreign
       Variable |       Obs        Mean    Std. Dev.       Min        Max
----------------+--------------------------------------------------------
            mpg |        22    24.77273    6.611187         14         41
         weight |        22    2315.909    433.0035       1760       3420
```

Domestic cars in our dataset average 19.8 miles per gallon, whereas foreign cars average 24.8.

Since by *varlist*: can be combined with summarize, it can also be combined with summarize, detail:

```
. by foreign: summarize mpg, detail
```

```
-> foreign = Domestic
                          Mileage (mpg)
-------------------------------------------------------------
       Percentiles      Smallest
 1%          12            12
 5%          14            12
10%          14            14         Obs                 52
25%        16.5           14         Sum of Wgt.         52

50%          19                      Mean           19.82692
                        Largest      Std. Dev.      4.743297
75%          22            28
90%          26            29         Variance       22.49887
95%          29            30         Skewness       .7712432
99%          34            34         Kurtosis       3.441459
```

```
-> foreign = Foreign
                          Mileage (mpg)
-------------------------------------------------------------
       Percentiles      Smallest
 1%          14            14
 5%          17            17
10%          17            17         Obs                 22
25%          21            18         Sum of Wgt.         22

50%         24.5                      Mean           24.77273
                        Largest      Std. Dev.      6.611187
75%          28            31
90%          35            35         Variance       43.70779
95%          35            35         Skewness        .657329
99%          41            41         Kurtosis        3.10734
```

◁

❑ Technical Note

summarize respects display formats if we specify the format option. When we type summarize price weight, we obtain

```
. summarize price weight
    Variable |      Obs        Mean    Std. Dev.       Min        Max
-------------+--------------------------------------------------------
       price |       74    6165.257    2949.496       3291      15906
      weight |       74    3019.459    777.1936       1760       4840
```

The display is accurate but is not as aesthetically pleasing as we may wish, particularly if we plan to use the output directly in published work. By placing formats on the variables, we can control how the table appears:

```
. format price weight %9.2fc
. summarize price weight, format
    Variable |      Obs        Mean    Std. Dev.       Min        Max
-------------+--------------------------------------------------------
       price |       74    6,165.26    2,949.50    3,291.00  15,906.00
      weight |       74    3,019.46      777.19    1,760.00   4,840.00
```

❑

If you specify a weight (see [U] **11.1.6 weight**), each observation is multiplied by the value of the weighting expression before the summary statistics are calculated so that the weighting expression is interpreted as the discrete density of each observation.

▷ Example 4

We have 1980 census data on each of the 50 states. Included in our variables is medage, the median age of the population of each state. If we type summarize medage, we obtain unweighted statistics:

```
. use http://www.stata-press.com/data/r10/census
(1980 Census data by state)
. summarize medage
    Variable |      Obs        Mean    Std. Dev.       Min        Max
-------------+--------------------------------------------------------
      medage |       50       29.54    1.693445        24.2       34.7
```

Also among our variables is pop, the population in each state. Typing summarize medage [w=pop] produces population-weighted statistics:

```
. summarize medage [w=pop]
(analytic weights assumed)
    Variable |      Obs      Weight        Mean   Std. Dev.       Min        Max
-------------+------------------------------------------------------------------
      medage |       50   225907472    30.11047     1.66933       24.2       34.7
```

The number listed under Weight is the sum of the weighting variable, pop, indicating that there are roughly 226 million people in the United States. The pop-weighted mean of medage is 30.11 (compared with 29.54 for the unweighted statistic), and the weighted standard deviation is 1.67 (compared with 1.69).

◁

▷ Example 5

We can obtain detailed summaries of weighted data as well. When we do this, *all* the statistics are weighted, including the percentiles.

```
. summarize medage [w=pop], detail
(analytic weights assumed)
```

                              Median age

	Percentiles	Smallest		
1%	27.1	24.2		
5%	27.7	26.1		
10%	28.2	27.1	Obs	50
25%	29.2	27.4	Sum of Wgt.	225907472
50%	29.9		Mean	30.11047
		Largest	Std. Dev.	1.66933
75%	30.9	32		
90%	32.1	32.1	Variance	2.786661
95%	32.2	32.2	Skewness	.5281972
99%	34.7	34.7	Kurtosis	4.494223

◁

❑ Technical Note

If you are writing a program and need to access the mean of a variable, the `meanonly` option provides for fast calls. For example, suppose that your program reads as follows:

```
program mean
        summarize '1', meanonly
        display "  mean = " r(mean)
end
```

The result of executing this is

```
. mean price
  mean = 6165.2568
```

❑

Saved Results

`summarize` saves the following in `r()`:

Scalars

r(N)	number of observations	r(p50)	50th percentile (detail only)
r(mean)	mean	r(p75)	75th percentile (detail only)
r(skewness)	skewness (detail only)	r(p90)	90th percentile (detail only)
r(min)	minimum	r(p95)	95th percentile (detail only)
r(max)	maximum	r(p99)	99th percentile (detail only)
r(sum_w)	sum of the weights	r(Var)	variance
r(p1)	1st percentile (detail only)	r(kurtosis)	kurtosis (detail only)
r(p5)	5th percentile (detail only)	r(sum)	sum of variable
r(p10)	10th percentile (detail only)	r(sd)	standard deviation
r(p25)	25th percentile (detail only)		

Methods and Formulas

Let x denote the variable on which we want to calculate summary statistics, and let x_i, $i = 1, \ldots, n$, denote an individual observation on x. Let v_i be the weight, and if no weight is specified, define $v_i = 1$ for all i.

Define V as the *sum of the weight*:

$$V = \sum_{i=1}^{n} v_i$$

Define w_i to be v_i normalized to sum to n, $w_i = v_i(n/V)$.

The *mean*, $\overline{x}$, is defined as

$$\overline{x} = \frac{1}{n} \sum_{i=1}^{n} w_i x_i$$

The *variance*, s^2, is defined as

$$s^2 = \frac{1}{n-1} \sum_{i=1}^{n} w_i (x_i - \overline{x})^2$$

The *standard deviation*, s, is defined as $\sqrt{s^2}$.

Define m_r as the rth moment about the mean $\overline{x}$:

$$m_r = \frac{1}{n} \sum_{i=1}^{n} w_i (x_i - \overline{x})^r$$

The *coefficient of skewness* is then defined as $m_3 m_2^{-3/2}$. The *coefficient of kurtosis* is defined as $m_4 m_2^{-2}$.

Let $x_{(i)}$ refer to the x in ascending order, and let $w_{(i)}$ refer to the corresponding weights of $x_{(i)}$. The four smallest values are $x_{(1)}$, $x_{(2)}$, $x_{(3)}$, and $x_{(4)}$. The four largest values are $x_{(n)}$, $x_{(n-1)}$, $x_{(n-2)}$, and $x_{(n-3)}$.

To obtain the pth *percentile*, which we will denote as $x_{[p]}$, let $P = np/100$. Let

$$W_{(i)} = \sum_{j=1}^{i} w_{(j)}$$

Find the first index i such that $W_{(i)} > P$. The pth *percentile* is then

$$x_{[p]} = \begin{cases} \dfrac{x_{(i-1)} + x_{(i)}}{2} & \text{if } W_{(i-1)} = P \\ x_{(i)} & \text{otherwise} \end{cases}$$

References

David, H. A. 2001. First (?) occurrence of common terms in statistics and probability. In *Annotated Readings in the History of Statistics*, ed. H. David and A. Edwards, 209–246. New York: Springer.

Galton, F. 1875. Statistics by intercomparison, with remarks on the law of frequency of error. *Philosophical Magazine* 49: 33–46.

Gleason, J. R. 1997. sg67: Univariate summaries with boxplots. *Stata Technical Bulletin* 36: 23–25. Reprinted in *Stata Technical Bulletin Reprints*, vol. 6, pp. 179–183.

——. 1999. sg67.1: Update to univar. *Stata Technical Bulletin* 51: 27–28. Reprinted in *Stata Technical Bulletin Reprints*, vol. 9, pp. 159–161.

Hald, A. 1998. *A History of Mathematical Statistics from 1750 to 1930*. New York: Wiley.

Hamilton, L. C. 1996. *Data Analysis for Social Scientists*. Belmont, CA: Wadsworth.

——. 2006. *Statistics with Stata (Updated for Version 9)*. Belmont, CA: Duxbury.

Kirkwood, B. R., and J. A. C. Sterne. 2003. *Essential Medical Statistics*. 2nd ed. Malden, MA: Blackwell.

Lauritzen, S. L. 2002. *Thiele: Pioneer in Statistics*. Oxford: Oxford University Press.

Plackett, R. L. 1958. The principle of the arithmetic mean. *Biometrika* 45: 130–135.

Stuart, A., and J. K. Ord. 1994. *Kendall's Advanced Theory of Statistics: Distribution Theory, Vol. 1*. 6th ed. London: Arnold.

Thiele, T. N. 1889. *Almindelig Iagttagelseslære: Sandsynlighedsregning og mindste Kvadraters Methode*. Kjøbenhavn: C.A. Reitzel. (English translation included in Lauritzen 2002.)

Weisberg, H. F. 1992. *Central Tendency and Variability*. Newbury Park, CA: Sage.

Also See

[R] **ameans** — Arithmetic, geometric, and harmonic means

[R] **centile** — Report centile and confidence interval

[R] **mean** — Estimate means

[R] **proportion** — Estimate proportions

[R] **ratio** — Estimate ratios

[R] **table** — Tables of summary statistics

[R] **tabstat** — Display table of summary statistics

[R] **tabulate, summarize()** — One- and two-way tables of summary statistics

[R] **total** — Estimate totals

[D] **codebook** — Describe data contents

[D] **describe** — Describe data in memory or in file

[D] **inspect** — Display simple summary of data's attributes

[ST] **stsum** — Summarize survival-time data

[SVY] **svy estimation** — Estimation commands for survey data

[XT] **xtsum** — Summarize xt data

Title

sunflower — Density-distribution sunflower plots

Syntax

sunflower *yvar* *xvar* $\left[\text{ }if\text{ }\right]$ $\left[\text{ }in\text{ }\right]$ $\left[\text{ }weight\text{ }\right]$ $\left[\text{ }, \text{ }options\text{ }\right]$

options	description
Main	
nograph	do not show graph
notable	do not show summary table; implied when by() is specified
marker_options	affect rendition of markers drawn at the plotted points
Bins/Petals	
binwidth(#)	width of the hexagonal bins
binar(#)	aspect ratio of the hexagonal bins
bin_options	affect rendition of hexagonal bins
light(#)	minimum observations for a light sunflower; default is light(3)
dark(#)	minimum observations for a dark sunflower; default is dark(13)
xcenter(#)	x-coordinate of the reference bin
ycenter(#)	y-coordinate of the reference bin
petalweight(#)	observations in a dark sunflower petal
petallength(#)	length of sunflower petal as a percentage
petal_options	affect rendition of sunflower petals
flowersonly	show petals only; do not render bins
nosinglepetal	suppress single petals
Add plots	
addplot(*plot*)	add other plots to generated graph
Y axis, X axis, Titles, Legend, Overall, By	
twoway_options	any options documented in [G] ***twoway_options***

(Continued on next page)

bin_options	description
[l|d] bstyle(*areastyle*)	overall look of hexagonal bins
[l|d] bcolor(*colorstyle*)	outline and fill color
[l|d] bfcolor(*colorstyle*)	fill color
[l|d] blstyle(*linestyle*)	overall look of outline
[l|d] blcolor(*colorstyle*)	outline color
[l|d] blwidth(*linewidthstyle*)	thickness of outline

petal_options	description
[l|d] flstyle(*linestyle*)	overall style of sunflower petals
[l|d] flcolor(*colorstyle*)	color of sunflower petals
[l|d] flwidth(*linewidthstyle*)	thickness of sunflower petals

All options are *rightmost*; see [G] **concept: repeated options**.
fweights are allowed; see [U] **11.1.6 weight**.

Description

sunflower draws density-distribution sunflower plots (Dupont and Plummer 2003). These plots are useful for displaying bivariate data whose density is too great for conventional scatterplots to be effective.

A sunflower is several line segments of equal length, called petals, that radiate from a central point. There are two varieties of sunflowers: light and dark. Each petal of a light sunflower represents 1 observation. Each petal of a dark sunflower represents several observations. Dark and light sunflowers represent high- and medium-density regions of the data, and marker symbols represent individual observations in low-density regions.

The plane defined by the variables *yvar* and *xvar* is divided into contiguous hexagonal bins. The number of observations contained within a bin determines how the bin will be represented.

1. When there are fewer than light(*#*) observations in a bin, each point is plotted using the usual marker symbols in a scatterplot.

2. Bins with at least light(*#*) but fewer than dark(*#*) observations are represented by a light sunflower.

3. Bins with at least dark(*#*) observations are represented by a dark sunflower.

Options

⌐ Main ⌐

nograph prevents the graph from being generated.

notable prevents the summary table from being displayed. This option is implied when the by() option is specified.

marker_options affect the rendition of markers drawn at the plotted points, including their shape, size, color, and outline; see [G] ***marker_options***.

Bins/Petals

binwidth(*#*) specifies the horizontal width of the hexagonal bins in the same units as *xvar*. By default,

$$binwidth = \texttt{max}(rbw, nbw)$$

where

$$rbw = \text{range of } xvar/40$$

$$nbw = \text{range of } xvar/\texttt{max(1,nb)}$$

and

$$nb = \texttt{int(min(sqrt(}n\texttt{),10 * log10(}n\texttt{)))}$$

where

$$n = \text{the number of observations in the dataset}$$

binar(*#*) specifies the aspect ratio for the hexagonal bins. The height of the bins is given by

$$binheight = binwidth \times \# \times 2/\sqrt{3}$$

where *binheight* and *binwidth* are specified in the units of *yvar* and *xvar*, respectively. The default is binar(*r*), where *r* results in the rendering of regular hexagons.

bin_options affect how the hexagonal bins are rendered.

lbstyle(*areastyle*) and dbstyle(*areastyle*) specify the look of the light and dark hexagonal bins, respectively. The options listed below allow you to change each attribute, but lbstyle() and dbstyle() provide the starting points. See [G] *areastyle* for a list of available area styles.

lbcolor(*colorstyle*) and dbcolor(*colorstyle*) specify one color to be used both to outline the shape and to fill the interior of the light and dark hexagonal bins, respectively. See [G] *colorstyle* for a list of color choices.

lbfcolor(*colorstyle*) and dbfcolor(*colorstyle*) specify the color to be used to fill the interior of the light and dark hexagonal bins, respectively. See [G] *colorstyle* for a list of color choices.

lblstyle(*linestyle*) and dblstyle(*linestyle*) specify the overall style of the line used to outline the area, which includes its pattern (solid, dashed, etc.), thickness, and color. The other options listed below allow you to change the line's attributes, but lblstyle() and dblstyle() are the starting points. See [G] *linestyle* for a list of choices.

lblcolor(*colorstyle*) and dblcolor(*colorstyle*) specify the color to be used to outline the light and dark hexagonal bins, respectively. See [G] *colorstyle* for a list of color choices.

lblwidth(*linewidthstyle*) and dblwidth(*linewidthstyle*) specify the thickness of the line to be used to outline the light and dark hexagonal bins, respectively. See [G] *linewidthstyle* for a list of choices.

light(*#*) specifies the minimum number of observations needed for a bin to be represented by a light sunflower. The default is light(3).

dark(*#*) specifies the minimum number of observations needed for a bin to be represented by a dark sunflower. The default is dark(13).

xcenter(#) and ycenter(#) specify the center of the reference bin. The default values are the median values of *xvar* and *yvar*, respectively. The centers of the other bins are implicitly defined by the location of the reference bin together with the common bin width and height.

petalweight(#) specifies the number of observations represented by each petal of a dark sunflower. The default value is chosen so that the maximum number of petals on a dark sunflower is 14.

petallength(#) specifies the length of petals in the sunflowers. The value specified is interpreted as a percentage of half the bin width. The default is 100%.

petal_options affect how the sunflower petals are rendered.

lflstyle(*linestyle*) and dflstyle(*linestyle*) specify the overall style of the light and dark sunflower petals, respectively.

lflcolor(*colorstyle*) and dflcolor(*colorstyle*) specify the color of the light and dark sunflower petals, respectively.

lflwidth(*linewidthstyle*) and dflwidth(*linewidthstyle*) specify the width of the light and dark sunflower petals, respectively.

flowersonly suppresses rendering of the bins. This option is equivalent to specifying lbcolor(none) and dbcolor(none).

nosinglepetal suppresses flowers from being drawn in light bins that contain only 1 observation and dark bins that contain as many observations as the petal weight (see the petalweight() option).

⌐‾‾ Add plots ⌐‾‾‾

addplot(*plot*) provides a way to add other plots to the generated graph; see [G] *addplot_option*.

⌐‾‾ Y axis, X axis, Titles, Legend, Overall, By ⌐‾‾‾‾‾‾‾‾‾‾‾‾‾‾‾‾‾‾‾‾‾‾‾‾‾

twoway_options are any of the options documented in [G] *twoway_options*. These include options for titling the graph (see [G] *title_options*), options for saving the graph to disk (see [G] *saving_option*), and the by() option (see [G] *by_option*).

Remarks

▷ Example 1

Using the auto dataset, we want to examine the relationship between weight and mpg. To do that, we type

```
. use http://www.stata-press.com/data/r10/auto
(1978 Automobile Data)

. sunflower mpg weight, binwid(500) petalw(2) dark(8)
Bin width         =        500
Bin height        =    8.38703
Bin aspect ratio  =   .0145268
Max obs in a bin  =         15
Light             =          3
Dark              =          8
X-center          =       3190
Y-center          =         20
Petal weight      =          2
```

flower type	petal weight	No. of petals	No. of flowers	estimated obs.	actual obs.
none				10	10
light	1	3	1	3	3
light	1	4	2	8	8
light	1	7	3	21	21
dark	2	4	1	8	8
dark	2	5	1	10	9
dark	2	8	1	16	15
				76	74

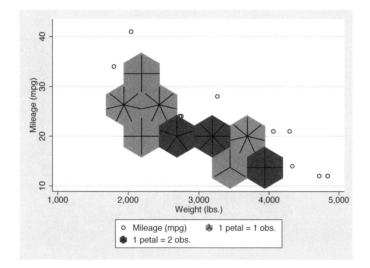

The three darkly shaded sunflowers immediately catch our eyes, indicating a group of eight cars that are heavy (nearly 4,000 pounds) and fuel inefficient and two groups of cars that get about 20 miles per gallon and weight in the neighborhood of 3,000 pounds, one with 10 cars and one with 8 cars. The lighter sunflowers with seven petals each indicate groups of seven cars that share similar weight and fuel economy characteristics. To obtain the number of cars in each group, we counted the number of petals in each flower and consulted the graph legend to see how many observations each petal represents.

◁

Methods and Formulas

sunflower is implemented as an ado-file.

Acknowledgments

We thank William D. Dupont and W. Dale Plummer, Jr. (Vanderbilt University), authors of the original sunflower command, for their assistance in producing this version.

References

Carr, D. B., R. J. Littlefield, W. L. Nicholson, and J. S. Littlefield. 1987. Scatterplot matrix techniques for large N. *Journal of the American Statistical Association* 82: 424–436.

Cleveland, W. S., and R. McGill. 1984. The many faces of a scatterplot. *Journal of the American Statistical Association* 79: 807–822.

Dupont, W. D., and W. D. Plummer, Jr. 2003. Density distribution sunflower plots. *Journal of Statistical Software* 8: 1–11.

——. 2005. Using density-distribution sunflower plots to explore bivariate relationships in dense data. *Stata Journal* 5: 371–384.

Huang, C., J. A. McDonald, and W. Stuetzle. 1997. Variable resolution bivariate plots. *Journal of Computational and Graphical Statistics* 6: 383–396.

Levy, D. 1999. *50 years of discovery: Medical milestones from the National Heart, Lung, and Blood Institute's Framingham Heart Study.* Hackensack, NJ: Center for Bio-Medical Communication.

Steichen, T. J., and N. J. Cox. 1999. flower: Stata module to draw sunflower plots. Stata program and help file downloadable from http://ideas.repec.org/c/boc/bocode/s393001.html. Accessed December 6, 2002.

Title

sureg — Zellner's seemingly unrelated regression

Syntax

Basic syntax

sureg (*depvar₁ varlist₁*) (*depvar₂ varlist₂*) ... (*depvarₙ varlistₙ*)

$[\,if\,]$ $[\,in\,]$ $[\,weight\,]$

Full syntax

sureg ($[\,eqname_1:\,]depvar_{1a}$ $[\,depvar_{1b}\,...\,=\,]varlist_1$ $[\,,\,\underline{no}constant\,]$)

($[\,eqname_2:\,]depvar_{2a}$ $[\,depvar_{2b}\,...\,=\,]varlist_2$ $[\,,\,\underline{no}constant\,]$)

...

($[\,eqname_N:\,]depvar_{Na}$ $[\,depvar_{Nb}\,...\,=\,]varlist_N$ $[\,,\,\underline{no}constant\,]$)

$[\,if\,]$ $[\,in\,]$ $[\,weight\,]$ $[\,,\,options\,]$

Explicit equation naming (*eqname:*) cannot be combined with multiple dependent variables in an equation specification.

options	description
Model	
<u>i</u>sure	iterate until estimates converge
<u>c</u>onstraints(*constraints*)	apply specified linear constraints
df adj.	
<u>sm</u>all	report small-sample statistics
dfk	use small-sample adjustment
dfk2	use alternate adjustment
Reporting	
<u>l</u>evel(*#*)	set confidence level; default is level(95)
<u>corr</u>	perform Breusch–Pagan test
Opt options	
optimization_options	control the optimization process; seldom used
† <u>no</u>header	suppress header table from above coefficient table
† <u>not</u>able	suppress coefficient table

†noheader and notable do not appear in the dialog box.

The *depvars* and the *varlists* may contain time-series operators; see [U] **11.4.3 Time-series varlists**.

bootstrap, by, jackknife, rolling, statsby, and xi are allowed; see [U] **11.1.10 Prefix commands**.

Weights are not allowed with the bootstrap prefix.

aweights are not allowed with the jackknife prefix.

aweights and fweights are allowed; see [U] **11.1.6 weight**.

See [U] **20 Estimation and postestimation commands** for more capabilities of estimation commands.

Description

sureg fits seemingly unrelated regression models (Zellner 1962, Zellner and Huang 1962, Zellner 1963). The acronyms SURE and SUR are often used for the estimator.

Options

_____ ⌐ Model ⌐ _____

isure specifies that sureg iterate over the estimated disturbance covariance matrix and parameter estimates until the parameter estimates converge. Under seemingly unrelated regression, this iteration converges to the maximum likelihood results. If this option is not specified, sureg produces two-step estimates.

constraints(*constraints*); see [R] **estimation options**.

_____ ⌐ df adj. ⌐ _____

small specifies that small-sample statistics be computed. It shifts the test statistics from chi-squared and Z statistics to F statistics and t statistics. Although the standard errors from each equation are computed using the degrees of freedom for the equation, the degrees of freedom for the t statistics are all taken to be those for the first equation.

dfk specifies the use of an alternate divisor in computing the covariance matrix for the equation residuals. As an asymptotically justified estimator, sureg by default uses the number of sample observations (n) as a divisor. When the dfk option is set, a small-sample adjustment is made and the divisor is taken to be $\sqrt{(n - k_i)(n - k_j)}$, where k_i and k_j are the numbers of parameters in equations i and j, respectively.

dfk2 specifies the use of an alternate divisor in computing the covariance matrix for the equation residuals. When the dfk2 option is set, the divisor is taken to be the mean of the residual degrees of freedom from the individual equations.

_____ ⌐ Reporting ⌐ _____

level(*#*); see [R] **estimation options**.

corr displays the correlation matrix of the residuals between equations and performs a Breusch–Pagan test for independent equations; that is, the disturbance covariance matrix is diagonal.

_____ ⌐ Opt options ⌐ _____

optimization_options control the iterative process that minimizes the sum of squared errors when isure is specified. These options are seldom used.

<u>iterate</u>(#) specifies the maximum number of iterations. When the number of iterations equals #, the optimizer stops and presents the current results, even if the convergence tolerance has not been reached. The default value of `iterate()` is the current value of `set maxiter`, which is `iterate(16000)` if `maxiter` has not been changed.

<u>trace</u> adds to the iteration log a display of the current parameter vector

`nolog` suppresses the display of the iteration log.

<u>tol</u>erance(#) specifies the tolerance for the coefficient vector. When the relative change in the coefficient vector from one iteration to the next is less than or equal to #, the optimization process is stopped. `tolerance(1e-6)` is the default.

The following options are available with `sureg` but are not shown in the dialog box:

`noheader` suppresses display of the table reporting F statistics, R-squared, and root mean squared error above the coefficient table.

`notable` suppresses display of the coefficient table.

Remarks

Seemingly unrelated regression models are so called because they appear to be joint estimates from several regression models, each with its own error term. The regressions are related because the (contemporaneous) errors associated with the dependent variables may be correlated.

▷ Example 1

When we fit models with the same set of right-hand-side variables, the seemingly unrelated regression results (in terms of coefficients and standard errors) are the same as fitting the models separately (using, say, `regress`). The same is true when the models are nested. Even in such cases, `sureg` is useful when we want to perform joint tests. For instance, let us assume that we think

$$\texttt{price} = \beta_0 + \beta_1 \texttt{foreign} + \beta_2 \texttt{length} + u_1$$
$$\texttt{weight} = \gamma_0 + \gamma_1 \texttt{foreign} + \gamma_2 \texttt{length} + u_2$$

Since the models have the same set of explanatory variables, we could estimate the two equations separately. Yet, we might still choose to estimate them with `sureg` because we want to perform the joint test $\beta_1 = \gamma_1 = 0$.

We use the `small` and `dfk` options to obtain small-sample statistics comparable with `regress` or `mvreg`.

Arnold Zellner (1927–) was born in New York. He studied physics at Harvard and economics at Berkeley, and then he taught economics at the Universities of Washington and Wisconsin before settling in Chicago in 1966. Among his many major contributions to econometrics and statistics are his work on seemingly unrelated regression, three-stage least squares, and Bayesian econometrics.

```
. use http://www.stata-press.com/data/r10/auto
(1978 Automobile Data)

. sureg (price foreign length) (weight foreign length), small dfk
Seemingly unrelated regression
```

Equation	Obs	Parms	RMSE	"R-sq"	F-Stat	P
price	74	2	2474.593	0.3154	16.35	0.0000
weight	74	2	250.2515	0.8992	316.54	0.0000

	Coef.	Std. Err.	t	P>|t|	[95% Conf. Interval]	
price						
foreign	2801.143	766.117	3.66	0.000	1286.674	4315.611
length	90.21239	15.83368	5.70	0.000	58.91219	121.5126
_cons	-11621.35	3124.436	-3.72	0.000	-17797.77	-5444.93
weight						
foreign	-133.6775	77.47615	-1.73	0.087	-286.8332	19.4782
length	31.44455	1.601234	19.64	0.000	28.27921	34.60989
_cons	-2850.25	315.9691	-9.02	0.000	-3474.861	-2225.639

These two equations have a common set of regressors, and we could have used a shorthand syntax to specify the equations:

```
. sureg (price weight = foreign length), small dfk
```

Here the results presented by sureg are the same as if we had estimated the equations separately:

```
. regress price foreign length
  (output omitted )
. regress weight foreign length
  (output omitted )
```

There is, however, a difference. We have allowed u_1 and u_2 to be correlated and have estimated the full variance–covariance matrix of the coefficients. sureg has estimated the correlations, but it does not report them unless we specify the corr option. We did not remember to specify corr when we fitted the model, but we can redisplay the results:

```
. sureg, notable noheader corr

Correlation matrix of residuals:
        price  weight
 price  1.0000
weight  0.5840  1.0000

Breusch-Pagan test of independence: chi2(1) =    25.237, Pr = 0.0000
```

The notable and noheader options prevented sureg from redisplaying the header and coefficient tables. We find that, for the same cars, the correlation of the residuals in the price and weight equations is .5840 and that we can reject the hypothesis that this correlation is zero.

We can test that the coefficients on foreign are jointly zero in both equations—as we set out to do—by typing test foreign; see [R] **test**. When we type a variable without specifying the equation, that variable is tested for zero in all equations in which it appears:

```
. test foreign
 ( 1)  [price]foreign = 0
 ( 2)  [weight]foreign = 0
        F(  2,   142) =    17.99
             Prob > F =    0.0000
```

◁

▷ Example 2

When the models do not have the same set of explanatory variables and are not nested, sureg
may lead to more efficient estimates than running the models separately as well as allowing joint
tests. This time, let us assume that we believe

$$\texttt{price} = \beta_0 + \beta_1\texttt{foreign} + \beta_2\texttt{mpg} + \beta_3\texttt{displ} + u_1$$

$$\texttt{weight} = \gamma_0 + \gamma_1\texttt{foreign} + \gamma_2\texttt{length} + u_2$$

To fit this model, we type

```
. sureg (price foreign mpg displ) (weight foreign length), corr
Seemingly unrelated regression
```

Equation	Obs	Parms	RMSE	"R-sq"	chi2	P
price	74	3	2165.321	0.4537	49.64	0.0000
weight	74	2	245.2916	0.8990	661.84	0.0000

	Coef.	Std. Err.	z	P>|z|	[95% Conf. Interval]	
price						
foreign	3058.25	685.7357	4.46	0.000	1714.233	4402.267
mpg	-104.9591	58.47209	-1.80	0.073	-219.5623	9.644042
displacement	18.18098	4.286372	4.24	0.000	9.779842	26.58211
_cons	3904.336	1966.521	1.99	0.047	50.0263	7758.645
weight						
foreign	-147.3481	75.44314	-1.95	0.051	-295.2139	.517755
length	30.94905	1.539895	20.10	0.000	27.93091	33.96718
_cons	-2753.064	303.9336	-9.06	0.000	-3348.763	-2157.365

```
Correlation matrix of residuals:

          price   weight
 price   1.0000
weight   0.3285   1.0000

Breusch-Pagan test of independence: chi2(1) =    7.984, Pr = 0.0047
```

(Continued on next page)

In comparison, if we had fitted the price model separately,

```
. regress price foreign mpg displ
```

Source	SS	df	MS		
Model	294104790	3	98034929.9		
Residual	340960606	70	4870865.81		
Total	635065396	73	8699525.97		

	Number of obs =	74
	F(3, 70) =	20.13
	Prob > F =	0.0000
	R-squared =	0.4631
	Adj R-squared =	0.4401
	Root MSE =	2207

price	Coef.	Std. Err.	t	P>|t|	[95% Conf. Interval]	
foreign	3545.484	712.7763	4.97	0.000	2123.897	4967.072
mpg	-98.88559	63.17063	-1.57	0.122	-224.8754	27.10426
displacement	22.40416	4.634239	4.83	0.000	13.16146	31.64686
_cons	2796.91	2137.873	1.31	0.195	-1466.943	7060.763

The coefficients are slightly different, but the standard errors are uniformly larger. This would still be true if we specified the dfk option to make a small-sample adjustment to the estimated covariance of the disturbances.

◁

❑ Technical Note

Constraints can be applied to SURE models using Stata's standard syntax for constraints. For a general discussion of constraints, see [R] **constraint**; for examples similar to seemingly unrelated regression models, see [R] **reg3**.

❑

Saved Results

sureg saves the following in e():

Scalars

e(N)	number of observations
e(k)	number of parameters in system
e(k_eq)	number of equations
e(mss_#)	model sum of squares for equation #
e(df_m#)	model degrees of freedom for equation #
e(rss_#)	residual sum of squares for equation #
e(df_r)	residual degrees of freedom
e(r2_#)	R-squared for equation #
e(F_#)	F statistic for equation # (small only)
e(rmse_#)	root mean squared error for equation #
e(dfk2_adj)	divisor used with VCE when dfk2 specified
e(ll)	log likelihood
e(chi2_#)	χ^2 for equation #
e(p_#)	significance for equation #
e(chi2_bp)	Breusch–Pagan χ^2
e(df_bp)	degrees of freedom for Breusch–Pagan χ^2 test
e(cons_#)	1 when equation # has a constant; 0, otherwise
e(ic)	number of iterations

Macros

e(cmd)	sureg
e(cmdline)	command as typed
e(depvar)	names of dependent variables
e(exog)	names of exogenous variables
e(eqnames)	names of equations
e(wtype)	weight type
e(wexp)	weight expression
e(method)	sure or isure
e(corr)	correlation structure
e(small)	small
e(dfk)	alternate divisor (dfk or dfk2 only)
e(predict)	program used to implement predict
e(properties)	b V

Matrices

e(b)	coefficient vector
e(V)	variance–covariance matrix of the estimators
e(Sigma)	$\hat{\Sigma}$ matrix

Functions

e(sample)	marks estimation sample

Methods and Formulas

sureg is implemented as an ado-file.

sureg uses the asymptotically efficient, feasible, generalized least-squares algorithm described in Greene (2003, 340–362). The computing formulas are given on page 342.

The R-squared reported is the percent of variance explained by the predictors. It may be used for descriptive purposes, but R-squared is not a well-defined concept when GLS is used.

sureg will refuse to compute the estimators if the same equation is named more than once or the covariance matrix of the residuals is singular.

The Breusch and Pagan (1980) χ^2 statistic—a Lagrange multiplier statistic—is given by

$$\lambda = T \sum_{m=1}^{M} \sum_{n=1}^{m-1} r_{mn}^2$$

where r_{mn} is the estimated correlation between the residuals of the M equations and T is the number of observations. It is distributed as χ^2 with $M(M-1)/2$ degrees of freedom.

References

Breusch, T., and A. Pagan. 1980. The LM test and its applications to model specification in econometrics. *Review of Economic Studies* 47: 239–254.

Greene, W. H. 2003. *Econometric Analysis*. 5th ed. Upper Saddle River, NJ: Prentice Hall.

McDowell, A. W. 2004. From the help desk: Seemingly unrelated regression with unbalanced equations. *Stata Journal* 4: 442–448.

Weesie, J. 1999. sg121: Seemingly unrelated estimation and the cluster-adjusted sandwich estimator. *Stata Technical Bulletin* 52: 34–47. Reprinted in *Stata Technical Bulletin Reprints*, vol. 9, pp. 231–248.

Zellner, A. 1962. An efficient method of estimating seemingly unrelated regressions and tests for aggregation bias. *Journal of the American Statistical Association* 57: 348–368.

———. 1963. Estimators for seemingly unrelated regression equations: Some exact finite sample results. *Journal of the American Statistical Association* 58: 977–992.

Zellner, A., and D. S. Huang. 1962. Further properties of efficient estimators for seemingly unrelated regression equations. *International Economic Review* 3: 300–313.

Also See

[R] **sureg postestimation** — Postestimation tools for sureg

[R] **mvreg** — Multivariate regression

[R] **nlsur** — Estimation of nonlinear systems of equations

[R] **reg3** — Three-stage estimation for systems of simultaneous equations

[R] **regress** — Linear regression

[U] **20 Estimation and postestimation commands**

Title

> **sureg postestimation** — Postestimation tools for sureg

Description

The following postestimation commands are available for `sureg`:

command	description
adjust	adjusted predictions of $\mathbf{x}\beta$
estat	AIC, BIC, VCE, and estimation sample summary
estimates	cataloging estimation results
lincom	point estimates, standard errors, testing, and inference for linear combinations of coefficients
mfx	marginal effects or elasticities
nlcom	point estimates, standard errors, testing, and inference for nonlinear combinations of coefficients
predict	predictions, residuals, influence statistics, and other diagnostic measures
predictnl	point estimates, standard errors, testing, and inference for generalized predictions
test	Wald tests for simple and composite linear hypotheses
testnl	Wald tests of nonlinear hypotheses

See the corresponding entries in the *Stata Base Reference Manual* for details.

Syntax for predict

predict [*type*] *newvar* [*if*] [*in*] [, equation(*eqno*[,*eqno*]) *statistic*]

statistic	description
Main	
xb	linear prediction; the default
stdp	standard error of the linear prediction
residuals	residuals
difference	difference between the linear predictions of two equations
stddp	standard error of the difference in linear predictions

These statistics are available both in and out of sample; type predict ... if e(sample) ... if wanted only for the estimation sample.

(Continued on next page)

Options for predict

┌─── Main ┐

equation(*eqno* [*,eqno*]) specifies to which equation(s) you are referring.

> equation() is filled in with one *eqno* for options xb, stdp, and residuals. equation(#1) would mean that the calculation is to be made for the first equation, equation(#2) would mean the second, and so on. You could also refer to the equations by their names. equation(income) would refer to the equation named income and equation(hours) to the equation named hours.

> If you do not specify equation(), the results are the same as if you specified equation(#1).

> difference and stddp refer to between-equation concepts. To use these options, you must specify two equations, for example, equation(#1,#2) or equation(income,hours). When two equations must be specified, equation() is required.

xb, the default, calculates the linear prediction (fitted values)—the prediction of $\mathbf{x}_j \mathbf{b}$ for the specified equation.

stdp calculates the standard error of the prediction for the specified equation. It can be thought of as the standard error of the predicted expected value or mean for the observation's covariate pattern. The standard error of the prediction is also referred to as the standard error of the fitted value.

residuals calculates the residuals.

difference calculates the difference between the linear predictions of two equations in the system. With equation(#1,#2), difference computes the prediction of equation(#1) minus the prediction of equation(#2).

stddp is allowed only after you have previously fitted a multiple-equation model. The standard error of the difference in linear predictions $(\mathbf{x}_{1j}\mathbf{b} - \mathbf{x}_{2j}\mathbf{b})$ between equations 1 and 2 is calculated.

For more information on using predict after multiple-equation estimation commands, see [R] **predict**.

Remarks

For an example of cross-equation testing of parameters using the test command, see example 1 in [R] **sureg**.

▷ Example 1

In example 1 of [R] **sureg**, we fitted a seemingly unrelated regressions model of price and weight. Here we obtain the fitted values.

```
. use http://www.stata-press.com/data/r10/auto
(1978 Automobile Data)
. sureg (price foreign length) (weight foreign length), small dfk
  (output omitted )
. predict phat, equation(price)
(option xb assumed; fitted values)
. predict what, equation(weight)
(option xb assumed; fitted values)
```

```
. summarize price phat weight what
    Variable |     Obs        Mean    Std. Dev.       Min        Max
-------------+--------------------------------------------------------
       price |      74    6165.257    2949.496       3291      15906
        phat |      74    6165.257    1656.407   1639.872   9398.138
      weight |      74    3019.459    777.1936       1760       4840
        what |      74    3019.459    736.9666   1481.199   4476.331
```

Just as in single-equation OLS regression, in a SURE model the sample mean of the fitted values for an equation equals the sample mean of the dependent variable.

◁

▷ Example 2

Suppose that for whatever reason we were interested in the difference between the predicted values of price and weight. predict has an option to compute this difference in one step:

```
. predict diff, equation(price, weight) difference
```

diff is the same as phat – what:

```
. generate mydiff = phat - what
. summarize diff mydiff
    Variable |     Obs        Mean    Std. Dev.       Min        Max
-------------+--------------------------------------------------------
        diff |      74    3145.797     1233.26  -132.2275   5505.914
      mydiff |      74    3145.797     1233.26  -132.2275   5505.914
```

◁

Methods and Formulas

All postestimation commands listed above are implemented as ado-files.

Also See

[R] **sureg** — Zellner's seemingly unrelated regression

[U] **20 Estimation and postestimation commands**

Title

swilk — Shapiro–Wilk and Shapiro–Francia tests for normality

Syntax

Shapiro–Wilk normality test

swilk *varlist* [*if*] [*in*] [, *options*]

Shapiro–Francia normality test

sfrancia *varlist* [*if*] [*in*]

options	description
Main	
generate(*newvar*)	create *newvar* containing W test coefficients
lnnormal	test for three-parameter log normality
noties	do not use average ranks for tied values

by is allowed with swilk and sfrancia; see [D] **by**.

Description

swilk performs the Shapiro–Wilk W test for normality, and sfrancia performs the Shapiro–Francia W' test for normality. swilk can be used with $4 \leq n \leq 2{,}000$ observations, and sfrancia can be used with $5 \leq n \leq 5{,}000$ observations; see [R] **sktest** for a test allowing more observations.

Options

Main

generate(*newvar*) creates new variable *newvar* containing the W test coefficients.

lnnormal specifies that the test be for three-parameter log normality, meaning that $\ln(X - k)$ is tested for normality, where k is calculated from the data as the value that makes the skewness coefficient zero. When simply testing $\ln(X)$ for normality, do not specify this option. See [R] **lnskew0** for estimation of k.

noties suppresses use of averaged ranks for tied values when calculating the W test coefficients.

Remarks

▷ Example 1

Using our automobile dataset, we will test whether the variables mpg and trunk are normally distributed:

```
. use http://www.stata-press.com/data/r10/auto
(1978 Automobile Data)
. swilk mpg trunk
```

	Shapiro-Wilk W test for normal data				
Variable	Obs	W	V	z	Prob>z
mpg	74	0.94821	3.335	2.627	0.00430
trunk	74	0.97921	1.339	0.637	0.26215

```
. sfrancia mpg trunk
```

	Shapiro-Francia W' test for normal data				
Variable	Obs	W'	V'	z	Prob>z
mpg	74	0.94872	3.629	2.490	0.00639
trunk	74	0.98446	1.100	0.190	0.42477

We can reject the hypothesis that mpg is normally distributed, but we cannot reject that trunk is normally distributed.

The values reported under W and W' are the Shapiro–Wilk and Shapiro–Francia test statistics. The tests also report V and V', which are more appealing indexes for departure from normality. The median values of V and V' are 1 for samples from normal populations. Large values indicate nonnormality. The 95% critical values of V (V'), which depend on the sample size, are between 1.2 and 2.4 (2.0 and 2.8); see Royston (1991a). There is no more information in V (V') than in W (W')—one is just the transform of the other.

◁

▷ Example 2

We have data on a variable called studytime, which we suspect is distributed lognormally:

```
. use http://www.stata-press.com/data/r10/cancer
(Patient Survival in Drug Trial)
. generate lnstudytime = ln(studytime)
. swilk lnstudytime
```

	Shapiro-Wilk W test for normal data				
Variable	Obs	W	V	z	Prob>z
lnstudytime	48	0.92731	3.311	2.547	0.00543

We can reject the lognormal assumption. We do *not* specify the lnnormal option when testing for log normality. The lnnormal option is for three-parameter log normality.

◁

▷ Example 3

Having discovered that $\ln(\text{studytime})$ is not distributed normally, we now test that $\ln(\text{studytime} - k)$ is normally distributed, where k is chosen so that the resulting skewness is zero. We obtain the estimate for k from lnskew0; see [R] **lnskew0**:

```
. lnskew0 lnstudytimek = studytime, level(95)
        Transform  |         k    [95% Conf. Interval]     Skewness
    ---------------+------------------------------------------------
     ln(studytim-k)|  -11.01181    -infinity  -.9477328    -.0000173
. swilk lnstudytimek, lnnormal
            Shapiro-Wilk W test for 3-parameter lognormal data
      Variable |     Obs        W          V          z     Prob>z
    -----------+------------------------------------------------
    lnstudytimek|      48   0.97064      1.337      1.261   0.10363
```

We cannot reject the hypothesis that $\ln(\text{studytime} + 11.01181)$ is distributed normally. We do specify the lnnormal option when using an estimated value of k.

◁

Saved Results

swilk and sfrancia save the following in r():

Scalars

r(N)	number of observations	r(W)	W or W'
r(p)	significance	r(V)	V or V'
r(z)	z statistic		

Methods and Formulas

swilk and sfrancia are implemented as ado-files.

The Shapiro–Wilk test is based on Shapiro and Wilk (1965) with a new approximation accurate for $4 \leq n \leq 2{,}000$ (Royston 1992). The calculations made by swilk are based on Royston (1982, 1992, 1993).

The Shapiro–Francia test (Shapiro and Francia 1972; Royston 1983) is an approximate test that is similar to the Shapiro–Wilk test for very large samples.

Samuel Sanford Shapiro (1930–) earned degrees in statistics and engineering from City College of New York, Columbia, and Rutgers. After employment in the U.S. army and industry, he joined the faculty at Florida International University in 1972. Shapiro has coauthored various texts in statistics and published several papers on distributional testing and other statistical topics.

Acknowledgment

swilk and sfrancia were written by Patrick Royston of the MRC Clinical Trials Unit, London.

References

Gould, W. W. 1992. sg3.7: Final summary of tests of normality. *Stata Technical Bulletin* 5: 10–11. Reprinted in *Stata Technical Bulletin Reprints*, vol. 1, pp. 114–115.

Royston, J. P. 1982. An extension of Shapiro and Wilk's *W* test for normality to large samples. *Applied Statistics* 31: 115–124.

——. 1983. A simple method for evaluating the Shapiro–Francia *W'* test of non-normality. *Applied Statistics* 32: 297–300.

——. 1991a. Estimating departure from normality. *Statistics in Medicine* 10: 1283–1293.

——. 1991b. sg3.2: Shapiro–Wilk and Shapiro–Francia tests. *Stata Technical Bulletin* 3: 19. Reprinted in *Stata Technical Bulletin Reprints*, vol. 1, p. 105.

——. 1992. Approximating the Shapiro–Wilk *W*-test for non-normality. *Statistics and Computing* 2: 117–119.

——. 1993. A toolkit for testing for non-normality in complete and censored samples. *Statistician* 42: 37–43.

Shapiro, S. S., and R. S. Francia. 1972. An approximate analysis of variance test for normality. *Journal of the American Statistical Association* 67: 215–216.

Shapiro, S. S., and M. B. Wilk. 1965. An analysis of variance test for normality (complete samples). *Biometrika* 52: 591–611.

Also See

[R] **lnskew0** — Find zero-skewness log or Box–Cox transform

[R] **lv** — Letter-value displays

[R] **sktest** — Skewness and kurtosis test for normality

Title

symmetry — Symmetry and marginal homogeneity tests

Syntax

Symmetry and marginal homogeneity tests

> symmetry *casevar controlvar* $\left[\,if\,\right]$ $\left[\,in\,\right]$ $\left[\,weight\,\right]$ $\left[\,,\,options\,\right]$

Immediate form of symmetry and marginal homogeneity tests

> symmi $\#_{11}$ $\#_{12}$ [...] \ $\#_{21}$ $\#_{22}$ [...] [\...] $\left[\,if\,\right]$ $\left[\,in\,\right]$ $\left[\,,\,options\,\right]$

options	description
Main	
notable	suppress output of contingency table
contrib	report contribution of each off-diagonal cell pair
exact	perform exact test of table symmetry
mh	perform two marginal homogeneity tests
trend	perform a test for linear trend in the (log) relative risk (RR)
cc	use continuity correction when calculating test for linear trend

fweights are allowed; see [U] **11.1.6 weight**.

Description

symmetry performs asymptotic symmetry and marginal homogeneity tests, as well as an exact symmetry test on $K \times K$ tables where there is a 1-to-1 matching of cases and controls (nonindependence). This testing is used to analyze matched-pair case–control data with multiple discrete levels of the exposure (outcome) variable. In genetics, the test is known as the transmission/disequilibrium test (TDT) and is used to test the association between transmitted and nontransmitted parental marker alleles to an affected child (Spieldman, McGinnis, and Ewens 1993). For 2×2 tables, the asymptotic test statistics reduce to the McNemar test statistic, and the exact symmetry test produces an exact McNemar test; see [ST] **epitab**. For many exposure variables, symmetry can optionally perform a test for linear trend in the log relative risk.

symmetry expects the data to be in the wide format; that is, each observation contains the matched case and control values in variables *casevar* and *controlvar*. Variables can be numeric or string.

symmi is the immediate form of symmetry. The symmi command uses the values specified on the command line; rows are separated by '\', and options are the same as for symmetry. See [U] **19 Immediate commands** for a general introduction to immediate commands.

394

Options

notable suppresses the output of the contingency table. By default, symmetry displays the $n \times n$ contingency table at the top of the output.

contrib reports the contribution of each off-diagonal cell pair to the overall symmetry χ^2.

exact performs an exact test of table symmetry. This option is recommended for sparse tables. CAUTION: the exact test requires substantial amounts of time and computer memory for large tables.

mh performs two marginal homogeneity tests that do not require the inversion of the variance–covariance matrix.

By default, symmetry produces the Stuart–Maxwell test statistic, which requires the inversion of the nondiagonal variance–covariance matrix, $\mathbf{V}$. When the table is sparse, the matrix may not be of full rank, and then the command substitutes a generalized inverse $\mathbf{V}^*$ for $\mathbf{V}^{-1}$. mh calculates optional marginal homogeneity statistics that do not require the inversion of the variance–covariance matrix. These tests may be preferred in certain situations. See *Methods and Formulas* and Bickenböller and Clerget-Darpoux (1995) for details on these test statistics.

trend performs a test for linear trend in the (log) relative risk (RR). This option is allowed only for numeric exposure (outcome) variables, and its use should be restricted to measurements on the ordinal or the interval scales.

cc specifies that the continuity correction be used when calculating the test for linear trend. This correction should be specified only when the levels of the exposure variable are equally spaced.

Remarks

symmetry and symmi may be used to analyze 1-to-1 matched case–control data with multiple discrete levels of the exposure (outcome) variable.

▷ Example 1

Consider a survey of 344 individuals (BMDP 1990, 267–270) who were asked in October 1986 whether they agreed with President Reagan's handling of foreign affairs. In January 1987, after the Iran-Contra affair became public, these same individuals were surveyed again and asked the same question. We would like to know if public opinion changed over this period.

We first describe the dataset and list a few observations.

```
. use http://www.stata-press.com/data/r10/iran
. describe
Contains data from http://www.stata-press.com/data/r10/iran.dta
  obs:           344
  vars:            2                          29 Jan 2007 02:37
  size:         2,064 (99.9% of memory free)

              storage   display     value
variable name  type    format      label     variable label

before         byte    %8.0g       vlab      Public Opinion before IC
after          byte    %8.0g       vlab      Public Opinion after IC

Sorted by:
```

```
. list in 1/5
```

	before	after
1.	agree	agree
2.	agree	disagree
3.	agree	unsure
4.	disagree	agree
5.	disagree	disagree

Each observation corresponds to one of the 344 individuals. The data are in wide form so that each observation has a before and an after measurement. We now perform the test without options.

```
. symmetry before after
```

Public Opinion before IC	Public Opinion after IC			
	agree	disagree	unsure	Total
agree	47	56	38	141
disagree	28	61	31	120
unsure	26	47	10	83
Total	101	164	79	344

	chi2	df	Prob>chi2
Symmetry (asymptotic)	14.87	3	0.0019
Marginal homogeneity (Stuart-Maxwell)	14.78	2	0.0006

The test first tabulates the data in a $K \times K$ table and then performs Bowker's (1948) test for table symmetry and the Stuart–Maxwell (Stuart 1955, Maxwell 1970) test for marginal homogeneity.

Both the symmetry test and the marginal homogeneity test are highly significant, thus indicating a shift in public opinion.

An exact test of symmetry is provided for use on sparse tables. This test is computationally intensive, so it should not be used on large tables. Since we are working on a fast computer, we will run the symmetry test again and this time include the exact option. We will suppress the output of the contingency table by specifying notable and include the contrib option so that we may further examine the cells responsible for the significant result.

```
. symmetry before after, contrib exact mh notable
```

Cells	Contribution to symmetry chi-squared
n1_2 & n2_1	9.3333
n1_3 & n3_1	2.2500
n2_3 & n3_2	3.2821

	chi2	df	Prob>chi2
Symmetry (asymptotic)	14.87	3	0.0019
Marginal homogeneity (Stuart-Maxwell)	14.78	2	0.0006
Marginal homogeneity (Bickenboller)	13.53	2	0.0012
Marginal homogeneity (no diagonals)	15.25	2	0.0005

Symmetry (exact significance probability)			0.0018

The largest contribution to the symmetry χ^2 is due to cells n_{12} and n_{21}. These correspond to changes between the agree and disagree categories. Of the 344 individuals, 56 (16.3%) changed from the agree to the disagree response, whereas only 28 (8.1%) changed in the opposite direction.

For these data, the results from the exact test are similar to those from the asymptotic test.

◁

▷ Example 2

Breslow and Day (1980, 163) reprinted data from Mack et al. (1976) from a case–control study of the effect of exogenous estrogen on the risk of endometrial cancer. The data consist of 59 elderly women diagnosed with endometrial cancer and 59 disease-free control subjects living in the same community as the cases. Cases and controls were matched on age, marital status, and time living in the community. The data collected included information on the daily dose of conjugated estrogen therapy. Breslow and Day analyzed these data by creating four levels of the dose variable. Here are the data as entered into a Stata dataset:

```
. use http://www.stata-press.com/data/r10/bd163
. list, noobs divider
```

case	control	count
0	0	6
0	0.1-0.299	2
0	0.3-0.625	3
0	0.626+	1
0.1-0.299	0	9
0.1-0.299	0.1-0.299	4
0.1-0.299	0.3-0.625	2
0.1-0.299	0.626+	1
0.3-0.625	0	9
0.3-0.625	0.1-0.299	2
0.3-0.625	0.3-0.625	3
0.3-0.625	0.626+	1
0.626+	0	12
0.626+	0.1-0.299	1
0.626+	0.3-0.625	2
0.626+	0.626+	1

This dataset is in a different format from that of the previous example. Instead of each observation representing one matched pair, each observation represents possibly multiple pairs indicated by the count variable. For instance, the first observation corresponds to six matched pairs where neither the case nor the control was on estrogen, the second observation corresponds to two matched pairs where the case was not on estrogen and the control was on 0.1 to 0.299 mg/day, etc.

To use symmetry to analyze this dataset, we must specify fweight to indicate that in our data there are observations corresponding to more than one matched pair.

```
. symmetry case control [fweight=count]
```

			control		
case	0	0.1-0.299	0.3-0.625	0.626+	Total
0	6	2	3	1	12
0.1-0.299	9	4	2	1	16
0.3-0.625	9	2	3	1	15
0.626+	12	1	2	1	16
Total	36	9	10	4	59

	chi2	df	Prob>chi2
Symmetry (asymptotic)	17.10	6	0.0089
Marginal homogeneity (Stuart-Maxwell)	16.96	3	0.0007

Both the test of symmetry and the test of marginal homogeneity are highly significant, thus leading us to reject the null hypothesis that there is no effect of exposure to estrogen on the risk of endometrial cancer.

Breslow and Day perform a test for trend assuming that the estrogen exposure levels were equally spaced by recoding the exposure levels as 1, 2, 3, and 4.

We can easily reproduce their results by recoding our data in this way and by specifying the trend option. Two new numeric variables were created, ca and co, corresponding to the variables case and control, respectively. Below we list some of the data and our results from symmetry:

```
. encode case, gen(ca)
. encode control, gen(co)
. label values ca
. label values co
. list in 1/4
```

	case	control	count	ca	co
1.	0	0	6	1	1
2.	0	0.1-0.299	2	1	2
3.	0	0.3-0.625	3	1	3
4.	0	0.626+	1	1	4

```
. symmetry ca co [fw=count], notable trend cc
```

	chi2	df	Prob>chi2
Symmetry (asymptotic)	17.10	6	0.0089
Marginal homogeneity (Stuart-Maxwell)	16.96	3	0.0007
Linear trend in the (log) RR	14.43	1	0.0001

We requested the continuity correction by specifying cc. Doing so is appropriate because our coded exposure levels are equally spaced.

The test for trend was highly significant, indicating an increased risk of endometrial cancer with increased dosage of conjugated estrogen.

You must be cautious: the way in which you code the exposure variable affects the linear trend statistic. If instead of coding the levels as 1, 2, 3, and 4, we had instead used 0, .2, .46, and .7

(roughly the midpoint in the range of each level), we would have obtained a χ^2 statistic of 11.19 for these data.

◁

Saved Results

symmetry saves the following in r():

Scalars

r(N_pair)	number of matched pairs
r(chi2)	asymptotic symmetry χ^2
r(df)	asymptotic symmetry degrees of freedom
r(p)	asymptotic symmetry p-value
r(chi2_sm)	MH (Stuart–Maxwell) χ^2
r(df_sm)	MH (Stuart–Maxwell) degrees of freedom
r(p_sm)	MH (Stuart–Maxwell) p-value
r(chi2_b)	MH (Bickenböller) χ^2
r(df_b)	MH (Bickenböller) degrees of freedom
r(p_b)	MH (Bickenböller) p-value
r(chi2_nd)	MH (no diagonals) χ^2
r(df_nd)	MH (no diagonals) degrees of freedom
r(p_nd)	MH (no diagonals) p-value
r(chi2_t)	χ^2 for linear trend
r(p_trend)	p-value for linear trend
r(p_exact)	exact symmetry p-value

Methods and Formulas

symmetry and symmi are implemented as ado-files.

Asymptotic tests

Consider a square table with K exposure categories, that is, K rows and K columns. Let n_{ij} be the count corresponding to row i and column j of the table, $N_{ij} = n_{ij} + n_{ji}$, for $i, j = 1, 2, \ldots, K$, and $n_{i.}$, and let $n_{.j}$ be the marginal totals for row i and column j, respectively. Asymptotic tests for symmetry and marginal homogeneity for this $K \times K$ table are calculated as follows:

The null hypothesis of complete symmetry $p_{ij} = p_{ji}$, $i \neq j$, is tested by calculating the test statistic (Bowker 1948),

$$T_{cs} = \sum_{i<j} \frac{(n_{ij} - n_{ji})^2}{n_{ij} + n_{ji}}$$

which is asymptotically distributed as χ^2 with $K(K-1)/2 - R$ degrees of freedom, where R is the number of off-diagonal cells with $N_{ij} = 0$.

The null hypothesis of marginal homogeneity, $p_{i.} = p_{.i}$, is tested by calculating the Stuart–Maxwell test statistic (Stuart 1955, Maxwell 1970),

$$T_{sm} = \mathbf{d}' \mathbf{V}^{-1} \mathbf{d}$$

where $\mathbf{d}$ is a column vector with elements equal to the differences $d_i = n_{i.} - n_{.i}$ for $i = 1, 2, \ldots, K$, and $\mathbf{V}$ is the variance–covariance matrix with elements

$$v_{ii} = n_{i.} + n_{.i} - 2n_{ii}$$
$$v_{ij} = -(n_{ij} + n_{ji}), \quad i \neq j$$

T_{sm} is asymptotically χ^2 with $K - 1$ degrees of freedom.

This test statistic properly accounts for the dependence between the table's rows and columns. When the matrix $\mathbf{V}$ is not of full rank, a generalized inverse $\mathbf{V}^*$ is substituted for $\mathbf{V}^{-1}$.

The Bickenböller and Clerget-Darpoux (1995) marginal homogeneity test statistic is calculated by

$$T_{\text{mh}} = \sum_i \frac{(n_{i.} - n_{.i})^2}{n_{i.} + n_{.i}}$$

This statistic is asymptotically distributed, under the assumption of marginal independence, as χ^2 with $K - 1$ degrees of freedom.

The marginal homogeneity (no diagonals) test statistic T_{mh}^0 is calculated in the same way as T_{mh}, except that the diagonal elements do not enter into the calculation of the marginal totals. Unlike the previous test statistic, T_{mh}^0 reduces to a McNemar test statistic for 2×2 tables. The test statistic $\{(K - 1)/2\} T_{\text{mh}}^0$ is asymptotically distributed as χ^2 with $K - 1$ degrees of freedom (Cleves, Olson, and Jacobs 1997; Spieldman and Ewens 1996).

Breslow and Day's test statistic for linear trend in the (log) of relative risk is

$$\frac{\left\{ \sum_{i<j} (n_{ij} - n_{ji})(X_j - X_i) - cc \right\}^2}{\sum_{i<j} (n_{ij} + n_{ji})(X_j - X_i)^2}$$

where the X_j are the doses associated with the various levels of exposure and cc is the continuity correction; it is asymptotically distributed as χ^2 with 1 degree of freedom.

The continuity correction option is applicable only when the levels of the exposure variable are equally spaced.

Exact symmetry test

The exact test is based on a permutation algorithm applied to the null distribution. The distribution of the off-diagonal elements n_{ij}, $i \neq j$, conditional on the sum of the complementary off-diagonal cells, $N_{ij} = n_{ij} + n_{ji}$, can be written as the product of $K(K - 1)/2$ binomial random variables,

$$P(\mathbf{n}) = \prod_{i<j} \binom{N_{ij}}{n_{ij}} \pi_{ij}^{n_{ij}} (1 - \pi_{ij})^{n_{ij}}$$

where $\mathbf{n}$ is a vector with elements n_{ij} and $\pi_{ij} = E(n_{ij}/N_{ij}|N_{ij})$. Under the null hypothesis of complete symmetry, $\pi_{ij} = \pi_{ji} = 1/2$, and thus the permutation distribution is given by

$$P_0(\mathbf{n}) = \prod_{i<j} \binom{N_{ij}}{n_{ij}} \left(\frac{1}{2}\right)^{N_{ij}}$$

The exact significance test is performed by evaluating

$$P_{\text{cs}} = \sum_{n \in p} P_0(\mathbf{n})$$

where $p = \{n : P_0(\mathbf{n}) < P_0(\mathbf{n}^*)\}$ and $\mathbf{n}^*$ is the observed contingency table data vector. The algorithm evaluates p_{cs} exactly. For information about permutation tests, see Good (2005 and 2006).

References

Bickenböller, H., and F. Clerget-Darpoux. 1995. Statistical properties of the allelic and genotypic transmission/disequilibrium test for multiallelic markers. *Genetic Epidemiology* 12: 865–870.

BMDP. 1990. *BMDP Statistical Software Manual.* Example 4F2.9. Los Angeles: BMDP Statistical Software.

Bowker, A. H. 1948. A test for symmetry in contingency tables. *Journal of the American Statistical Association* 43: 572–574.

Breslow, N. E., and N. E. Day. 1980. *Statistical Methods in Cancer Research: The Analysis of Case–Control Studies,* vol. 1, 182–198. Lyon: IARC.

Cleves, M. A. 1997. sg74: Symmetry and marginal homogeneity test/TDT. *Stata Technical Bulletin* 40: 23–27. Reprinted in *Stata Technical Bulletin Reprints,* vol. 7, pp. 193–197.

——. 1999. sg110: Hardy–Weinberg equilibrium test and allele frequency estimation. *Stata Technical Bulletin* 48: 34–37. Reprinted in *Stata Technical Bulletin Reprints,* vol. 8, pp. 280–284.

Cleves, M. A., J. M. Olson, and K. B. Jacobs. 1997. Exact transmission–disequilibrium tests with multiallelic markers. *Genetic Epidemiology* 14: 337–347.

Cui, J. 2000. sg150: Hardy–Weinberg equilibrium test in case–control studies. *Stata Technical Bulletin* 57: 17–19. Reprinted in *Stata Technical Bulletin Reprints,* vol. 10, pp. 218–220.

Good, P. I. 2005. *Permutation, Parametric, and Bootstrap Tests of Hypotheses.* 3rd ed. New York: Springer.

——. 2006. *Resampling Methods: A Practical Guide to Data Analysis.* 3rd ed. Boston: Birkhäuser.

Mack, T. M., M. C. Pike, B. E. Henderson, R. I. Pfeffer, V. R. Gerkins, B. S. Arthur, and S. E. Brown. 1976. Estrogens and endometrial cancer in a retirement community. *New England Journal of Medicine* 294: 1262–1267.

Mander, A. 2000. sbe38: Haplotype frequency estimation using an EM algorithm and log-linear modeling. *Stata Technical Bulletin* 57: 5–7. Reprinted in *Stata Technical Bulletin Reprints,* vol. 10, pp. 104–107.

Maxwell, A. E. 1970. Comparing the classification of subjects by two independent judges. *British Journal of Psychiatry* 116: 651–655.

Spieldman, R. S., and W. J. Ewens. 1996. The TDT and other family-based tests for linkage disequilibrium and association. *American Journal of Human Genetics* 59: 983–989.

Spieldman, R. S., R. E. McGinnis, and W. J. Ewens. 1993. Transmission test for linkage disequilibrium: The insulin gene region and insulin-dependence diabetes mellitus. *American Journal of Human Genetics* 52: 506–516.

Stuart, A. 1955. A test for homogeneity of the marginal distributions in a two-way classification. *Biometrika* 42: 412–416.

Also See

[ST] **epitab** — Tables for epidemiologists

Title

table — Tables of summary statistics

Syntax

table *rowvar* [*colvar* [*supercolvar*]] [*if*] [*in*] [*weight*] [, *options*]

options	description
Main	
contents(*clist*)	contents of table cells; select up to five statistics; default is contents(freq)
by(*superrowvarlist*)	superrow variables
Options	
cellwidth(#)	cell width
csepwidth(#)	column-separation width
stubwidth(#)	stub width
scsepwidth(#)	supercolumn-separation width
center	center-align table cells; default is right-align
left	left-align table cells; default is right-align
cw	perform casewise deletion
row	add row totals
col	add column totals
scol	add supercolumn totals
concise	suppress rows with all missing entries
missing	show missing statistics with period
replace	replace current data with table statistics
name(*string*)	name new variables with prefix *string*
format(%*fmt*)	display format for numbers in cells; default is format(%9.0g)

by is allowed; see [D] **by**.

fweights, aweights, iweights, and pweights are allowed; see [U] **11.1.6 weight**.

sd (standard deviation) is not allowed with pweights.

where the elements of *clist* may be

freq	frequency	median *varname*	median
mean *varname*	mean of *varname*	p1 *varname*	1st percentile
sd *varname*	standard deviation	p2 *varname*	2nd percentile
sum *varname*	sum	...	3rd–49th percentiles
rawsum *varname*	sums ignoring optionally specified weight	p50 *varname*	50th percentile (median)
count *varname*	count of nonmissing observations	...	51st–97th percentiles
n *varname*	same as count	p98 *varname*	98th percentile
max *varname*	maximum	p99 *varname*	99th percentile
min *varname*	minimum	iqr *varname*	interquartile range

Rows, columns, supercolumns, and superrows are thus defined as

row 1	.
row 2	.

	supercol 1		supercol 2	
	col 1	col 2	col 1	col 2
row 1	.	.	.	.
row 2	.	.	.	.

	col 1	col 2
row 1	.	.
row 2	.	.

	supercol 1		supercol 2	
	col 1	col 2	col 1	col 2
superrow 1:				
row 1	.	.	.	.
row 2	.	.	.	.
superrow 2:				
row 1	.	.	.	.
row 2	.	.	.	.

Description

table calculates and displays tables of statistics.

Options

_____ Main _____

contents(*clist*) specifies the contents of the table's cells; if not specified, contents(freq) is used by default. contents(freq) produces a table of frequencies. contents(mean mpg) produces a table of the means of variable mpg. contents(freq mean mpg sd mpg) produces a table of frequencies together with the mean and standard deviation of variable mpg. Up to five statistics may be specified.

by(*superrowvarlist*) specifies that numeric or string variables be treated as superrows. Up to four variables may be specified in *superrowvarlist*. The by() option may be specified with the by prefix.

_____ Options _____

cellwidth(#) specifies the width of the cell in units of digit widths; 10 means the space occupied by 10 digits, which is 1234567890. The default cellwidth() is not a fixed number, but a number chosen by table to spread the table out while presenting a reasonable number of columns across the page.

csepwidth(#) specifies the separation between columns in units of digit widths. The default is not a fixed number, but a number chosen by table according to what it thinks looks best.

stubwidth(#) specifies the width, in units of digit widths, to be allocated to the left stub of the table. The default is not a fixed number, but a number chosen by table according to what it thinks looks best.

scsepwidth(#) specifies the separation between supercolumns in units of digit widths. The default is not a fixed number, but a number chosen by table to present the results best.

center specifies that results be centered in the table's cells. The default is to right-align results. For centering to work well, you typically need to specify a display format as well. center format(%9.2f) is popular.

left specifies that column labels be left-aligned. The default is to right-align column labels to distinguish them from supercolumn labels, which are left-aligned. If you specify left, both column and supercolumn labels are left-aligned.

cw specifies casewise deletion. If cw is not specified, all observations possible are used to calculate each of the specified statistics. cw is relevant only when you request a table containing statistics on multiple variables. For instance, contents(mean mpg mean weight) would produce a table reporting the means of variables mpg and weight. Consider an observation in which mpg is known but weight is missing. By default, that observation will be used in the calculation of the mean of mpg. If you specify cw, the observation will be excluded in the calculation of the means of both mpg and weight.

row specifies that a row be added to the table reflecting the total across the rows.

col specifies that a column be added to the table reflecting the total across columns.

scol specifies that a supercolumn be added to the table reflecting the total across supercolumns.

concise specifies that rows with all missing entries not be displayed.

missing specifies that missing statistics be shown in the table as periods (Stata's missing-value indicator). The default is that missing entries be left blank.

replace specifies that the data in memory be replaced with data containing 1 observation per cell (row, column, supercolumn, and superrow) and with variables containing the statistics designated in contents().

This option is rarely specified. If you do not specify this option, the data in memory remain unchanged.

If you do specify this option, the first statistic will be named table1, the second table2, and so on. For instance, if contents(mean mpg sd mpg) was specified, the means of mpg would be in variable table1 and the standard deviations in table2.

name(*string*) is relevant only if you specify replace. name() allows changing the default stub name that replace uses to name the new variables associated with the statistics. If you specify name(stat), the first statistic will be placed in variable stat1, the second in stat2, and so on.

format(%*fmt*) specifies the display format for presenting numbers in the table's cells. format(%9.0g) is the default; format(%9.2f) and format(%9.2fc) are popular alternatives. The width of the format you specify does not matter, except that %*fmt* must be valid. The width of the cells is chosen by table to present the results best. Option cellwidth() allows you to override table's choice.

Limits

Up to four variables may be specified in the by(), so with the three row, column, and supercolumn variables, seven-way tables may be displayed.

Up to five statistics may be displayed in each cell of the table.

The sum of the number of rows, columns, supercolumns, and superrows is called the number of margins. A table may contain up to 3,000 margins. Thus a one-way table may contain 3,000 rows. A two-way table could contain 2,998 rows and two columns, 2,997 rows and three columns, ..., 1,500 rows and 1,500 columns, ..., two rows and 2,998 columns. A three-way table is similarly limited by the sum of the number of rows, columns, and supercolumns. A $r \times c \times d$ table is feasible if $r + c + d \leq 3{,}000$. The limit is set in terms of the sum of the rows, columns, supercolumns, and superrows, and not, as you might expect, in terms of their product.

Remarks

Remarks are presented under the following headings:

One-way tables
Two-way tables
Three-way tables
Four-way and higher-dimensional tables

One-way tables

▷ Example 1

From the automobile dataset, here is a simple one-way table:

```
. use http://www.stata-press.com/data/r10/auto
(1978 Automobile Data)
. table rep78, contents(mean mpg)
```

Repair Record 1978	mean(mpg)
1	21
2	19.125
3	19.4333
4	21.6667
5	27.3636

We are not limited to including only one statistic:

```
. table rep78, c(n mpg  mean mpg  sd mpg  median mpg)
```

Repair Record 1978	N(mpg)	mean(mpg)	sd(mpg)	med(mpg)
1	2	21	4.24264	21
2	8	19.125	3.758324	18
3	30	19.4333	4.141325	19
4	18	21.6667	4.93487	22.5
5	11	27.3636	8.732385	30

We abbreviated `contents()` as `c()`. The `format()` option will allow us to better format the numbers in the table:

```
. table rep78, c(n mpg  mean mpg  sd mpg  median mpg) format(%9.2f)
```

Repair Record 1978	N(mpg)	mean(mpg)	sd(mpg)	med(mpg)
1	2	21.00	4.24	21.00
2	8	19.12	3.76	18.00
3	30	19.43	4.14	19.00
4	18	21.67	4.93	22.50
5	11	27.36	8.73	30.00

The `center` option will center the results under the headings:

```
. table rep78, c(n mpg  mean mpg  sd mpg  median mpg) format(%9.2f) center
```

Repair Record 1978	N(mpg)	mean(mpg)	sd(mpg)	med(mpg)
1	2	21.00	4.24	21.00
2	8	19.12	3.76	18.00
3	30	19.43	4.14	19.00
4	18	21.67	4.93	22.50
5	11	27.36	8.73	30.00

◁

Two-way tables

▷ Example 2

In example 1, when we typed 'table rep78, ...', we obtained a one-way table. If we were to type 'table rep78 foreign, ...', we would obtain a two-way table:

```
. table rep78 foreign, c(mean mpg)
```

Repair Record 1978	Car type Domestic	Foreign
1	21	
2	19.125	
3	19	23.3333
4	18.4444	24.8889
5	32	26.3333

Note the missing cells. Certain combinations of repair record and car type do not exist in our dataset.

As with one-way tables, we can specify a display format for the cells and center the numbers within the cells if we wish.

```
. table rep78 foreign, c(mean mpg) format(%9.2f) center
```

Repair Record 1978	Car type Domestic	Foreign
1	21.00	
2	19.12	
3	19.00	23.33
4	18.44	24.89
5	32.00	26.33

We can obtain row totals by specifying the `row` option and obtain column totals by specifying the `col` option. We specify both below:

```
. table rep78 foreign, c(mean mpg) format(%9.2f) center row col
```

Repair Record 1978	Car type Domestic	Foreign	Total
1	21.00		21.00
2	19.12		19.12
3	19.00	23.33	19.43
4	18.44	24.89	21.67
5	32.00	26.33	27.36
Total	19.54	25.29	21.29

`table` can display multiple statistics within cells, but once we move beyond one-way tables, the table becomes busy:

```
. table foreign rep78, c(mean mpg  n mpg) format(%9.2f) center
```

Car type	Repair Record 1978				
	1	2	3	4	5
Domestic	21.00	19.12	19.00	18.44	32.00
	2	8	27	9	2
Foreign			23.33	24.89	26.33
			3	9	9

This two-way table with two statistics per cell works well in this case. That was, in part, helped along by our interchanging the rows and columns. We turned the table around by typing `table foreign rep78` rather than `table rep78 foreign`.

Another way to display two-way tables is to specify a row and superrow rather than a row and column. We do that below and display three statistics per cell:

```
. table foreign, by(rep78) c(mean mpg  sd mpg  n mpg) format(%9.2f) center
```

Repair Record 1978 and Car type	mean(mpg)	sd(mpg)	N(mpg)
1			
Domestic	21.00	4.24	2
Foreign			
2			
Domestic	19.12	3.76	8
Foreign			
3			
Domestic	19.00	4.09	27
Foreign	23.33	2.52	3
4			
Domestic	18.44	4.59	9
Foreign	24.89	2.71	9
5			
Domestic	32.00	2.83	2
Foreign	26.33	9.37	9

◁

Three-way tables

▷ Example 3

We have data on the prevalence of byssinosis, a form of pneumoconiosis to which workers exposed to cotton dust are susceptible. The dataset is on 5,419 workers in a large cotton mill. We know whether each worker smokes, his or her race, and the dustiness of the work area. The categorical variables are

smokes	Smoker or nonsmoker in the last five years.
race	White or other.
workplace	1 (most dusty), 2 (less dusty), 3 (least dusty).

Moreover, this dataset includes a frequency-weight variable pop. Here is a three-way table showing the fraction of workers with byssinosis:

```
. use http://www.stata-press.com/data/r10/byssin
(Byssinosis incidence)
. table workplace smokes race [fw=pop], c(mean prob)
```

Dustiness of workplace	Race and Smokes			
	——— other ———		——— white ———	
	no	yes	no	yes
least	.0107527	.0101523	.0081549	.0162774
less	.02	.0081633	.0136612	.0143149
most	.0820896	.1679105	.0833333	.2295082

This table would look better if we showed the fraction to four digits:

```
. table workplace smokes race [fw=pop], c(mean prob) format(%9.4f)
```

Dustiness of workplace	Race and Smokes ——— other ——— no	yes	——— white ——— no	yes
least	0.0108	0.0102	0.0082	0.0163
less	0.0200	0.0082	0.0137	0.0143
most	0.0821	0.1679	0.0833	0.2295

In this table, the rows are the dustiness of the workplace, the columns are whether the worker smokes, and the supercolumns are the worker's race.

Now we request that the table include the supercolumn totals by specifying the option sctotal, which we can abbreviate as sc:

```
. table workplace smokes race [fw=pop], c(mean prob) format(%9.4f) sc
```

Dustiness of workplace	Race and Smokes ——— other ——— no	yes	——— white ——— no	yes	——— Total ——— no	yes
least	0.0108	0.0102	0.0082	0.0163	0.0090	0.0145
less	0.0200	0.0082	0.0137	0.0143	0.0159	0.0123
most	0.0821	0.1679	0.0833	0.2295	0.0826	0.1929

The supercolumn total is the total over race and is divided into its columns based on smokes. Here is the table with the column rather than the supercolumn totals:

```
. table workplace smokes race [fw=pop], c(mean prob) format(%9.4f) col
```

Dustiness of workplace	Race and Smokes ——— other ——— no	yes	Total	——— white ——— no	yes	Total
least	0.0108	0.0102	0.0104	0.0082	0.0163	0.0129
less	0.0200	0.0082	0.0135	0.0137	0.0143	0.0140
most	0.0821	0.1679	0.1393	0.0833	0.2295	0.1835

Here is the table with both column and supercolumn totals:

```
. table workplace smokes race [fw=pop], c(mean prob) format(%9.4f) sc col
```

Dustiness of workplace	Race and Smokes ——— other ——— no	yes	Total	——— white ——— no	yes	Total	——— Total ——— no	yes	Total
least	0.0108	0.0102	0.0104	0.0082	0.0163	0.0129	0.0090	0.0145	0.0122
less	0.0200	0.0082	0.0135	0.0137	0.0143	0.0140	0.0159	0.0123	0.0138
most	0.0821	0.1679	0.1393	0.0833	0.2295	0.1835	0.0826	0.1929	0.1570

table is struggling to keep this table from becoming too wide—notice how it divided the words in the title in the top-left stub. Here, if the table had more columns, or, if we demanded more digits, table would be forced to segment the table and present it in pieces, which it would do:

```
. table workplace smokes race [fw=pop], c(mean prob) format(%9.6f) sc col
```

Dustiness of workplace	Race and Smokes					
	other			white		
	no	yes	Total	no	yes	Total
least	0.010753	0.010152	0.010417	0.008155	0.016277	0.012949
less	0.020000	0.008163	0.013483	0.013661	0.014315	0.014035
most	0.082090	0.167910	0.139303	0.083333	0.229508	0.183521

Dustiness of workplace	Race and Smokes		
	Total		
	no	yes	Total
least	0.008990	0.014471	0.012174
less	0.015901	0.012262	0.013846
most	0.082569	0.192905	0.156951

Here three digits is probably enough, so here is the table including all the row, column, and supercolumn totals:

```
. table workplace smokes race [fw=pop], c(mean prob) format(%9.3f)
> sc col row
```

Dustiness of workplace	Race and Smokes								
	other			white			Total		
	no	yes	Total	no	yes	Total	no	yes	Total
least	0.011	0.010	0.010	0.008	0.016	0.013	0.009	0.014	0.012
less	0.020	0.008	0.013	0.014	0.014	0.014	0.016	0.012	0.014
most	0.082	0.168	0.139	0.083	0.230	0.184	0.083	0.193	0.157
Total	0.025	0.048	0.038	0.014	0.035	0.026	0.018	0.039	0.030

We can show multiple statistics:

```
. table workplace smokes race [fw=pop], c(mean prob  n prob) format(%9.3f)
> sc col row
```

Dustiness of workplace	Race and Smokes								
	other			white			Total		
	no	yes	Total	no	yes	Total	no	yes	Total
least	0.011	0.010	0.010	0.008	0.016	0.013	0.009	0.014	0.012
	465	591	1,056	981	1,413	2,394	1,446	2,004	3,450
less	0.020	0.008	0.013	0.014	0.014	0.014	0.016	0.012	0.014
	200	245	445	366	489	855	566	734	1,300
most	0.082	0.168	0.139	0.083	0.230	0.184	0.083	0.193	0.157
	134	268	402	84	183	267	218	451	669
Total	0.025	0.048	0.038	0.014	0.035	0.026	0.018	0.039	0.030
	799	1,104	1,903	1,431	2,085	3,516	2,230	3,189	5,419

◁

Four-way and higher-dimensional tables

▷ Example 4

Let's pretend that our byssinosis dataset also recorded each worker's sex (it does not, and we have made up this extra information). We obtain a four-way table just as we would a three-way table, but we specify the fourth variable as a superrow by including it in the by() option:

```
. use http://www.stata-press.com/data/r10/byssin1
(Byssinosis incidence)
. table workplace smokes race [fw=pop], by(sex) c(mean prob) format(%9.3f)
> sc col row
```

Sex and Dustiness of workplace	Race and Smokes								
	── other ──			── white ──			── Total ──		
	no	yes	Total	no	yes	Total	no	yes	Total
Female									
least	0.006	0.009	0.008	0.009	0.021	0.016	0.009	0.018	0.014
less	0.020	0.008	0.010	0.015	0.015	0.015	0.016	0.012	0.014
most	0.057	0.154	0.141				0.057	0.154	0.141
Total	0.017	0.051	0.043	0.011	0.020	0.016	0.012	0.032	0.024
Male									
least	0.013	0.011	0.012	0.006	0.007	0.006	0.009	0.008	0.009
less	0.020	0.000	0.019	0.000	0.013	0.011	0.016	0.013	0.014
most	0.091	0.244	0.136	0.083	0.230	0.184	0.087	0.232	0.167
Total	0.029	0.041	0.033	0.020	0.056	0.043	0.025	0.052	0.039

If our dataset also included work group and we wanted a five-way table, we could include both the sex and work-group variables in the by() option. You may include up to four variables in by(), and so produce up to 7-way tables.

◁

Methods and Formulas

table is implemented as an ado-file. The contents of cells are calculated by collapse and are displayed by tabdisp; see [D] **collapse** and [P] **tabdisp**.

Also See

[R] **adjust** — Tables of adjusted means and proportions

[R] **summarize** — Summary statistics

[R] **tabstat** — Display table of summary statistics

[R] **tabulate oneway** — One-way tables of frequencies

[R] **tabulate twoway** — Two-way tables of frequencies

[D] **collapse** — Make dataset of summary statistics

[P] **tabdisp** — Display tables

Title

tabstat — Display table of summary statistics

Syntax

tabstat *varlist* [*if*] [*in*] [*weight*] [, *options*]

options	description
Main	
by(*varname*)	group statistics by variable
<u>s</u>tatistics(*statname*[...])	report specified statistics
Options	
<u>label</u>width(#)	width for by() variable labels; default is labelwidth(16)
<u>var</u>width(#)	variable width; default is varwidth(12)
columns(<u>v</u>ariables)	display variables in table columns; the default
columns(<u>s</u>tatistics)	display statistics in table columns
<u>f</u>ormat[(%fmt)]	display format for statistics; default format is %9.0g
<u>case</u>wise	perform casewise deletion of observations
<u>noto</u>tal	do not report overall statistics; use with by()
<u>m</u>issing	report statistics for missing values of by() variable
<u>nose</u>parator	do not use separator line between by() categories
<u>long</u>stub	make left table stub wider
save	save summary statistics in r()

by is allowed; see [D] **by**.
aweights and fweights are allowed; see [U] **11.1.6 weight**.

Description

tabstat displays summary statistics for a series of numeric variables in one table, possibly broken down on (conditioned by) another variable.

Without the by() option, tabstat is a useful alternative to summarize because it allows you to specify the list of statistics to be displayed.

With the by() option, tabstat resembles tabulate used with its summarize() option in that both report statistics of *varlist* for the different values of *varname*. tabstat allows more flexibility in terms of the statistics presented and the format of the table.

tabstat is sensitive to the linesize; it widens the table if possible and wraps if necessary.

Options

Main

by(*varname*) specifies that the statistics be displayed separately for each unique value of *varname*; *varname* may be numeric or string. For instance, `tabstat height` would present the overall mean of height. `tabstat height, by(sex)` would present the mean height of males, and of females, and the overall mean height. Do not confuse the `by()` option with the `by` prefix; both may be specified.

statistics(*statname* $[\dots]$) specifies the statistics to be displayed; the default is equivalent to specifying `statistics(mean)`. (`stats()` is a synonym for `statistics()`.) Multiple statistics may be specified and are separated by white space, such as `statistics(mean sd)`. Available statistics are

statname	definition	*statname*	definition
mean	mean	p1	1st percentile
count	count of nonmissing observations	p5	5th percentile
n	same as count	p10	10th percentile
sum	sum	p25	25th percentile
max	maximum	median	median (same as p50)
min	minimum	p50	50th percentile (same as median)
range	range = max − min	p75	75th percentile
sd	standard deviation	p90	90th percentile
variance	variance	p95	95th percentile
cv	coefficient of variation (sd/mean)	p99	99th percentile
semean	standard error of mean (sd/$\sqrt{n}$)	iqr	interquartile range = p75 − p25
skewness	skewness	q	equivalent to specifying p25 p50 p75
kurtosis	kurtosis		

Options

labelwidth(#) specifies the maximum width to be used within the stub to display the labels of the by() variable. The default is `labelwidth(16)`. $8 \le \# \le 32$.

varwidth(#) specifies the maximum width to be used within the stub to display the names of the variables. The default is `varwidth(12)`. `varwidth()` is effective only with `columns(statistics)`. Setting `varwidth()` implies `longstub`. $8 \le \# \le 16$.

columns(variables | statistics) specifies whether to display variables or statistics in the columns of the table. `columns(variables)` is the default when more than one variable is specified.

format and format(%*fmt*) specify how the statistics are to be formatted. The default is to use a %9.0g format.

format specifies that each variable's statistics be formatted with the variable's display format; see [D] **format**.

format(%*fmt*) specifies the format to be used for all statistics. The maximum width of the specified format should not exceed nine characters.

casewise specifies casewise deletion of observations. Statistics are to be computed for the sample that is not missing for any of the variables in *varlist*. The default is to use all the nonmissing values for each variable.

nototal is for use with by(); it specifies that the overall statistics not be reported.

missing specifies that missing values of the by() variable be treated just like any other value and that statistics should be displayed for them. The default is not to report the statistics for the by()== *missing* group. If the by() variable is a string variable, by()=="" is considered to mean missing.

noseparator specifies that a separator line between the by() categories not be displayed.

longstub specifies that the left stub of the table be made wider so that it can include names of the statistics or variables in addition to the categories of by(*varname*). The default is to describe the statistics or variables in a header. longstub is ignored if by(*varname*) is not specified.

save specifies that the summary statistics be returned in r(). The overall (unconditional) statistics are returned in matrix r(StatTotal) (rows are statistics, columns are variables). The conditional statistics are returned in the matrices r(Stat1), r(Stat2), ..., and the names of the corresponding variables are returned in the macros r(name1), r(name2), ...

Remarks

This command is probably most easily understood by going through a series of examples.

▷ Example 1

We have data on the price, weight, mileage rating, and repair record of 22 foreign and 52 domestic 1978 automobiles. We want to summarize these variables for the different origins of the automobiles.

```
. use http://www.stata-press.com/data/r10/auto
(1978 Automobile Data)

. tabstat price weight mpg rep78, by(foreign)

Summary statistics: mean
  by categories of: foreign (Car type)
```

foreign	price	weight	mpg	rep78
Domestic	6072.423	3317.115	19.82692	3.020833
Foreign	6384.682	2315.909	24.77273	4.285714
Total	6165.257	3019.459	21.2973	3.405797

More summary statistics can be requested via the option statistics(). The group totals can be suppressed with the nototal option.

```
. tabstat price weight mpg rep78, by(foreign) stat(mean sd min max) nototal

Summary statistics: mean, sd, min, max
  by categories of: foreign (Car type)
```

foreign	price	weight	mpg	rep78
Domestic	6072.423	3317.115	19.82692	3.020833
	3097.104	695.3637	4.743297	.837666
	3291	1800	12	1
	15906	4840	34	5
Foreign	6384.682	2315.909	24.77273	4.285714
	2621.915	433.0035	6.611187	.7171372
	3748	1760	14	3
	12990	3420	41	5

Although the header of the table describes the statistics running vertically in the "cells", the table may become hard to read, especially with many variables or statistics. The option `longstub` specifies that a column be added describing the contents of the cells. The option `format` can be issued to specify that `tabstat` display the statistics by using the display format of the variables rather than the overall default %9.0g.

```
. tabstat price weight mpg rep78, by(foreign) stat(mean sd min max) long format
```

foreign	stats	price	weight	mpg	rep78
Domestic	mean	6,072.4	3,317.1	19.8269	3.02083
	sd	3,097.1	695.364	4.7433	.837666
	min	3,291	1,800	12	1
	max	15,906	4,840	34	5
Foreign	mean	6,384.7	2,315.9	24.7727	4.28571
	sd	2,621.9	433.003	6.61119	.717137
	min	3,748	1,760	14	3
	max	12,990	3,420	41	5
Total	mean	6,165.3	3,019.5	21.2973	3.4058
	sd	2,949.5	777.194	5.7855	.989932
	min	3,291	1,760	12	1
	max	15,906	4,840	41	5

We can specify a layout of the table in which the statistics run horizontally and the variables run vertically by specifying the `col(statistics)` option.

```
. tabstat price weight mpg rep78, by(foreign) stat(min mean max) col(stat) long
```

foreign	variable	min	mean	max
Domestic	price	3291	6072.423	15906
	weight	1800	3317.115	4840
	mpg	12	19.82692	34
	rep78	1	3.020833	5
Foreign	price	3748	6384.682	12990
	weight	1760	2315.909	3420
	mpg	14	24.77273	41
	rep78	3	4.285714	5
Total	price	3291	6165.257	15906
	weight	1760	3019.459	4840
	mpg	12	21.2973	41
	rep78	1	3.405797	5

Finally, `tabstat` can also be used to enhance `summarize` so we can specify the statistics to be displayed. For instance, we can display the number of observations, the mean, the coefficient of variation, and the 25%, 50%, and 75% quantiles for a list of variables.

```
. tabstat price weight mpg rep78, stat(n mean cv q) col(stat)
```

variable	N	mean	cv	p25	p50	p75
price	74	6165.257	.478406	4195	5006.5	6342
weight	74	3019.459	.2573949	2240	3190	3600
mpg	74	21.2973	.2716543	18	20	25
rep78	69	3.405797	.290661	3	3	4

Since we did not specify the by() option, these statistics were not displayed for the subgroups of the data formed by the categories of the by() variable.   ◁

Methods and Formulas

tabstat is implemented as an ado-file.

Acknowledgments

The tabstat command was written by Jeroen Weesie and Vincent Buskens of Utrecht University in The Netherlands.

Also See

[R] **summarize** — Summary statistics

[R] **table** — Tables of summary statistics

[R] **tabulate, summarize()** — One- and two-way tables of summary statistics

[D] **collapse** — Make dataset of summary statistics

Title

> **tabulate oneway** — One-way tables of frequencies

Syntax

One-way tables of frequencies

> ta̲bulate *varname* $\lceil$ *if* $\rceil$ $\lceil$ *in* $\rceil$ $\lceil$ *weight* $\rceil$ $\lceil$, *tabulate1_options* $\rceil$

One-way table for each variable—a convenience tool

> tab1 *varlist* $\lceil$ *if* $\rceil$ $\lceil$ *in* $\rceil$ $\lceil$ *weight* $\rceil$ $\lceil$, *tab1_options* $\rceil$

tabulate1_options	description
Main	
subpop(*varname*)	exclude observations for which *varname* = 0
mi̲ssing	treat missing values like other values
nofreq	do not display frequencies
nol̲abel	display numeric codes rather than value labels
pl̲ot	produce a bar chart of the relative frequencies
sort	display the table in descending order of frequency
Advanced	
generate(*stubname*)	create indicator variables for *stubname*
matcell(*matname*)	save frequencies in *matname*; programmer's option
matrow(*matname*)	save unique values of *varname* in *matname*; programmer's option

tab1_options	description
Main	
subpop(*varname*)	exclude observations for which *varname* = 0
mi̲ssing	treat missing values like other values
nofreq	do not display frequencies
nol̲abel	display numeric codes rather than value labels
pl̲ot	produce a bar chart of the relative frequencies
sort	display the table in descending order of frequency

by is allowed with tabulate and tab1; see [D] **by**.

fweights, aweights, and iweights are allowed by tabulate. fweights are allowed by tab1. See
   [U] **11.1.6 weight**.

Description

tabulate produces one-way tables of frequency counts.

For information about two-way tables of frequency counts along with various measures of association, including the common Pearson χ^2, the likelihood-ratio χ^2, Cramér's V, Fisher's exact test, Goodman and Kruskal's gamma, and Kendall's τ_b, see [R] **tabulate twoway**.

tab1 produces a one-way tabulation for each variable specified in *varlist*.

Also see [R] **table** and [R] **tabstat** if you want one-, two-, or n-way tables of frequencies and a wide variety of summary statistics. See [R] **tabulate, summarize()** for a description of tabulate with the summarize() option; it produces tables (breakdowns) of means and standard deviations. table is better than tabulate, summarize(), but tabulate, summarize() is faster. See [ST] **epitab** for 2×2 tables with statistics of interest to epidemiologists.

Options

_____ Main _____

subpop(*varname*) excludes observations for which *varname* = 0 in tabulating frequencies. The mathematical results of tabulate ..., subpop(myvar) are the same as tabulate ... if myvar !=0, but the table may be presented differently. The identities of the rows and columns will be determined from all the data, including the myvar = 0 group, so there may be entries in the table with frequency 0.

Consider tabulating answer, a variable that takes on values 1, 2, and 3, but consider tabulating it just for the male==1 subpopulation. Assume that answer is never 2 in this group. tabulate answer if male==1 produces a table with two rows: one for answer 1 and one for answer 3. There will be no row for answer 2 because answer 2 was never observed. tabulate answer, subpop(male) produces a table with three rows. The row for answer 2 will be shown as having 0 frequency.

missing requests that missing values be treated like other values in calculations of counts, percentages, and other statistics.

nofreq suppresses the printing of the frequencies.

nolabel causes the numeric codes to be displayed rather than the value labels.

plot produces a bar chart of the relative frequencies in a one-way table. (Also see [R] **histogram**.)

sort puts the table in descending order of frequency (and ascending order of the variable within equal values of frequency).

_____ Advanced _____

generate(*stubname*) creates a set of indicator variables (*stubname*1, *stubname*2, ...) reflecting the observed values of the tabulated variable. The generate() option may not be used with the by prefix.

matcell(*matname*) saves the reported frequencies in *matname*. This option is for use by programmers.

matrow(*matname*) saves the numeric values of the $r \times 1$ row stub in *matname*. This option is for use by programmers. matrow() may not be specified if the row variable is a string.

Limits

One-way tables may have a maximum of 12,000 rows (Stata/MP and Stata/SE), 3,000 rows (Stata/IC), or 500 rows (Small Stata).

Remarks

Remarks are presented under the following headings:

> tabulate
> tab1

For each value of a specified variable, `tabulate` reports the number of observations with that value. The number of times a value occurs is called its *frequency*.

tabulate

▷ Example 1

We have data summarizing the speed limit and the accident rate per million vehicle miles along various Minnesota highways in 1973. The variable containing the speed limit is called `spdlimit`. If we `summarize` the variable, we obtain its mean and standard deviation:

```
. use http://www.stata-press.com/data/r10/hiway
(Minnesota Highway Data, 1973)

. summarize spdlimit
```

Variable	Obs	Mean	Std. Dev.	Min	Max
spdlimit	39	55	5.848977	40	70

The average speed limit is 55 miles per hour. We can learn more about this variable by tabulating it:

```
. tabulate spdlimit
```

Speed limit	Freq.	Percent	Cum.
40	1	2.56	2.56
45	3	7.69	10.26
50	7	17.95	28.21
55	15	38.46	66.67
60	11	28.21	94.87
65	1	2.56	97.44
70	1	2.56	100.00
Total	39	100.00	

We see that one highway has a speed limit of 40 miles per hour, three have speed limits of 45, 7 of 50, and so on. The column labeled `Percent` shows the percentage of highways in the dataset that have the indicated speed limit. For instance, 38.46% of highways in our dataset have a speed limit of 55 miles per hour. The final column shows the cumulative percentage. We see that 66.67% of highways in our dataset have a speed limit of 55 miles per hour or less.

◁

▷ Example 2

The `plot` option places a sideways histogram alongside the table:

```
. tabulate spdlimit, plot
Speed limit |        Freq.
------------+------------------------------------------
         40 |            1 |*
         45 |            3 |***
         50 |            7 |*******
         55 |           15 |***************
         60 |           11 |***********
         65 |            1 |*
         70 |            1 |*
------------+------------------------------------------
      Total |           39
```

Of course, `graph` can produce better-looking histograms; see [R] **histogram**.

◁

▷ Example 3

`tabulate` labels tables using *variable* and *value labels* if they exist. To demonstrate how this works, let's add a new variable to our dataset that categorizes `spdlimit` into three categories. We will call this new variable `spdcat`:

```
. generate spdcat=recode(spdlimit,50,60,70)
```

The `recode()` function divides `spdlimit` into 50 miles per hour or below, 51–60, and above 60; see [D] **functions**. We specified the breakpoints in the arguments (`spdlimit,50,60,70`). The first argument is the variable to be recoded. The second argument is the first breakpoint, the third argument is the second breakpoint, and so on. We can specify as many breakpoints as we wish.

`recode()` used our arguments not only as the breakpoints but also to label the results. If `spdlimit` is less than or equal to 50, `spdcat` is set to 50; if `spdlimit` is between 51 and 60, `spdcat` is 60; otherwise, `spdcat` is arbitrarily set to 70. (See [U] **25 Dealing with categorical variables**.)

Since we just created the variable `spdcat`, it is not yet labeled. When we make a table using this variable, `tabulate` uses the variable's name to label it:

```
. tabulate spdcat
     spdcat |      Freq.     Percent        Cum.
------------+-----------------------------------
         50 |         11       28.21       28.21
         60 |         26       66.67       94.87
         70 |          2        5.13      100.00
------------+-----------------------------------
      Total |         39      100.00
```

Even through the table is not well labeled, `recode()`'s coding scheme provides us with clues as to the table's meaning. The first line of the table corresponds to 50 miles per hour and below, the next to 51 through 60 miles per hour, and the last to above 60 miles per hour.

We can improve this table by labeling the values and variables:

```
. label define scat 50 "40 to 50" 60 "55 to 60" 70 "Above 60"
. label values spdcat scat
. label variable spdcat "Speed Limit Category"
```

We define a *value label* called scat that attaches labels to the numbers 50, 60, and 70 using the label define command; see [U] **12.6.3 Value labels**. We label the value 50 as '40 to 50', since we looked back at our original tabulation in the first example and saw that the speed limit was never less than 40. Similarly, we could have labeled the last category '65 to 70' since the speed limit is never greater than 70 miles per hour.

Next, we requested that Stata label the values of the new variable spdcat using the value label scat. Finally, we labeled our variable Speed Limit Category. We are now ready to tabulate the result:

```
. tabulate spdcat
```

Speed Limit Category	Freq.	Percent	Cum.
40 to 50	11	28.21	28.21
55 to 60	26	66.67	94.87
Above 60	2	5.13	100.00
Total	39	100.00	

◁

▷ Example 4

If we have missing values in our dataset, tabulate ignores them unless we explicitly indicate otherwise. We have no missing data in our example, so let's add some:

```
. replace spdcat=. in 39
(1 real change made, 1 to missing)
```

We changed the first observation on spdcat to *missing*. Let's now tabulate the result:

```
. tabulate spdcat
```

Speed Limit Category	Freq.	Percent	Cum.
40 to 50	11	28.95	28.95
55 to 60	26	68.42	97.37
Above 60	1	2.63	100.00
Total	38	100.00	

Comparing this output with that in the previous example, we see that the total frequency count is now one less than it was—38 rather than 39. Also the 'Above 60' category now has only one observation where it used to have two, so we evidently changed a road with a high speed limit.

We want tabulate to treat missing values just as it treats numbers, so we specify the missing option:

```
. tabulate spdcat, missing
```

Speed Limit Category	Freq.	Percent	Cum.
40 to 50	11	28.21	28.21
55 to 60	26	66.67	94.87
Above 60	1	2.56	97.44
.	1	2.56	100.00
Total	39	100.00	

We now see our missing value—the last category, labeled '.', shows a frequency count of 1. The table sum is once again 39.

Let's put our dataset back as it was originally:

```
. replace spdcat=70 in 39
(1 real change made)
```

◁

❑ Technical Note

tabulate also can automatically create indicator variables from categorical variables. We will briefly review that capability here, but see [U] **25 Dealing with categorical variables** for a complete description. Let's begin by describing our highway dataset:

```
. describe
Contains data from http://www.stata-press.com/data/r10/hiway.dta
  obs:            39                          Minnesota Highway Data, 1973
  vars:            2                          16 Nov 2006 12:39
  size:          507 (99.9% of memory free)

              storage  display    value
variable name  type    format    label     variable label

spdlimit       byte    %8.0g               Speed limit
rate           float   %9.0g     rcat      Accident rate per million
                                             vehicle miles
spdcat         float   %9.0g     scat      Speed Limit Category

Sorted by:
     Note:  dataset has changed since last saved
```

Our dataset contains three variables. We will type `tabulate spdcat, generate(spd)`, `describe` our data, and then explain what happened.

```
. tabulate spdcat, generate(spd)
```

Speed Limit Category	Freq.	Percent	Cum.
40 to 50	11	28.21	28.21
55 to 60	26	66.67	94.87
Above 60	2	5.13	100.00
Total	39	100.00	

```
. describe

Contains data from http://www.stata-press.com/data/r10/hiway.dta
  obs:            39                          Minnesota Highway Data, 1973
  vars:            6                          16 Nov 2006 12:39
  size:          624 (99.9% of memory free)
```

variable name	storage type	display format	value label	variable label
spdlimit	byte	%8.0g		Speed limit
rate	float	%9.0g	rcat	Accident rate per million vehicle miles
spdcat	float	%9.0g	scat	Speed Limit Category
spd1	byte	%8.0g		spdcat==40 to 50
spd2	byte	%8.0g		spdcat==55 to 60
spd3	byte	%8.0g		spdcat==Above 60

```
Sorted by:
     Note:  dataset has changed since last saved
```

When we typed `tabulate` with the `generate()` option, Stata responded by producing a one-way frequency table, so it appeared that the option did nothing. Yet when we `describe` our dataset, we find that we now have *six* variables instead of the original three. The new variables are named `spd1`, `spd2`, and `spd3`.

When we specify the `generate()` option, we are telling Stata to not only produce the table but also create a set of indicator variables that correspond to that table. Stata adds a numeric suffix to the name we specify in the parentheses. `spd1` refers to the first line of the table, `spd2` to the second line, and so on. Also Stata labels the variables so that we know what they mean. `spd1` is an indicator variable that is *true* (takes on the value 1) when `spdcat` is between 40 and 50; otherwise, it is zero. (There is an exception: if `spdcat` is missing, so are the `spd1`, `spd2`, and `spd3` variables. This did not happen in our dataset.)

We want to prove our claim. Since we have not yet introduced two-way tabulations, we will use the `summarize` statement:

```
. summarize spdlimit if spd1==1
```

Variable	Obs	Mean	Std. Dev.	Min	Max
spdlimit	11	47.72727	3.437758	40	50

```
. summarize spdlimit if spd2==1
```

Variable	Obs	Mean	Std. Dev.	Min	Max
spdlimit	26	57.11538	2.519157	55	60

```
. summarize spdlimit if spd3==1
```

Variable	Obs	Mean	Std. Dev.	Min	Max
spdlimit	2	67.5	3.535534	65	70

Notice the indicated minimum and maximum in each of the tables above. When we restrict the sample to `spd1`, `spdlimit` is between 40 and 50; when we restrict the sample to `spd2`, `spdlimit` is between 55 and 60; when we restrict the sample to `spd3`, `spdlimit` is between 65 and 70.

Thus `tabulate` provides an easy way to create indicator (sometimes called dummy) variables. We could now use these variables in, for instance, regression analysis. See [U] **25 Dealing with categorical variables** for an example.

❑

tab1

tab1 is a convenience tool. Typing

```
. tab1 myvar thisvar thatvar, plot
```

is equivalent to typing

```
. tabulate myvar, plot
. tabulate thisvar, plot
. tabulate thatvar, plot
```

Saved Results

tabulate and tab1 save the following in r():

Scalars

r(N)   number of observations   r(r)   number of rows

Methods and Formulas

tab1 is implemented as an ado-file.

Reference

Harrison, D. A. 2006. Stata tip 34: Tabulation by listing. *Stata Journal* 6: 425–427.

Also See

[R] **table** — Tables of summary statistics

[R] **tabstat** — Display table of summary statistics

[R] **tabulate twoway** — Two-way tables of frequencies

[R] **tabulate, summarize()** — One- and two-way tables of summary statistics

[D] **collapse** — Make dataset of summary statistics

[ST] **epitab** — Tables for epidemiologists

[SVY] **svy: tabulate oneway** — One-way tables for survey data

[SVY] **svy: tabulate twoway** — Two-way tables for survey data

[XT] **xttab** — Tabulate xt data

[U] **12.6.3 Value labels**

[U] **25 Dealing with categorical variables**

Title

> **tabulate twoway** — Two-way tables of frequencies

Syntax

Two-way tables

> $\underline{\text{ta}}\text{bulate}$ *varname*$_1$ *varname*$_2$ $\begin{bmatrix} if \end{bmatrix}$ $\begin{bmatrix} in \end{bmatrix}$ $\begin{bmatrix} weight \end{bmatrix}$ $\begin{bmatrix} , options \end{bmatrix}$

Two-way tables for all possible combinations—a convenience tool

> tab2 *varlist* $\begin{bmatrix} if \end{bmatrix}$ $\begin{bmatrix} in \end{bmatrix}$ $\begin{bmatrix} weight \end{bmatrix}$ $\begin{bmatrix} , options \end{bmatrix}$

Immediate form of two-way tabulations

> tabi $\#_{11}$ $\#_{12}$ $\begin{bmatrix} \dots \end{bmatrix}$ \ $\#_{21}$ $\#_{22}$ $\begin{bmatrix} \dots \end{bmatrix}$ $\begin{bmatrix} \setminus \dots \end{bmatrix}$ $\begin{bmatrix} , \text{replace } options \end{bmatrix}$

options	description
Main	
$\underline{\text{chi}}2$	report Pearson's χ^2
$\underline{\text{e}}\text{xact}\begin{bmatrix} (\#) \end{bmatrix}$	report Fisher's exact test
gamma	report Goodman and Kruskal's gamma
lrchi2	report likelihood-ratio χ^2
$\underline{\text{t}}\text{aub}$	report Kendall's τ_b
V	report Cramér's V
$\underline{\text{cc}}\text{hi}2$	report Pearson's χ^2 in each cell
$\underline{\text{col}}\text{umn}$	report relative frequency within its column of each cell
$\underline{\text{r}}\text{ow}$	report relative frequency within its row of each cell
$\underline{\text{cl}}\text{rchi}2$	report likelihood-ratio χ^2 in each cell
$\underline{\text{cell}}$	report the relative frequency of each cell
expected	report expected frequency in each cell
$\underline{\text{nof}}\text{req}$	do not display frequencies
$\underline{\text{m}}\text{issing}$	treat missing values like other values
$\underline{\text{wrap}}$	do not wrap wide tables
$\begin{bmatrix} \text{no} \end{bmatrix}\text{key}$	report/suppress cell contents key
$\underline{\text{nol}}\text{abel}$	display numeric codes rather than value labels
$\underline{\text{nolog}}$	do not display enumeration log for Fisher's exact test
Advanced	
matcell(*matname*)	save frequencies in *matname*; programmer's option
matrow(*matname*)	save unique values of *varname*$_1$ in *matname*; programmer's option
matcol(*matname*)	save unique values of *varname*$_2$ in *matname*; programmer's option
† $\underline{\text{all}}$	equivalent to specifying chi2 lrchi2 V gamma taub

†all does not appear in the dialog box.

by is allowed with tabulate and tab2; see [D] **by**.

fweights, aweights, and iweights are allowed by tabulate. fweights are allowed by tab2.
    See [U] **11.1.6 weight**.

Description

tabulate produces two-way tables of frequency counts, along with various measures of association, including the common Pearson's χ^2, the likelihood-ratio χ^2, Cramér's V, Fisher's exact test, Goodman and Kruskal's gamma, and Kendall's τ_b.

Line size is respected. That is, if you resize the Results window before running tabulate, the resulting two-way tabulation will take advantage of the available horizontal space. Stata for Unix(console) users can instead use the set linesize command to take advantage of this feature.

tab2 produces all possible two-way tabulations of the variables specified in *varlist*.

tabi displays the $r \times c$ table, using the values specified; rows are separated by '\'. If no options are specified, it is as if exact were specified for 2×2 tables and chi2 were specified otherwise. See [U] **19 Immediate commands** for a general description of immediate commands. Specifics for tabi can be found toward the end of [R] **tabulate twoway**.

See [R] **tabulate oneway** if you want one-way tables of frequencies. See [R] **table** and [R] **tabstat** if you want one-, two-, or n-way tables of frequencies and a wide variety of summary statistics. See [R] **tabulate, summarize()** for a description of tabulate with the summarize() option; it produces tables (breakdowns) of means and standard deviations. table is better than tabulate, summarize(), but tabulate, summarize() is faster. See [ST] **epitab** for 2×2 tables with statistics of interest to epidemiologists.

Options

_____| Main |_____

chi2 calculates and displays Pearson's χ^2 for the hypothesis that the rows and columns in a two-way table are independent. chi2 may not be specified if aweights or iweights are specified.

exact $\big[(\#)\big]$ displays the significance calculated by Fisher's exact test and may be applied to $r \times c$ as well as to 2×2 tables. For 2×2 tables, both one- and two-sided probabilities are displayed. For $r \times c$ tables, one-sided probabilities are displayed. The optional positive integer # is a multiplier on the amount of memory that the command is permitted to consume. The default is 1. This option should not be necessary for reasonable $r \times c$ tables. If the command terminates with error 910, try exact(2).

gamma displays Goodman and Kruskal's gamma, along with its asymptotic standard error. gamma is appropriate only when both variables are ordinal. gamma may not be specified if aweights or iweights are specified.

lrchi2 displays the likelihood-ratio χ^2 statistic. The request is ignored if any cell of the table contains no observations. lrchi2 may not be specified if aweights or iweights are specified.

taub displays Kendall's τ_b along with its asymptotic standard error. taub is appropriate only when both variables are ordinal. taub may not be specified if aweights or iweights are specified.

V (note capitalization) displays Cramér's V. V may not be specified if aweights or iweights are specified.

cchi2 displays each cell's contribution to Pearson's chi-squared in a two-way table.

column displays the relative frequency of each cell within its column in a two-way table.

row displays the relative frequency of each cell within its row in a two-way table.

clrchi2 displays each cell's contribution to the likelihood-ratio chi-squared in a two-way table.

cell displays the relative frequency of each cell in a two-way table.

expected displays the expected frequency of each cell in a two-way table.

nofreq suppresses the printing of the frequencies.

missing requests that missing values be treated like other values in calculations of counts, percentages, and other statistics.

wrap requests that Stata take no action on wide, two-way tables to make them readable. Unless wrap is specified, wide tables are broken into pieces to enhance readability.

[no]key suppresses or forces the display of a key above two-way tables. The default is to display the key if more than one cell statistic is requested, and otherwise to omit it. key forces the display of the key. nokey suppresses its display.

nolabel causes the numeric codes to be displayed rather than the value labels.

nolog suppresses the display of the log for Fisher's exact test. Using Fisher's exact test requires counting all tables that have a probability exceeding that of the observed table given the observed row and column totals. The log counts down each stage of the network computations, starting from the number of columns and counting down to 1, displaying the number of nodes in the network at each stage. A log is not displayed for 2×2 tables.

$\overline{\qquad\quad \boxed{\text{Advanced}}\ \rule[0.5ex]{0pt}{0pt} \qquad\qquad\qquad\qquad\qquad\qquad\qquad\qquad\qquad\qquad\qquad\qquad\qquad\qquad}$

matcell(*matname*) saves the reported frequencies in *matname*. This option is for use by programmers.

matrow(*matname*) saves the numeric values of the $r \times 1$ row stub in *matname*. This option is for use by programmers. matrow() may not be specified if the row variable is a string.

matcol(*matname*) saves the numeric values of the $1 \times c$ column stub in *matname*. This option is for use by programmers. matcol() may not be specified if the column variable is a string.

replace indicates that the immediate data specified as arguments to the tabi command be left as the current data in place of whatever data were there.

The following option is available with tabulate but is not shown in the dialog box:

all is equivalent to specifying chi2 lrchi2 V gamma taub. Note the omission of exact. When all is specified, no may be placed in front of the other options. all noV requests all measures of association but Cramér's V (and Fisher's exact). all exact requests all association measures, including Fisher's exact test. all may not be specified if aweights or iweights are specified.

Limits

Two-way tables may have a maximum of 1,200 rows and 80 columns (Stata/MP and Stata/SE), 300 rows and 20 columns (Stata/IC), or 160 rows and 20 columns (Small Stata). If larger tables are needed, see [R] **table**.

Remarks

Remarks are presented under the following headings:

tabulate
Measures of association
N-way tables
Weighted data
Tables with immediate data
tab2

For each value of a specified variable (or a set of values for a pair of variables), `tabulate` reports the number of observations with that value. The number of times a value occurs is called its *frequency*.

tabulate

▷ Example 1

`tabulate` will make two-way tables if we specify two variables following the word `tabulate`. In our highway dataset, we have a variable called `rate` that divides the accident rate into three categories: below 4, 4–7, and above 7 per million vehicle miles. Let's make a table of the speed limit category and the accident-rate category:

```
. use http://www.stata-press.com/data/r10/hiway2
(Minnesota Highway Data, 1973)

. tabulate spdcat rate
```

Speed Limit Category	Accident rate per million vehicle miles			Total
	Below 4	4-7	Above 7	
40 to 50	3	5	3	11
55 to 60	19	6	1	26
Above 60	2	0	0	2
Total	24	11	4	39

The table indicates that three stretches of highway have an accident rate below 4 and a speed limit of 40 to 50 miles per hour. The table also shows the row and column sums (called the *marginals*). The number of highways with a speed limit of 40 to 50 miles per hour is 11, which is the same result we obtained in our previous one-way tabulations.

Stata can present this basic table in several ways—16, to be precise—and we will show just a few below. It might be easier to read the table if we included the row percentages. For instance, of 11 highways in the lowest speed limit category, three are also in the lowest accident-rate category. Three-elevenths amounts to some 27.3%. We can ask Stata to fill in this information for us by using the `row` option:

```
. tabulate spdcat rate, row
```

┌─────────────┐
│ Key   │
├─────────────┤
│ *frequency* │
│ *row percentage* │
└─────────────┘

Speed Limit Category	Accident rate per million vehicle miles			Total
	Below 4	4-7	Above 7	
40 to 50	3	5	3	11
	27.27	45.45	27.27	100.00
55 to 60	19	6	1	26
	73.08	23.08	3.85	100.00
Above 60	2	0	0	2
	100.00	0.00	0.00	100.00
Total	24	11	4	39
	61.54	28.21	10.26	100.00

The number listed below each frequency is the percentage of cases that each cell represents out of its row. That is easy to remember because we see 100% listed in the "Total" column. The bottom row is also informative. We see that 61.54% of all the highways in our dataset fall into the lowest accident-rate category, that 28.21% are in the middle category, and that 10.26% are in the highest.

tabulate can calculate column percentages and cell percentages, as well. It does so when we specify the column or cell options, respectively. We can even specify them together. Below is a table that includes everything:

(Continued on next page)

```
. tabulate spdcat rate, row column cell
```

Key
frequency
row percentage
column percentage
cell percentage

Speed Limit Category	Accident rate per million vehicle miles			Total
	Below 4	4-7	Above 7	
40 to 50	3	5	3	11
	27.27	45.45	27.27	100.00
	12.50	45.45	75.00	28.21
	7.69	12.82	7.69	28.21
55 to 60	19	6	1	26
	73.08	23.08	3.85	100.00
	79.17	54.55	25.00	66.67
	48.72	15.38	2.56	66.67
Above 60	2	0	0	2
	100.00	0.00	0.00	100.00
	8.33	0.00	0.00	5.13
	5.13	0.00	0.00	5.13
Total	24	11	4	39
	61.54	28.21	10.26	100.00
	100.00	100.00	100.00	100.00
	61.54	28.21	10.26	100.00

The number at the top of each cell is the frequency count. The second number is the row percentage—they sum to 100% going across the table. The third number is the column percentage—they sum to 100% going down the table. The bottom number is the cell percentage—they sum to 100% going down all the columns and across all the rows. For instance, highways with a speed limit above 60 miles per hour and in the lowest accident rate category account for 100% of highways with a speed limit above 60 miles per hour; 8.33% of highways in the lowest accident-rate category; and 5.13% of all our data.

A fourth option, `nofreq`, tells Stata not to print the frequency counts. To construct a table consisting of only row percentages, we type

```
. tabulate spdcat rate, row nofreq
```

Speed Limit Category	Accident rate per million vehicle miles			Total
	Below 4	4-7	Above 7	
40 to 50	27.27	45.45	27.27	100.00
55 to 60	73.08	23.08	3.85	100.00
Above 60	100.00	0.00	0.00	100.00
Total	61.54	28.21	10.26	100.00

Measures of association

▷ Example 2

tabulate will calculate the Pearson χ^2 test for the independence of the rows and columns if we specify the chi2 option. Suppose that we have 1980 census data on 956 cities in the United States and wish to compare the age distribution across regions of the country. Assume that agecat is the median age in each city and that region denotes the region of the country in which the city is located.

```
. use http://www.stata-press.com/data/r10/citytemp2
(City Temperature Data)
. tabulate region agecat, chi2
```

Census Region	agecat 19-29	30-34	35+	Total
NE	46	83	37	166
N Cntrl	162	92	30	284
South	139	68	43	250
West	160	73	23	256
Total	507	316	133	956

```
              Pearson chi2(6) =  61.2877   Pr = 0.000
```

We obtain the standard two-way table and, at the bottom, a summary of the χ^2 test. Stata informs us that the χ^2 associated with this table has 6 degrees of freedom and is 61.29. The observed differences are significant.

The table is, perhaps, easier to understand if we suppress the frequencies and print just the row percentages:

```
. tabulate region agecat, row nofreq chi2
```

Census Region	agecat 19-29	30-34	35+	Total
NE	27.71	50.00	22.29	100.00
N Cntrl	57.04	32.39	10.56	100.00
South	55.60	27.20	17.20	100.00
West	62.50	28.52	8.98	100.00
Total	53.03	33.05	13.91	100.00

```
              Pearson chi2(6) =  61.2877   Pr = 0.000
```

◁

▷ Example 3

We have data on dose level and outcome for a set of patients and wish to evaluate the association between the two variables. We can obtain all the association measures by specifying the all and exact options:

(Continued on next page)

```
. use http://www.stata-press.com/data/r10/dose

. tabulate dose function, all exact

Enumerating sample-space combinations:
stage 3:  enumerations = 1
stage 2:  enumerations = 9
stage 1:  enumerations = 0
```

		Function		
Dosage	< 1 hr	1 to 4	4+	Total
1/day	20	10	2	32
2/day	16	12	4	32
3/day	10	16	6	32
Total	46	38	12	96

```
            Pearson chi2(4) =    6.7780   Pr = 0.148
    likelihood-ratio chi2(4) =   6.9844   Pr = 0.137
                Cramer's V =     0.1879
                     gamma =     0.3689   ASE = 0.129
             Kendall's tau-b =   0.2378   ASE = 0.086
             Fisher's exact =              0.145
```

We find evidence of association but not enough to be truly convincing.

If we had not also specified the **exact** option, we would not have obtained Fisher's exact test. Stata can calculate this statistic both for 2×2 tables and for $r \times c$. For 2×2 tables, the calculation is almost instant. On more general tables, however, the calculation can take longer.

We carefully constructed our example so that **all** would be meaningful. Kendall's τ_b and Goodman and Kruskal's gamma are relevant only when both dimensions of the table can be ordered, say, from low to high or from worst to best. The other statistics, however, are applicable in all cases.

◁

❑ Technical Note

Be careful when attempting to compute the p-value for Fisher's exact test since the number of tables that contribute to the p-value can be extremely large and a solution may not be feasible. The errors that are indicative of this situation are errors 910, exceeded memory limitations, and 1401, integer overflow due to large row-margin frequencies. If execution terminates because of memory limitations, use **exact(2)** to permit the algorithm to consume twice the memory, **exact(3)** for three times the memory, etc. The default memory usage should be sufficient for reasonable tables.

❑

N-way tables

If you need more than two-way tables, your best alternative to is use **table**, not **tabulate**; see [R] **table**.

The technical note below shows you how to use **tabulate** to create a sequence of two-way tables that together form, in effect, a three-way table, but using **table** is easy and produces prettier results:

```
. use http://www.stata-press.com/data/r10/birthcat
(City data)
. table birthcat region agecat, c(freq)
```

| birthcat | agecat and Census Region | | | | | | | |
| | 19-29 | | | | 30-34 | | | |
	NE N Cntrl	South	West		NE N Cntrl	South	West
29-136	11	23	11	11	34	27	10	8
137-195	31	97	65	46	48	58	45	42
196-529	4	38	59	91	1	3	12	21

| birthcat | agecat and Census Region | | | |
| | 35+ | | | |
	NE N Cntrl	South	West
29-136	34	26	27	18
137-195	3	4	7	4
196-529			4	

❑ Technical Note

We can make *n*-way tables by combining the by *varlist*: prefix with `tabulate`. Continuing with the dataset of 956 cities, say that we want to make a table of age category by birth-rate category by region of the country. The birth-rate category variable is named `birthcat` in our dataset. To make separate tables for each age category, we would type

```
. by agecat, sort: tabulate birthcat region
```

```
-> agecat = 19-29
```

| birthcat | Census Region | | | | Total |
	NE	N Cntrl	South	West	
29-136	11	23	11	11	56
137-195	31	97	65	46	239
196-529	4	38	59	91	192
Total	46	158	135	148	487

```
-> agecat = 30-34
```

| birthcat | Census Region | | | | Total |
	NE	N Cntrl	South	West	
29-136	34	27	10	8	79
137-195	48	58	45	42	193
196-529	1	3	12	21	37
Total	83	88	67	71	309

```
-> agecat = 35+
```

birthcat	NE	N Cntrl	South	West	Total
		Census Region			
29-136	34	26	27	18	105
137-195	3	4	7	4	18
196-529	0	0	4	0	4
Total	37	30	38	22	127

❑

Weighted data

▷ Example 4

tabulate can process weighted as well as unweighted data. As with all Stata commands, we indicate the weight by specifying the [*weight*] modifier; see [U] **11.1.6 weight**.

Continuing with our dataset of 956 cities, we also have a variable called pop, the population of each city. We can make a table of region by age category, weighted by population, by typing

```
. tabulate region agecat [freq=pop]
```

Census Region	agecat 19-29	30-34	35+	Total
NE	4,721,387	10,421,387	5,323,610	20,466,384
N Cntrl	16,901,550	8,964,756	4,015,593	29,881,899
South	13,894,254	7,686,531	4,141,863	25,722,648
West	16,698,276	7,755,255	2,375,118	26,828,649
Total	52,215,467	34,827,929	15,856,184	102899580

If we specify the cell, column, or row options, they will also be appropriately weighted. Below we repeat the table, suppressing the counts and substituting row percentages:

```
. tabulate region agecat [freq=pop], nofreq row
```

Census Region	agecat 19-29	30-34	35+	Total
NE	23.07	50.92	26.01	100.00
N Cntrl	56.56	30.00	13.44	100.00
South	54.02	29.88	16.10	100.00
West	62.24	28.91	8.85	100.00
Total	50.74	33.85	15.41	100.00

◁

Tables with immediate data

▷ Example 5

tabi ignores the dataset in memory and uses as the table the values that we specify on the command line:

```
. tabi 30 18 \ 38 14
```

		col		
row	1		2	Total
1	30		18	48
2	38		14	52
Total	68		32	100

```
              Fisher's exact =                    0.289
      1-sided Fisher's exact =                    0.179
```

We may specify any of the options of tabulate and are not limited to 2×2 tables:

```
. tabi 30 18 38 \ 13 7 22, chi2 exact
Enumerating sample-space combinations:
stage 3:  enumerations = 1
stage 2:  enumerations = 3
stage 1:  enumerations = 0
```

		col			
row	1		2	3	Total
1	30		18	38	86
2	13		7	22	42
Total	43		25	60	128

```
        Pearson chi2(2) =   0.7967   Pr = 0.671
          Fisher's exact =              0.707
. tabi 30 13 \ 18 7 \ 38 22, all exact col
```

Key
frequency
column percentage

```
Enumerating sample-space combinations:
stage 3:  enumerations = 1
stage 2:  enumerations = 3
stage 1:  enumerations = 0
```

		col		
row	1		2	Total
1	30		13	43
	34.88		30.95	33.59
2	18		7	25
	20.93		16.67	19.53
3	38		22	60
	44.19		52.38	46.88
Total	86		42	128
	100.00		100.00	100.00

```
        Pearson chi2(2) =   0.7967   Pr = 0.671
likelihood-ratio chi2(2) =   0.7985   Pr = 0.671
             Cramer's V =   0.0789
                  gamma =   0.1204   ASE = 0.160
        Kendall's tau-b =   0.0630   ASE = 0.084
          Fisher's exact =                0.707
```

For 2×2 tables, both one- and two-sided Fisher's exact probabilities are displayed; this is true of both `tabulate` and `tabi`. See *Cumulative incidence data* and *Case–control data* in [ST] **epitab** for more discussion on the relationship between one- and two-sided probabilities.

◁

❑ Technical Note

`tabi`, as with all immediate commands, leaves any data in memory undisturbed. With the `replace` option, however, the data in memory are replaced by the data from the table:

```
. tabi 30 18 \ 38 14, replace
           |         col
       row |         1          2 |     Total
-----------+----------------------+----------
         1 |        30         18 |        48
         2 |        38         14 |        52
-----------+----------------------+----------
     Total |        68         32 |       100

           Fisher's exact =                 0.289
    1-sided Fisher's exact =                 0.179
. list

     +--------------------+
     | row    col    pop  |
     |--------------------|
  1. |  1      1      30  |
  2. |  1      2      18  |
  3. |  2      1      38  |
  4. |  2      2      14  |
     +--------------------+
```

With this dataset, you could recreate the above table by typing

```
. tabulate row col [freq=pop], exact
           |         col
       row |         1          2 |     Total
-----------+----------------------+----------
         1 |        30         18 |        48
         2 |        38         14 |        52
-----------+----------------------+----------
     Total |        68         32 |       100

           Fisher's exact =                 0.289
    1-sided Fisher's exact =                 0.179
```

❑

tab2

`tab2` is a convenience tool. Typing

. `tab2 myvar thisvar thatvar, chi2`

is equivalent to typing

. `tabulate myvar thisvar, chi2`
. `tabulate myvar thatvar, chi2`
. `tabulate thisvar thatvar, chi2`

Saved Results

`tabulate`, `tab2`, and `tabi` save the following in `r()`:

Scalars

r(N)	number of observations	r(p_exact)	Fisher's exact p
r(r)	number of rows	r(chi2_lr)	likelihood-ratio χ^2
r(c)	number of columns	r(p_lr)	significance of likelihood-ratio χ^2
r(chi2)	Pearson's χ^2	r(CramersV)	Cramér's V
r(p)	significance of Pearson's χ^2	r(ase_gam)	ASE of gamma
r(gamma)	gamma	r(ase_taub)	ASE of τ_b
r(p1_exact)	one-sided Fisher's exact p	r(taub)	τ_b

r(p1_exact) is defined only for 2×2 tables. Also, the `matrow()`, `matcol()`, and `matcell()` options allow you to obtain the row values, column values, and frequencies, respectively.

Methods and Formulas

`tab2` and `tabi` are implemented as ado-files.

Let n_{ij}, $i = 1, \ldots, I$, $j = 1, \ldots, J$, be the number of observations in the ith row and jth column. If the data are not weighted, n_{ij} is just a count. If the data are weighted, n_{ij} is the sum of the weights of all data corresponding to the (i, j) cell.

Define the row and column marginals as

$$ n_{i\cdot} = \sum_{j=1}^{J} n_{ij} \qquad\qquad n_{\cdot j} = \sum_{i=1}^{I} n_{ij} $$

and let $n = \sum_i \sum_j n_{ij}$ be the overall sum. Also define the concordance and discordance as

$$ A_{ij} = \sum_{k>i}\sum_{l>j} n_{kl} + \sum_{k<i}\sum_{l<j} n_{kl} \qquad\qquad D_{ij} = \sum_{k>i}\sum_{l<j} n_{kl} + \sum_{k<i}\sum_{l>j} n_{kl} $$

along with twice the number of concordances $P = \sum_i \sum_j n_{ij} A_{ij}$ and twice the number of discordances $Q = \sum_i \sum_j n_{ij} D_{ij}$.

The Pearson χ^2 statistic with $(I-1)(J-1)$ degrees of freedom (so called because it is based on Pearson 1900; see Conover 1999, 240 and Fienberg 1980, 9) is defined as

$$X^2 = \sum_i \sum_j \frac{(n_{ij} - m_{ij})^2}{m_{ij}}$$

where $m_{ij} = n_i . n_{.j}/n$.

The likelihood-ratio χ^2 statistic with $(I-1)(J-1)$ degrees of freedom (Fienberg 1980, 40) is defined as

$$G^2 = 2 \sum_i \sum_j n_{ij}\ln(n_{ij}/m_{ij})$$

Cramér's V (Cramér 1946; also see Agresti 1984, 23–24) is a measure of association designed so that the attainable upper bound is 1. For 2×2 tables, $-1 \le V \le 1$, and otherwise, $0 \le V \le 1$.

$$V = \begin{cases} (n_{11}n_{22} - n_{12}n_{21})/(n_1.n_2.n_{.1}n_{.2})^{1/2} & \text{for } 2 \times 2 \\ \{(X^2/n)/\min(I-1, J-1)\}^{1/2} & \text{otherwise} \end{cases}$$

Gamma (Goodman and Kruskal 1954, 1959, 1963, 1972; also see Agresti 1984, 159–161) ignores tied pairs and is based only on the number of concordant and discordant pairs of observations, $-1 \le \gamma \le 1$,

$$\gamma = (P-Q)/(P+Q)$$

with asymptotic variance

$$16 \sum_i \sum_j n_{ij}(QA_{ij} - PD_{ij})^2/(P+Q)^4$$

Kendall's τ_b (Kendall 1945; also see Agresti 1984, 161–163), $-1 \le \tau_b \le 1$, is similar to gamma, except that it uses a correction for ties,

$$\tau_b = (P-Q)/(w_r w_c)^{1/2}$$

with asymptotic variance

$$\frac{\sum_i \sum_j n_{ij}(2w_r w_c d_{ij} + \tau_b v_{ij})^2 - n^3 \tau_b^2 (w_r + w_c)^2}{(w_r w_c)^4}$$

where

$$w_r = n^2 - \sum_i n_{i\cdot}^2$$

$$w_c = n^2 - \sum_j n_{\cdot j}^2$$

$$d_{ij} = A_{ij} - D_{ij}$$

$$v_{ij} = n_{i\cdot} w_c + n_{\cdot j} w_r$$

Fisher's exact test (Fisher 1935 and Finney 1948; see Zelterman and Louis 1992, 293–301, for the 2×2 case) yields the probability of observing a table that gives at least as much evidence of association as the one actually observed under the assumption of no association. Holding row and column marginals fixed, the hypergeometric probability P of every possible table A is computed, and the

$$P = \sum_{T \in A} \Pr(T)$$

where A is the set of all tables with the same marginals as the observed table, $T^\star$, such that $\Pr(T) \le \Pr(T^\star)$. For 2×2 tables, the one-sided probability is calculated by further restricting A to tables in the same tail as $T^\star$. The first algorithm extending this calculation to $r \times c$ tables was Pagano and Halvorsen (1981); the one implemented here is the FEXACT algorithm by Mehta and Patel (1986). This is a search-tree clipping method originally published by Mehta and Patel (1983) with further refinements by Joe (1988) and Clarkson, Fan, and Joe (1993). Fisher's exact test is a permutation test. For more information on permutation tests, see Good (2005 and 2006).

References

Agresti, A. 1984. *Analysis of Ordinal Categorical Data*. New York: Wiley.

Campbell, M. J., and D. Machin. 1999. *Medical Statistics: A Commonsense Approach*. 3rd ed. New York: Wiley.

Clarkson, D. B., Y. Fan, and H. Joe. 1993. A remark on Algorithm 643: FEXACT: An algorithm for performing Fisher's exact test in $r \times c$ contingency tables. *ACM Transactions on Mathematical Software* 19: 484–488.

Conover, W. J. 1999. *Practical Nonparametric Statistics*. 3rd ed. New York: Wiley.

Cox, N. J. 1996. sg57: An immediate command for two-way tables. *Stata Technical Bulletin* 33: 7–9. Reprinted in *Stata Technical Bulletin Reprints*, vol. 6, pp. 140–143.

———. 1999. sg113: Tabulation of modes. *Stata Technical Bulletin* 50: 26–27. Reprinted in *Stata Technical Bulletin Reprints*, vol. 9, pp. 180–181.

———. 2003. sg113_1: Software update: Tabulation of modes. *Stata Journal* 3: 211.

Cramér, H. 1946. *Mathematical Methods of Statistics*. Princeton: Princeton University Press.

Fienberg, S. E. 1980. *The Analysis of Cross-Classified Categorical Data*. 2nd ed. Cambridge, MA: MIT Press.

Finney, D. J. 1948. The Fisher–Yates test of significance in 2×2 contingency tables. *Biometrika* 35: 145–156.

Fisher, R. A. 1935. The logic of inductive inference. *Journal of the Royal Statistical Society, Series A* 98: 39–54.

Good, P. 2005. *Permutation, Parametric, and Bootstrap Tests of Hypotheses: A Practical Guide to Resampling Methods for Testing Hypotheses*. 3rd ed. New York: Springer.

———. 2006. *Resampling Methods: A Practical Guide to Data Analysis*. 3rd ed. Boston: Birkhäuser.

Goodman, L. A., and W. H. Kruskal. 1954. Measures of association for cross classifications. *Journal of the American Statistical Association* 49: 732–764.

———. 1959. Measures of association for cross classifications II: Further discussion and references. *Journal of the American Statistical Association* 54: 123–163.

——. 1963. Measures of association for cross classifications III: Approximate sampling theory. *Journal of the American Statistical Association* 58: 310–364.

——. 1972. Measures of association for cross classifications IV: Simplification of asymptotic variances. *Journal of the American Statistical Association* 67: 415–421.

Harrison, D. A. 2006. Stata tip 34: Tabulation by listing. *Stata Journal* 6: 425–427.

Joe, H. 1988. Extreme probabilities for contingency tables under row and column independence with application to Fisher's exact test. *Communications in Statistics—Theory and Methods* 17: 3677–3685.

Judson, D. H. 1992. sg12: Extended tabulate utilities. *Stata Technical Bulletin* 10: 22–23. Reprinted in *Stata Technical Bulletin Reprints*, vol. 2, pp. 140–141.

Kendall, M. G. 1945. The treatment of ties in rank problems. *Biometrika* 33: 239–251.

Mehta, C. R., and N. R. Patel. 1983. A network algorithm for performing Fisher's exact test in $r \times c$ contingency tables. *Journal of the American Statistical Association* 78: 427–434.

——. 1986. Algorithm 643 FEXACT: A FORTRAN subroutine for Fisher's exact test on unordered $r \times c$ contingency tables. *ACM Transactions on Mathematical Software* 12: 154–161.

Newson, R. 2002. Parameters behind "nonparametric" statistics: Kendall's tau, Somers' D, and median differences. *Stata Journal* 2: 45–64.

Pagano, M., and K. Halvorsen. 1981. An algorithm for finding the exact significance levels of $r \times c$ tables. *Journal of the American Statistical Association* 76: 931–934.

Pearson, K. 1900. On the criterion that a given system of deviations from the probable in the case of a correlated system of variables is such that it can be reasonably supposed to have arisen from random sampling. *Philosophical Magazine*, Series 5, 50: 157–175.

Weesie, J. 2001. dm91: Patterns of missing values. *Stata Technical Bulletin* 61: 5–7. Reprinted in *Stata Technical Bulletin Reprints*, vol. 10, pp. 49–51.

Wolfe, R. 1999. sg118: Partitions of Pearson's χ^2 for analyzing two-way tables that have ordered columns. *Stata Technical Bulletin* 51: 37–40. Reprinted in *Stata Technical Bulletin Reprints*, vol. 9, pp. 203–207.

Zelterman, D., and T. A. Louis. 1992. Contingency tables in medical studies. In *Medical Uses of Statistics*, 2nd ed., ed. J. C. Bailar III and F. Mosteller, 293–310. Boston: Dekker.

Also See

Title

> **tabulate, summarize()** — One- and two-way tables of summary statistics

Syntax

$$\underline{\text{ta}}\text{bulate } varname_1 \; \big[\, varname_2 \,\big] \; \big[\, if \,\big] \; \big[\, in \,\big] \; \big[\, weight \,\big] \; \big[\, , \; options \,\big]$$

options	description
Main	
<u>su</u>mmarize(*varname₃*)	report summary statistics for *varname₃*
[<u>no</u>]<u>means</u>	include or suppress means
[<u>no</u>]<u>standard</u>	include or suppress standard deviations
[<u>no</u>]<u>freq</u>	include or suppress frequencies
[<u>no</u>]<u>obs</u>	include or suppress number of observations
<u>nol</u>abel	show numeric codes, not labels
<u>wr</u>ap	do not break wide tables
<u>mi</u>ssing	treat missing values of *varname₁* and *varname₂* as categories

by is allowed; see [D] **by**.

aweights and fweights are allowed; see [U] **11.1.6 weight**.

Description

tabulate, summarize() produces one- and two-way tables (breakdowns) of means and standard deviations. See [R] **tabulate oneway** and [R] **tabulate twoway** for one- and two-way frequency tables. See [R] **table** for a more flexible command that produces one-, two-, and n-way tables of frequencies and a wide variety of summary statistics. **table** is better, but **tabulate, summarize()** is faster. Also see [R] **tabstat** for yet another alternative.

Options

⌐ Main ⌐

summarize(*varname₃*) identifies the name of the variable for which summary statistics are to be reported. If you do not specify this option, a table of frequencies is produced; see [R] **tabulate oneway** and [R] **tabulate twoway**. The description here concerns **tabulate** when this option is specified.

(Continued on next page)

441

[no]means includes or suppresses only the means from the table.

The summarize() table normally includes the mean, standard deviation, frequency, and, if the data are weighted, number of observations. Individual elements of the table may be included or suppressed by the [no]means, [no]standard, [no]freq, and [no]obs options. For example, typing

           . tabulate category, summarize(myvar) means standard

produces a summary table by category containing only the means and standard deviations of myvar. You could also achieve the same result by typing

           . tabulate category, summarize(myvar) nofreq

[no]standard includes or suppresses only the standard deviations from the table; see [no]means option above.

[no]freq includes or suppresses only the frequencies from the table; see [no]means option above.

[no]obs includes or suppresses only the reported number of observations from the table. If the data are not weighted, the number of observations is identical to the frequency, and by default only the frequency is reported. If the data are weighted, the frequency refers to the sum of the weights. See [no]means option above.

nolabel causes the numeric codes to be displayed rather than the label values.

wrap requests that no action be taken on wide tables to make them readable. Unless wrap is specified, wide tables are broken into pieces to enhance readability.

missing requests that missing values of *varname*$_1$ and *varname*$_2$ be treated as categories rather than as observations to be omitted from the analysis.

Remarks

tabulate with the summarize() option produces one- and two-way tables of summary statistics. When combined with the by prefix, it can produce n-way tables as well.

Remarks are presented under the following headings:

           One-way tables
           Two-way tables

One-way tables

▷ Example 1

We have data on 74 automobiles. Included in our dataset are the variables foreign, which marks domestic and foreign cars, and mpg, the car's mileage rating. Typing tabulate foreign displays a breakdown of the number of observations we have by the values of the foreign variable.

```
. use http://www.stata-press.com/data/r10/auto
(1978 Automobile Data)

. tabulate foreign
```

Car type	Freq.	Percent	Cum.
Domestic	52	70.27	70.27
Foreign	22	29.73	100.00
Total	74	100.00	

We discover that we have 52 domestic cars and 22 foreign cars in our dataset. If we add the summarize(*varname*) option, however, tabulate produces a table of summary statistics for *varname*:

```
. tabulate foreign, summarize(mpg)
             |     Summary of Mileage (mpg)
   Car type  |       Mean    Std. Dev.        Freq.
-------------+-----------------------------------
   Domestic  |   19.826923    4.7432972          52
    Foreign  |   24.772727    6.6111869          22
-------------+-----------------------------------
      Total  |   21.297297    5.7855032          74
```

We also discover that the average gas mileage for domestic cars is about 20 mpg and the average foreign is almost 25 mpg. Overall, the average is 21 mpg in our dataset.

◁

❑ Technical Note

We might now wonder if the difference in gas mileage between foreign and domestic cars is statistically significant. We can use the oneway command to find out; see [R] **oneway**. To obtain an analysis-of-variance table of mpg on foreign, we type

```
. oneway mpg foreign
                        Analysis of Variance
    Source              SS         df       MS            F     Prob > F
------------------------------------------------------------------------
Between groups      378.153515      1    378.153515     13.18    0.0005
Within groups       2065.30594     72    28.6848048
------------------------------------------------------------------------
    Total           2443.45946     73    33.4720474

Bartlett's test for equal variances:  chi2(1) =    3.4818   Prob>chi2 = 0.062
```

The F statistic is 13.18, and the difference between foreign and domestic cars' mileage ratings is significant at the 0.05% level.

There are several ways that we could have statistically compared mileage ratings—see, for instance, [R] **anova**, [R] **oneway**, [R] **regress**, and [R] **ttest**—but oneway seemed the most convenient.

❑

Two-way tables

▷ Example 2

tabulate, summarize can be used to obtain two-way as well as one-way breakdowns. For instance, we obtained summary statistics on mpg decomposed by foreign by typing tabulate foreign, summarize(mpg). We can specify up to two variables before the comma:

(*Continued on next page*)

```
. generate wgtcat = autocode(weight,4,1760,4840)

. tabulate wgtcat foreign, summarize(mpg)
```

Means, Standard Deviations and Frequencies of Mileage (mpg)

wgtcat	Car type Domestic	Foreign	Total
2530	28.285714	27.0625	27.434783
	3.0937725	5.9829619	5.2295149
	7	16	23
3300	21.75	19.6	21.238095
	2.4083189	3.4351128	2.7550819
	16	5	21
4070	17.26087	14	17.125
	1.8639497	0	1.9406969
	23	1	24
4840	14.666667	.	14.666667
	3.32666	.	3.32666
	6	0	6
Total	19.826923	24.772727	21.297297
	4.7432972	6.6111869	5.7855032
	52	22	74

In addition to the means, standard deviations, and frequencies for each weight–mileage cell, also reported are the summary statistics by weight, by mileage, and overall. For instance, the last row of the table reveals that the average mileage of domestic cars is 19.83 and that of foreign cars is 24.77—domestic cars yield poorer mileage than foreign cars. But we now see that domestic cars yield better gas mileage within weight class—the reason domestic cars yield poorer gas mileage is because they are, on average, heavier.

◁

▷ Example 3

If we do not specify the statistics to be included in a table, tabulate reports the mean, standard deviation, and frequency. We can specify the statistics that we want to see using the means, standard, and freq options:

```
. tabulate wgtcat foreign, summarize(mpg) means
```

Means of Mileage (mpg)

wgtcat	Car type Domestic	Foreign	Total
2530	28.285714	27.0625	27.434783
3300	21.75	19.6	21.238095
4070	17.26087	14	17.125
4840	14.666667	.	14.666667
Total	19.826923	24.772727	21.297297

When we specify one or more of the means, standard, and freq options, only those statistics are displayed. Thus we could obtain a table containing just the means and standard deviations by typing means standard after the summarize(mpg) option. We can also suppress selected statistics by placing no in front of the option name. Another way of obtaining only the means and standard deviations is to add the nofreq option:

```
. tabulate wgtcat foreign, summarize(mpg) nofreq
```
                Means and Standard Deviations of Mileage (mpg)

wgtcat	Car type Domestic	Foreign	Total
2530	28.285714 3.0937725	27.0625 5.9829619	27.434783 5.2295149
3300	21.75 2.4083189	19.6 3.4351128	21.238095 2.7550819
4070	17.26087 1.8639497	14 0	17.125 1.9406969
4840	14.666667 3.32666	. .	14.666667 3.32666
Total	19.826923 4.7432972	24.772727 6.6111869	21.297297 5.7855032

◁

Also See

[R] **adjust** — Tables of adjusted means and proportions

[R] **table** — Tables of summary statistics

[R] **tabstat** — Display table of summary statistics

[R] **tabulate oneway** — One-way tables of frequencies

[R] **tabulate twoway** — Two-way tables of frequencies

[D] **collapse** — Make dataset of summary statistics

[SVY] **svy: tabulate oneway** — One-way tables for survey data

[SVY] **svy: tabulate twoway** — Two-way tables for survey data

[U] **12.6 Dataset, variable, and value labels**

[U] **25 Dealing with categorical variables**

Title

test — Test linear hypotheses after estimation

Syntax

Basic syntax   (see [R] **anova postestimation** for `test` after anova;

         see [MV] **manova postestimation** for `test` after manova)

`test` *coeflist*   *(Syntax 1)*

`test` *exp=exp* $\left[$ =... $\right]$ *(Syntax 2)*

`test` [*eqno*] $\left[$: *varlist* $\right]$ *(Syntax 3)*

`test` [*eqno=eqno* $\left[$ =... $\right]$] $\left[$: *varlist* $\right]$ *(Syntax 4)*

`testparm` *varlist* $\left[$, <u>e</u>qual <u>eq</u>uation(*eqno*) $\right]$

Full syntax

`test` (*spec*) $\left[$ (*spec*) ... $\right]$ $\left[$, *test_options* $\right]$

test_options	description
Options	
<u>m</u>test $\left[$ (*opt*) $\right]$	test each condition separately
coef	report estimated constrained coefficients
<u>ac</u>cumulate	test hypothesis jointly with previously tested hypotheses
<u>not</u>est	suppress the output
common	test only variables common to all the equations
<u>cons</u>tant	include the constant in coefficients to be tested
<u>nosvy</u>adjust	carry out the Wald test as $W/k \sim F(k, d)$; for use with svy estimation commands
<u>min</u>imum	perform test with the constant, drop terms until the test becomes nonsingular, and test without the constant on the remaining terms; highly technical
† matvlc(*matname*)	save the variance–covariance matrix; programmer's option

† matvlc(*matname*) does not appear in the dialog box.

varlist and *varname* may contain time-series operators; see [U] **11.4.3 Time-series varlists**.

Syntax 1 tests that coefficients are 0.

Syntax 2 tests that linear expressions are equal.

Syntax 3 tests that coefficients in *eqno* are 0.

Syntax 4 tests equality of coefficients between equations.

446

spec is one of

> *coeflist*
> *exp=exp* $\big[$ *=exp* $\big]$
> $[eqno]$ $\big[$: *varlist* $\big]$
> $[eqno_1{=}eqno_2$ $\big[$ =... $\big]\big]$ $\big[$: *varlist* $\big]$

coeflist is

> *varlist*
> $[eqno]varname$ $\big[$ $[eqno]varname$... $\big]$
> $[eqno]$ _b $[varname]$ $\big[$ $[eqno]$ _b $[varname]$... $\big]$

exp is a linear expression containing

> *varname*
> _b $[varname]$
> $[eqno]varname$
> $[eqno]$ _b $[varname]$

eqno is

> # #
> *name*

Distinguish between [ ], which are to be typed, and $\big[\ \big]$, which indicate optional arguments.

Although not shown in the syntax diagram, parentheses around *spec* are required only with multiple specifications. Also, the diagram does not show that test may be called without arguments to redisplay the results from the last test.

Description

test performs Wald tests for simple and composite linear hypotheses about the parameters of the most recently fitted model.

test supports svy estimators, carrying out an adjusted Wald test by default in such cases. test can be used with svy estimation results, see [SVY] **svy postestimation**.

testparm provides a useful alternative to test that permits *varlist* rather than a list of coefficients (which is often nothing more than a list of variables), allowing the use of standard Stata notation, including '-' and '*', which are given the *exp*ression interpretation by test.

test and testparm perform Wald tests. For likelihood-ratio tests, see [R] **lrtest**. For Wald-type tests of nonlinear hypotheses, see [R] **testnl**. To display estimates for one-dimensional linear or nonlinear expressions of coefficients, see [R] **lincom** and [R] **nlcom**.

See [R] **anova postestimation** for test after anova.

See [MV] **manova postestimation** for test after manova.

Options for testparm

equal tests that the variables appearing in *varlist*, which also appear in the previously fitted model, are equal to each other rather than jointly equal to zero.

equation(*eqno*) is relevant only for multiple-equation models, such as mvreg, mlogit, and heckman. It specifies the equation for which the all-zero or all-equal hypothesis is tested. equation(#1) specifies that the test be conducted regarding the first equation #1. equation(price) specifies that the test concern the equation named price.

Options for test

<u>Options</u>

mtest [(*opt*)] specifies that tests be performed for each condition separately. *opt* specifies the method for adjusting *p*-values for multiple testing. Valid values for *opt* are

<u>bon</u>ferroni	Bonferroni's method
<u>ho</u>lm	Holm's method
<u>si</u>dak	Šidák's method
<u>noadj</u>ust	no adjustment is to be made

Specifying mtest without an argument is equivalent to mtest(noadjust).

coef specifies that the constrained coefficients be displayed.

accumulate allows a hypothesis to be tested jointly with the previously tested hypotheses.

notest suppresses the output. This option is useful when you are interested only in the joint test of several hypotheses, specified in a subsequent call of test, accumulate.

common specifies that when you use the [*eqno1=eqno2* [=...]] form of *spec*, the variables common to the equations *eqno1*, *eqno2*, etc., be tested. The default action is to complain if the equations have variables not in common.

constant specifies that _cons be included in the list of coefficients to be tested when using the [*eqno1=eqno2* [=...]] or [*eqno*] forms of *spec*. The default is not to include _cons.

nosvyadjust is for use with svy estimation commands. It specifies that the Wald test be carried out as $W/k \sim F(k, d)$ rather than as $(d - k + 1)W/(kd) \sim F(k, d - k + 1)$, where k = the dimension of the test and d = the total number of sampled PSUs minus the total number of strata.

minimum is a highly technical option. It first performs the test with the constant added. If this test is singular, coefficients are dropped until the test becomes nonsingular. Then the test without the constant is performed with the remaining terms.

The following option is available with test but is not shown in the dialog box:

matvlc(*matname*), a programmer's option, saves the variance–covariance matrix of the linear combinations involved in the suite of tests. For the test of the linear constraints $Lb = c$, *matname* contains LVL', where **V** is the estimated variance–covariance matrix of b.

Remarks

Remarks are presented under the following headings:

Introductory examples
Special syntaxes after multiple-equation estimation
Constrained coefficients
Multiple testing

Introductory examples

test performs F or χ^2 tests of linear restrictions applied to the most recently fitted model (e.g., regress or svy: regress in the linear regression case; logit, stcox, svy: logit ... in the single-equation maximum-likelihood case; and mlogit, mvreg, streg ... in the multiple-equation maximum-likelihood case). test may be used after *any* estimation command, although for maximum likelihood techniques, test produces a Wald test that depends only on the estimate of the covariance matrix—you may prefer to use the more computationally expensive likelihood-ratio test; see [U] **20 Estimation and postestimation commands** and [R] **lrtest**.

There are several variations on the syntax for test. The second syntax,

$$\text{test } exp=exp\big[=\dots\big]$$

is allowed after any form of estimation, although it is not that useful after anova unless you wish to concoct your own test. The anova case is discussed in [R] **anova**, so we will ignore it. Putting aside anova, after fitting a model of *depvar* on x1, x2, and x3, typing test x1+x2=x3 tests the restriction that the coefficients on x1 and x2 sum to the coefficient on x3. The expressions can be arbitrarily complicated; for instance, typing test x1+2*(x2+x3)=x2+3*x3 is the same as typing test x1+x2=x3.

As a convenient shorthand, test also allows you to specify equality for multiple expressions; for example, test x1+x2 = x3+x4 = x5+x6 tests that the three specified pairwise sums of coefficients are equal.

test understands that when you type x1, you are referring to the coefficient on x1. You could also more explicitly type test _b[x1]+_b[x2]=_b[x3]; or you could test _coef[x1]+_coef[x2]=_coef[x3], or test [#1]x1+[#1]x2= [#1]x3, or many other things since there is more than one way to refer to an estimated coefficient; see [U] **13.5 Accessing coefficients and standard errors**. The shorthand involves less typing. On the other hand, you must be more explicit after estimation of multiple-equation models since there may be more than one coefficient associated with an independent variable. You might type, for instance, test [#2]x1+[#2]x2=[#2]x3 to test the constraint in equation 2 or, more readably, test [ford]x1+[ford]x2=[ford]x3, meaning that Stata will test the constraint on the equation corresponding to ford, which might be equation 2. ford would be an equation name after, say, sureg, or, after mlogit, ford would be one of the outcomes. For mlogit, you could also type test [2]x1+[2]x2=[2]x3—note the lack of the #—meaning not equation 2, but the equation corresponding to the numeric outcome 2. You can even test constraints across equations: test [ford]x1+[ford]x2=[buick]x3.

The syntax,

$$\text{test } coeflist$$

is available after all estimation commands except anova and is a convenient way to test that multiple coefficients are zero following estimation. A *coeflist* can simply be a list of variable names,

$$\text{test } varname \big[\, varname \dots \big]$$

and it is most often specified that way. After you have fitted a model of depvar on x1, x2, and x3, typing test x1 x3 tests that the coefficients on x1 and x3 are jointly zero. After multiple-equation estimation, this would test that the coefficients on x1 and x3 are zero in all equations that contain them. You can also be more explicit and type, for instance, test [ford]x1 [ford]x3 to test that the coefficients on x1 and x3 are zero in the equation for ford.

In the multiple-equation case, there are more alternatives. You could also test that the coefficients on x1 and x3 are zero in the equation for ford by typing test [ford]: x1 x3. You could test that all coefficients except the coefficient on the constant are zero in the equation for ford by typing test [ford]. You could test that the coefficients on x1 and x3 in the equation for ford are equal to the corresponding coefficients in the equation corresponding to buick by typing test[ford=buick]: x1 x3. You could test that all the corresponding coefficients except the constant in three equations are equal by typing test [ford=buick=volvo].

testparm is much like the first syntax of test and cannot be used after anova. Its usefulness will be demonstrated below.

The examples below use regress, but what is said applies equally after any single-equation estimation command (such as logistic, etc.). It also applies after multiple-equation estimation commands as long as references to coefficients are qualified with an equation name or number in square brackets placed before them. The convenient syntaxes for dealing with tests of many coefficients in multiple-equation models are demonstrated in *Special syntaxes after multiple-equation estimation* below.

▷ Example 1

We have 1980 census data on the 50 states recording the birth rate in each state (brate), the median age (medage) and its square (medagesq), and the region of the country in which each state is located.

The variable reg1 is 1 if the state is located in the Northeast and zero otherwise, whereas reg2 marks the North Central, reg3 the South, and reg4 the West. We estimate the following regression:

```
. use http://www.stata-press.com/data/r10/census4
(birth rate, median age)

. regress brate medage medagesq reg2-reg4
```

Source	SS	df	MS		
Model	38803.419	5	7760.68381		
Residual	3393.40095	44	77.1227489		
Total	42196.82	49	861.159592		

	Number of obs =	50
	F(5, 44) =	100.63
	Prob > F =	0.0000
	R-squared =	0.9196
	Adj R-squared =	0.9104
	Root MSE =	8.782

brate	Coef.	Std. Err.	t	P>|t|	[95% Conf. Interval]	
medage	-109.0957	13.52452	-8.07	0.000	-136.3526	-81.83886
medagesq	1.635208	.2290536	7.14	0.000	1.173581	2.096835
reg2	15.00284	4.252068	3.53	0.001	6.433365	23.57233
reg3	7.366435	3.953336	1.86	0.069	-.6009898	15.33386
reg4	21.39679	4.650602	4.60	0.000	12.02412	30.76946
_cons	1947.61	199.8405	9.75	0.000	1544.858	2350.362

test can now be used to perform a variety of statistical tests. We can test the hypothesis that the coefficient on reg3 is zero by typing

```
. test reg3=0

 ( 1)  reg3 = 0

       F(  1,    44) =    3.47
            Prob > F =    0.0691
```

The F statistic with 1 numerator and 44 denominator degrees of freedom is 3.47. The significance level of the test is 6.91%—we can reject the hypothesis at the 10% level but not at the 5% level.

This result from `test` is identical to one presented in the output from `regress`, which indicates that the t statistic on the `reg3` coefficient is 1.863 and that its significance level is 0.069. The t statistic presented in the output can be used to test the hypothesis that the corresponding coefficient is zero, although it states the test in slightly different terms. The F distribution with 1 numerator degree of freedom is, however, identical to the t^2 distribution. We note that $1.863^2 \approx 3.47$ and that the significance levels in each test agree, although one extra digit is presented by the `test` command.

◁

❑ Technical Note

After all estimation commands, including those that use the maximum likelihood method, the test that one variable is zero is identical to that reported by the command's output. The tests are performed in the same way—using the estimated covariance matrix—and are known as Wald tests. If the estimation command reports significance levels and confidence intervals using z rather than t statistics, `test` reports results using the χ^2 rather than the F statistic.

❑

▷ Example 2

If that were all `test` could do, it would be useless. We can use `test`, however, to perform other tests. For instance, we can `test` the hypothesis that the coefficient on `reg2` is 21 by typing

```
. test reg2=21
( 1)  reg2 = 21
      F(  1,    44) =    1.99
           Prob > F =    0.1654
```

We find that we cannot reject that hypothesis, or at least we cannot reject it at any significance level below 16.5%.

◁

▷ Example 3

The previous test is useful, but we could almost as easily perform it by hand using the results presented in the regression output if we were well read on our statistics. We could type

```
. display Ftail(1,44,((_coef[reg2]-21)/4.252068)^2)
.16544972
```

So, now let's `test` something a bit more difficult; whether the coefficient on `reg2` is the same as the coefficient on `reg4`:

```
. test reg2=reg4
( 1)  reg2 - reg4 = 0
      F(  1,    44) =    2.84
           Prob > F =    0.0989
```

We find that we cannot reject the equality hypothesis at the 5% level, but we can at the 10% level.

◁

▷ Example 4

When we tested the equality of the reg2 and reg4 coefficients, Stata rearranged our algebra. When Stata displayed its interpretation of the specified test, it indicated that we were testing whether reg2 *minus* reg4 is zero. The rearrangement is innocuous and, in fact, allows Stata to perform much more complicated algebra, for instance,

```
. test 2*(reg2-3*(reg3-reg4))=reg3+reg2+6*(reg4-reg3)
 ( 1)   reg2 - reg3 = 0
        F(  1,     44) =     5.06
             Prob > F =     0.0295
```

Although we requested what appeared to be a lengthy hypothesis, once Stata simplified the algebra, it realized that all we wanted to do was test whether the coefficient on reg2 is the same as the coefficient on reg3.

◁

❑ Technical Note

Stata's ability to simplify and test complex hypotheses is limited to *linear* hypotheses. If you attempt to test a nonlinear hypothesis, you will be told that it is not possible:

```
. test reg2/reg3=reg2+reg3
not possible with test
r(131);
```

To test a nonlinear hypothesis, see [R] **testnl**.

❑

▷ Example 5

The real power of test is demonstrated when we test *joint* hypotheses. Perhaps we wish to test whether the region variables, taken as a whole, are significant by testing whether the coefficients on reg2, reg3, and reg4 are simultaneously zero. test allows us to specify multiple conditions to be tested, each embedded within parentheses.

```
. test (reg2=0) (reg3=0) (reg4=0)
 ( 1)   reg2 = 0
 ( 2)   reg3 = 0
 ( 3)   reg4 = 0
        F(  3,     44) =     8.85
             Prob > F =     0.0001
```

test displays the set of conditions and reports an F statistic of 8.85. test also reports the degrees of freedom of the test to be 3, the "dimension" of the hypothesis, and the residual degrees of freedom, 44. The significance level of the test is close to 0, so we can strongly reject the hypothesis of no difference between the regions.

An alternative method to specify simultaneous hypotheses uses the convenient shorthand of conditions with multiple equality operators.

```
. test reg2=reg3=reg4=0
 ( 1)   reg2 - reg3 = 0
 ( 2)   reg2 - reg4 = 0
 ( 3)   reg2 = 0
        F(  3,     44) =     8.85
             Prob > F =     0.0001
```

◁

❑ Technical Note

Another method to test simultaneous hypotheses is to specify a `test` for each constraint and `accumulate` it with the previous constraints:

```
. test reg2=0

 ( 1)   reg2 = 0

        F(  1,    44) =    12.45
             Prob > F =     0.0010
. test reg3=0, accumulate

 ( 1)   reg2 = 0
 ( 2)   reg3 = 0

        F(  2,    44) =     6.42
             Prob > F =     0.0036
. test reg4=0, accumulate

 ( 1)   reg2 = 0
 ( 2)   reg3 = 0
 ( 3)   reg4 = 0

        F(  3,    44) =     8.85
             Prob > F =     0.0001
```

We tested the hypothesis that the coefficient on `reg2` was zero by typing `test reg2=0`. We then tested whether the coefficient on `reg3` was also zero by typing `test reg3=0, accumulate`. The `accumulate` option told Stata that this was not the start of a new test but a continuation of a previous one. Stata responded by showing us the two equations and reporting an F statistic of 6.42. The significance level associated with those two coefficients being zero is 0.36%.

When we added the last constraint `test reg4=0, accumulate`, we discovered that the three region variables are significant. If all we wanted was the overall significance and we did not want to bother seeing the interim results, we could have used the `notest` option:

```
. test reg2=0, notest

 ( 1)   reg2 = 0
. test reg3=0, accumulate notest

 ( 1)   reg2 = 0
 ( 2)   reg3 = 0
. test reg4=0, accumulate

 ( 1)   reg2 = 0
 ( 2)   reg3 = 0
 ( 3)   reg4 = 0

        F(  3,    44) =     8.85
             Prob > F =     0.0001
```

❑

▷ Example 6

Since tests that coefficients are zero are so common in applied statistics, the `test` command has a more convenient syntax to accommodate this case.

```
. test reg2 reg3 reg4

 ( 1)   reg2 = 0
 ( 2)   reg3 = 0
 ( 3)   reg4 = 0

        F(  3,    44) =     8.85
             Prob > F =     0.0001
```

◁

▷ Example 7

We will now show how to use `testparm`. In its first syntax, `test` accepts a list of variable names but not a *varlist*.

```
. test reg2-reg4
- not found
r(111);
```

In a *varlist*, `reg2-reg3` means variables `reg2` and `reg3` and all the variables between, yet we received an error. `test` is confused because the `-` has two meanings: it means subtraction in an expression and "through" in a *varlist*. Similarly, '`*`' means "any set of characters" in a *varlist* and multiplication in an expression. `testparm` avoids this confusion—it allows only a *varlist*.

```
. testparm reg2-reg4

 ( 1)   reg2 = 0
 ( 2)   reg3 = 0
 ( 3)   reg4 = 0

       F( 3,   44) =    8.85
            Prob > F =    0.0001
```

`testparm` has another advantage. We have five variables in our dataset that start with the characters `reg`: `region`, `reg1`, `reg2`, `reg3`, and `reg4`. `reg*` thus means those five variables:

```
. describe reg*
```

variable name	storage type	display format	value label	variable label
region	int	%8.0g	region	Census Region
reg1	byte	%9.0g		region==NE
reg2	byte	%9.0g		region==N Cntrl
reg3	byte	%9.0g		region==South
reg4	byte	%9.0g		region==West

We cannot type `test reg*` because, in an expression, '`*`' means multiplication, but here is what would happen if we attempted to test all the variables that begin with `reg`:

```
. test region reg1 reg2 reg3 reg4
region not found
r(111);
```

The variable `region` was not included in our model, so it was not found. However, with `testparm`,

```
. testparm reg*

 ( 1)   reg2 = 0
 ( 2)   reg3 = 0
 ( 3)   reg4 = 0

       F( 3,   44) =    8.85
            Prob > F =    0.0001
```

That is, `testparm` took `reg*` to mean all variables that start with `reg` that were in our model.

◁

❑ Technical Note

Actually, `reg*` means what it always does—all variables in our dataset that begin with `reg`—in this case, `region reg1 reg2 reg3 reg4`. `testparm` just ignores any variables you specify that are not in the model.

❑

▷ Example 8

We just used `test` (`testparm`, actually, but it does not matter) to test the hypothesis that `reg2`, `reg3`, and `reg4` are jointly zero. We can review the results of our last test by typing `test` without arguments:

```
. test
 ( 1)  reg2 = 0
 ( 2)  reg3 = 0
 ( 3)  reg4 = 0
      F(  3,    44) =    8.85
           Prob > F =    0.0001
```

◁

❑ Technical Note

`test` does not care how we build joint hypotheses; we may freely mix different forms of syntax. (We can even start with `testparm`, but we cannot use it thereafter because it does not have an `accumulate` option.)

Say that we type `test reg2 reg3 reg4` to test that the coefficients on our region dummies are jointly zero. We could then add a fourth constraint, say, that `medage` = 100, by typing `test medage=100, accumulate`. Or, if we had introduced the `medage` constraint first (our first `test` command had been `test medage=100`), we could then add the region dummy test by typing `test reg2 reg3 reg4, accumulate` or `test (reg2=0) (reg3=0) (reg4=0), accumulate`.

Remember that all previous tests are cleared when we do not specify the `accumulate` option. No matter what tests we performed in the past, if we type `test medage medagesq`, omitting the `accumulate` option, we would test that `medage` and `medagesq` are jointly zero.

❑

▷ Example 9

Let's `test` the hypothesis that all the included regions have the *same* coefficient—that the Northeast is significantly different from the rest of the nation:

```
. test reg2=reg3=reg4
 ( 1)  reg2 - reg3 = 0
 ( 2)  reg2 - reg4 = 0
      F(  2,    44) =    8.23
           Prob > F =    0.0009
```

We find that they are not all the same. The syntax `reg2=reg3=reg4` with multiple = operators is just a convenient shorthand for typing that the first expression equals the second expression and that the first expression equals the third expression,

```
. test (reg2=reg3) (reg2=reg4)
```

We performed the test for equality of the three regions by imposing two constraints: region 2 has the same coefficient as region 3, and region 2 has the same coefficient as region 4. Alternatively, we could have tested that the coefficients on regions 2 and 3 are the same and that the coefficients on regions 3 and 4 are the same. We would obtain the same results in either case.

To test for equality of the three regions, we might, likely by mistake, type equality constraints for *all* pairs of regions:

```
. test (reg2=reg3) (reg2=reg4) (reg3=reg4)
( 1)   reg2 - reg3 = 0
( 2)   reg2 - reg4 = 0
( 3)   reg3 - reg4 = 0
       Constraint 3 dropped
       F(  2,    44) =     8.23
            Prob > F =     0.0009
```

Equality of regions 2 and 3 and of regions 2 and 4, however, implies equality of regions 3 and 4. test recognized that the last constraint is implied by the other constraints and hence dropped it.

◁

❑ Technical Note

Generally, Stata uses = for assignment, as in gen *newvar* = *exp*, and == as the operator for testing equality in expressions. For your convenience, test allows both = and == to be used.

❑

▷ Example 10

The test for the equality of the regions is also possible with the testparm command. When we include the equal option, testparm tests that the coefficients of all the variables specified are equal:

```
. testparm reg*, equal
( 1) - reg2 + reg3 = 0
( 2) - reg2 + reg4 = 0
      F(  2,    44) =     8.23
           Prob > F =     0.0009
```

We can also obtain the equality test by accumulating single equality tests.

```
. test reg2=reg4, notest
( 1)   reg2 - reg4 = 0
. test reg3=reg4, accum
( 1)   reg2 - reg4 = 0
( 2)   reg3 - reg4 = 0
      F(  2,    44) =     8.23
           Prob > F =     0.0009
```

◁

❑ Technical Note

If we specify a set of inconsistent constraints, test will tell us by dropping the constraint or constraints that led to the inconsistency. For instance, let's test that the coefficients on region 2 and region 4 are the same, add the test that the coefficient on region 2 is 20, and finally add the test that the coefficient on region 4 is 21:

```
. test (reg2==reg4) (reg2=20) (reg4=21)
( 1)   reg2 - reg4 = 0
( 2)   reg2 = 20
( 3)   reg4 = 21
       Constraint 2 dropped
       F(  2,    44) =     1.82
            Prob > F =     0.1737
```

test informed us that it was dropping constraint 2. All three equations cannot be simultaneously true, so test drops whatever it takes to get back to something that makes sense.

❑

Special syntaxes after multiple-equation estimation

Everything said above about tests after single-equation estimation applies to tests after multiple-equation estimation, as long as you remember to specify the equation name. To demonstrate, let's estimate a seemingly unrelated regression by using sureg; see [R] **sureg**.

```
. use http://www.stata-press.com/data/r10/auto
(1978 Automobile Data)
. sureg (price foreign mpg displ) (weight foreign length)
Seemingly unrelated regression
```

Equation	Obs	Parms	RMSE	"R-sq"	chi2	P
price	74	3	2165.321	0.4537	49.64	0.0000
weight	74	2	245.2916	0.8990	661.84	0.0000

	Coef.	Std. Err.	z	P>|z|	[95% Conf. Interval]	
price						
foreign	3058.25	685.7357	4.46	0.000	1714.233	4402.267
mpg	-104.9591	58.47209	-1.80	0.073	-219.5623	9.644042
displacement	18.18098	4.286372	4.24	0.000	9.779842	26.58211
_cons	3904.336	1966.521	1.99	0.047	50.0263	7758.645
weight						
foreign	-147.3481	75.44314	-1.95	0.051	-295.2139	.517755
length	30.94905	1.539895	20.10	0.000	27.93091	33.96718
_cons	-2753.064	303.9336	-9.06	0.000	-3348.763	-2157.365

To test the significance of foreign in the price equation, we could type

```
. test [price]foreign
 ( 1)  [price]foreign = 0
           chi2(  1) =    19.89
         Prob > chi2 =     0.0000
```

which is the same result reported by sureg: $4.460^2 \approx 19.89$. To test foreign in both equations, we could type

```
. test [price]foreign [weight]foreign
 ( 1)  [price]foreign = 0
 ( 2)  [weight]foreign = 0
           chi2(  2) =    31.61
         Prob > chi2 =     0.0000
```

or

```
. test foreign
 ( 1)  [price]foreign = 0
 ( 2)  [weight]foreign = 0
           chi2(  2) =    31.61
         Prob > chi2 =     0.0000
```

This last syntax—typing the variable name by itself—tests the coefficients in all equations in which they appear. The variable `length` appears in only the `weight` equation, so typing

```
. test length
( 1)   [weight]length = 0
            chi2(  1) =   403.94
          Prob > chi2 =    0.0000
```

yields the same result as typing `test [weight]length`. We may also specify a linear expression rather than a list of coefficients:

```
. test mpg=displ
( 1)   [price]mpg - [price]displ = 0
            chi2(  1) =     4.85
          Prob > chi2 =    0.0277
```

or

```
. test [price]mpg = [price]displ
( 1)   [price]mpg - [price]displ = 0
            chi2(  1) =     4.85
          Prob > chi2 =    0.0277
```

A variation on this syntax can be used to test cross-equation constraints:

```
. test [price]foreign = [weight]foreign
( 1)   [price]foreign - [weight]foreign = 0
            chi2(  1) =    23.07
          Prob > chi2 =    0.0000
```

Typing an equation name in square brackets by itself tests all the coefficients except the intercept in that equation:

```
. test [price]
( 1)   [price]foreign = 0
( 2)   [price]mpg = 0
( 3)   [price]displacement = 0
            chi2(  3) =    49.64
          Prob > chi2 =    0.0000
```

Typing an equation name in square brackets, a colon, and a list of variable names tests those variables in the specified equation:

```
. test [price]: foreign displ
( 1)   [price]foreign = 0
( 2)   [price]displacement = 0
            chi2(  2) =    25.19
          Prob > chi2 =    0.0000
```

`test [`*eqname*$_1$`=`*eqname*$_2$`]` tests that all the coefficients in the two equations are equal. We cannot use that syntax here because there are different variables in the model:

```
. test [price=weight]
variables differ between equations
(to test equality of coefficients in common, specify option -common-)
r(111);
```

The option `common` specifies a test of the equality coefficients common to the equations `price` and `weight`,

```
. test [price=weight], common
 ( 1)  [price]foreign - [weight]foreign = 0
           chi2(  1) =    23.07
         Prob > chi2 =    0.0000
```

By default, `test` does not include the constant, the coefficient of the constant variable `_cons`, in the test. The option `cons` specifies that the constant be included.

```
. test [price=weight], common cons
 ( 1)  [price]foreign - [weight]foreign = 0
 ( 2)  [price]_cons - [weight]_cons = 0
           chi2(  2) =    51.23
         Prob > chi2 =  0.0000
```

We can also use a modification of this syntax with the model if we also type a colon and the names of the variables we want to test:

```
. test [price=weight]: foreign
 ( 1)  [price]foreign - [weight]foreign = 0
           chi2(  1) =    23.07
         Prob > chi2 =    0.0000
```

We have only one variable in common between the two equations, but if there had been more, we could have listed them.

Finally, a simultaneous test of multiple constraints may be specified just as after single-equation estimation.

```
. test ([price]: foreign) ([weight]: foreign)
 ( 1)  [price]foreign = 0
 ( 2)  [weight]foreign = 0
           chi2(  2) =    31.61
         Prob > chi2 =  0.0000
```

`test` can also test for equality of coefficients across more than two equations. For instance, `test [eq1=eq2=eq3]` specifies a test that the coefficients in the three equations eq1, eq2, and eq3 are equal. This requires that the same variables be included in the three equations. If some variables are entered only in some of the equations, you can type `test [eq1=eq2=eq3], common` to test that the coefficients of the variables common to all three equations are equal. Alternatively, you can explicitly list the variables for which equality of coefficients across the equations is to be tested. For instance, `test [eq1=eq2=eq3]: time money` tests that the coefficients of the variables `time` and `money` do not differ between the equations.

❑ Technical Note

`test [eq1=eq2=eq3], common` tests the equality of the coefficients common to all equations, but it does *not* test the equality of all common coefficients. Consider the case where

eq1	contains the variables var1 var2 var3
eq2	contains the variables var1 var2 var4
eq3	contains the variables var1 var3 var4

Obviously, only `var1` is common to all three equations. Thus `test [eq1=eq2=eq3], common` tests that the coefficients of `var1` do not vary across the equations, so it is equivalent to `test [eq1=eq2=eq3]: var1`. To perform a test of the coefficients of variables common to two equations, you could explicitly list the constraints to be tested,

```
. test ([eq1=eq2=eq3]:var1) ([eq1=eq2]:var2) ([eq1=eq3]:var3) ([eq2=eq3]:var4)
```

or use `test` with the `accumulate` option, and maybe also with the `notest` option, to form the appropriate joint hypothesis:

```
. test [eq1=eq2], common notest
. test [eq1=eq3], common accumulate notest
. test [eq2=eq3], common accumulate
```

❑

Constrained coefficients

If the test indicates that the data do not allow you to conclude that the constraints are not satisfied, you may want to inspect the constrained coefficients. The option `coef` specified that the constrained results, estimated by GLS, are shown.

```
. test [price=weight], common coef
( 1)  [price]foreign - [weight]foreign = 0
          chi2(  1) =    23.07
        Prob > chi2 =    0.0000
```

Constrained coefficients

	Coef.	Std. Err.	z	P>|z|	[95% Conf. Interval]	
price						
foreign	-216.4015	74.06083	-2.92	0.003	-361.558	-71.2449
mpg	-121.5717	58.36972	-2.08	0.037	-235.9742	-7.169116
displacement	7.632566	3.681114	2.07	0.038	.4177148	14.84742
_cons	7312.856	1834.034	3.99	0.000	3718.215	10907.5
weight						
foreign	-216.4015	74.06083	-2.92	0.003	-361.558	-71.2449
length	30.34875	1.534815	19.77	0.000	27.34057	33.35693
_cons	-2619.719	302.6632	-8.66	0.000	-3212.928	-2026.51

The constrained coefficient of `foreign` is −216.40 with standard error 74.06 in equations `price` and `weight`. The other coefficients and their standard errors are affected by imposing the equality constraint of the two coefficients of `foreign` because the unconstrained estimates of these two coefficients were correlated with the estimates of the other coefficients.

❑ Technical Note

The two-step constrained coefficients b_c displayed by `test, coef` are asymptotically equivalent to the one-stage constrained estimates that are computed by specifying the constraints during estimation using the `constraint()` option of estimation commands (Gourieroux and Monfort 1998, chap. 10). Generally, one-step constrained estimates have better small-sample properties. For inspection and interpretation, however, two-step constrained estimates are a convenient alternative. Moreover, some estimation commands (e.g., `stcox`, `logit`, many `xt`-estimators, ...) do not have a `constraint()` option.

❑

Multiple testing

When performing the test of a joint hypothesis, you might want to inspect the underlying 1-degree-of-freedom hypotheses. Which constraint "is to blame"? `test` displays the univariate as well as the simultaneous test if the option `mtest` is specified. For example,

```
. test [price=weight], common cons mtest
 ( 1)  [price]foreign - [weight]foreign = 0
 ( 2)  [price]_cons - [weight]_cons = 0
```

	chi2	df	p
(1)	23.07	1	0.0000 #
(2)	11.17	1	0.0008 #
all	51.23	2	0.0000

```
                    # unadjusted p-values
```

Both coefficients seem to contribute to the highly significant result. The 1-degree-of-freedom test shown here is identical to those if `test` had been invoked to test just this simple hypotheses. There is, of course, a real risk in inspecting these simple hypotheses. Especially in high-dimensional hypotheses, you may easily find one hypothesis that happens to be significant. Multiple testing procedures are designed to provide some safeguard against this risk. p-values of the univariate hypotheses are modified so that the probability of falsely rejecting one of the null hypotheses is bounded. `test` provides the methods due to Bonferroni, Šidák, and Holm.

```
. test [price=weight], common cons mtest(b)
 ( 1)  [price]foreign - [weight]foreign = 0
 ( 2)  [price]_cons - [weight]_cons = 0
```

	chi2	df	p
(1)	23.07	1	0.0000 #
(2)	11.17	1	0.0017 #
all	51.23	2	0.0000

```
                # Bonferroni adjusted p-values
```

Saved Results

`test` and `testparm` save the following in `r()`:

Scalars

r(p)	two-sided p-value	r(chi2)	χ^2
r(F)	F statistic	r(ss)	sum of squares (test)
r(df)	test constraints degrees of freedom	r(rss)	residual sum of squares
r(df_r)	residual degrees of freedom	r(drop)	1 if constraints were dropped, 0
r(dropped_i)	index of ith constraint dropped		otherwise

Macros

r(mtmethod)	method of adjustment for multiple testing

Matrices

r(mtest)	multiple test results

r(ss) and r(rss) are defined only when `test` is used for testing effects after `anova`.

Methods and Formulas

`test` and `testparm` are implemented as ado-files.

`test` and `testparm` perform Wald tests. Let the estimated coefficient vector be $\mathbf{b}$ and the estimated variance–covariance matrix be $\mathbf{V}$. Let $\mathbf{Rb} = \mathbf{r}$ denote the set of q linear hypotheses to be tested jointly.

The Wald test statistic is (Judge et al. 1985, 20–28)

$$W = (\mathbf{Rb} - \mathbf{r})'(\mathbf{RVR}')^{-1}(\mathbf{Rb} - \mathbf{r})$$

If the estimation command reports its significance levels using Z statistics, a chi-squared distribution with q degrees of freedom,

$$W \sim \chi_q^2$$

is used for computation of the significance level of the hypothesis test.

If the estimation command reports its significance levels using t statistics with d degrees of freedom, an F statistic,

$$F = \frac{1}{q}W$$

is computed, and an F distribution with q numerator degrees of freedom and d denominator degrees of freedom computes the significance level of the hypothesis test.

The two-step constrained estimates b_c displayed by `test` with the option `coef` are the GLS estimates of the unconstrained estimates b subject to the specified constraints $Rb = c$ (Gourieroux and Monfort 1998, chap. 10),

$$\mathbf{b_c} = \mathbf{b} - \mathbf{R}'(\mathbf{RVR}')^{-1}\mathbf{R}(\mathbf{Rb} - \mathbf{r})$$

with variance–covariance matrix

$$\mathbf{V_c} = \mathbf{V} - \mathbf{VR}'(\mathbf{RVR}')^{-1}\mathbf{RV}$$

If `test` displays a Wald test for joint (simultaneous) hypotheses, it can also display all 1-degree-of-freedom tests, with p-values adjusted for multiple testing. Let $p_1, p_2, \ldots, p_k$ be the unadjusted p-values of these 1-degree-of-freedom tests. The Bonferroni-adjusted p-values are defined as $p_i^b = \min(1, kp_i)$. The Šidák-adjusted p-values are $p_i^s = 1 - (1 - p_i)^k$. Holm's method for adjusting p-values is defined as $p_i^h = \min(1, k_i p_i)$, where k_i is the number of p-values at least as large as p_i. Note that $p_i^h < p_i^b$, reflecting that Holm's method is strictly less conservative than the widely used Bonferroni method.

If `test` is used after a `svy` command, it carries out an adjusted Wald test—this adjustment should not be confused with the adjustment for multiple testing. Both adjustments may actually be combined. Specifically, the survey adjustment uses an approximate F statistic $(d-k+1)W/(kd)$, where W is the Wald test statistic, k is the dimension of the hypothesis test, and $d =$ the total number of sampled PSUs minus the total number of strata. Under the null hypothesis, $(d-k+1)F/(kd) \sim F(k, d-k+1)$, where $F(k, d-k+1)$ is an F distribution with k numerator degrees of freedom and $d-k+1$ denominator degrees of freedom. If `svynoadjust` is specified, the p-value is computed using $W/k \sim F(k, d)$.

See Korn and Graubard (1990) for a detailed description of the Bonferroni adjustment technique and for a discussion of the relative merits of it and of the adjusted and unadjusted Wald tests.

Acknowledgment

The svy adjustment code was adopted from another command developed in collaboration with John L. Eltinge, Bureau of Labor Statistics.

References

Beale, E. M. L. 1960. Confidence regions in nonlinear estimation. *Journal of the Royal Statistical Society, Series B* 22: 41–88.

Eltinge, J. L., and W. M. Sribney. 1996. svy5: Estimates of linear combinations and hypothesis tests for survey data. *Stata Technical Bulletin* 31: 31–42. Reprinted in *Stata Technical Bulletin Reprints*, vol. 6, pp. 246–259.

Gourieroux, C., and A. Monfort. 1995/1998. *Statistics and Econometric Models*. 2 vols. Cambridge: Cambridge University Press.

Holm, C. 1979. A simple sequentially rejective multiple test procedure. *Scandinavian Journal of Statistics* 6: 65–70.

Judge, G. G., W. E. Griffiths, R. C. Hill, H. Lütkepohl, and T.-C. Lee. 1985. *The Theory and Practice of Econometrics*. 2nd ed. New York: Wiley.

Korn, E. L., and B. I. Graubard. 1990. Simultaneous testing of regression coefficients with complex survey data: Use of Bonferroni t statistics. *American Statistician* 44: 270–276.

Weesie, J. 1999. sg100: Two-stage linear constrained estimation. *Stata Technical Bulletin* 47: 24–30. Reprinted in *Stata Technical Bulletin Reprints*, vol. 8, pp. 217–225.

Also See

Title

testnl — Test nonlinear hypotheses after estimation

Syntax

testnl *exp=exp*$\left[$*=exp...*$\right]$ $\left[$, *options*$\right]$

testnl (*exp=exp*$\left[$*=exp...*$\right]$) $\left[$(*exp=exp*$\left[$*=exp...*$\right]$) ...$\right]$ $\left[$, *options*$\right]$

options	description
<u>m</u>test$\left[$(*opt*)$\right]$	test each condition separately
<u>nosvya</u>djust	carry out the Wald test as $W/k \sim F(k, d)$; for use with svy estimation commands
<u>iter</u>ate(#)	use maximum # of iterations to find the optimal step size

The second syntax means that if more than one constraint is specified, each must be surrounded by parentheses.

Description

testnl tests (linear or nonlinear) hypotheses about the estimated parameters from the most recently fitted model.

testnl produces Wald-type tests of smooth nonlinear (or linear) hypotheses about the estimated parameters from the most recently fitted model. The *p*-values are based on the "delta method", an approximation appropriate in large samples.

testnl can be used with svy estimation results; see [SVY] **svy postestimation**.

The format (*exp₁=exp₂=exp₃* ...) for a simultaneous-equality hypothesis is just a convenient shorthand for a list ($exp_1=exp_2$) ($exp_1=exp_3$), etc.

testnl may also be used to test linear hypotheses. test is faster if you want to test only linear hypotheses. testnl is the only option for testing linear and nonlinear hypotheses simultaneously.

Options

mtest$\left[$(*opt*)$\right]$ specifies that tests be performed for each condition separately. *opt* specifies the method for adjusting *p*-values for multiple testing. Valid values for *opt* are

<u>b</u>onferroni	Bonferroni's method
<u>h</u>olm	Holm's method
<u>s</u>idak	Šidák's method
<u>noad</u>just	no adjustment is to be made

Specifying mtest without an argument is equivalent to specifying mtest(noadjust).

nosvyadjust is for use with svy estimation commands. It specifies that the Wald test be carried out as $W/k \sim F(k, d)$ rather than as $(d - k + 1)W/(kd) \sim F(k, d - k + 1)$, where $k =$ the dimension of the test and $d =$ the total number of sampled PSUs minus the total number of strata.

iterate(#) specifies the maximum number of iterations used to find the optimal step size in the calculation of numerical derivatives of the test expressions. By default, the maximum number of iterations is 100, but convergence is usually achieved after only a few iterations. You should rarely have to use this option.

Remarks

Remarks are presented under the following headings:

> Introduction
> Using testnl to perform linear tests
> Specifying constraints
> Dropped constraints
> Output
> Multiple constraints
> Manipulability

Introduction

▷ Example 1

We have just estimated the parameters of an earnings model on cross-sectional time-series data using one of Stata's more sophisticated estimators:

```
. use http://www.stata-press.com/data/r10/earnings
(NLS Women 14-24 in 1968)
. xtgee ln_w grade age age2, corr(exchangeable) nolog
GEE population-averaged model          Number of obs      =       1326
Group variable:                idcode  Number of groups   =        269
Link:                        identity  Obs per group: min =          1
Family:                      Gaussian                 avg =        4.9
Correlation:             exchangeable                 max =          9
                                       Wald chi2(3)       =     327.33
Scale parameter:               .0976738  Prob > chi2      =     0.0000
```

ln_wage	Coef.	Std. Err.	z	P>|z|	[95% Conf. Interval]	
grade	.0749686	.0066111	11.34	0.000	.062011	.0879261
age	.1080806	.0235861	4.58	0.000	.0618526	.1543086
age2	-.0016253	.0004739	-3.43	0.001	-.0025541	-.0006966
_cons	-.8788933	.2830899	-3.10	0.002	-1.433739	-.3240473

An implication of this model is that peak earnings occur at age -_b[age]/(2*_b[age2]), which here is equal to 33.2. Say that we have a theory that peak earnings should occur at age 16 + 1/_b[grade].

```
. testnl -_b[age]/(2*_b[age2]) = 16 + 1/_b[grade]
  (1)  -_b[age]/(2*_b[age2]) = 16 + 1/_b[grade]
              chi2(1) =        1.71
          Prob > chi2 =      0.1914
```

These data do not reject our theory.   ◁

Using testnl to perform linear tests

testnl may be used to test linear constraints, but test is faster; see [R] **test**. You could type

```
. testnl _b[x4] = _b[x1]
```

but it would take less computer time if you typed

```
. test _b[x4] = _b[x1]
```

Specifying constraints

The constraints to be tested can be formulated in many different ways. You could type

```
. testnl _b[mpg]*_b[weight] = 1
```

or

```
. testnl _b[mpg] = 1/_b[weight]
```

or you could express the constraint any other way you wished. (To say that testnl allows constraints to be specified in different ways does not mean that the test itself does not depend on the formulation. This point is briefly discussed later.) In formulating the constraints, you must, however, exercise one caution: users of test often refer to the coefficient on a variable by specifying the variable name. For example,

```
. test mpg = 0
```

More formally, they should type

```
. test _b[mpg] = 0
```

but test allows the _b[] surrounding the variable name to be omitted. testnl does not allow this shorthand. Typing

```
. testnl mpg=0
```

specifies the constraint that the value of variable mpg in the first observation is zero. If you make this mistake, sometimes testnl will catch it:

```
. testnl mpg=0
equation (1) contains reference to X rather than _b[X]
r(198);
```

In other cases, testnl may not catch the mistake; then the constraint will be dropped because it does not make sense:

```
. testnl mpg=0
  Constraint (1) dropped
```

(There are reasons other than this for constraints being dropped.) The worst case, however, is

```
. testnl _b[weight]*mpg = 1
```

when what you mean is not that _b[weight] equals the reciprocal of the value of mpg in the first observation, but rather that

```
. testnl _b[weight]*_b[mpg] = 1
```

Sometimes this mistake will be caught by the "contains reference to X rather than _b[X]" error, and sometimes it will not. Be careful.

`testnl`, like `test`, can be used after any Stata estimation command, including the survey estimators. When you use it after a multiple-equation command, such as `mlogit` or `heckman`, you refer to coefficients by using Stata's standard syntax: [*eqname*]_b[*varname*].

Stata's single-equation estimation output looks like this:

	Coef ...		
weight	12.27 ...	<-	coefficient is _b[weight]
mpg	3.21 ...		

Stata's multiple-equation output looks like this:

	Coef ...		
cat1	...		
weight	12.27 ...	<-	coefficient is [cat1]_b[weight]
mpg	3.21 ...		
8	...		
weight	5.83 ...	<-	coefficient is [8]_b[weight]
mpg	7.43 ...		

Dropped constraints

`testnl` automatically drops constraints when

1. They are nonbinding, e.g., _b[mpg]=_b[mpg]. More subtle cases include

    ```
    _b[mpg]*_b[weight] = 4
    _b[weight] = 2
    _b[mpg] = 2
    ```

    In this example, the third constraint is nonbinding since it is implied by the first two.

2. They are contradictory, e.g., _b[mpg]=2 and _b[mpg]=3. More subtle cases include

    ```
    _b[mpg]*_b[weight] = 4
    _b[weight] = 2
    _b[mpg] = 3
    ```

    The third constraint contradicts the first two.

Output

`testnl` reports the constraints being tested followed by a F or χ^2 test:

```
. use http://www.stata-press.com/data/r10/auto2
(1978 Automobile Data)

. regress price mpg weight weightsq foreign
(output omitted )
```

```
. testnl (39*_b[mpg]^2 = _b[foreign]) (_b[mpg]/_b[weight] = 4)

 (1)  39*_b[mpg]^2 = _b[foreign]
 (2)  _b[mpg]/_b[weight] = 4

              F(2, 69) =         0.08
              Prob > F =       0.9195

. logit foreign price weight mpg
 (output omitted )

. testnl (45*_b[mpg]^2 = _b[price]) (_b[mpg]/_b[weight] = 4)

 (1)  45*_b[mpg]^2 = _b[price]
 (2)  _b[mpg]/_b[weight] = 4

            chi2(2) =         2.45
        Prob > chi2 =       0.2945
```

Multiple constraints

▷ Example 2

We illustrate the simultaneous test of a series of constraints using simulated data on labor-market promotion in a given year. We fit a probit model with separate effects for education, experience, and experience-squared for men and women.

```
. use http://www.stata-press.com/data/r10/promotion
. generate yedu_m  = yedu  * (male==1)
. generate yexp_m  = yexp  * (male==1)
. generate yexp2_m = yexp2 * (male==1)
. generate yedu_f  = yedu  * (male==0)
. generate yexp_f  = yexp  * (male==0)
. generate yexp2_f = yexp2 * (male==0)
. probit promo male yedu_m yexp_m yexp2_m  yedu_f yexp_f yexp2_f, nolog
```

Probit regression				Number of obs	=	775
				LR chi2(7)	=	424.42
				Prob > chi2	=	0.0000
Log likelihood = -245.42768				Pseudo R2	=	0.4637

promo	Coef.	Std. Err.	z	P>|z|	[95% Conf. Interval]	
male	.6489974	.203739	3.19	0.001	.2496763	1.048318
yedu_m	1.390517	.1527288	9.10	0.000	1.091174	1.68986
yexp_m	1.422539	.1544255	9.21	0.000	1.11987	1.725207
yexp2_m	-.3749457	.1160113	-3.23	0.001	-.6023236	-.1475677
yedu_f	.9730237	.1056136	9.21	0.000	.7660248	1.180023
yexp_f	.4559544	.0901169	5.06	0.000	.2793285	.6325803
yexp2_f	-.1027149	.0573059	-1.79	0.073	-.2150325	.0096026
_cons	.9872018	.1148215	8.60	0.000	.7621559	1.212248

```
Note: 1 failure and 2 successes completely determined.
```

The effects of human capital seem to differ between men and women. A formal test confirms this.

```
. test (_b[yedu_m]=_b[yedu_f]) (_b[yexp_m]=_b[yexp_f]) (_b[yexp2_m]=_b[yexp2_f])

 ( 1)  yedu_m - yedu_f = 0
 ( 2)  yexp_m - yexp_f = 0
 ( 3)  yexp2_m - yexp2_f = 0

        chi2(  3) =    35.43
      Prob > chi2 =     0.0000
```

How do we interpret this gender difference? It has repeatedly been stressed (see, for example, Long 1997, 47–50; Allison 1999) that comparison of groups in binary-response models, and similarly in other latent-variable models, is hampered by an identification problem: with β the regression coefficients for the latent variable and σ the standard deviation of the latent residual, only the β/σ are identified. In fact, in terms of the latent regression, the probit coefficients should be interpreted as β/σ, not as the β. If we cannot claim convincingly that the residual standard deviation σ does not vary between the sexes, equality of the regression coefficients β implies that the coefficients of the probit model for men and women are *proportional* but not necessarily equal. This is a nonlinear hypothesis in terms of the probit coefficients, not a linear one.

```
. testnl _b[yedu_m]/_b[yedu_f] = _b[yexp_m]/_b[yexp_f] = _b[yexp2_m]/_b[yexp2_f]

 (1)  _b[yedu_m]/_b[yedu_f] = _b[yexp_m]/_b[yexp_f]
 (2)  _b[yedu_m]/_b[yedu_f] = _b[yexp2_m]/_b[yexp2_f]

              chi2(2) =        9.21
          Prob > chi2 =      0.0100
```

We conclude that we find fairly strong evidence against the proportionality of the coefficients, and hence we have to conclude that success in the labor market is produced in different ways by men and women. (But remember, these were simulated data.)

◁

▷ Example 3

The syntax for specifying the equality of multiple expressions is just a convenient shorthand for specifying a series of constraints, namely, that the first expression equals the second expression, the first expression also equals the third expression, etc. The Wald test performed and the output of `testnl` are the same whether we use the shorthand or we specify the series of constraints. The lengthy specification as a series of constraints can be simplified using the continuation symbols `///`.

```
. testnl (_b[yedu_m]/_b[yedu_f] = _b[yexp_m]/_b[yexp_f]) ///
         (_b[yedu_m]/_b[yedu_f] = _b[yexp2_m]/_b[yexp2_f])

 (1)  _b[yedu_m]/_b[yedu_f] = _b[yexp_m]/_b[yexp_f]
 (2)  _b[yedu_m]/_b[yedu_f] = _b[yexp2_m]/_b[yexp2_f]

              chi2(2) =        9.21
          Prob > chi2 =      0.0100
```

Having established differences between men and women, we would like to do multiple testing between the ratios. Since we did not specify hypotheses in advance, we prefer to adjust the p-values of tests using, in this case, Bonferroni's method.

```
. testnl _b[yedu_m]/_b[yedu_f] = _b[yexp_m]/_b[yexp_f] = ///
                       _b[yexp2_m]/_b[yexp2_f], mtest(b)

 (1)  _b[yedu_m]/_b[yedu_f] = _b[yexp_m]/_b[yexp_f]
 (2)  _b[yedu_m]/_b[yedu_f] = _b[yexp2_m]/_b[yexp2_f]
```

	chi2	df	p
(1)	6.89	1	0.0173 #
(2)	0.93	1	0.6713 #
all	9.21	2	0.0100

    # Bonferroni adjusted p-values

◁

Manipulability

Although `testnl` allows you to specify constraints in different ways that are mathematically equivalent, as noted above, this does not mean that the tests are the same. This difference is known as the manipulability of the Wald test for nonlinear hypotheses; also see [R] **boxcox**. The test might even be significant for one formulation but not significant for another formulation that is mathematically equivalent. Trying out different specifications to find a formulation with the desired p-value is totally inappropriate, though it may actually be fun to try. There is no variance under representation because the nonlinear Wald test is actually a standard Wald test for a linearization of the constraint, which depends on the particular specification. We note that the likelihood-ratio test is not manipulable in this sense.

From a statistical point of view, it is best to choose a specification of the constraints that is as linear is possible. Doing so usually improves the accuracy of the approximation of the null-distribution of the test by a χ^2 or F distribution. The example above used the nonlinear Wald test to test whether the coefficients of human capital variables for men were proportional to those of women. A specification of proportionality of coefficients in terms of ratios of coefficients is fairly nonlinear if the coefficients in the denominator are close to 0. A more linear version of the test results from a bilinear formulation. Thus instead of

```
. testnl _b[yedu_m]/_b[yedu_f] = _b[yexp_m]/_b[yexp_f]
   (1)  _b[yedu_m]/_b[yedu_f] = _b[yexp_m]/_b[yexp_f]
              chi2(1) =        6.89
           Prob > chi2 =       0.0087
```

perhaps

```
. testnl _b[yedu_m]*_b[yexp_f] =  _b[yedu_f]*_b[yexp_m]
   (1)  _b[yedu_m]*_b[yexp_f] =  _b[yedu_f]*_b[yexp_m]
              chi2(1) =        13.95
           Prob > chi2 =       0.0002
```

is better, and in fact it has been suggested that the latter version of the test is more reliable. This assertion is confirmed by performing simulations and is in line with theoretical results of Philips and Park (1988). There is strong evidence against the proportionality of human capital effects between men and women, implying for this example that differences in the residual variances between the sexes can be ruled out as the explanation of the sex differences in the analysis of labor market participation.

Saved Results

`testnl` saves the following in `r()`:

Scalars

r(df)	degrees of freedom
r(df_r)	residual degrees of freedom
r(chi2)	χ^2
r(p)	significance
r(F)	F statistic

Matrices

r(G)	derivatives of $R(\mathbf{b})$ with respect to $\mathbf{b}$; see *Methods and Formulas* below.
r(R)	$R(\mathbf{b})-\mathbf{q}$; see *Methods and Formulas* below.

Methods and Formulas

`testnl` is implemented as an ado-file.

After fitting a model, define $\mathbf{b}$ as the resulting $1 \times k$ parameter vector and $\mathbf{V}$ as the $k \times k$ covariance matrix. The (linear or nonlinear) hypothesis is given by $R(\mathbf{b}) = \mathbf{q}$, where R is a function returning a $j \times 1$ vector. The Wald test formula is (Greene 2003, 487)

$$ W = \left\{ R(\mathbf{b}) - \mathbf{q} \right\}' \left(\mathbf{G}\mathbf{V}\mathbf{G}' \right)^{-1} \left\{ R(\mathbf{b}) - \mathbf{q} \right\} $$

where $\mathbf{G}$ is the derivative matrix of $R(\mathbf{b})$ with respect to $\mathbf{b}$. W is distributed as χ^2 if $\mathbf{V}$ is an asymptotic covariance matrix. $F = W/j$ is distributed as F for linear regression.

The adjustment methods for multiple testing are described in [R] **test**. The adjustment for survey design effects is described in [SVY] **svy postestimation**.

References

Allison, P. D. 1999. Comparing logit and probit coefficients across groups. *Sociological Methods and Research* 28: 186–208.

Gould, W. W. 1996. crc43: Wald test of nonlinear hypotheses after model estimation. *Stata Technical Bulletin* 29: 2–4. Reprinted in *Stata Technical Bulletin Reprints*, vol. 5, pp. 15–18.

Greene, W. H. 2003. *Econometric Analysis*. 5th ed. Upper Saddle River, NJ: Prentice Hall.

Long, J. S. 1997. *Regression Models for Categorical and Limited Dependent Variables*. Thousand Oaks, CA: Sage.

Phillips, P. C. B., and J. Y. Park. 1988. On the formulation of Wald tests of nonlinear restrictions. *Econometrica* 56: 1065–1083.

Also See

[R] **lincom** — Linear combinations of estimators

[R] **lrtest** — Likelihood-ratio test after estimation

[R] **nlcom** — Nonlinear combinations of estimators

[R] **test** — Test linear hypotheses after estimation

[U] **13.5 Accessing coefficients and standard errors**

[U] **20 Estimation and postestimation commands**

Title

tetrachoric — Tetrachoric correlations for binary variables

Syntax

tetrachoric *varlist* [*if*] [*in*] [*weight*] [, *options*]

options	description
Main	
stats(*statlist*)	list of statistics; select up to 4 statistics; default is stats(rho)
edwards	use the noniterative Edwards and Edwards estimator; default is the maximum likelihood estimator
print(#)	significance level for displaying coefficients
star(#)	significance level for displaying with a star
bonferroni	use Bonferroni-adjusted significance level
sidak	use Šidák-adjusted significance level
pw	calculate all the pairwise correlation coefficients by using all available data (pairwise deletion)
zeroadjust	adjust frequencies when one cell has a zero count
matrix	display output in matrix form
notable	suppress display of correlations
posdef	modify correlation matrix to be positive (semi)definite

statlist	description
rho	tetrachoric correlation coefficient
se	standard error of rho
obs	number of observations
p	exact two-sided significance level

by is allowed; see [D] **by**.

fweights are allowed; see [U] **11.1.6 weight**.

Description

tetrachoric computes estimates of the tetrachoric correlation coefficients of the binary variables in *varlist*. All these variables should be 0, 1, or missing values.

Tetrachoric correlations assume a latent bivariate-normal distribution (X_1, X_2) for each pair of variables (v_1, v_2), with a threshold model for the manifest variables, $v_i = 1$ if and only if $X_i > 0$. The means and variances of the latent variables are not identified, but the correlation, r, of X_1 and X_2 can be estimated from the joint distribution of v_1 and v_2 and is called the tetrachoric correlation coefficients.

tetrachoric computes pairwise estimates of the tetrachoric correlations by the (iterative) maximum likelihood estimator obtained from bivariate probit without explanatory variables (see [R] **biprobit**) by using the Edwards and Edwards (1984) noniterative estimator as the initial value.

The pairwise correlation matrix is returned as r(Rho) and can be used to perform a factor analysis or a principal components analysis of binary variables using the factormat or pcamat commands; see [MV] **factor** and [MV] **pca**.

Options

_____ Main _____

stats(*statlist*) specifies the statistics to be displayed in the matrix of output. stats(rho) is the default. Up to 4 statistics may be specified. stats(rho se p obs) would display the tetrachoric correlation coefficient, its standard error, the significance level, and the number of observations. If *varlist* contains only two variables, all statistics are shown in tabular form. stats(), print(), and star() have no effect unless the matrix option is also specified.

edwards specifies that the noniterative Edwards and Edwards estimator be used. The default is the maximum likelihood estimator. If you analyze many binary variables, you may want to use the fast noniterative estimator proposed by Edwards and Edwards (1984). However, if you have skewed variables, the approximation does not perform well.

print(#) specifies the maximum significance level of correlation coefficients to be printed. Correlation coefficients with larger significance levels are left blank in the matrix. Typing tetrachoric . . . , print(.10) would list only those correlation coefficients that are significant at the 10% level or lower.

star(#) specifies the maximum significance level of correlation coefficients to be marked with a star. Typing tetrachoric . . . , star(.05) would "star" all correlation coefficients significant at the 5% level or lower.

bonferroni makes the Bonferroni adjustment to calculated significance levels. This option affects printed significance levels and the print() and star() options. Thus tetrachoric . . . , print(.05) bonferroni prints coefficients with Bonferroni-adjusted significance levels of .05 or less.

sidak makes the Šidák adjustment to calculated significance levels. This option affects printed significance levels and the print() and star() options. Thus tetrachoric . . . , print(.05) sidak prints coefficients with Šidák-adjusted significance levels of .05 or less.

pw specifies that the tetrachoric correlation be calculated by using all available data. By default, tetrachoric uses casewise deletion, where observations are ignored if any of the specified variables in *varlist* are missing.

zeroadjust specifies that when one of the cells has a zero count, a frequency adjustment be applied in such a way as to increase the zero to one-half and maintain row and column totals.

matrix forces tetrachoric to display the statistics as a matrix, even if *varlist* contains only two variables. matrix is implied if more than two variables are specified.

notable suppresses the output.

posdef modifies the correlation matrix so that it is positive semidefinite, i.e., a proper correlation matrix. The modified result is the correlation matrix associated with the least-squares approximation of the tetrachoric correlation matrix by a positive-semidefinite matrix. If the correlation matrix is modified, the standard errors and significance levels are not displayed and are returned in r().

Remarks

Remarks are presented under the following headings:

Association in 2-by-2 tables
Factor analysis of dichotomous variables
Tetrachoric correlations with simulated data

Association in 2-by-2 tables

Although a wide variety of measures of association in cross tabulations have been proposed, such measures are essentially equivalent (monotonically related) in the special case of 2×2 tables—there is only 1 degree of freedom for nonindependence. Still, some measures have more desirable properties than others. Here we compare two measures: the standard Pearson correlation coefficient and the tetrachoric correlation coefficient. Given asymmetric row or column margins, Pearson correlations are limited to a range smaller than -1 to 1, although tetrachoric correlations can still span the range from -1 to 1. To illustrate, consider the following set of tables for two binary variables, X and Z:

	$Z = 0$	$Z = 1$	
$X = 0$	$20 - a$	$10 + a$	30
$X = 1$	a	$10 - a$	10
	20	20	40

For a equal to 0, 1, 2, 5, 8, 9, and 10, the Pearson and tetrachoric correlations for the above table are

a	0	1	2	5	8	9	10
Pearson	0.577	0.462	0.346	0	−0.346	−0.462	−0.577
Tetrachoric	1.000	0.792	0.607	0	−0.607	−0.792	−1.000

The restricted range for the Pearson correlation is especially unfortunate when you try to analyze the association between binary variables by using models developed for continuous data, such as factor analysis and principal component analysis.

The tetrachoric correlation of two variables (Y_1, Y_2) can be thought of as the Pearson correlation of two latent bivariate-normal distributed variables (Y_1^*, Y_2^*) with threshold measurement models $Y_i = (Y_i^* > c_i)$ for unknown cutpoints c_i. Or equivalently, $Y_i = (Y_i^{**} > 0)$ where the latent bivariate-normal (Y_1^{**}, Y_2^{**}) are shifted versions of (Y_1^*, Y_2^*) so that the cutpoints are zero. Obviously, you must judge whether assuming underlying latent variables is meaningful for the data. If this assumption is justified, tetrachoric correlations have two advantages. First you have an intuitive understanding of the size of correlations that are substantively interesting in your field of research, and this intuition is based on correlations that range from -1 to 1. Second since the tetrachoric correlation for binary variables estimates the Pearson correlation of the latent continuous variables (assumed multivariate-normal distributed), you can use the tetrachoric correlations to analyze multivariate relationships between the dichotomous variables. When doing so, remember that you must interpret the model in terms of the underlying continuous variables.

▷ Example 1

To illustrate tetrachoric correlations, we examine three binary variables from the `familyvalues` dataset (described in example 2).

```
. use http://www.stata-press.com/data/r10/familyvalues
(Attitudes on gender, relationships and family)
. tabulate RS075 RS076
```

fam att: women in charge bad	fam att: trad division of labor		Total
	0	1	
0	1,564	979	2,543
1	119	632	751
Total	1,683	1,611	3,294

```
. correlate  RS074 RS075 RS076
(obs=3291)
```

	RS074	RS075	RS076
RS074	1.0000		
RS075	0.0396	1.0000	
RS076	0.1595	0.3830	1.0000

```
. tetrachoric RS074 RS075 RS076
(obs=3291)
```

	RS074	RS075	RS076
RS074	1.0000		
RS075	0.0689	1.0000	
RS076	0.2480	0.6427	1.0000

As usual, the tetrachoric correlation coefficients are larger (in absolute value) and more dispersed than the Pearson correlations.

◁

Factor analysis of dichotomous variables

▷ Example 2

Factor analysis is a popular model for measuring latent continuous traits. The standard estimators are appropriate only for continuous unimodal data. Because of the skewness implied by Bernoulli-distributed variables (especially when the probability is distributed unevenly), a factor analysis of a Pearson correlation matrix can be rather misleading when used in this context. A factor analysis of a matrix of tetrachoric correlations is more appropriate under these conditions (Uebersax 2000). We illustrate this with data on gender, relationship, and family attitudes of spouses using the Households in The Netherlands survey 1995 (Weesie et al. 1995). For attitude variables, it seems reasonable to assume that agreement or disagreement is just a coarse measurement of more nuanced underlying attitudes.

To demonstrate, we examine a few of the variables from the `familyvalues` dataset.

```
. use http://www.stata-press.com/data/r10/familyvalues
(Attitudes on gender, relationships and family)
. describe RS056-RS063
```

variable name	storage type	display format	value label	variable label
RS056	byte	%9.0g		fam att: should be together
RS057	byte	%9.0g		fam att: should fight for relat
RS058	byte	%9.0g		fam att: should avoid conflict
RS059	byte	%9.0g		fam att: woman better nurturer
RS060	byte	%9.0g		fam att: both spouses money goo
RS061	byte	%9.0g		fam att: woman techn school goo
RS062	byte	%9.0g		fam att: man natural breadwinne
RS063	byte	%9.0g		fam att: common leisure good

```
. summarize RS056-RS063
```

Variable	Obs	Mean	Std. Dev.	Min	Max
RS056	3298	.5630685	.4960816	0	1
RS057	3296	.5400485	.4984692	0	1
RS058	3283	.6387451	.4804374	0	1
RS059	3308	.654474	.4756114	0	1
RS060	3302	.3906723	.487975	0	1
RS061	3293	.7102946	.4536945	0	1
RS062	3307	.5857272	.4926705	0	1
RS063	3298	.5379018	.498637	0	1

```
. correlate RS056-RS063
(obs=3221)
```

	RS056	RS057	RS058	RS059	RS060	RS061	RS062
RS056	1.0000						
RS057	0.1350	1.0000					
RS058	0.2377	0.0258	1.0000				
RS059	0.1816	0.0097	0.2550	1.0000			
RS060	-0.1020	-0.0538	-0.0424	0.0126	1.0000		
RS061	-0.1137	0.0610	-0.1375	-0.2076	0.0706	1.0000	
RS062	0.2014	0.0285	0.2273	0.4098	-0.0793	-0.2873	1.0000
RS063	0.2057	0.1460	0.1049	0.0911	0.0179	-0.0233	0.0975

	RS063
RS063	1.0000

Skewness in these data is relatively modest. For comparison, here are the tetrachoric correlations:

```
. tetrachoric RS056-RS063
(obs=3221)
```

	RS056	RS057	RS058	RS059	RS060	RS061	RS062
RS056	1.0000						
RS057	0.2114	1.0000					
RS058	0.3716	0.0416	1.0000				
RS059	0.2887	0.0158	0.4007	1.0000			
RS060	-0.1620	-0.0856	-0.0688	0.0208	1.0000		
RS061	-0.1905	0.1011	-0.2382	-0.3664	0.1200	1.0000	
RS062	0.3135	0.0452	0.3563	0.6109	-0.1267	-0.4845	1.0000
RS063	0.3187	0.2278	0.1677	0.1467	0.0286	-0.0388	0.1538

	RS063
RS063	1.0000

Again we see that the tetrachoric correlations are generally larger in absolute value than the Pearson correlations. The bivariate probit and Edwards and Edwards estimators (option edwards) implemented in tetrachoric may return a correlation matrix that is not positive semidefinite—a mathematical property of any real correlation matrix. Positive definiteness is required by commands for analyses of correlation matrices, such as factormat and pcamat; see [MV] **factor** and [MV] **pca**. The posdef option of tetrachoric tests for positive definiteness and projects the estimated correlation matrix to a positive-semidefinite matrix if needed.

```
. tetrachoric RS056-RS063, notable posdef
. matrix C = r(corr)
```

This time, we suppressed the display of the correlations with the notable option and requested that the correlation matrix be positive semidefinite with the posdef option. Had the correlation matrix not been positive definite, tetrachoric would have displayed a warning message and then adjusted the matrix to be positive semidefinite. We placed the resulting tetrachoric correlation matrix into a matrix, C, so that we can perform a factor analysis upon it.

tetrachoric with the posdef option asserted that C was positive definite since no warning message was displayed. We can verify this by using a familiar characterization of symmetric positive-definite matrices: all eigenvalues are real and positive.

```
. matrix symeigen eigenvectors eigenvalues = C
. matrix list eigenvalues
eigenvalues[1,8]
          e1         e2         e3         e4         e5         e6         e7
r1  2.5974789  1.3544664  1.0532476  .77980391  .73462018  .57984565  .54754512
          e8
r1  .35299228
```

We can proceed with a factor analysis on the matrix C. We use factormat and select iterated principal factors as the estimation method; see [MV] **factor**.

(Continued on next page)

```
. factormat C, n(3221) ipf factor(2)
(obs=3221)
```

Factor analysis/correlation   Number of obs   =   3221
   Method: iterated principal factors   Retained factors =   2
   Rotation: (unrotated)   Number of params =   15

Factor	Eigenvalue	Difference	Proportion	Cumulative
Factor1	2.06855	1.40178	0.7562	0.7562
Factor2	0.66677	0.47180	0.2438	1.0000
Factor3	0.19497	0.06432	0.0713	1.0713
Factor4	0.13065	0.10967	0.0478	1.1191
Factor5	0.02098	0.10085	0.0077	1.1267
Factor6	-0.07987	0.01037	-0.0292	1.0975
Factor7	-0.09024	0.08626	-0.0330	1.0645
Factor8	-0.17650	.	-0.0645	1.0000

LR test: independent vs. saturated: chi2(28) = 4620.01 Prob>chi2 = 0.0000

Factor loadings (pattern matrix) and unique variances

Variable	Factor1	Factor2	Uniqueness
RS056	0.5528	0.4120	0.5247
RS057	0.1124	0.4214	0.8098
RS058	0.5333	0.0718	0.7105
RS059	0.6961	-0.1704	0.4865
RS060	-0.1339	-0.0596	0.9785
RS061	-0.5126	0.2851	0.6560
RS062	0.7855	-0.2165	0.3361
RS063	0.2895	0.3919	0.7626

◁

▷ Example 3

We noted in example 2 that the matrix of estimates of the tetrachoric correlation coefficients need not be positive definite. Here is an example.

```
. use http://www.stata-press.com/data/r10/familyvalues
(Attitudes on gender, relationships and family)
. tetrachoric RS056-RS063 in 1/20, posdef
(obs=18)
```

matrix with tetrachoric correlations is not positive (semi-)definite;
   it has 2 negative eigenvalues
   maxdiff(corr,adj-corr) = 0.2346
   (adj-corr: tetrachoric correlations adjusted to be positive (semi)-definite)

adj-corr	RS056	RS057	RS058	RS059	RS060	RS061	RS062
RS056	1.0000						
RS057	0.5284	1.0000					
RS058	0.3012	0.2548	1.0000				
RS059	0.3251	0.2791	0.0550	1.0000			
RS060	-0.5197	-0.4222	-0.7163	0.0552	1.0000		
RS061	0.3448	0.4815	-0.0958	-0.1857	-0.0980	1.0000	
RS062	0.1066	-0.0375	0.0072	0.3909	-0.2333	-0.7654	1.0000
RS063	0.3830	0.4939	0.4336	0.0075	-0.8937	-0.0337	0.4934

adj-corr	RS063
RS063	1.0000

```
. mata
                                                    ── mata (type end to exit) ──
: C2 = st_matrix("r(corr)")
: eigenvecs = .
: eigenvals = .
: symeigensystem(C2, eigenvecs, eigenvals)
: eigenvals
                    1              2              3              4

    1     3.156592567    2.065279398    1.324911199    .7554904485

                    5              6              7              8

    1     .4845368741    .2131895139    2.02944e-16    -1.11650e-17

: end
```

The estimated tetrachoric correlation matrix is rank-2 deficient. With this C2 matrix, we can only use models of correlation that allow for singular cases.

◁

Tetrachoric correlations with simulated data

▷ Example 4

We use drawnorm (see [D] **drawnorm**) to generate a sample of 1,000 observations from a bivariate normal distribution with means -1 and 1, unit variances, and correlation 0.4.

```
. clear
. set seed 314159265
. matrix m = (1, -1)
. matrix V = (1, 0.4 \ 0.4, 1)
. drawnorm c1 c2, n(1000) means(m) cov(V)
(obs 1000)
```

Now consider the measurement model assumed by the tetrachoric correlations. We only observe whether c1 and c2 are greater than zero,

```
. generate d1 = (c1 > 0)
. generate d2 = (c2 > 0)
. tabulate d1 d2
```

	d2		
d1	0	1	Total
0	161	7	168
1	687	145	832
Total	848	152	1,000

We want to estimate the correlation of c1 and c2 from the binary variables d1 and d2. Pearson's correlation of the binary variables d1 and d2 is 0.138—a seriously biased estimate of the underlying correlation $\rho = 0.4$.

```
. correlate d1 d2
(obs=1000)

              |      d1        d2
 -------------+------------------
           d1 |  1.0000
           d2 |  0.1381    1.0000
```

The tetrachoric correlation coefficient of d1 and d2 estimates the Pearson correlation of the latent continuous variables, c1 and c2.

```
. tetrachoric d1 d2
   Number of obs =      1000
  Tetrachoric rho =       0.3947
        Std error =       0.0750

  Test of Ho: d1 and d2 are independent
  2-sided exact P =       0.0000
```

The estimate of the tetrachoric correlation of d1 and d2, 0.3947, is much closer to the underlying correlation, .4, between c1 and c2.

◁

Saved Results

tetrachoric saves the following in r():

Scalars

r(rho)	tetrachoric correlation coefficient between variables 1 and 2
r(N)	number of observations
r(nneg)	number of negative eigenvalues (posdef only)
r(se_rho)	standard error of r(rho)
r(p)	exact two-sided significance level

Macros

r(method)	estimator used

Matrices

r(Rho)	tetrachoric correlation matrix
r(Se_Rho)	standard errors of r(Rho)
r(corr)	synonym for r(Rho)
r(Nobs)	number of observations used in computing correlation
r(P)	exact two-sided significance level matrix

Methods and Formulas

tetrachoric is implemented as an ado-file.

tetrachoric provides two estimators for the tetrachoric correlation ρ of two binary variables with the frequencies n_{ij}, $i, j = 0, 1$. tetrachoric defaults to the slower (iterative) maximum likelihood estimator obtained from bivariate probit without explanatory variables (see [R] **biprobit**) by using the Edwards and Edwards noniterative estimator as the initial value. A fast (noniterative) estimator is also available by specifying the edwards option (Edwards and Edwards 1984; Digby 1983)

$$\widehat{\rho} = \frac{\alpha - 1}{\alpha + 1}$$

where

$$\alpha = \left(\frac{n_{00}n_{11}}{n_{01}n_{10}} \right)^{\pi/4} \quad (\pi = 3.14\ldots)$$

if all $n_{ij} > 0$. If $n_{00} = 0$ or $n_{11} = 0$, $\widehat{\rho} = -1$; if $n_{01} = 0$ or $n_{10} = 0$, $\widehat{\rho} = 1$.

The asymptotic variance of the Edwards and Edwards estimator of the tetrachoric correlation is easily obtained by the delta method,

$$\text{avar}(\widehat{\rho}) = \left(\frac{\pi\alpha}{2(1+\alpha)^2} \right)^2 \left(\frac{1}{n_{00}} + \frac{1}{n_{01}} + \frac{1}{n_{10}} + \frac{1}{n_{11}} \right)$$

provided all $n_{ij} > 0$, otherwise it is left undefined (missing). The Edwards and Edwards estimator is fast, but may be inaccurate if the margins are very skewed.

`tetrachoric` reports exact p-values for statistical independence, computed by the `exact` option of [R] **tabulate twoway**.

References

Brown, M. B. 1977. The tetrachoric correlation and its asymptotic standard error. *Applied Statistics* 26: 343–351.

Brown, M. B., and J. K. Benedetti. 1977. On the mean and variance of the tetrachoric correlation coefficient. *Psychometrika* 42: 347–355.

Digby, P. G. N. 1983. Approximating the tetrachoric correlation coefficient. *Biometrics* 39: 753–757.

Edwards, J. H., and A. W. F. Edwards. 1984. Approximating the tetrachoric correlation coefficient. *Biometrics* 40: 563.

Golub, G. H., and C. F. Van Loan. 1996. *Matrix Computations.* 3rd ed. Baltimore: Johns Hopkins University Press.

Uebersax, J. 2000. Estimating a latent trait model by factor analysis of tetrachoric correlations. http://ourworld.compuserve.com/homepages/jsuebersax/irt.htm

Weesie, J., M. Kalmijn, W. Bernasco, and D. Giesen. 1995. Households in The Netherlands 1995. Datafile, ISCORE, University of Utrecht, Utrecht, Netherlands.

Also See

[MV] **factor** — Factor analysis

[MV] **pca** — Principal components analysis

[R] **tabulate twoway** — Two-way tables of frequencies

[R] **biprobit** — Bivariate probit regression

[R] **correlate** — Correlations (covariances) of variables or coefficients

[R] **spearman** — Spearman's and Kendall's correlations

Title

> **tobit** — Tobit regression

Syntax

$$\underline{\text{tobit}} \; depvar \; \left[indepvars\right] \; \left[if\right] \; \left[in\right] \; \left[weight\right], \; \text{ll}\left[(\#)\right] \; \text{ul}\left[(\#)\right] \; \left[options\right]$$

options	description
Model	
* ll $\left[(\#)\right]$	left-censoring limit
* ul $\left[(\#)\right]$	right-censoring limit
<u>off</u>set(*varname*)	include *varname* in model with coefficient constrained to 1
SE/Robust	
vce(*vcetype*)	*vcetype* may be oim, <u>r</u>obust, <u>cl</u>uster *clustvar*, <u>boot</u>strap, or jackknife
Reporting	
<u>l</u>evel(#)	set confidence level; default is level(95)
Max options	
maximize_options	control the maximization process; seldom used

* You must specify at least one of ll $\left[(\#)\right]$ or ul $\left[(\#)\right]$.

bootstrap, by, jackknife, nestreg, rolling, statsby, stepwise, svy, and xi are allowed; see
    [U] **11.1.10 Prefix commands**.

Weights are not allowed with the bootstrap prefix.

aweights are not allowed with the jackknife prefix.

vce() and weights are not allowed with the svy prefix.

aweights, fweights, pweights, and iweights are allowed; see [U] **11.1.6 weight**.

See [U] **20 Estimation and postestimation commands** for more capabilities of estimation commands.

Description

tobit fits a model of *depvar* on *indepvars* where the censoring values are fixed.

Options

> **Model**

ll $\left[(\#)\right]$ and ul $\left[(\#)\right]$ indicate the lower and upper limits for censoring, respectively. You may specify one or both. Observations with *depvar* $\leq$ ll() are left-censored; observations with *depvar* $\geq$ ul() are right-censored; and remaining observations are not censored. You do not have to specify the censoring values at all. It is enough to type ll, ul, or both. When you do not specify a censoring value, tobit assumes that the lower limit is the minimum observed in the data (if ll is specified) and the upper limit is the maximum (if ul is specified).

offset(*varname*); see [R] **estimation options**.

vce(*vcetype*) specifies the type of standard error reported, which includes types that are derived
from asymptotic theory, that are robust to some kinds of misspecification, that allow for intragroup
correlation, and that use bootstrap or jackknife methods; see [R] *vce_option*.

level(*#*); see [R] **estimation options**.

maximize_options: iterate(*#*), [no]log, trace, tolerance(*#*), ltolerance(*#*); see [R] **maximize**. These options are seldom used.

Unlike most maximum likelihood commands, tobit defaults to nolog—it suppresses the iteration
log.

Remarks

Tobit estimation was originally developed by Tobin (1958). A consumer durable was purchased if
a consumer's desire was high enough, where desire was measured by the dollar amount spent by the
purchaser. If no purchase was made, the measure of desire was censored at zero.

▷ Example 1

We will demonstrate tobit with an artificial example, which in the process will allow us to
emphasize the assumptions underlying the estimation. We have a dataset containing the mileage
ratings and weights of 74 cars. There are no censored variables in this dataset, but we are going to
create one. Before that, however, the relationship between mileage and weight in our complete data is

```
. use http://www.stata-press.com/data/r10/auto
(1978 Automobile Data)

. generate wgt = weight/1000

. regress mpg wgt
```

Source	SS	df	MS		Number of obs =	74
					F(1, 72) =	134.62
Model	1591.99024	1	1591.99024		Prob > F =	0.0000
Residual	851.469221	72	11.8259614		R-squared =	0.6515
					Adj R-squared =	0.6467
Total	2443.45946	73	33.4720474		Root MSE =	3.4389

mpg	Coef.	Std. Err.	t	P>|t|	[95% Conf. Interval]	
wgt	-6.008687	.5178782	-11.60	0.000	-7.041058	-4.976316
_cons	39.44028	1.614003	24.44	0.000	36.22283	42.65774

(We divided weight by 1,000 simply to make discussing the resulting coefficients easier. We find
that each additional 1,000 pounds of weight reduces mileage by 6 mpg.)

mpg in our data ranges from 12 to 41. Let us now pretend that our data were censored in the sense
that we could not observe a mileage rating below 17 mpg. If the true mpg is 17 or less, all we know
is that the mpg is less than or equal to 17:

```
. replace mpg=17 if mpg<=17
(14 real changes made)

. tobit mpg wgt, ll
```

Tobit regression

				Number of obs	=	74
				LR chi2(1)	=	72.85
				Prob > chi2	=	0.0000
Log likelihood = -164.25438				Pseudo R2	=	0.1815

mpg	Coef.	Std. Err.	t	P>|t|	[95% Conf. Interval]	
wgt	-6.87305	.7002559	-9.82	0.000	-8.268658	-5.477442
_cons	41.49856	2.05838	20.16	0.000	37.39621	45.6009
/sigma	3.845701	.3663309			3.115605	4.575797

```
  Obs. summary:       18  left-censored observations at mpg<=17
                      56       uncensored observations
                       0  right-censored observations
```

The `replace` before estimation was not really necessary—we remapped all the mileage ratings below 17 to 17 merely to reassure you that `tobit` was not somehow using uncensored data. We typed `ll` after `tobit` to inform `tobit` that the data were left-censored. `tobit` found the minimum of `mpg` in our data and assumed that was the censoring point. We could also have dispensed with `replace` and typed `ll(17)`, informing `tobit` that all values of the dependent variable 17 and below are really censored at 17. In either case, at the bottom of the table, we are informed that there are, as a result, 18 left-censored observations.

On these data, our estimate is now a reduction of 6.9 mpg per 1,000 extra pounds of weight as opposed to 6.0. The parameter reported as `/sigma` is the estimated standard error of the regression; the resulting 3.8 is comparable to the estimated root mean squared error reported by `regress` of 3.4.

◁

❑ Technical Note

You would never want to throw away information by purposely censoring variables. The `regress` estimates are in every way preferable to those of `tobit`. Our example is designed solely to illustrate the relationship between `tobit` and `regress`. If you have uncensored data, use `regress`. If your data are censored, you have no choice but to use `tobit`.

❑

▷ Example 2

`tobit` can also fit models that are censored from above. This time, let's assume that we do not observe the actual mileage rating of cars yielding 24 mpg or better—we know only that it is at least 24. (Also assume that we have undone the change to `mpg` we made in the previous example.)

```
. use http://www.stata-press.com/data/r10/auto, clear
(1978 Automobile Data)

. generate wgt = weight/100

. regress mpg wgt
  (output omitted )
```

```
. tobit mpg wgt, ul(24)
```

Tobit regression

				Number of obs	=	74
				LR chi2(1)	=	90.72
				Prob > chi2	=	0.0000
Log likelihood = -129.8279				Pseudo R2	=	0.2589

mpg	Coef.	Std. Err.	t	P>|t|	[95% Conf. Interval]	
wgt	-.5080645	.043493	-11.68	0.000	-.5947459	-.4213831
_cons	36.08037	1.432056	25.19	0.000	33.22628	38.93445
/sigma	2.385357	.2444604			1.898148	2.872566

```
Obs. summary:          0  left-censored observations
                      51      uncensored observations
                      23 right-censored observations at mpg>=24
```

◁

▷ Example 3

tobit can also fit models that are censored from both sides (the so-called two-limit tobit):

```
. tobit mpg wgt, ll(17) ul(24)
```

Tobit regression

				Number of obs	=	74
				LR chi2(1)	=	77.60
				Prob > chi2	=	0.0000
Log likelihood = -104.25976				Pseudo R2	=	0.2712

mpg	Coef.	Std. Err.	t	P>|t|	[95% Conf. Interval]	
wgt	-.5764448	.0724542	-7.96	0.000	-.7208457	-.4320438
_cons	38.07468	2.255917	16.88	0.000	33.57865	42.57072
/sigma	2.886337	.3952143			2.098676	3.673998

```
Obs. summary:         18  left-censored observations at mpg<=17
                      33      uncensored observations
                      23 right-censored observations at mpg>=24
```

◁

James Tobin (1918–2002) was an American economist who after education and research at Harvard moved to Yale, where he was on the faculty from 1950 to 1988. He made many outstanding contributions to economics and was awarded the Nobel Prize in 1981 "for his analysis of financial markets and their relations to expenditure decisions, employment, production and prices." He appeared thinly disguised as a character in Herman Wouk's novel *The Caine Mutiny* (1951) who thwarts the ambition of Willie Keith to be first in his class at midshipman school: "A mandarin-like midshipman named Tobit, with a domed forehead, measured quiet speech, and a mind like a sponge, was ahead of the field by a spacious percentage."

Saved Results

tobit saves the following in e():

Scalars

e(N)	number of observations	e(df_r)	residual degrees of freedom
e(N_unc)	number of uncensored obs.	e(r2_p)	pseudo-R-squared
e(N_lc)	number of left-censored obs.	e(chi2)	χ^2
e(N_rc)	number of right-censored obs.	e(ll)	log likelihood
e(llopt)	contents of ll(), if specified	e(ll_0)	log likelihood, constant-only model
e(ulopt)	contents of ul(), if specified	e(N_clust)	number of clusters
e(k_aux)	number of auxiliary parameters	e(converged)	1 if converged, 0 otherwise
e(df_m)	model degrees of freedom		

Macros

e(cmd)	tobit	e(chi2type)	LR; type of model χ^2 test
e(cmdline)	command as typed	e(vce)	*vcetype* specified in vce()
e(depvar)	name of dependent variable	e(vcetype)	title used to label Std. Err.
e(censored)	variable specified in censored()	e(crittype)	optimization criterion
e(wtype)	weight type	e(properties)	b V
e(wexp)	weight expression	e(footnote)	program used to implement the footnote display
e(title)	title in estimation output		
e(clustvar)	name of cluster variable	e(predict)	program used to implement predict
e(offset)	offset		

Matrices

e(b)	coefficient vector	e(V)	variance–covariance matrix of the estimators

Functions

e(sample)	marks estimation sample

Methods and Formulas

tobit is implemented as an ado-file.

See the *Methods and Formulas* of [R] **intreg**.

See Tobin (1958) for the original derivation of the tobit model. An introductory description of the tobit model can be found in, for instance, Johnston and DiNardo (1997, 436–441), Kmenta (1997, 562–566), Long (1997, 196–210), and Maddala (1992, 338–342).

References

Amemiya, T. 1973. Regression analysis when the dependent variable is truncated normal. *Econometrica* 41: 997–1016.

——. 1984. Tobit models: A survey. *Journal of Econometrics* 24: 3–61.

Cong, R. 2000. sg144: Marginal effects of the tobit model. *Stata Technical Bulletin* 56: 27–34. Reprinted in *Stata Technical Bulletin Reprints*, vol. 10, pp. 189–197.

Drukker, D. M. 2002. Bootstrapping a conditional moments test for normality after tobit estimation. *Stata Journal* 2: 125–139.

Goldberger, A. S. 1983. Abnormal selection bias. In *Studies in Econometrics, Time Series, and Multivariate Statistics*, ed. S. Karlin, T. Amemiya, and L. A. Goodman, 67–84. New York: Academic Press.

Hurd, M. 1979. Estimation in truncated samples when there is heteroscedasticity. *Journal of Econometrics* 11: 247–258.

Johnston, J., and J. DiNardo. 1997. *Econometric Methods*. 4th ed. New York: McGraw–Hill.

Kendall, M. G., and A. Stuart. 1973. *The Advanced Theory of Statistics: Inference and Relationship, Vol. 2*. New York: Hafner.

Kmenta, J. 1997. *Elements of Econometrics*. 2nd ed. Ann Arbor: University of Michigan Press.

Long, J. S. 1997. *Regression Models for Categorical and Limited Dependent Variables*. Thousand Oaks, CA: Sage.

Maddala, G. S. 1992. *Introduction to Econometrics*. 2nd ed. New York: Macmillan.

McDonald, J. A., and R. Moffitt. 1980. The use of tobit analysis. *Review of Economics and Statistics* 62: 318–321.

Shiller, R. J. 1999. The ET interview: Professor James Tobin. *Econometric Theory* 15: 867–900.

Stewart, M. B. 1983. On least-squares estimation when the dependent variable is grouped. *Review of Economic Studies* 50: 737–753.

Tobin, J. 1958. Estimation of relationships for limited dependent variables. *Econometrica* 26: 24–36.

Also See

[R] **tobit postestimation** — Postestimation tools for tobit

[R] **intreg** — Interval regression

[R] **cnreg** — Censored-normal regression

[R] **heckman** — Heckman selection model

[R] **ivtobit** — Tobit model with endogenous regressors

[R] **regress** — Linear regression

[R] **truncreg** — Truncated regression

[XT] **xtintreg** — Random-effects interval-data regression models

[XT] **xttobit** — Random-effects tobit models

[U] **20 Estimation and postestimation commands**

Title

> **tobit postestimation** — Postestimation tools for tobit

Description

The following postestimation commands are available for `tobit`:

command	description
adjust	adjusted predictions of $\mathbf{x}\boldsymbol{\beta}$
estat	AIC, BIC, VCE, and estimation sample summary
estat (svy)	postestimation statistics for survey data
estimates	cataloging estimation results
hausman	Hausman's specification test
lincom	point estimates, standard errors, testing, and inference for linear combinations of coefficients
linktest	link test for model specification
lrtest[1]	likelihood-ratio test
mfx	marginal effects or elasticities
nlcom	point estimates, standard errors, testing, and inference for nonlinear combinations of coefficients
predict	predictions, residuals, influence statistics, and other diagnostic measures
predictnl	point estimates, standard errors, testing, and inference for generalized predictions
test	Wald tests for simple and composite linear hypotheses
testnl	Wald tests of nonlinear hypotheses

[1] `lrtest` is not appropriate with `svy` estimation results.

See the corresponding entries in the *Stata Base Reference Manual* for details, but see [SVY] **estat** for details about `estat` (svy).

Syntax for predict

> predict $\left[\,type\,\right]$ *newvar* $\left[\,if\,\right]$ $\left[\,in\,\right]$ $\left[\,,\ statistic\ \underline{\text{nooff}}\text{set}\,\right]$

> predict $\left[\,type\,\right]$ $\left\{\,stub* \,|\, newvar_{\text{reg}}\ newvar_{\text{sigma}}\,\right\}$ $\left[\,if\,\right]$ $\left[\,in\,\right]$, $\underline{\text{sc}}\text{ores}$

statistic	description
xb	linear prediction; the default
stdp	standard error of the linear prediction
stdf	standard error of the forecast
pr(a,b)	$\Pr(a < y_j < b)$
e(a,b)	$E(y_j \,|\, a < y_j < b)$
ystar(a,b)	$E(y_j^*)$, $y_j^* = \max\{a, \min(y_j, b)\}$

These statistics are available both in and out of sample; type `predict ... if e(sample) ...` if wanted only for the estimation sample.

`stdf` is not allowed with `svy` estimation results.

where a and b may be numbers or variables; a missing $(a \geq .)$ means $-\infty$, and b missing $(b \geq .)$ means $+\infty$; see [U] **12.2.1 Missing values**.

Options for predict

xb, the default, calculates the linear prediction.

stdp calculates the standard error of the prediction, which can be thought of as the standard error of the predicted expected value or mean for the observation's covariate pattern. The standard error of the prediction is also referred to as the standard error of the fitted value.

stdf calculates the standard error of the forecast, which is the standard error of the point prediction for 1 observation. It is commonly referred to as the standard error of the future or forecast value. By construction, the standard errors produced by stdf are always larger than those produced by stdp; see [R] **regress** *Methods and Formulas*.

pr(a,b) calculates $\Pr(a < \mathbf{x}_j \mathbf{b} + u_j < b)$, the probability that $y_j | \mathbf{x}_j$ would be observed in the interval (a, b).

a and b may be specified as numbers or variable names; *lb* and *ub* are variable names;
pr(20,30) calculates $\Pr(20 < \mathbf{x}_j \mathbf{b} + u_j < 30)$;
pr(*lb*,*ub*) calculates $\Pr(lb < \mathbf{x}_j \mathbf{b} + u_j < ub)$; and
pr(20,*ub*) calculates $\Pr(20 < \mathbf{x}_j \mathbf{b} + u_j < ub)$.

a missing $(a \geq .)$ means $-\infty$; pr(.,30) calculates $\Pr(-\infty < \mathbf{x}_j \mathbf{b} + u_j < 30)$;
pr(*lb*,30) calculates $\Pr(-\infty < \mathbf{x}_j \mathbf{b} + u_j < 30)$ in observations for which $lb \geq .$
and calculates $\Pr(lb < \mathbf{x}_j \mathbf{b} + u_j < 30)$ elsewhere.

b missing $(b \geq .)$ means $+\infty$; pr(20,.) calculates $\Pr(+\infty > \mathbf{x}_j \mathbf{b} + u_j > 20)$;
pr(20,*ub*) calculates $\Pr(+\infty > \mathbf{x}_j \mathbf{b} + u_j > 20)$ in observations for which $ub \geq .$
and calculates $\Pr(20 < \mathbf{x}_j \mathbf{b} + u_j < ub)$ elsewhere.

e(a,b) calculates $E(\mathbf{x}_j \mathbf{b} + u_j \mid a < \mathbf{x}_j \mathbf{b} + u_j < b)$, the expected value of $y_j | \mathbf{x}_j$ conditional on $y_j | \mathbf{x}_j$ being in the interval (a, b), meaning that $y_j | \mathbf{x}_j$ is censored.
a and b are specified as they are for pr().

ystar(a,b) calculates $E(y_j^*)$, where $y_j^* = a$ if $\mathbf{x}_j \mathbf{b} + u_j \leq a$, $y_j^* = b$ if $\mathbf{x}_j \mathbf{b} + u_j \geq b$, and $y_j^* = \mathbf{x}_j \mathbf{b} + u_j$ otherwise, meaning that y_j^* is truncated. a and b are specified as they are for pr().

nooffset is relevant only if you specified offset(*varname*). It modifies the calculations made by predict so that they ignore the offset variable; the linear prediction is treated as $\mathbf{x}_j \mathbf{b}$ rather than as $\mathbf{x}_j \mathbf{b} + \text{offset}_j$.

scores calculates equation-level score variables.

The first new variable will contain $\partial \ln L / \partial (\mathbf{x}_j \boldsymbol{\beta})$.

The second new variable will contain $\partial \ln L / \partial \sigma$.

Remarks

Following Cong (2000), write the tobit model as

$$y_i^* = \begin{cases} y_i, & \text{if } a < y_i < b \\ a, & \text{if } y_i \leq a \\ b, & \text{if } y_i \geq b \end{cases}$$

y_i is a latent variable; instead, we observe y_i^*, which is bounded between a and b if y_i is outside those bounds.

There are four types of marginal effects that may be of interest in the tobit model, depending on the application:

1. The β coefficients themselves measure how the unobserved variable y_i changes with respect to changes in the regressors.

2. The marginal effects of the truncated expected value $E[y_i * | a < y_i^* < b]$ measure the changes in y_i with respect to changes in the regressors among the subpopulation for which y_i is not at a boundary.

3. The marginal effects of the censored expected value $E[y_i^*]$ describe how the observed variable y_i^* changes with respect to the regressors.

4. The marginal effects of $\Pr[a < y_i^* < b]$ describe how the probability of being uncensored changes with respect to the regressors.

In the next example, we show how to obtain each of these.

▷ Example 1

In example 3 of [R] **tobit**, we fitted a two-limit tobit model of mpg on wgt. The marginal effects of y_i are simply the β's reported by tobit:

```
. use http://www.stata-press.com/data/r10/auto
(1978 Automobile Data)

. generate wgt = weight/100

. tobit mpg wgt, ll(17) ul(24)
```

Tobit regression

					Number of obs	=	74
					LR chi2(1)	=	77.60
					Prob > chi2	=	0.0000
Log likelihood = -104.25976					Pseudo R2	=	0.2712

mpg	Coef.	Std. Err.	t	P>|t|	[95% Conf. Interval]	
wgt	-.5764448	.0724542	-7.96	0.000	-.7208457	-.4320438
_cons	38.07468	2.255917	16.88	0.000	33.57865	42.57072
/sigma	2.886337	.3952143			2.098676	3.673998

Obs. summary:	18	left-censored observations at mpg<=17
	33	uncensored observations
	23	right-censored observations at mpg>=24

If mpg were not censored at 17 and 24, then a 100-pound increase in a car's weight (which is a 1-unit increase in wgt) would lower fuel economy by 0.58 miles per gallon.

To get the marginal effect on the truncated expected value, we type

```
. mfx compute, predict(e(17,24))
```

Marginal effects after tobit
```
    y  = E(mpg|17<mpg<24) (predict, e(17,24))
       =    20.56787
```

variable	dy/dx	Std. Err.	z	P>|z|	[95% C.I.]		X
wgt	-.2311626	.04275	-5.41	0.000	-.314955	-.14737	30.1946

For cars whose fuel economy lies between 17 and 24 miles per gallon, a 100-pound increase in the car's weight lowers fuel economy by about 0.23 miles per gallon.

To get the marginal effect on the censored expected value, we type

```
. mfx compute, predict(ystar(17,24))

Marginal effects after tobit
     y  = E(mpg*|17<mpg<24) (predict, ystar(17,24))
        =  20.631014
```

variable	dy/dx	Std. Err.	z	P>|z|	[95% C.I.]	X
wgt	-.4461247	.04788	-9.32	0.000	-.539975 -.352274	30.1946

The observed fuel economy (which is bounded between 17 and 24) will decline by 0.45 miles per gallon for a 100-pound increase in weight.

Finally, what happens to the probability of being censored as we change wgt? We type

```
. mfx compute, predict(pr(17,24))

Marginal effects after tobit
     y  = Pr(17<mpg<24) (predict, pr(17,24))
        =  .77392453
```

variable	dy/dx	Std. Err.	z	P>|z|	[95% C.I.]	X
wgt	.0054246	.01263	0.43	0.668	-.019326 .030175	30.1946

If we increase the weight of a car by 100 pounds, the probability that the fuel economy lies between 17 and 24 miles per gallon increases by 0.005, or, equivalently, the probability that the fuel economy is less than 17 or greater than 24 miles per gallon decreases by 0.005. This marginal effect is small, in part, because a large fraction (77%) of cars are predicted to lie within those bounds.

◁

Methods and Formulas

All postestimation commands listed above are implemented as ado-files.

Reference

Cong, R. 2000. sg144: Marginal effects of the tobit model. *Stata Technical Bulletin* 56: 27–34. Reprinted in *Stata Technical Bulletin Reprints*, vol. 10, pp. 189–197.

Also See

[R] **tobit** — Tobit regression

[U] **20 Estimation and postestimation commands**

Title

> **total** — Estimate totals

Syntax

> `total` *varlist* $\left[\textit{if}\right]$ $\left[\textit{in}\right]$ $\left[\textit{weight}\right]$ $\left[\textit{, options}\right]$

options	description
if/in/over	
over(*varlist*[, <u>nolabel</u>])	group over subpopulations defined by *varlist*; optionally, suppress group labels
SE/Cluster	
vce(*vcetype*)	*vcetype* may be `analytic`, <u>cl</u>uster *clustvar*, <u>boot</u>strap, or `jackknife`
Reporting	
<u>level</u>(#)	set confidence level; default is `level(95)`
<u>nohe</u>ader	suppress table header
<u>nol</u>egend	suppress table legend

bootstrap, jackknife, rolling, statsby, and svy are allowed; see [U] **11.1.10 Prefix commands**.
Weights are not allowed with the bootstrap prefix.
vce() and weights are not allowed with the svy prefix.
fweights, pweights, and iweights are allowed; see [U] **11.1.6 weight**.
See [U] **20 Estimation and postestimation commands** for more capabilities of estimation commands.

Description

total produces estimates of totals, along with standard errors.

Options

> **if/in/over**

over(*varlist* [, <u>nolabel</u>]) specifies that estimates be computed for multiple subpopulations, which are identified by the different values of the variables in *varlist*.

When this option is supplied with one variable name, such as over(*varname*), the value labels of *varname* are used to identify the subpopulations. If *varname* does not have labeled values (or there are unlabeled values), the values themselves are used, provided that they are nonnegative integers. Noninteger values, negative values, and labels that are not valid Stata names are substituted with a default identifier.

When over() is supplied with multiple variable names, each subpopulation is assigned a unique default identifier.

nolabel specifies that value labels attached to the variables identifying the subpopulations be ignored.

SE/Cluster

vce(*vcetype*) specifies the type of standard error reported, which includes types that are derived from asymptotic theory, that allow for intragroup correlation, and that use bootstrap or jackknife methods; see [R] *vce_option*.

vce(analytic), the default, uses the analytically derived variance estimator associated with the sample total.

Reporting

level(*#*); see [R] **estimation options**.

noheader prevents the table header from being displayed. This option implies nolegend.

nolegend prevents the table legend identifying the subpopulations from being displayed.

Remarks

▷ Example 1

Suppose that we collected data on incidence of heart attacks. The variable heartatk indicates whether a person ever had a heart attack (1 means yes; 0 means no). We can then estimate the total number of persons who have had heart attacks for each sex in the population represented by the data we collected.

```
. use http://www.stata-press.com/data/r10/total
. total heartatk [pw=swgt], over(sex)
Total estimation                    Number of obs   =    4946
        Male: sex = Male
      Female: sex = Female
```

Over	Total	Std. Err.	[95% Conf. Interval]	
heartatk				
Male	944559	104372.3	739943	1149175
Female	581590	82855.59	419156.3	744023.7

◁

(Continued on next page)

Saved Results

total saves the following in e():

Scalars

e(N)	number of observations	e(k_eq)	number of equations in e(b)
e(N_over)	number of subpopulations	e(df_r)	sample degrees of freedom
e(N_clust)	number of clusters		

Macros

e(cmd)	total	e(over)	*varlist* from over()
e(cmdline)	command as typed	e(over_labels)	labels from over() variables
e(varlist)	*varlist*	e(over_namelist)	names from e(over_labels)
e(wtype)	weight type	e(vce)	*vcetype* specified in vce()
e(wexp)	weight expression	e(vcetype)	title used to label Std. Err.
e(title)	title in estimation output	e(estat_cmd)	program used to implement estat
e(cluster)	name of cluster variable	e(properties)	b V

Matrices

e(b)	vector of total estimates
e(V)	(co)variance estimates
e(_N)	vector of numbers of nonmissing observations
e(error)	error code corresponding to e(b)

Functions

e(sample)	marks estimation sample

Methods and Formulas

total is implemented as an ado-file.

The total estimator

Let y denote the variable on which to calculate the total and $y_j, j = 1, \ldots, n$, denote an individual observation on y. Let w_j be the frequency weight (or iweight), and if no weight is specified, define $w_j = 1$ for all j. See the next section for pweighted data. The sum of the weights is an estimate of the population size:

$$\widehat{N} = \sum_{j=1}^{n} w_j$$

If the population values of y are denoted by $Y_j, j = 1, \ldots, N$, the associated population total is

$$Y = \sum_{j=1}^{N} Y_j = N\overline{y}$$

where $\overline{y}$ is the population mean. The total is estimated as

$$\widehat{Y} = \widehat{N}\overline{y}$$

The variance estimator for the total is

$$\widehat{V}(\widehat{Y}) = \widehat{N}^2 \widehat{V}(\overline{y})$$

where $\widehat{V}(\overline{y})$ is the variance estimator for the mean; see [R] **mean**. The standard error of the total is the square root of the variance.

If x, x_j, $\overline{x}$, and $\widehat{X}$ are similarly defined for another variable (observed jointly with y), the covariance estimator between $\widehat{X}$ and $\widehat{Y}$ is

$$\widehat{\mathrm{Cov}}(\widehat{X}, \widehat{Y}) = \widehat{N}^2 \widehat{\mathrm{Cov}}(\overline{x}, \overline{y})$$

where $\widehat{\mathrm{Cov}}(\overline{x}, \overline{y})$ is the covariance estimator between two means; see [R] **mean**.

Survey data

See [SVY] **variance estimation** and [SVY] **poststratification** for discussions that provide background information for the following formulas.

The survey total estimator

Let Y_j be a survey item for the jth individual in the population, where $j = 1, \ldots, M$ and M is the size of the population. The associated population total for the item of interest is

$$Y = \sum_{j=1}^{M} Y_j$$

Let y_j be the survey item for the jth sampled individual from the population, where $j = 1, \ldots, m$ and m is the number of observations in the sample.

The estimator $\widehat{Y}$ for the population total Y is

$$\widehat{Y} = \sum_{j=1}^{m} w_j y_j$$

where w_j is a sampling weight. The estimator for the number of individuals in the population is

$$\widehat{M} = \sum_{j=1}^{m} w_j$$

The score variable for the total estimator is the variable itself,

$$z_j(\widehat{Y}) = y_j$$

The poststratified total estimator

Let P_k denote the set of sampled observations that belong to poststratum k and define $I_{P_k}(j)$ to indicate if the jth observation is a member of poststratum k; where $k = 1, \ldots, L_P$ and L_P is the number of poststrata. Also let M_k denote the population size for poststratum k. P_k and M_k are identified by specifying the `poststrata()` and `postweight()` options on `svyset`.

The estimator for the poststratified total is

$$\widehat{Y}^P = \sum_{k=1}^{L_P} \frac{M_k}{\widehat{M}_k} \widehat{Y}_k = \sum_{k=1}^{L_P} \frac{M_k}{\widehat{M}_k} \sum_{j=1}^{m} I_{P_k}(j)\, w_j y_j$$

where

$$\widehat{M}_k = \sum_{j=1}^{m} I_{P_k}(j) w_j$$

The score variable for the poststratified total is

$$z_j(\widehat{Y}^P) = \sum_{k=1}^{L_P} I_{P_k}(j) \frac{M_k}{\widehat{M}_k} \left(y_j - \frac{\widehat{Y}_k}{\widehat{M}_k} \right)$$

Subpopulation estimation

Let S denote the set of sampled observations that belong to the subpopulation of interest, and define $I_S(j)$ to indicate if the jth observation falls within the subpopulation.

The estimator for the subpopulation total is

$$\widehat{Y}^S = \sum_{j=1}^{m} I_S(j)\, w_j y_j$$

and its score variable is

$$z_j(\widehat{Y}^S) = I_S(j)\, y_j$$

The estimator for the poststratified subpopulation total is

$$\widehat{Y}^{PS} = \sum_{k=1}^{L_P} \frac{M_k}{\widehat{M}_k} \widehat{Y}_k^S = \sum_{k=1}^{L_P} \frac{M_k}{\widehat{M}_k} \sum_{j=1}^{m} I_{P_k}(j) I_S(j)\, w_j y_j$$

and its score variable is

$$z_j(\widehat{Y}^{PS}) = \sum_{k=1}^{L_P} I_{P_k}(j) \frac{M_k}{\widehat{M}_k} \left\{ I_S(j)\, y_j - \frac{\widehat{Y}_k^S}{\widehat{M}_k} \right\}$$

References

Cochran, W. G. 1977. *Sampling Techniques*. 3rd ed. New York: Wiley.

Stuart, A., and J. K. Ord. 1994. *Kendall's Advanced Theory of Statistics: Distribution Theory, Vol. I.* 6th ed. London: Arnold.

Also See

[R] **total postestimation** — Postestimation tools for total

[R] **mean** — Estimate means

[R] **proportion** — Estimate proportions

[R] **ratio** — Estimate ratios

[SVY] **direct standardization** — Direct standardization of means, proportions, and ratios

[SVY] **poststratification** — Poststratification for survey data

[SVY] **subpopulation estimation** — Subpopulation estimation for survey data

[SVY] **svy estimation** — Estimation commands for survey data

[SVY] **variance estimation** — Variance estimation for survey data

[U] **20 Estimation and postestimation commands**

Title

total postestimation — Postestimation tools for total

Description

The following postestimation commands are available for `total`:

command	description
estat	VCE
estat (svy)	postestimation statistics for survey data
estimates	cataloging estimation results
lincom	point estimates, standard errors, testing, and inference for linear combinations of coefficients
nlcom	point estimates, standard errors, testing, and inference for nonlinear combinations of coefficients
test	Wald tests for simple and composite linear hypotheses
testnl	Wald tests of nonlinear hypotheses

See the corresponding entries in the *Stata Base Reference Manual* for details, but see [SVY] **estat** for details about `estat` (svy).

Remarks

▷ Example 1

Continuing with our data on incidence of heart attacks from example 1 in [R] **total**, we want to test whether there are twice as many heart attacks among men than women in the population.

```
. use http://www.stata-press.com/data/r10/total
. total heartatk [pw=swgt], over(sex)
(output omitted )
. test _b[Male] = 2*_b[Female]
( 1)  [heartatk]Male - 2 [heartatk]Female = 0
      F(  1,  4945) =     1.25
            Prob > F =    0.2643
```

Thus we do not reject our hypothesis that the total number of heart attacks for men is twice that for women in the population.

◁

Methods and Formulas

All postestimation commands listed above are implemented as ado-files.

Also See

[R] **total** — Estimate totals

[U] **20 Estimation and postestimation commands**

Title

translate — Print and translate logs

Syntax

Print log and SMCL files

> print *filename* $\left[\right.$, like(*ext*) name(*windowname*) *override_options* $\left.\right]$

Translate log files to SMCL files and vice versa

> translate *filename*$_{in}$ *filename*$_{out}$ $\left[\right.$, translator(*tname*) name(*windowname*)

> *override_options* replace $\left.\right]$

View translator parameter settings

> translator query $\left[\right.$*tname*$\left.\right]$

Change translator parameter settings

> translator set $\left[\right.$*tname setopt setval*$\left.\right]$

Return translator parameter settings to default values

> translator reset *tname*

List current mappings from one extension to another

> transmap query $\left[\right.$.*ext*$\left.\right]$

Specify that files with one extension be treated the same as files with another extension

> transmap define .*ext*$_{new}$.*ext*$_{old}$

filename in print, in addition to being a filename to be printed, may be specified as @Results to mean the Results window and @Viewer to mean the Viewer window.

filename$_{in}$ in translate may be specified just as *filename* in print.

tname in translator specifies the name of a translator; see the translator() option under *Options for translate*.

Description

print prints log, SMCL, and text files. Although there is considerable flexibility in how print (and translate, which print uses) can be set to work, they have already been set up and should just work:

```
. print mylog.smcl
. print mylog.log
```

Unix users may discover that they need to do a bit of setup before `print` works; see *Printing files, Unix* below. International Unix users may also wish to modify the default paper size. All users can tailor `print` and `translate` to their needs.

print may also be used to print the current contents of the Results window or the Viewer. For instance, the current contents of the Results window could be printed by typing

    . print @Results

`translate` translates log and SMCL files from one format to another, the other typically being suitable for printing. `translate` can also translate SMCL logs (logs created by typing, say, `log using mylog`) to ASCII text:

    . translate mylog.smcl mylog.log

You can use `translate` to recover a log when you have forgotten to start one. You may type

    . translate @Results mylog.txt

to capture as ASCII text what is currently shown in the Results window.

This entry provides a general overview of `print` and `translate` and covers in detail the printing and translation of text (nongraphic) files.

Options for print

like(*ext*) specifies how the file should be translated to a form suitable for printing. The default is to determine the translation method from the extension of *filename*. Thus `mylog.smcl` is translated according to the rule for translating `smcl` files, `myfile.txt` is translated according to the rule for translating `txt` files, and so on. (These rules are, in fact, `translate`'s smcl2prn and txt2prn translators, but put that aside for the moment.)

Rules for the following extensions are predefined:

.txt	Assume input file contains ASCII text
.log	Assume input file contains Stata log ASCII text
.smcl	Assume input file contains SMCL

To print a file that has an extension different from those listed above, you can define a new extension, but you do not have to do that. Assume that you wish to print the file `read.me`, which you know to contain ASCII text. If you were just to type `print read.me`, you would be told that Stata cannot translate .me files. (You would actually be told that the translator for me2prn was not found.) You could type `print read.me, like(txt)` to tell `print` to print `read.me` like a .txt file.

On the other hand, you could type

    . transmap define .me .txt

to tell Stata that .me files are always to be treated like .txt files. If you did that, Stata would remember the new rule, even in future sessions.

When you specify option like(), you override the recorded rules. So, if you were to type `print mylog.smcl, like(txt)`, the file would be printed as ASCII text (meaning that all the SMCL commands would show).

name(*windowname*) specifies which window to print when printing a Viewer. The default is for Stata to print the topmost Viewer (Unix(GUI) users: see the second technical note in *Printing files, Unix*). The name() option is ignored when printing the Results window.

The window name is located inside parentheses in the window title. For example, if the title for a Viewer window is *Viewer (#1) [help print]*, the name for the window is #1.

override_options refers to translate's override options. print uses translate to translate the file into a format suitable for sending to the printer. To find out what you can override for *X* files (files ending in *.X*), type translator query *X*2prn. For instance, to find out what you can override when printing .smcl files, type

```
. translator query smcl2prn
  (output omitted )
```

In the omitted output (which varies across operating systems), you might learn that there is an rmargin # tunable value, which specifies the right margin in inches. You could specify the *override_option* rmargin(#) to temporarily override the default value, or you could type translator set smcl2prn rmargin # beforehand to permanently reset the value.

Alternatively, on some computers with some translators, you might discover that nothing can be set.

Options for translate

translator(*tname*) specifies the name of the translator to be used to translate the file. The available translators are

tname	input	output
smcl2ps	SMCL	PostScript
log2ps	Stata ASCII text log	PostScript
txt2ps	generic ASCII text file	PostScript
Viewer2ps	Viewer window	PostScript
Results2ps	Results window	PostScript
smcl2prn	SMCL	default printer format
log2prn	Stata ASCII text log	default printer format
txt2prn	generic ASCII text log	default printer format
Results2prn	Results window	default printer format
Viewer2prn	Viewer window	default printer format
smcl2txt	SMCL	generic ASCII text file
smcl2log	SMCL	Stata ASCII text log
Results2txt	Results window	generic ASCII text file
Viewer2txt	Viewer window	generic ASCII text file
smcl2pdf	SMCL	PDF (Macintosh only)
log2pdf	Stata ASCII text log	PDF (Macintosh only)
txt2pdf	generic ASCII text log	PDF (Macintosh only)
Results2pdf	Results window	PDF (Macintosh only)
Viewer2pdf	Viewer window	PDF (Macintosh only)

If translator() is not specified, translate determines which translator to use from extensions of the filenames specified. Typing translate myfile.smcl myfile.ps would use the smcl2ps translator. Typing translate myfile.smcl myfile.ps, translate(smcl2prn) would override the default and use the smcl2prn translator.

Actually, when you type translate *a.b c.d*, translate looks up *.b* in the transmap extension-synonym table. If *.b* is not found, the translator *b*2*d* is used. If *.b* is found in the table, the mapped extension is used (call it *b′*), and then the translator *b′*2*d* is used. For example,

command	translator used
. translate myfile.smcl myfile.ps	smcl2ps
. translate myfile.odd myfile.ps	odd2ps, which does not exist, so error
. transmap define .odd .txt	
. translate myfile.odd myfile.ps	txt2ps

You can list the mappings that `translate` uses by typing `transmap query`.

name(*windowname*) specifies which window to translate when translating a Viewer. The default is for Stata to translate the topmost Viewer. The `name()` option is ignored when translating the Results window.

The window name is located inside parentheses in the window title. For example, if the title for a Viewer window is *Viewer (#1) [help print]*, the name for the window is #1.

override_options override any of the default options of the specified or implied translator. To find out what you can override for, say, `log2ps`, type

```
. translator query log2ps
  (output omitted )
```

In the omitted output, you might learn that there is an `rmargin #` tunable value, which, for `log2ps`, specifies the right margin in inches. You could specify the *override_option* `rmargin(#)` to temporarily override the default value or type `translator set log2ps rmargin #` beforehand to permanently reset the value.

replace specifies that *filename*$_{out}$ be replaced if it already exists.

Remarks

Remarks below are presented under the headings

> *Printing files*
> *Printing files, Windows and Macintosh*
> *Printing files, Unix*
> *Translating files from one format to another*

Printing files

Printing should be easy; just type

```
. print mylog.smcl
. print mylog.log
```

You can use `print` to print SMCL files, ASCII text files, and even the contents of the Results and Viewer windows:

```
. print @Results
. print @Viewer
. print @Viewer, name(#2)
```

For information about printing and translating graph files, see [G] **graph print** and see [G] **graph export**.

Printing files, Windows and Macintosh

When you type `print`, you are using the same facility that you would be using if you had selected **Print** from the **File** menu. If you try to print a file that Stata does not know about, Stata will complain:

```
. print read.me
translator me2prn not found
(perhaps you need to specify the like() option)
r(111);
```

Then you could type

```
. print read.me, like(txt)
```

to indicate that you wanted `read.me` sent to the printer in the same fashion as if the file were named `readme.txt`, or you could type

```
. transmap define .me .txt
. print read.me
```

Here you are telling Stata once and for all that you want files ending in `.me` to be treated in the same way as files ending in `.txt`. Stata will remember this mapping, even across sessions. To clear the `.me` mapping, type

```
. transmap define .me
```

To see all the mappings, type

```
. transmap query
```

To print to a file, use the `translate` command, not `print`:

```
. translate mylog.smcl mylog.prn
```

`translate` prints to a file by using the Windows print driver when the new filename ends in `.prn`. Under Macintosh, the `prn` translators are the same as the `pdf` translators. We suggest that you simply use the `.pdf` file extension when printing to a file.

Printing files, Unix

Stata assumes that you have a PostScript printer attached to your Unix computer and that the Unix command `lpr(1)` can be used to send PostScript files to it, but you can change this. On your Unix system, typing

```
mycomputer$ lpr < filename
```

may not be sufficient to print PostScript files. For instance, perhaps on your system you would need to type

```
mycomputer$ lpr -Plexmark < filename
```

or

```
mycomputer$ lpr -Plexmark filename
```

or something else. To set the print command to be `lpr -Plexmark` *filename* and to state that the printer expects to receive PostScript files, type

```
. printer define prn ps "lpr -Plexmark @"
```

To set the print command to `lpr -Plexmark <` *filename* and to state that the printer expects to receive ASCII text files, type

```
. printer define prn txt "lpr -Plexmark < @"
```

That is, just type the command necessary to send files to your printer and include an @ sign where the filename should be substituted. Two file formats are available: `ps` and `txt`. The default setting, as shipped from the factory, is

```
. printer define prn ps "lpr < @"
```

We will return to the `printer` command in the technical note that follows because it has some other capabilities you should know about.

In any case, after you redefine the default printer, the following should just work:

```
. print mylog.smcl
. print mylog.log
```

If you try to print a file that Stata does not know about, it will complain:

```
. print read.me
translator me2prn not found
r(111);
```

Here you could type

```
. print read.me, like(txt)
```

to indicate that you wanted `read.me` sent to the printer in the same fashion as if the file were named `readme.txt`, or you could type

```
. transmap define .me .txt
. print read.me
```

Here you are telling Stata once and for all that you want files ending in `.me` to be treated in the same way as files ending in `.txt`. Stata will remember this setting for `.me`, even across sessions.

If you want to clear the `.me` setting, type

```
. transmap define .me
```

If you want to see all your settings, type

```
. transmap query
```

❑ Technical Note

The syntax of the `printer` command is

> printer <u>def</u>ine *printername* $\big[$ { ps | txt } "*Unix command with @*" $\big]$

> printer <u>q</u>uery $\big[$ *printername* $\big]$

You may define multiple printers. By default, `print` uses the printer named `prn`, but `print` has the syntax

> print *filename* $\big[$, like(*ext*) printer(*printername*) *override_options* $\big]$

so, if you define multiple printers, you may route your output to them.

For instance, if you have a second printer on your system, you might type

```
. printer define lexmark ps "lpr -Plexmark < @"
```

After doing that, you could type

```
. print myfile.smcl, printer(lexmark)
```

Any printers that you set will be remembered even across sessions. You can delete printers:

```
. printer define lexmark
```

You can list all the defined printers by typing `printer query`, and you can list the definition of a particular printer, say, `prn`, by typing `printer query prn`.

The default printer `prn` we have predefined for you is

```
. printer define prn ps "lpr < @"
```

meaning that we assume that it is a PostScript printer and that the Unix command `lpr(1)`, without options, is sufficient to cause files to print. Feel free to change the default definition. If you change it, the change will be remembered across sessions.
❑

❑ Technical Note

Unix(GUI) users should note that X-Windows does not have the concept of a window z-order, which prevents Stata from determining which window is the topmost window. Instead, Stata determines which window is topmost based on which window has the focus. However, some window managers will set the focus to a window without bringing the window to the top. What Stata considers the topmost window may not appear topmost visually. For this reason, you should always use the `name()` option to ensure that the correct window is printed.
❑

❑ Technical Note

When you select the Results window to print from the **Print** menu or toolbar button, the result is the same as if you were to issue the `print` command. When you select a Viewer window to print from the **Print** menu or toolbar button, the result is the same as if you were to issue the `print` command with a `name()` option.
❑

The translation to PostScript format is done by `translate` and, in particular, is performed by the translators `smcl2ps`, `log2ps`, and `txt2ps`. There are many tunable parameters in each of these translators. You can display the current values of these tunable parameters for, say, `smcl2ps` by typing

```
. translator query smcl2ps
  (output omitted )
```

and you can set any of the tunable parameters (for instance, setting `smcl2ps`'s `rmargin` value to 1) by typing

```
. translator set smcl2ps rmargin 1
  (output omitted )
```

Any settings you make will be remembered across sessions. You can reset `smcl2ps` to be as it was when Stata was shipped by typing

```
. translator reset smcl2ps
```

Translating files from one format to another

If you have a SMCL log, which you might have created by previously typing `log using mylog`, you can translate it to an ASCII text log by typing

```
. translate myfile.smcl myfile.log
```

and you can translate it to a PostScript file by typing

```
. translate myfile.smcl myfile.ps
```

`translate` translates files from one format to another, and, in fact, `print` uses `translate` to produce a file suitable for sending to the printer.

When you type

```
. translate a.b c.d
```

`translate` looks for the predefined translator *b2d* and uses that to perform the translation. If there is a `transmap` synonym for *b*, however, the mapped value b' is used: $b'2d$.

Only certain translators exist, and they are listed under the description of the `translate()` option in *Options for translate* above, or you can type

```
. translator query
```

for a complete (and perhaps more up-to-date) list.

Anyway, `translate` forms the name *b2d* or $b'2d$, and if the translator does not exist, `translate` issues an error. With the `translator()` option, you can specify exactly which translator to use, and then it does not matter how your files are named.

The only other thing to know is that some translators have tunable parameters that affect how they perform their translation. You can type

```
. translator query translator_name
```

to find out what those parameters are. Some translators have no tunable parameters, and some have many:

```
. translator query smcl2ps
```

header	on					
headertext						
logo	on					
user						
projecttext						
cmdnumber	on					
fontsize	9		lmargin	1.00		
pagesize	letter		rmargin	1.00		
pagewidth	8.50		tmargin	1.00		
pageheight	11.00		bmargin	1.00		
scheme	monochrome					
cust1_result_color	0	0	0	cust2_result_color	0	0	0
cust1_standard_color	0	0	0	cust2_standard_color	0	0	0
cust1_error_color	0	0	0	cust2_error_color	255	0	0
cust1_input_color	0	0	0	cust2_input_color	0	0	0
cust1_link_color	0	0	0	cust2_link_color	0	0	255
cust1_hilite_color	0	0	0	cust2_hilite_color	0	0	0
cust1_result_bold	on			cust2_result_bold	on		
cust1_standard_bold	off			cust2_standard_bold	off		

cust1_error_bold	on	cust2_error_bold	on
cust1_input_bold	off	cust2_input_bold	off
cust1_link_bold	off	cust2_link_bold	off
cust1_hilite_bold	on	cust2_hilite_bold	on
cust1_link_underline	on	cust2_link_underline	on
cust1_hilite_underline	off	cust2_hilite_underline	off

You can temporarily override any setting by specifying option *setopt*(*setval*) on the `translate` (or `print`) command. For instance, you can type

>   . translate ..., ... cmdnumber(off)

or you can reset the value permanently by typing

>   . translator set smcl2ps *setopt* *setval*

For instance,

>   . translator set smcl2ps cmdnumber off

If you reset a value, Stata will remember the change, even in future sessions.

Windows and Macintosh users: The `smcl2ps` (and the other `*2ps` translators) are not used by `print`, even when you have a PostScript printer attached to your computer. Instead, the Windows or Macintosh print driver is used. Resetting `smcl2ps` values will not affect printing; instead, you change the defaults in the Printers Control Panel in Windows and by selecting **Page Setup...** from the **File** menu in Macintosh. You can, however, `translate` files yourself using the `smcl2ps` translator and the other `*2ps` translators.

Saved Results

`transmap query` .*ext* saves in macro `r(suffix)` the mapped extension (without the leading period) or saves *ext* if the *ext* is not mapped.

`translator query` *translatorname* saves *setval* in macro `r(`*setopt*`)` for every *setopt*, *setval* pair.

`printer query` *printername* (Unix only) saves in macro `r(suffix)` the "filetype" of the input that the printer expects (currently "ps" or "txt") and, in macro `r(command)`, the command to send output to the printer.

Methods and Formulas

`print` is implemented as an ado-file.

Also See

[R] **log** — Echo copy of session to file or device

[G] **graph export** — Export current graph

[G] **graph print** — Print current graph

[P] **smcl** — Stata Markup and Control Language

[U] **15 Printing and preserving output**

Title

> **treatreg** — Treatment-effects model

Syntax

Basic syntax

> treatreg *depvar* [*indepvars*], <u>treat</u>(*depvar_t* = *indepvars_t*) [<u>two</u>step]

Full syntax for maximum likelihood estimates only

> treatreg *depvar* [*indepvars*] [*if*] [*in*] [*weight*] ,
>
>   <u>treat</u>(*depvar_t* = *indepvars_t* [, <u>noc</u>onstant]) [*treatreg_ml_options*]

Full syntax for two-step consistent estimates only

> treatreg *depvar* [*indepvars*] [*if*] [*in*] ,
>
>   <u>treat</u>(*depvar_t* = *indepvars_t* [, <u>noc</u>onstant]) <u>two</u>step [*treatreg_ts_options*]

treatreg_ml_options	description
Model	
* <u>treat</u>()	equation for treatment effects
<u>noc</u>onstant	suppress constant term
<u>cons</u>traints(*constraints*)	apply specified linear constraints
<u>col</u>linear	keep collinear variables
SE/Robust	
vce(*vcetype*)	*vcetype* may be oim, <u>r</u>obust, <u>c</u>luster *clustvar*, opg, <u>boot</u>strap, or <u>jack</u>knife
Reporting	
<u>l</u>evel(#)	set confidence level; default is level(95)
<u>first</u>	report first-step probit estimates
noskip	perform likelihood-ratio test
<u>hazard</u>(*newvar*)	create *newvar* containing hazard from treatment equation
Max options	
maximize_options	control maximization process; seldom used

* <u>treat</u>(*depvar_t* =*indepvars_t* [, <u>noc</u>onstant]) is required.

509

treatreg_ts_options	description
Model	
`treat()`	equation for treatment effects
`twostep`	produce two-step consistent estimate
`noconstant`	suppress constant term
SE	
`vce(`*vcetype*`)`	*vcetype* may be `conventional`, `bootstrap`, or `jackknife`
Reporting	
`level(#)`	set confidence level; default is `level(95)`
`first`	report first-step probit estimates
`hazard(`*newvar*`)`	create *newvar* containing hazard from treatment equation

* `treat(`*depvar_t* =*indepvars_t* [, `noconstant`]`)` and `twostep` are required.

depvar, *indepvars*, *depvar_t*, and *indepvars_t* may contain time-series operators; see [U] **11.4.3 Time-series varlists**.
`bootstrap`, `by`, `jackknife`, `rolling`, `statsby`, `svy`, and `xi` are allowed; see [U] **11.1.10 Prefix commands**.
Weights are not allowed with the `bootstrap` prefix.
`aweights` are not allowed with the `jackknife` prefix.
`twostep`, `vce()`, `first`, `noskip`, `hazard()`, and weights are not allowed with the `svy` prefix.
`pweights`, `aweights`, `fweights`, and `iweights` are allowed with maximum likelihood estimation;
    see [U] **11.1.6 weight**. No weights are allowed if `twostep` is specified.
See [U] **20 Estimation and postestimation commands** for more capabilities of estimation commands.

Description

treatreg fits a treatment-effects model by using either a two-step consistent estimator or full maximum likelihood. The treatment-effects model considers the effect of an endogenously chosen binary treatment on another endogenous continuous variable, conditional on two sets of independent variables.

Options for maximum likelihood estimates

⌐ **Model** ⌐

`treat(`*depvar_t* = *indepvars_t* [, `noconstant`]`)` specifies the variables and options for the treatment equation. It is an integral part of specifying a treatment-effects model and is required.

`noconstant`, `constraints(`*constraints*`)`, `collinear`; see [R] **estimation options**.

⌐ **SE/Robust** ⌐

`vce(`*vcetype*`)` specifies the type of standard error reported, which includes types that are derived from asymptotic theory, that are robust to some kinds of misspecification, that allow for intragroup correlation, and that use bootstrap or jackknife methods; see [R] ***vce_option***.

⌐ **Reporting** ⌐

`level(#)`; see [R] **estimation options**.

first specifies that the first-step probit estimates of the treatment equation be displayed before estimation.

noskip specifies that a full maximum-likelihood model with only a constant for the regression equation be fitted. This model is not displayed but is used as the base model to compute a likelihood-ratio test for the model test statistic displayed in the estimation header. By default, the overall model test statistic is an asymptotically equivalent Wald test that all the parameters in the regression equation are zero (except the constant). For many models, this option can substantially increase estimation time.

hazard(*newvar*) will create a new variable containing the hazard from the treatment equation. The hazard is computed from the estimated parameters of the treatment equation.

⌐ Max options ⌐

maximize_options: <u>diff</u>icult, <u>tech</u>nique(*algorithm_spec*), <u>iter</u>ate(*#*), [<u>no</u>]log, <u>tra</u>ce, gradient, showstep, <u>hess</u>ian, <u>shownr</u>tolerance, <u>tol</u>erance(*#*), <u>ltol</u>erance(*#*), <u>gtol</u>erance(*#*), <u>nrtol</u>erance(*#*), <u>nonrtol</u>erance, from(*init_specs*); see [R] **maximize**. These options are seldom used.

Setting the optimization type to technique(bhhh) resets the default *vcetype* to vce(opg).

Options for two-step consistent estimates

⌐ Model ⌐

treat(*depvar_t* = *indepvars_t*[, <u>noc</u>onstant]) specifies the variables and options for the treatment equation. It is an integral part of specifying a treatment-effects model and is required.

twostep specifies that two-step consistent estimates of the parameters, standard errors, and covariance matrix be produced, instead of the default maximum likelihood estimates.

noconstant; see [R] **estimation options**.

⌐ SE ⌐

vce(*vcetype*) specifies the type of standard error reported, which includes types that are derived from asymptotic theory and that use bootstrap or jackknife methods; see [R] **vce_option**.

vce(conventional), the default, uses the conventionally derived variance estimator for generalized least-squares regression.

⌐ Reporting ⌐

level(*#*); see [R] **estimation options**.

first specifies that the first-step probit estimates of the treatment equation be displayed before estimation.

hazard(*newvar*) will create a new variable containing the hazard from the treatment equation. The hazard is computed from the estimated parameters of the treatment equation.

Remarks

The treatment-effects model estimates the effect of an endogenous binary treatment, z_j, on a continuous, fully observed variable y_j, conditional on the independent variables x_j and w_j. The primary interest is in the regression function

$$y_j = \mathbf{x}_j\boldsymbol{\beta} + \delta z_j + \epsilon_j$$

where z_j is an endogenous dummy variable indicating whether the treatment is assigned or not. The binary decision to obtain the treatment z_j is modeled as the outcome of an unobserved latent variable, z_j^*. It is assumed that z_j^* is a linear function of the exogenous covariates $\mathbf{w}_j$ and a random component u_j. Specifically,

$$z_j^* = \mathbf{w}_j\boldsymbol{\gamma} + u_j$$

and the observed decision is

$$z_j = \begin{cases} 1, & \text{if } z_j^* > 0 \\ 0, & \text{otherwise} \end{cases}$$

where ϵ and u are bivariate normal with mean zero and covariance matrix

$$\begin{bmatrix} \sigma & \rho \\ \rho & 1 \end{bmatrix}$$

There are many variations of this model in the literature. Maddala (1983) derives the maximum likelihood and two-step estimators of the version implemented here and also gives a brief review of several empirical applications of this model. Barnow, Cain, and Goldberger (1981) provide another useful derivation of this model. Barnow, Cain, and Goldberger (1981) concentrate on deriving the conditions for which the self-selection bias of the simple OLS estimator of the treatment effect, δ, is nonzero and of a specific sign.

▷ Example 1

We will illustrate **treatreg** with part of the Mroz dataset distributed with Berndt (1991). This dataset contains 753 observations on women's labor supply. Our subsample is of 250 observations, with 150 market laborers and 100 nonmarket laborers.

```
. use http://www.stata-press.com/data/r10/labor
. describe
Contains data from http://www.stata-press.com/data/r10/labor.dta
  obs:           250
  vars:           15                          18 Apr 2007 05:01
  size:        16,000 (98.3% of memory free)

              storage  display    value
variable name  type    format     label      variable label

lfp           float   %9.0g                  1 if woman worked in 1975
whrs          float   %9.0g                  wife's hours of work
k16           float   %9.0g                  # of children younger than 6
k618          float   %9.0g                  # of children between 6 and 18
wa            float   %9.0g                  wife's age
we            float   %9.0g                  wife's education attainment
ww            float   %9.0g                  wife's wage
hhrs          float   %9.0g                  husband's hours worked in 1975
ha            float   %9.0g                  husband's age
he            float   %9.0g                  husband's educational attainment
hw            float   %9.0g                  husband's wage
faminc        float   %9.0g                  family income
wmed          float   %9.0g                  wife's mother's educational
                                               attainment
wfed          float   %9.0g                  wife's father's educational
                                               attainment
cit           float   %9.0g                  1 if live in large city

Sorted by:
```

```
. summarize
```

Variable	Obs	Mean	Std. Dev.	Min	Max
lfp	250	.6	.4908807	0	1
whrs	250	799.84	915.6035	0	4950
kl6	250	.236	.5112234	0	3
k618	250	1.364	1.370774	0	8
wa	250	42.92	8.426483	30	60
we	250	12.352	2.164912	5	17
ww	250	2.27523	2.59775	0	14.631
hhrs	250	2234.832	600.6702	768	5010
ha	250	45.024	8.171322	30	60
he	250	12.536	3.106009	3	17
hw	250	7.494435	4.636192	1.0898	40.509
faminc	250	23062.54	12923.98	3305	91044
wmed	250	9.136	3.536031	0	17
wfed	250	8.608	3.751082	0	17
cit	250	.624	.4853517	0	1

We will assume that the wife went to college if her educational attainment is more than 12 years. Let wc be the dummy variable indicating whether the individual went to college. With this definition, our sample contains the following distribution of college education:

```
. generate wc = 0
. replace wc = 1 if we > 12
(69 real changes made)
. tab wc
```

wc	Freq.	Percent	Cum.
0	181	72.40	72.40
1	69	27.60	100.00
Total	250	100.00	

We will model the wife's wage as a function of her age, whether the family was living in a big city, and whether she went to college. An ordinary least-squares estimation produces the following results:

```
. regress ww wa cit wc
```

Source	SS	df	MS		Number of obs =	250
					F(3, 246) =	4.82
Model	93.2398568	3	31.0799523		Prob > F =	0.0028
Residual	1587.08776	246	6.45157627		R-squared =	0.0555
					Adj R-squared =	0.0440
Total	1680.32762	249	6.74830369		Root MSE =	2.54

ww	Coef.	Std. Err.	t	P>|t|	[95% Conf. Interval]	
wa	-.0104985	.0192667	-0.54	0.586	-.0484472	.0274502
cit	.1278922	.3389058	0.38	0.706	-.5396351	.7954195
wc	1.332192	.3644344	3.66	0.000	.6143819	2.050001
_cons	2.278337	.8432385	2.70	0.007	.6174488	3.939225

Is 1.332 a consistent estimate of the marginal effect of a college education on wages? If individuals choose whether or not to attend college and the error term of the model that gives rise to this choice is correlated with the error term in the wage equation, then the answer is no. (See Barnow, Cain, and Goldberger 1981 for a good discussion of the existence and sign of selectivity bias.) We might suspect

that individuals with higher abilities, either innate or due to the circumstances of their birth, would be more likely to go to college and to earn higher wages. Such ability is, of course, unobserved. Furthermore, if the error term in our model for going to college is correlated with ability, and the error term in our wage equation is correlated with ability, the two terms should be positively correlated. These conditions make the problem of signing the selectivity bias equivalent to an omitted-variable problem. In the case at hand, because we would expect the correlation between the omitted variable and a college education to be positive, we suspect that OLS is biased upward.

To account for the bias, we fit the treatment-effects model. We model the wife's college decision as a function of her mother's and her father's educational attainment. Thus we are interested in fitting the model

$$ww = \beta_0 + \beta_1 wa + \beta_2 cit + \delta wc + \epsilon$$

$$wc^* = \gamma_0 + \gamma_1 wmed + \gamma_2 wfed + u$$

where

$$wc = \begin{cases} 1, & wc^* > 0, \text{ i.e., wife went to college} \\ 0, & \text{otherwise} \end{cases}$$

and where ϵ and u have a bivariate normal distribution with zero mean and covariance matrix

$$\begin{bmatrix} \sigma & \rho \\ \rho & 1 \end{bmatrix}$$

The following output gives the maximum likelihood estimates of the parameters of this model:

```
. treatreg ww wa cit, treat(wc=wmed wfed)
Iteration 0:   log likelihood = -707.07237
Iteration 1:   log likelihood = -707.07215
Iteration 2:   log likelihood = -707.07215

Treatment-effects model -- MLE              Number of obs   =        250
                                            Wald chi2(3)    =       4.11
Log likelihood = -707.07215                 Prob > chi2     =     0.2501
```

	Coef.	Std. Err.	z	P>|z|	[95% Conf. Interval]	
ww						
wa	-.0110424	.0199652	-0.55	0.580	-.0501735	.0280887
cit	.127636	.3361938	0.38	0.704	-.5312917	.7865638
wc	1.271327	.7412951	1.72	0.086	-.1815842	2.724239
_cons	2.318638	.9397573	2.47	0.014	.4767478	4.160529
wc						
wmed	.1198055	.0320056	3.74	0.000	.0570757	.1825352
wfed	.0961886	.0290868	3.31	0.001	.0391795	.1531977
_cons	-2.631876	.3309128	-7.95	0.000	-3.280453	-1.983299
/athrho	.0178668	.1899898	0.09	0.925	-.3545063	.3902399
/lnsigma	.9241584	.0447455	20.65	0.000	.8364588	1.011858
rho	.0178649	.1899291			-.3403659	.371567
sigma	2.519747	.1127473			2.308179	2.750707
lambda	.0450149	.4786442			-.8931105	.9831404

```
LR test of indep. eqns. (rho = 0):   chi2(1) =     0.01   Prob > chi2 = 0.9251
```

In the input, we specified that the continuous dependent variable, ww (wife's wage), is a linear function of cit and wa. Note the syntax for the treatment variable. The treatment wc is not included in the first variable list; it is specified in the treat() option. In this example, wmed and wfed are specified as the exogenous covariates in the treatment equation.

The output has the form of many two-equation estimators in Stata. We note that our conjecture that the OLS estimate was biased upward is verified. But it is perhaps more interesting that the size of the bias is negligible, and the likelihood-ratio test at the bottom of the output indicates that we cannot reject the null hypothesis that the two error terms are uncorrelated. This result might be due to several specification errors. We ignored the selectivity bias due to the endogeneity of entering the labor market. We have also written both the wage equation and the college-education equation in linear form, ignoring any higher power terms or interactions.

The results for the two ancillary parameters require explanation. For numerical stability during optimization, treatreg does not directly estimate ρ or σ. Instead, treatreg estimates the inverse hyperbolic tangent of ρ,

$$\text{atanh } \rho = \frac{1}{2} \ln \left(\frac{1 + \rho}{1 - \rho} \right)$$

and $\ln\sigma$. Also, treatreg reports $\lambda = \rho\sigma$, along with an estimate of the standard error of the estimate and confidence interval.

◁

▷ Example 2

Stata also produces a two-step estimator of the model with the twostep option. Maximum likelihood estimation of the parameters can be time consuming with large datasets, and the two-step estimates may provide a good alternative in such cases. Continuing with the women's wage model, we can obtain the two-step estimates with consistent covariance estimates by typing

```
. treatreg ww wa cit, treat(wc=wmed wfed) twostep
Treatment-effects model -- two-step estimates    Number of obs   =        250
                                                 Wald chi2(3)    =       3.67
                                                 Prob > chi2     =     0.2998
```

	Coef.	Std. Err.	z	P>|z|	[95% Conf. Interval]	
ww						
wa	-.0111623	.020152	-0.55	0.580	-.0506594	.0283348
cit	.1276102	.33619	0.38	0.704	-.53131	.7865305
wc	1.257995	.8007428	1.57	0.116	-.3114319	2.827422
_cons	2.327482	.9610271	2.42	0.015	.4439031	4.21106
wc						
wmed	.1198888	.0319859	3.75	0.000	.0571976	.1825801
wfed	.0960764	.0290581	3.31	0.001	.0391236	.1530292
_cons	-2.631496	.3308344	-7.95	0.000	-3.279919	-1.983072
hazard						
lambda	.0548738	.5283928	0.10	0.917	-.9807571	1.090505
rho	0.02178					
sigma	2.5198211					
lambda	.05487379	.5283928				

The reported `lambda` (λ) is the parameter estimate on the hazard from the augmented regression, which is derived in Maddala (1983) and presented in *Methods and Formulas* below.

◁

❏ Technical Note

The difference in expected earnings between participants and nonparticipants is

$$E\left(y_j \mid z_j = 1\right) - E\left(y_j \mid z_j = 0\right) \;=\; \delta + \rho\sigma \left[\frac{\phi(\mathbf{w}_j\boldsymbol{\gamma})}{\Phi(\mathbf{w}_j\boldsymbol{\gamma})\left\{1 - \Phi(\mathbf{w}_j\boldsymbol{\gamma})\right\}} \right]$$

where ϕ is the standard normal density and Φ is the standard normal cumulative distribution function. If the correlation between the error terms, ρ, is zero, the problem reduces to one estimable by OLS and the difference is simply δ. Since ρ is positive in the example, least squares overestimates the treatment effect, δ.

❏

(Continued on next page)

Saved Results

treatreg saves the following in e():

Scalars

e(N)	number of observations	e(selambda)	standard error of λ
e(k)	number of parameters	e(sigma)	estimate of sigma
e(k_eq)	number of equations	e(chi2)	χ^2
e(k_eq_model)	number of equations in model Wald test	e(chi2_c)	χ^2 for comparison test
		e(p_c)	p-value for comparison test
e(k_aux)	number of auxiliary parameters	e(p)	significance
e(k_dv)	number of dependent variables	e(rho)	ρ
e(df_m)	model degrees of freedom	e(rank)	rank of e(V)
e(ll)	log likelihood	e(ic)	number of iterations
e(N_clust)	number of clusters	e(rc)	return code
e(lambda)	λ	e(converged)	1 if converged, 0 otherwise

Macros

e(cmd)	treatreg	e(diparm#)	display transformed parameter #
e(cmdline)	command as typed	e(opt)	type of optimization
e(depvar)	name of dependent variable	e(method)	ml or twostep
e(hazard)	variable containing hazard	e(ml_method)	type of ml method
e(wtype)	weight type	e(user)	name of likelihood-evaluator program
e(wexp)	weight expression		
e(title)	title in estimation output	e(technique)	maximization technique
e(clustvar)	name of cluster variable	e(crittype)	optimization criterion
e(chi2type)	Wald or LR; type of model χ^2 test	e(properties)	b V
e(chi2_ct)	Wald or LR; type of model χ^2 test corresponding to e(chi2_c)	e(footnote)	program used to implement the footnote display
e(vce)	vcetype specified in vce()	e(predict)	program used to implement predict
e(vcetype)	title used to label Std. Err.		

Matrices

e(b)	coefficient vector	e(V)	variance–covariance matrix of the estimators
e(ilog)	iteration log (up to 20 iterations)		
e(gradient)	gradient vector		

Functions

e(sample)	marks estimation sample

Methods and Formulas

treatreg is implemented as an ado-file. Maddala (1983, 117–122) derives both the maximum likelihood and the two-step estimator implemented here. Greene (2003, 787–789) also provides an introduction to the treatment-effects model.

The primary regression equation of interest is

$$y_j = \mathbf{x}_j\boldsymbol{\beta} + \delta z_j + \epsilon_j$$

where z_j is a binary decision variable that is assumed to stem from an unobservable latent variable:

$$z_j^* = \mathbf{w}_j\boldsymbol{\gamma} + u_j$$

The decision to obtain the treatment is made according to the rule

$$
z_j = \begin{cases} 1, & \text{if } z_j^* > 0 \\ 0, & \text{otherwise} \end{cases}
$$

where ϵ and u are bivariate normal with mean zero and covariance matrix

$$
\begin{bmatrix} \sigma & \rho \\ \rho & 1 \end{bmatrix}
$$

The likelihood function for this model is given in Maddala (1983, 122). Greene (2000, 180) discusses the standard method of reducing a bivariate normal to a function of a univariate normal and the correlation ρ. The following is the log likelihood for observation j,

$$
\ln L_j = \begin{cases} \ln \Phi \left\{ \dfrac{\mathbf{w}_j \boldsymbol{\gamma} + (y_j - \mathbf{x}_j \boldsymbol{\beta} - \delta)\rho/\sigma}{\sqrt{1 - \rho^2}} \right\} - \dfrac{1}{2} \left(\dfrac{y_j - \mathbf{x}_j \boldsymbol{\beta} - \delta}{\sigma} \right)^2 - \ln(\sqrt{2\pi}\sigma), & z_j = 1 \\[3ex] \ln \Phi \left\{ \dfrac{-\mathbf{w}_j \boldsymbol{\gamma} - (y_j - \mathbf{x}_j \boldsymbol{\beta})\rho/\sigma}{\sqrt{1 - \rho^2}} \right\} - \dfrac{1}{2} \left(\dfrac{y_j - \mathbf{x}_j \boldsymbol{\beta}}{\sigma} \right)^2 - \ln(\sqrt{2\pi}\sigma), & z_j = 0 \end{cases}
$$

where $\Phi()$ is the cumulative distribution function of the standard normal distribution.

In the maximum likelihood estimation, σ and ρ are not directly estimated. Rather $\ln \sigma$ and $\operatorname{atanh} \rho$ are directly estimated, where

$$
\operatorname{atanh} \rho = \frac{1}{2} \ln \left(\frac{1 + \rho}{1 - \rho} \right)
$$

The standard error of $\lambda = \rho\sigma$ is approximated through the delta method, which is given by

$$
\operatorname{Var}(\lambda) \approx \mathbf{D} \operatorname{Var} \left\{ (\operatorname{atanh} \rho \ \ln\sigma) \right\} \mathbf{D}'
$$

where $\mathbf{D}$ is the Jacobian of λ with respect to $\operatorname{atanh} \rho$ and $\ln \sigma$.

Maddala (1983, 120–122) also derives the two-step estimator. In the first stage, probit estimates are obtained of the treatment equation

$$
\Pr(z_j = 1 \mid \mathbf{w}_j) = \Phi(\mathbf{w}_j \boldsymbol{\gamma})
$$

From these estimates, the hazard, h_j, for each observation j is computed as

$$
h_j = \begin{cases} \phi(\mathbf{w}_j \widehat{\boldsymbol{\gamma}}) / \Phi(\mathbf{w}_j \widehat{\boldsymbol{\gamma}}), & z_j = 1 \\[2ex] -\phi(\mathbf{w}_j \widehat{\boldsymbol{\gamma}}) / \left\{ 1 - \Phi(\mathbf{w}_j \widehat{\boldsymbol{\gamma}}) \right\}, & z_j = 0 \end{cases}
$$

where ϕ is the standard normal density function. If

$$
d_j = h_j (h_j + \widehat{\boldsymbol{\gamma}} \mathbf{w}_j)
$$

then

$$E\left(y_j \mid z_j\right) = \mathbf{x}_j\boldsymbol{\beta} + \delta\mathbf{z}_j + \rho\sigma h_j$$
$$\mathrm{Var}\left(y_j \mid z_j\right) = \sigma^2\left(1 - \rho^2 d_j\right)$$

The two-step parameter estimates of $\boldsymbol{\beta}$ and δ are obtained by augmenting the regression equation with the hazard h. Thus the regressors become $\left[\,\mathbf{x}\ \mathbf{z}\ h\,\right]$, and the additional parameter estimate β_h is obtained on the variable containing the hazard. A consistent estimate of the regression disturbance variance is obtained using the residuals from the augmented regression and the parameter estimate on the hazard

$$\widehat{\sigma}^{\,2} = \frac{\mathbf{e}'\mathbf{e} + \beta_h^2 \sum_{j=1}^{N} d_j}{N}$$

The two-step estimate of ρ is then

$$\widehat{\rho} = \frac{\beta_h}{\widehat{\sigma}}$$

To understand how the consistent estimates of the coefficient covariance matrix based on the augmented regression are derived, let $\mathbf{A} = \left[\,\mathbf{x}\ \mathbf{z}\ h\,\right]$ and $\mathbf{D}$ be a square diagonal matrix of size N with $\left(1 - \widehat{\rho}^{\,2} d_j\right)$ on the diagonal elements. The conventional VCE is

$$\mathbf{V}_{\mathrm{twostep}} = \widehat{\sigma}^{\,2}(\mathbf{A}'\mathbf{A})^{-1}(\mathbf{A}'\mathbf{D}\mathbf{A} + \mathbf{Q})(\mathbf{A}'\mathbf{A})^{-1}$$

where

$$\mathbf{Q} = \widehat{\rho}^{\,2}(\mathbf{A}'\mathbf{D}\mathbf{A})\mathbf{V}_{\mathbf{p}}(\mathbf{A}'\mathbf{D}\mathbf{A})$$

and $\mathbf{V}_{\mathbf{p}}$ is the variance–covariance estimate from the probit estimation of the treatment equation.

References

Barnow, B., G. Cain, and A. S. Goldberger. 1981. Issues in the analysis of selectivity bias. In *Evaluation Studies Review Annual*, vol. 5, ed. E. Stromsdorfer and G. Farkas, 43–59. Beverly Hills: Sage.

Berndt, E. R. 1991. *The Practice of Econometrics*. New York: Addison–Wesley.

Cong, R., and D. M. Drukker. 2000. sg141: Treatment-effects model. *Stata Technical Bulletin* 55: 25–33. Reprinted in *Stata Technical Bulletin Reprints*, vol. 10, pp. 159–169.

Greene, W. H. 2000. *Econometric Analysis*. 4th ed. Upper Saddle River, NJ: Prentice Hall.

———. 2003. *Econometric Analysis*. 5th ed. Upper Saddle River, NJ: Prentice Hall.

Maddala, G. S. 1983. *Limited-Dependent and Qualitative Variables in Econometrics*. Cambridge: Cambridge University Press.

Also See

Title

> **treatreg postestimation** — Postestimation tools for treatreg

Description

The following postestimation commands are available for `treatreg`:

command	description
adjust[1]	adjusted predictions of $\mathbf{x}\boldsymbol{\beta}$
estat[2]	AIC, BIC, VCE, and estimation sample summary
estat (svy)	postestimation statistics for survey data
estimates	cataloging estimation results
lincom	point estimates, standard errors, testing, and inference for linear combinations of coefficients
lrtest[3]	likelihood-ratio test
mfx	marginal effects or elasticities
nlcom	point estimates, standard errors, testing, and inference for nonlinear combinations of coefficients
predict	predictions, residuals, influence statistics, and other diagnostic measures
predictnl	point estimates, standard errors, testing, and inference for generalized predictions
suest[2]	seemingly unrelated estimation
test	Wald tests for simple and composite linear hypotheses
testnl	Wald tests of nonlinear hypotheses

[1] `adjust` is not appropriate with time-series operators.

[2] `estat ic` and `suest` are not appropriate after `treatreg, twostep`.

[3] `lrtest` is not appropriate with svy estimation results.

See the corresponding entries in the *Stata Base Reference Manual* for details, but see
[SVY] **estat** for details about `estat` (svy).

Syntax for predict

After ML or twostep

> predict [*type*] *newvar* [*if*] [*in*] [, *statistic*]

After ML

> predict [*type*] { *stub** | *newvar*$_{\text{reg}}$ *newvar*$_{\text{treat}}$ *newvar*$_{\text{athrho}}$ *newvar*$_{\text{lnsigma}}$ }
>
> [*if*] [*in*] , <u>sc</u>ores

statistic	description
Main	
xb	linear prediction; the default
stdp	standard error of the prediction
stdf	standard error of the forecast
yctrt	$E(y_j \mid \text{treatment} = 1)$
ycntrt	$E(y_j \mid \text{treatment} = 0)$
ptrt	$\Pr(\text{treatment} = 1)$
xbtrt	linear prediction for treatment equation
stdptrt	standard error of the linear prediction for treatment equation

These statistics are available both in and out of sample; type predict ... if e(sample) ... if wanted only for the estimation sample.

stdf is not allowed with svy estimation results.

Options for predict

⌐ Main ⌐

xb, the default, calculates the linear prediction, $\mathbf{x}_j\mathbf{b}$.

stdp calculates the standard error of the prediction, which can be thought of as the standard error of the predicted expected value or mean for the observation's covariate pattern. The standard error of the prediction is also referred to as the standard error of the fitted value.

stdf calculates the standard error of the forecast, which is the standard error of the point prediction for one observation. It is commonly referred to as the standard error of the future or forecast value. By construction, the standard errors produced by stdf are always larger than those produced by stdp; see [R] **regress postestimation** *Methods and Formulas*.

yctrt calculates the expected value of the dependent variable conditional on the presence of the treatment: $E(y_j \mid \text{treatment} = 1)$.

ycntrt calculates the expected value of the dependent variable conditional on the absence of the treatment: $E(y_j \mid \text{treatment} = 0)$.

ptrt calculates the probability of the presence of the treatment:
$\Pr(\text{treatment} = 1) = \Pr(\mathbf{w}_j\boldsymbol{\gamma} + u_j > 0)$.

xbtrt calculates the linear prediction for the treatment equation.

stdptrt calculates the standard error of the linear prediction for the treatment equation.

scores, not available with twostep, calculates equation-level score variables.

The first new variable will contain $\partial \ln L/\partial(\boldsymbol{x}_j\boldsymbol{\beta})$.

The second new variable will contain $\partial \ln L/\partial(\boldsymbol{w}_j\boldsymbol{\gamma})$.

The third new variable will contain $\partial \ln L/\partial \operatorname{atanh} \rho$.

The fourth new variable will contain $\partial \ln L/\partial \ln \sigma$.

Remarks

▷ Example 1

The default statistic produced by predict after treatreg is the expected value of the dependent variable from the underlying distribution of the regression model. For example 1 in [R] **treatreg**, this model is

$$ \mathtt{ww} = \beta_0 + \beta_1 \mathtt{wa} + \beta_2 \mathtt{cit} + \delta \mathtt{wc} + \epsilon $$

Several other interesting aspects of the treatment effects model can be explored with predict. We continue with our wage model, the wife's expected wage, conditional on attending college, can be obtained with the yctrt option. The wife's expected wages, conditional on not attending college, can be obtained with the ycntrt option. Thus the difference in expected wages between participants and nonparticipants is the difference between yctrt and ycntrt. For the case at hand, we have the following calculation:

```
. predict wwctrt, yctrt
. predict wwcntrt, ycntrt
. generate diff = wwctrt - wwcntrt
. summarize diff
```

Variable	Obs	Mean	Std. Dev.	Min	Max
diff	250	1.356912	.0134202	1.34558	1.420173

◁

Methods and Formulas

All postestimation commands listed above are implemented as ado-files.

Also See

[R] **treatreg** — Treatment-effects model

[U] **20 Estimation and postestimation commands**

Title

truncreg — Truncated regression

Syntax

truncreg *depvar* [*indepvars*] [*if*] [*in*] [*weight*] [, *options*]

options	description
Model	
<u>nocon</u>stant	suppress constant term
ll(*varname* | #)	lower limit for left truncation
ul(*varname* | #)	upper limit for right truncation
<u>off</u>set(*varname*)	include *varname* in model with coefficient constrained to 1
<u>constr</u>aints(*constraints*)	apply specified linear constraints
<u>coll</u>inear	keep collinear variables
SE/Robust	
vce(*vcetype*)	*vcetype* may be oim, <u>r</u>obust, <u>c</u>luster *clustvar*, opg, <u>boot</u>strap, or <u>jack</u>knife
Reporting	
<u>level</u>(#)	set confidence level; default is level(95)
noskip	perform likelihood-ratio test
Max options	
maximize_options	control maximization process; seldom used

depvar and *indepvars* may contain time-series operators; see [U] **11.4.3 Time-series varlists**.

bootstrap, by, jackknife, rolling, statsby, svy, and xi are allowed; see [U] **11.1.10 Prefix commands**.

Weights are not allowed with the bootstrap prefix.

aweights are not allowed with the jackknife prefix.

vce(), noskip, and weights are not allowed with the svy prefix.

aweights, fweights, iweights, and pweights are allowed; see [U] **11.1.6 weight**.

See [U] **20 Estimation and postestimation commands** for more capabilities of estimation commands.

Description

truncreg fits a regression model of *depvar* on *varlist* from a sample drawn from a restricted part of the population. Under the normality assumption for the whole population, the error terms in the truncated regression model have a truncated normal distribution, which is a normal distribution that has been scaled upward so that the distribution integrates to one over the restricted range.

Options

<u>Model</u>

noconstant; see [R] **estimation options**.

ll(*varname* | #) and ul(*varname* | #) indicate the lower and upper limits for truncation, respectively. You may specify one or both. Observations with *depvar* ≤ ll() are left-truncated, observations with *depvar* ≥ ul() are right-truncated, and the remaining observations are not truncated. See [R] **tobit** for a more detailed description.

offset(*varname*), constraints(*constraints*), collinear; see [R] **estimation options**.

    SE/Robust
---

vce(*vcetype*) specifies the type of standard error reported, which includes types that are derived from asymptotic theory, that are robust to some kinds of misspecification, that allow for intragroup correlation, and that use bootstrap or jackknife methods; see [R] **vce_option**.

    Reporting
---

level(#); see [R] **estimation options**.

noskip specifies that a full maximum-likelihood model with only a constant for the regression equation be fitted. This model is not displayed but is used as the base model to compute a likelihood-ratio test for the model test statistic displayed in the estimation header. By default, the overall model test statistic is an asymptotically equivalent Wald test of all the parameters in the regression equation being zero (except the constant). For many models, this option can substantially increase estimation time.

    Max options
---

maximize_options: difficult, technique(*algorithm_spec*), iterate(#), [no]log, trace, gradient, showstep, hessian, shownrtolerance, tolerance(#), ltolerance(#), gtolerance(#), nrtolerance(#), nonrtolerance, from(*init_specs*); see [R] **maximize**. These options are seldom used, but you may use the ltol(#) option to relax the convergence criterion; the default is 1e-6 during specification searches.

Setting the optimization type to technique(bhhh) resets the default *vcetype* to vce(opg).

Remarks

Truncated regression fits a model of a dependent variable on independent variables from a restricted part of a population. Truncation is essentially a characteristic of the distribution from which the sample data are drawn. If x has a normal distribution with mean μ and standard deviation σ, the density of the truncated normal distribution is

$$f\left(x \mid a < x < b\right) = \frac{f(x)}{\Phi\left(\frac{b-\mu}{\sigma}\right) - \Phi\left(\frac{a-\mu}{\sigma}\right)}$$

$$= \frac{\frac{1}{\sigma}\phi\left(\frac{x-\mu}{\sigma}\right)}{\Phi\left(\frac{b-\mu}{\sigma}\right) - \Phi\left(\frac{a-\mu}{\sigma}\right)}$$

where ϕ and Φ are the density and distribution functions of the standard normal distribution.

Compared with the mean of the untruncated variable, the mean of the truncated variable is greater if the truncation is from below, and the mean of the truncated variable is smaller if the truncation is from above. Moreover, truncation reduces the variance compared with the variance in the untruncated distribution.

▷ Example 1

We will demonstrate `truncreg` with part of the Mroz dataset distributed with Berndt (1991). This dataset contains 753 observations on women's labor supply. Our subsample is of 250 observations, with 150 market laborers and 100 nonmarket laborers.

```
. use http://www.stata-press.com/data/r10/laborsub
. describe
Contains data from http://www.stata-press.com/data/r10/laborsub.dta
  obs:            250
  vars:             6                          25 Sep 2006 18:36
  size:         2,750 (99.6% of memory free)
```

variable name	storage type	display format	value label	variable label
lfp	byte	%9.0g		1 if woman worked in 1975
whrs	int	%9.0g		Wife's hours of work
k16	byte	%9.0g		# of children younger than 6
k618	byte	%9.0g		# of children between 6 and 18
wa	byte	%9.0g		Wife's age
we	byte	%9.0g		Wife's educational attainment

```
Sorted by:
. summarize, sep(0)
```

Variable	Obs	Mean	Std. Dev.	Min	Max
lfp	250	.6	.4908807	0	1
whrs	250	799.84	915.6035	0	4950
k16	250	.236	.5112234	0	3
k618	250	1.364	1.370774	0	8
wa	250	42.92	8.426483	30	60
we	250	12.352	2.164912	5	17

We first perform ordinary least-squares estimation on the market laborers.

```
. regress whrs k16 k618 wa we if whrs > 0
```

Source	SS	df	MS
Model	7326995.15	4	1831748.79
Residual	94793104.2	145	653745.546
Total	102120099	149	685369.794

```
Number of obs =     150
F(  4,   145) =    2.80
Prob > F      =  0.0281
R-squared     =  0.0717
Adj R-squared =  0.0461
Root MSE      =  808.55
```

whrs	Coef.	Std. Err.	t	P>|t|	[95% Conf. Interval]
k16	-421.4822	167.9734	-2.51	0.013	-753.4748 -89.48953
k618	-104.4571	54.18616	-1.93	0.056	-211.5538 2.639666
wa	-4.784917	9.690502	-0.49	0.622	-23.9378 14.36797
we	9.353195	31.23793	0.30	0.765	-52.38731 71.0937
_cons	1629.817	615.1301	2.65	0.009	414.0371 2845.597

Now we use `truncreg` to perform truncated regression with truncation from below zero.

```
. truncreg whrs kl6 k618 wa we, ll(0)
(note: 100 obs. truncated)

Fitting full model:

Iteration 0:   log likelihood = -1205.6992
Iteration 1:   log likelihood = -1200.9873
Iteration 2:   log likelihood = -1200.9159
Iteration 3:   log likelihood = -1200.9157
Iteration 4:   log likelihood = -1200.9157

Truncated regression
Limit:   lower =           0                    Number of obs =      150
         upper =        +inf                    Wald chi2(4)  =    10.05
Log likelihood = -1200.9157                     Prob > chi2   =   0.0395
```

whrs	Coef.	Std. Err.	z	P>|z|	[95% Conf.	Interval]
kl6	-803.0042	321.3614	-2.50	0.012	-1432.861	-173.1474
k618	-172.875	88.72898	-1.95	0.051	-346.7806	1.030579
wa	-8.821123	14.36848	-0.61	0.539	-36.98283	19.34059
we	16.52873	46.50375	0.36	0.722	-74.61695	107.6744
_cons	1586.26	912.355	1.74	0.082	-201.9233	3374.442
/sigma	983.7262	94.44303	10.42	0.000	798.6213	1168.831

If we assume that our data were censored, the tobit model is

```
. tobit whrs kl6 k618 wa we, ll(0)

Tobit regression                              Number of obs   =      250
                                              LR chi2(4)      =    23.03
                                              Prob > chi2     =   0.0001
Log likelihood = -1367.0903                   Pseudo R2       =   0.0084
```

whrs	Coef.	Std. Err.	t	P>|t|	[95% Conf.	Interval]
kl6	-827.7657	214.7407	-3.85	0.000	-1250.731	-404.8008
k618	-140.0192	74.22303	-1.89	0.060	-286.2129	6.174547
wa	-24.97919	13.25639	-1.88	0.061	-51.08969	1.131317
we	103.6896	41.82393	2.48	0.014	21.31093	186.0683
_cons	589.0001	841.5467	0.70	0.485	-1068.556	2246.556
/sigma	1309.909	82.73335			1146.953	1472.865

```
Obs. summary:          100  left-censored observations at whrs<=0
                       150      uncensored observations
                         0 right-censored observations
```

◁

❑ Technical Note

Whether truncated regression is more appropriate than the ordinary least-squares estimation depends on the purpose of that estimation. If we are interested in the mean of wife's working hours conditional on the subsample of market laborers, least-squares estimation is appropriate. However if we are interested in the mean of wife's working hours regardless of market or nonmarket labor status, least-squares estimates could be seriously misleading.

Truncation and censoring are different concepts. A sample has been censored if no observations have been systematically excluded but some of the information contained in them has been suppressed. In a truncated distribution, only the part of the distribution above (or below, or between) the truncation points is relevant to our computations. We need to scale it up by the probability that an observation falls in the range that interests us to make the distribution integrate to one. The censored distribution used by tobit, however, is a mixture of discrete and continuous distributions. Instead of rescaling over the observable range, we simply assign the full probability from the censored regions to the censoring points. The truncated regression model is sometimes less well behaved than the tobit model. Davidson and MacKinnon (1993) provide an example where truncation results in more inconsistency than censoring.

❏

Saved Results

truncreg saves the following in e():

Scalars

e(N)	number of observations	e(ll_0)	log likelihood, constant-only model
e(N_bf)	number of obs. before truncation	e(N_clust)	number of clusters
e(chi2)	model χ^2	e(sigma)	estimate of sigma
e(k_eq)	number of equations in e(b)	e(p)	significance
e(k_eq_model)	number of equations in model Wald test	e(rank)	rank of e(V)
		e(ic)	number of iterations
e(k_aux)	number of auxiliary parameters	e(rc)	return code
e(df_m)	model degrees of freedom	e(converged)	1 if converged, 0 otherwise
e(ll)	log likelihood		

Macros

e(cmd)	truncreg	e(vce)	*vcetype* specified in vce()
e(cmdline)	command as typed	e(vcetype)	title used to label Std. Err.
e(llopt)	contents of ll(), if specified	e(opt)	type of optimization
e(ulopt)	contents of ul(), if specified	e(ml_method)	type of ml method
e(depvar)	name of dependent variable	e(chi2type)	Wald or LR; type of model χ^2 test
e(wtype)	weight type	e(user)	name of likelihood-evaluator program
e(wexp)	weight expression	e(technique)	maximization technique
e(title)	title in estimation output	e(crittype)	optimization criterion
e(clustvar)	name of cluster variable	e(properties)	b V
e(offset)	offset	e(predict)	program used to implement predict

Matrices

e(b)	coefficient vector	e(V)	variance–covariance matrix of the estimators
e(ilog)	iteration log (up to 20 iterations)		
e(gradient)	gradient vector	e(means)	means of independent variables
		e(dummy)	indicator for dummy variables

Functions

e(sample)	marks estimation sample

Methods and Formulas

truncreg is implemented as an ado-file. Greene (2003, 756–761) and Davidson and MacKinnon (1993, 534–537) provide introductions to the truncated regression model.

Let $\mathbf{y} = \mathbf{X}\boldsymbol{\beta} + \boldsymbol{\epsilon}$ be the model. $\mathbf{y}$ represents continuous outcomes either observed or not observed. Our model assumes that $\boldsymbol{\epsilon} \sim N(\mathbf{0}, \sigma^2 \mathbf{I})$.

Let a be the lower limit and b be the upper limit. The log likelihood is

$$\ln L = -\frac{n}{2} \log(2\pi\sigma^2) - \frac{1}{2\sigma^2} \sum_{j=1}^{n} (y_j - \mathbf{x}_j\boldsymbol{\beta})^2 - \sum_{j=1}^{n} \log \left\{ \Phi\left(\frac{b - \mathbf{x}_j\boldsymbol{\beta}}{\sigma}\right) - \Phi\left(\frac{a - \mathbf{x}_j\boldsymbol{\beta}}{\sigma}\right) \right\}$$

References

Berndt, E. R. 1991. *The Practice of Econometrics: Classic and Contemporary.* New York: Addison–Wesley.

Cong, R. 1999. sg122: Truncated regression. *Stata Technical Bulletin* 52: 47–52. Reprinted in *Stata Technical Bulletin Reprints*, vol. 9, pp. 248–255.

Davidson, R., and J. G. MacKinnon. 1993. *Estimation and Inference in Econometrics.* New York: Oxford University Press.

Greene, W. H. 2003. *Econometric Analysis.* 5th ed. Upper Saddle River, NJ: Prentice Hall.

Also See

Title

truncreg postestimation — Postestimation tools for truncreg

Description

The following postestimation commands are available for truncreg:

command	description
adjust[1]	adjusted predictions of $\mathbf{x}\boldsymbol{\beta}$
estat	AIC, BIC, VCE, and estimation sample summary
estat (svy)	postestimation statistics for survey data
estimates	cataloging estimation results
lincom	point estimates, standard errors, testing, and inference for linear combinations of coefficients
lrtest[2]	likelihood-ratio test
mfx	marginal effects or elasticities
nlcom	point estimates, standard errors, testing, and inference for nonlinear combinations of coefficients
predict	predictions, residuals, influence statistics, and other diagnostic measures
predictnl	point estimates, standard errors, testing, and inference for generalized predictions
suest	seemingly unrelated estimation
test	Wald tests for simple and composite linear hypotheses
testnl	Wald tests of nonlinear hypotheses

[1] adjust is not appropriate with time-series operators.

[2] lrtest is not appropriate with svy estimation results.

See the corresponding entries in the *Stata Base Reference Manual* for details, but see [SVY] **estat** for details about estat (svy).

Syntax for predict

predict [*type*] *newvar* [*if*] [*in*] [, *statistic* <u>nooff</u>set]

predict [*type*] { *stub** | *newvar*_reg *newvar*_lnsigma } [*if*] [*in*] , <u>scores</u>

options	description
Main	
xb	linear prediction; the default
stdp	standard error of the prediction
stdf	standard error of the forecast
<u>pr</u>(a,b)	$\Pr(a < y_j < b)$
e(a,b)	$E(y_j \mid a < y_j < b)$
<u>ystar</u>(a,b)	$E(y_j^*)$, $y_j^* = \max\{a, \min(y_j, b)\}$

These statistics are available both in and out of sample; type predict ... if e(sample) ... if wanted only for the estimation sample.

stdf is not allowed with svy estimation results.

where a and b may be numbers or variables; a missing ($a \geq .$) means $-\infty$, and b missing ($b \geq .$) means $+\infty$; see [U] **12.2.1 Missing values**.

Options for predict

⌐ Main ⌐

xb, the default, calculates the linear prediction.

stdp calculates the standard error of the prediction, which can be thought of as the standard error of the predicted expected value or mean for the observation's covariate pattern. The standard error of the prediction is also referred to as the standard error of the fitted value.

stdf calculates the standard error of the forecast, which is the standard error of the point prediction for 1 observation. It is commonly referred to as the standard error of the future or forecast value. By construction, the standard errors produced by stdf are always larger than those produced by stdp; see [R] **regress** *Methods and Formulas*.

pr(a,b) calculates $\Pr(a < \mathbf{x}_j\mathbf{b} + u_j < b)$, the probability that $y_j|\mathbf{x}_j$ would be observed in the interval (a, b).

a and b may be specified as numbers or variable names; *lb* and *ub* are variable names;
pr(20,30) calculates $\Pr(20 < \mathbf{x}_j\mathbf{b} + u_j < 30)$;
pr(*lb*,*ub*) calculates $\Pr(lb < \mathbf{x}_j\mathbf{b} + u_j < ub)$; and
pr(20,*ub*) calculates $\Pr(20 < \mathbf{x}_j\mathbf{b} + u_j < ub)$.

a missing ($a \geq .$) means $-\infty$; pr(.,30) calculates $\Pr(-\infty < \mathbf{x}_j\mathbf{b} + u_j < 30)$;
pr(*lb*,30) calculates $\Pr(-\infty < \mathbf{x}_j\mathbf{b} + u_j < 30)$ in observations for which *lb* $\geq .$
and calculates $\Pr(lb < \mathbf{x}_j\mathbf{b} + u_j < 30)$ elsewhere.

b missing ($b \geq .$) means $+\infty$; pr(20,.) calculates $\Pr(+\infty > \mathbf{x}_j\mathbf{b} + u_j > 20)$;
pr(20,*ub*) calculates $\Pr(+\infty > \mathbf{x}_j\mathbf{b} + u_j > 20)$ in observations for which *ub* $\geq .$
and calculates $\Pr(20 < \mathbf{x}_j\mathbf{b} + u_j < ub)$ elsewhere.

e(a,b) calculates $E(\mathbf{x}_j\mathbf{b} + u_j \mid a < \mathbf{x}_j\mathbf{b} + u_j < b)$, the expected value of $y_j|\mathbf{x}_j$ conditional on $y_j|\mathbf{x}_j$ being in the interval (a, b), meaning that $y_j|\mathbf{x}_j$ is censored.
a and b are specified as they are for pr().

ystar(a,b) calculates $E(y_j^*)$, where $y_j^* = a$ if $\mathbf{x}_j\mathbf{b} + u_j \leq a$, $y_j^* = b$ if $\mathbf{x}_j\mathbf{b} + u_j \geq b$, and $y_j^* = \mathbf{x}_j\mathbf{b} + u_j$ otherwise, meaning that y_j^* is truncated. a and b are specified as they are for pr().

nooffset is relevant only if you specified offset(*varname*). It modifies the calculations made by predict so that they ignore the offset variable; the linear prediction is treated as $\mathbf{x}_j\mathbf{b}$ rather than as $\mathbf{x}_j\mathbf{b} + \text{offset}_j$.

scores calculates equation-level score variables.

The first new variable will contain $\partial \ln L / \partial(\mathbf{x}_j\boldsymbol{\beta})$.

The second new variable will contain $\partial \ln L / \partial \sigma$.

Methods and Formulas

All postestimation commands listed above are implemented as ado-files.

Also See

[R] **truncreg** — Truncated regression

[U] **20 Estimation and postestimation commands**

Title

ttest — Mean-comparison tests

Syntax

One-sample mean-comparison test

ttest *varname* == # [*if*] [*in*] [, level(#)]

Two-sample mean-comparison test (unpaired)

ttest *varname*$_1$ == *varname*$_2$ [*if*] [*in*], unpaired [unequal welch level(#)]

Two-sample mean-comparison test (paired)

ttest *varname*$_1$ == *varname*$_2$ [*if*] [*in*] [, level(#)]

Two-group mean-comparison test

ttest *varname* [*if*] [*in*], by(*groupvar*) [*options*$_1$]

Immediate form of one-sample mean-comparison test

ttesti #$_{obs}$ #$_{mean}$ #$_{sd}$ #$_{val}$ [, level(#)]

Immediate form of two-sample mean-comparison test

ttesti #$_{obs1}$ #$_{mean1}$ #$_{sd1}$ #$_{obs2}$ #$_{mean2}$ #$_{sd2}$ [, *options*$_2$]

options$_1$	description
Main	
* by(*groupvar*)	variable defining the groups
unequal	unpaired data have unequal variances
welch	use Welch's approximation
level(#)	set confidence level; default is level(95)

*by(*groupvar*) is required.

options$_2$	description
Main	
unequal	unpaired data have unequal variances
welch	use Welch's approximation
level(#)	set confidence level; default is level(95)

by is allowed with test; see [D] by.

532

Description

ttest performs t tests on the equality of means. In the first form, ttest tests that *varname* has a mean of #. In the second form, ttest tests that *varname*$_1$ and *varname*$_2$ have the same mean, assuming *unpaired* data. In the third form, ttest tests that *varname*$_1$ and *varname*$_2$ have the same mean, assuming *paired* data. In the fourth form, ttest tests that *varname* has the same mean within the two groups defined by *groupvar*.

ttesti is the immediate form of ttest; see [U] **19 Immediate commands**.

For the equivalent of a two-sample t test with sampling weights (pweights), use the svy: mean command with the over() option, and then use lincom; see [R] **mean** and [SVY] **svy postestimation**.

Options

┌─── Main ───

by(*groupvar*) specifies the *groupvar* that defines the two groups that ttest will use to test the hypothesis that their means are equal. Specifying by(*groupvar*) implies an unpaired (two sample) t test. Do not confuse the by() option with the by prefix; you can specify both.

unpaired specifies that the data be treated as unpaired. The unpaired option is used when the two sets of values to be compared are in different variables.

unequal specifies that the unpaired data not be assumed to have equal variances.

welch specifies that the approximate degrees of freedom for the test be obtained from Welch's formula (1947) rather than from Satterthwaite's approximation formula (1946), which is the default when unequal is specified. Specifying welch implies unequal.

level(#) specifies the confidence level, as a percentage, for confidence intervals. The default is level(95) or as set by set level; see [U] **20.7 Specifying the width of confidence intervals**.

Remarks

▷ Example 1: One-sample mean-comparison test

In the first form, ttest tests whether the mean of the sample is equal to a known constant under the assumption of unknown variance. Assume that we have a sample of 74 automobiles. We know each automobile's average mileage rating and wish to test whether the overall average for the sample is 20 miles per gallon.

```
. use http://www.stata-press.com/data/r10/auto
(1978 Automobile Data)

. ttest mpg==20

One-sample t test
```

Variable	Obs	Mean	Std. Err.	Std. Dev.	[95% Conf. Interval]
mpg	74	21.2973	.6725511	5.785503	19.9569 22.63769

```
         mean = mean(mpg)                                        t =    1.9289
Ho: mean = 20                                degrees of freedom =        73

    Ha: mean < 20               Ha: mean != 20               Ha: mean > 20
 Pr(T < t) = 0.9712        Pr(|T| > |t|) = 0.0576        Pr(T > t) = 0.0288
```

The test indicates that the underlying mean is not 20 with a significance level of 5.8%.

◁

▷ Example 2: Two-sample mean-comparison test

We are testing the effectiveness of a new fuel additive. We run an experiment with 12 cars. We run the cars without and with the fuel treatment. The results of the experiment are as follows:

Without Treatment	With Treatment	Without Treatment	With Treatment
20	24	18	17
23	25	24	28
21	21	20	24
25	22	24	27
18	23	23	21
17	18	19	23

By creating two variables called mpg1 and mpg2 representing mileage without and with the treatment, respectively, we can test the equality of means by typing

```
. use http://www.stata-press.com/data/r10/fuel
. ttest mpg1==mpg2
Paired t test
```

Variable	Obs	Mean	Std. Err.	Std. Dev.	[95% Conf. Interval]	
mpg1	12	21	.7881701	2.730301	19.26525	22.73475
mpg2	12	22.75	.9384465	3.250874	20.68449	24.81551
diff	12	-1.75	.7797144	2.70101	-3.46614	-.0338602

```
    mean(diff) = mean(mpg1 - mpg2)                              t =  -2.2444
Ho: mean(diff) = 0                            degrees of freedom =        11

Ha: mean(diff) < 0            Ha: mean(diff) != 0            Ha: mean(diff) > 0
Pr(T < t) = 0.0232         Pr(|T| > |t|) = 0.0463          Pr(T > t) = 0.9768
```

We find that the means are statistically different from each other at any level greater than 4.6%.

◁

▷ Example 3: Group mean-comparison test

Let's pretend that the preceding data were collected by running 24 cars: 12 cars with the additive and 12 without. Although we might be tempted to enter the data in the same way, we should resist (see the technical note below). Instead, we enter the data as 24 observations on mpg with an additional variable, treated, taking on 1 if the car received the fuel treatment and 0 otherwise:

```
. use http://www.stata-press.com/data/r10/fuel3

. ttest mpg, by(treated)
```

Two-sample t test with equal variances

Group	Obs	Mean	Std. Err.	Std. Dev.	[95% Conf. Interval]	
0	12	21	.7881701	2.730301	19.26525	22.73475
1	12	22.75	.9384465	3.250874	20.68449	24.81551
combined	24	21.875	.6264476	3.068954	20.57909	23.17091
diff		-1.75	1.225518		-4.291568	.7915684

```
      diff = mean(0) - mean(1)                             t =  -1.4280
Ho: diff = 0                             degrees of freedom =       22

   Ha: diff < 0                 Ha: diff != 0                  Ha: diff > 0
Pr(T < t) = 0.0837       Pr(|T| > |t|) = 0.1673        Pr(T > t) = 0.9163
```

This time we do not find a statistically significant difference.

If we were not willing to assume that the variances were equal and wanted to use Welch's formula, we could type

```
. ttest mpg, by(treated) welch
```

Two-sample t test with unequal variances

Group	Obs	Mean	Std. Err.	Std. Dev.	[95% Conf. Interval]	
0	12	21	.7881701	2.730301	19.26525	22.73475
1	12	22.75	.9384465	3.250874	20.68449	24.81551
combined	24	21.875	.6264476	3.068954	20.57909	23.17091
diff		-1.75	1.225518		-4.28369	.7836902

```
      diff = mean(0) - mean(1)                             t =  -1.4280
Ho: diff = 0                     Welch's degrees of freedom =  23.2465

   Ha: diff < 0                 Ha: diff != 0                  Ha: diff > 0
Pr(T < t) = 0.0833       Pr(|T| > |t|) = 0.1666        Pr(T > t) = 0.9167
```

◁

❑ Technical Note

In two-group randomized designs, subjects will sometimes refuse the assigned treatment but still be measured for an outcome. In this case, take care to specify the group properly. You might be tempted to let *varname* contain missing where the subject refused and thus let ttest drop such observations from the analysis. Zelen (1979) argues that it would be better to specify that the subject belongs to the group in which he or she was randomized, even though such inclusion will dilute the measured effect.

❑

❑ Technical Note

There is a second, inferior way to organize the data in the preceding example. Remember, we ran a test on 24 cars, 12 without the additive and 12 with. Nevertheless, we could have entered the data in the same way as we did when we had 12 cars, each run without and with the additive, by creating two variables—mpg1 and mpg2.

This method is inferior because it suggests a connection that is not there. For the 12-car experiment, there was most certainly a connection—it was the same car. In the 24-car experiment, however, it is arbitrary which mpg results appear next to which. Nevertheless, if our data are organized like this, ttest can accommodate us.

```
. use http://www.stata-press.com/data/r10/fuel

. ttest mpg1==mpg2, unpaired
Two-sample t test with equal variances
```

Variable	Obs	Mean	Std. Err.	Std. Dev.	[95% Conf. Interval]	
mpg1	12	21	.7881701	2.730301	19.26525	22.73475
mpg2	12	22.75	.9384465	3.250874	20.68449	24.81551
combined	24	21.875	.6264476	3.068954	20.57909	23.17091
diff		-1.75	1.225518		-4.291568	.7915684

```
    diff = mean(mpg1) - mean(mpg2)                              t =  -1.4280
Ho: diff = 0                                degrees of freedom =        22

    Ha: diff < 0              Ha: diff != 0                Ha: diff > 0
 Pr(T < t) = 0.0837    Pr(|T| > |t|) = 0.1673        Pr(T > t) = 0.9163
```

❑

▷ Example 4

ttest can be used to test the equality of a pair of means; see [R] **oneway** for testing the equality of more than two means.

Suppose that we have data on the 50 states. The dataset contains the median age of the population (medage) and the region of the country (region) for each state. Region 1 refers to the Northeast, region 2 to the North Central, region 3 to the South, and region 4 to the West. Using oneway, we can test the equality of all four means.

```
. use http://www.stata-press.com/data/r10/census
(1980 Census data by state)

. oneway medage region
                        Analysis of Variance
    Source              SS          df      MS            F     Prob > F
```

Source	SS	df	MS	F	Prob > F
Between groups	46.3961903	3	15.4653968	7.56	0.0003
Within groups	94.1237947	46	2.04616945		
Total	140.519985	49	2.8677548		

```
Bartlett's test for equal variances:   chi2(3) =  10.5757  Prob>chi2 = 0.014
```

We find that the means are different, but we are interested only in testing whether the means for the East (region==1) and West (region==4) are different. We could use oneway,

```
. oneway medage region if region==1 | region==4
                            Analysis of Variance
        Source              SS         df       MS            F     Prob > F

Between groups         46.241247        1    46.241247      20.02    0.0002
Within groups         46.1969169       20   2.30984584

        Total         92.4381638       21   4.40181733
Bartlett's test for equal variances:  chi2(1) =   2.4679  Prob>chi2 = 0.116
```

or we could use ttest:

```
. ttest medage if region==1 | region==4, by(region)
Two-sample t test with equal variances

    Group |   Obs        Mean    Std. Err.   Std. Dev.   [95% Conf. Interval]

       NE |     9    31.23333    .3411581    1.023474    30.44662    32.02005
     West |    13    28.28462    .4923577    1.775221    27.21186    29.35737

 combined |    22    29.49091    .4473059    2.098051    28.56069    30.42113

     diff |          2.948718    .6590372                 1.57399    4.323445

    diff = mean(NE) - mean(West)                              t =    4.4743
Ho: diff = 0                                degrees of freedom =        20

    Ha: diff < 0              Ha: diff != 0               Ha: diff > 0
 Pr(T < t) = 0.9999      Pr(|T| > |t|) = 0.0002      Pr(T > t) = 0.0001
```

The significance levels of both tests are the same.

◁

Immediate form

▷ Example 5

ttesti is like ttest, except that we specify summary statistics rather than variables as arguments. For instance, we are reading an article that reports the mean number of sunspots per month as 62.6 with a standard deviation of 15.8. There are 24 months of data. We wish to test whether the mean is 75:

```
. ttesti 24 62.6 15.8 75
One-sample t test

          |   Obs        Mean    Std. Err.   Std. Dev.   [95% Conf. Interval]

        x |    24        62.6    3.225161        15.8    55.92825    69.27175

    mean = mean(x)                                            t =   -3.8448
Ho: mean = 75                              degrees of freedom =        23

    Ha: mean < 75            Ha: mean != 75              Ha: mean > 75
 Pr(T < t) = 0.0004      Pr(|T| > |t|) = 0.0008      Pr(T > t) = 0.9996
```

◁

▷ Example 6

There is no immediate form of ttest with paired data since the test is also a function of the covariance, a number unlikely to be reported in any published source. For nonpaired data, however, we might type

```
. ttesti 20 20 5  32 15 4
```

Two-sample t test with equal variances

	Obs	Mean	Std. Err.	Std. Dev.	[95% Conf.	Interval]
x	20	20	1.118034	5	17.65993	22.34007
y	32	15	.7071068	4	13.55785	16.44215
combined	52	16.92308	.6943785	5.007235	15.52905	18.3171
diff		5	1.256135		2.476979	7.523021

```
       diff = mean(x) - mean(y)                                    t =    3.9805
Ho: diff = 0                                    degrees of freedom =        50

    Ha: diff < 0                  Ha: diff != 0                    Ha: diff > 0
 Pr(T < t) = 0.9999       Pr(|T| > |t|) = 0.0002            Pr(T > t) = 0.0001
```

If we had typed ttesti 20 20 5 32 15 4, unequal, the test would have assumed unequal variances.

◁

Saved Results

ttest and ttesti save the following in r():

Scalars

r(N_1)	sample size n_1	r(t)	t statistic
r(N_2)	sample size n_2	r(sd_1)	standard deviation for first variable
r(p_l)	lower one-sided p-value	r(sd_2)	standard deviation for second variable
r(p_u)	upper one-sided p-value	r(mu_1)	$\bar{x}_1$ mean for population 1
r(p)	two-sided p-value	r(mu_2)	$\bar{x}_2$ mean for population 2
r(se)	estimate of standard error	r(df_t)	degrees of freedom

Methods and Formulas

ttest and ttesti are implemented as ado-files.

See, for instance, Hoel (1984, 140–161) or Dixon and Massey (1983, 121–130) for an introduction and explanation of the calculation of these tests.

The test for $\mu = \mu_0$ for unknown σ is given by

$$t = \frac{(\bar{x} - \mu_0)\sqrt{n}}{s}$$

The statistic is distributed as Student's t with $n - 1$ degrees of freedom (Gosset 1908).

The test for $\mu_x = \mu_y$ when σ_x and σ_y are unknown but $\sigma_x = \sigma_y$ is given by

$$
t = \frac{\overline{x} - \overline{y}}{\left\{ \frac{(n_x - 1)s_x^2 + (n_y - 1)s_y^2}{n_x + n_y - 2} \right\}^{1/2} \left(\frac{1}{n_x} + \frac{1}{n_y} \right)^{1/2}}
$$

The result is distributed as Student's t with $n_x + n_y - 2$ degrees of freedom.

You could perform `ttest` (without the `unequal` option) in a regression setting given that regression assumes a homoskedastic error model. To compare with the `ttest` command, denote the underlying observations on x and y by x_j, $j = 1, \ldots, n_x$ and y_j, $j = 1, \ldots, n_y$. In a regression framework, typing `ttest` without the `unequal` option is equivalent to

creating a new variable z_j that represents the stacked observations on x and y (so that $z_j = x_j$ for $j = 1, \ldots, n_x$ and $z_{n_x+j} = y_j$ for $j = 1, \ldots, n_y$)

and then estimating the equation $z_j = \beta_0 + \beta_1 d_j + \epsilon_j$, where $d_j = 0$ for $j = 1, \ldots, n_x$ and $d_j = 1$ for $j = n_x + 1, \ldots, n_x + n_y$ (i.e., $d_j = 0$ when the z observations represent x, and $d_j = 1$ when the z observations represent y).

The estimated value of β_1, b_1, will equal $\overline{y} - \overline{x}$, and the reported t statistic will be the same t statistic as given by the formula above.

The test for $\mu_x = \mu_y$ when σ_x and σ_y are unknown and $\sigma_x \neq \sigma_y$ is given by

$$
t = \frac{\overline{x} - \overline{y}}{\left(s_x^2/n_x + s_y^2/n_y \right)^{1/2}}
$$

The result is distributed as Student's t with ν degrees of freedom, where ν is given by (with Satterthwaite's formula)

$$
\frac{\left(s_x^2/n_x + s_y^2/n_y \right)^2}{\dfrac{\left(s_x^2/n_x \right)^2}{n_x - 1} + \dfrac{\left(s_y^2/n_y \right)^2}{n_y - 1}}
$$

With Welch's formula (1947), the number of degrees of freedom is given by

$$
-2 + \frac{\left(s_x^2/n_x + s_y^2/n_y \right)^2}{\dfrac{\left(s_x^2/n_x \right)^2}{n_x + 1} + \dfrac{\left(s_y^2/n_y \right)^2}{n_y + 1}}
$$

The test for $\mu_x = \mu_y$ for matched observations (also known as paired observations, correlated pairs, or permanent components) is given by

$$
t = \frac{\overline{d}\sqrt{n}}{s_d}
$$

where $\overline{d}$ represents the mean of $x_i - y_i$ and s_d represents the standard deviation. The test statistic t is distributed as Student's t with $n - 1$ degrees of freedom.

You can also use `ttest` without the `unpaired` option in a regression setting since a paired comparison includes the assumption of constant variance. The `ttest` with an unequal variance assumption does not lend itself to an easy representation in regression settings and is not discussed here. $(x_j - y_j) = \beta_0 + \epsilon_j$.

William Sealy Gosset (1876–1937) was born in Canterbury, England. He studied chemistry and mathematics at Oxford and worked as a chemist with the brewers Guinness in Dublin. Gosset became interested in statistical problems, which he discussed with Karl Pearson and later with Fisher and Neyman, and published several important papers under the pseudonym "Student", including that on the test that usually bears his name.

References

Boland, P. J. 2000. William Sealy Gosset—alias 'Student' 1876–1937. In *Creators of Mathematics: The Irish Connection*, ed. K. Houston, 105–112. Dublin: University College Dublin Press.

Dixon, W. J., and F. J. Massey, Jr. 1983. *Introduction to Statistical Analysis*. 4th ed. New York: McGraw–Hill.

Gleason, J. R. 1999. sg101: Pairwise comparisons of means, including the Tukey wsd method. *Stata Technical Bulletin* 47: 31–37. Reprinted in *Stata Technical Bulletin Reprints*, vol. 8, pp. 225–233.

Gosset, W. S. [Student, pseud.] 1908. The probable error of a mean. *Biometrika* 6: 1–25.

——. 1943. *"Student's" Collected Papers*, ed. E. S. Pearson and J. Wishart. London: Biometrika Office, University College.

Hoel, P. G. 1984. *Introduction to Mathematical Statistics*. 5th ed. New York: Wiley.

Pearson, E. S., and R. L. Plackett. 1990. *'Student': A Statistical Biography of William Sealy Gosset*. Oxford: Oxford University Press.

Preece, D. A. 1982. t is for trouble (and textbooks): A critique of some examples of the paired-samples t-test. *Statistician* 31: 169–195.

Satterthwaite, F. E. 1946. An approximate distribution of estimates of variance components. *Biometrics Bulletin* 2: 110–114.

Senn, S. J., and W. Richardson. 1994. The first *t*-test. *Statistics in Medicine* 13: 785–803.

Welch, B. L. 1947. The generalization of student's problem when several different population variances are involved. *Biometrika* 34: 28–35.

Zelen, M. 1979. A new design for randomized clinical trials. *New England Journal of Medicine* 300: 1242–1245.

Also See

[R] **bitest** — Binomial probability test

[R] **ci** — Confidence intervals for means, proportions, and counts

[R] **mean** — Estimate means

[R] **oneway** — One-way analysis of variance

[R] **prtest** — One- and two-sample tests of proportions

[R] **sdtest** — Variance-comparison tests

[MV] **hotelling** — Hotelling's T-squared generalized means test

Title

> **update** — Update Stata

Syntax

Report on update level of currently installed Stata

    update

Set update source

    update from *location*

Compare update level of currently installed Stata with that of source

    update query [, from(*location*)]

Update ado-files from source if necessary

    update ado [, from(*location*) into(*dirname*)]

Update executable from source if necessary

    update executable [, from(*location*) into(*dirname*) force]

Perform update ado and update executable if necessary

    update all [, from(*location*)]

Complete installation of newly updated executable

    update swap [, clear]

Set automatic updates (Windows and Macintosh only)

    <u>set</u> update_query { on | off }
    <u>set</u> update_interval #
    <u>set</u> update_prompt { on | off }

Description

The update command reports on the current update level and installs official updates to Stata. Official updates are updates to Stata as it was originally shipped from StataCorp, not the additions to Stata published in, for instance, the *Stata Journal* (SJ) or *Stata Technical Bulletin* (STB). Those additions are installed using the net command and updated using the adoupdate command; see [R] **net** and [R] **adoupdate**.

update without arguments reports on the update level of the currently installed Stata.

update from sets an update source, where *location* is a directory name or URL. If you are on the Internet, type update from http://www.stata.com.

update query compares the update level of the currently installed Stata with that available from the update source and displays a report.

update ado compares the update level of the official ado-files of the currently installed Stata with those available from the update source. If the currently installed ado-files need updating, update ado copies and installs those files from the update source that are necessary to bring the ado-files up to date. update ado is seldom used by itself, it is called automatically by update all.

update executable compares the update level of the currently installed Stata executable with that available from the update source. If the currently installed Stata needs updating, update executable copies the new executable from the update source, but the last step of the installation—erasing the old executable and renaming the new executable—must be performed by the update swap command. update executable displays instructions on how to do this. update executable is seldom used by itself, it is called automatically by update all.

update all does the same as update ado followed by update executable. This is what you should usually type to check for and install updates.

update swap completes the installation of a newly updated executable by automatically swapping the newly downloaded executable with the currently running executable. During this process, Stata will shut down to perform the swap before automatically launching the new executable. This command is necessary after either the update executable or the update all command when a new executable is downloaded.

set update_query determines if update query is to be automatically performed when Stata is launched. Only Windows and Macintosh platforms can be set for automatic updating.

set update_interval # sets the number of days to elapse before performing the next automatic update query. The default # is 7. The interval starts from the last time an update query was performed (automatically or manually). Only Windows and Macintosh platforms can be set for automatic updating.

set update_prompt determines whether a dialog is to be displayed before performing an automatic update query. The dialog allows you to perform an update query now, perform one the next time Stata is launched, perform one after the next interval has passed, or disable automatic update query. Only Windows and Macintosh platforms can be set for automatic updating.

Options

from(*location*) specifies the location of the update source. You can specify the from() option on the individual update commands or use the update from command. Which you do makes no difference. You typically do not need to use this option.

into(*dirname*) specifies the name of the directory into which the updates are to be copied. *dirname* can be a directory name or a sysdir codeword, such as UPDATES or STATA; see [P] **sysdir**. This option is seldom used.

For update ado, the default is into(UPDATES), the official update directory. Network computer managers might want to specify into() if they want to download the updates but leave the last step—copying the files into the official directory—to do themselves.

For update executable, the default is into(STATA), the official Stata directory. Network computer managers might want to specify into() so that they can copy the update into a more accessible directory. Then the last step of copying the new executable over the existing executable would be left for them to perform.

force is used with update executable to force downloading a new executable even if, based on the date comparison, Stata does not think it is necessary. There is seldom a reason to specify this option. There is no such option for update ado because, if you wanted to force the reinstallation of all ado-file updates, you would need only to erase the UPDATES directory. You can type sysdir to see where the UPDATES directory is on your computer; see [P] **sysdir**.

clear specifies that it is okay for update swap to shut down this Stata and start the new one, even if data in memory have not been saved to disk.

Remarks

update updates the two official components of Stata—its binary executable and its ado-files—from the official source: http://www.stata.com. Jumping ahead of the story, the easiest thing to do if you are connected to the Internet is to type

        . update all

and follow the instructions. If Stata is up to date, update all will do nothing. Otherwise, it will download whatever is necessary and display detailed instructions on what, if anything, needs to be done next. If you first want to know what update all will do, type

        . update query

update query will present a report comparing what you have installed with what is available and will recommend that you do nothing or that you type update all. You should never update your ado-files without also updating your executable (and vice versa) if both ado-files and executable updates are available, which is why update all will be recommended no matter what type of updates are available.

If you want only a report on what you have installed without comparing with what is available, type

        . update

update will show you what you have installed and where it is and will recommend that you type update query to compare that with what is available.

Before doing any of this, you can type

        . update from http://www.stata.com

but that is not really necessary because http://www.stata.com is the default location.

In addition to using the update command, you may select **Official Updates** from the **Help** menu. The menu item does the same thing as the command, but it does not provide the file redirection option into(), which managers of networked computers may wish to use so that they can download the updates and then copy the files to the official locations for themselves.

For examples of using `update`, see

Windows	[GSW] **20 Updating and extending Stata—Internet functionality** and
	[U] **28 Using the Internet to keep up to date**
Macintosh	[GSM] **20 Updating and extending Stata—Internet functionality** and
	[U] **28 Using the Internet to keep up to date**
Unix	[GSU] **20 Updating and extending Stata—Internet functionality** and
	[U] **28 Using the Internet to keep up to date**

Notes for multiuser system administrators

There are two types of updates: ado-file updates and the binary executable update. Stata's periodic updates are usually ado-file updates, but sometimes they are both.

By default, `update` installs the ado-file updates, which may include lots of small files. `update` is careful about how it does this. First it downloads all the files you need to a temporary place. Next it closes the connection to http://www.stata.com and checks the files to make sure that they are complete. Only after all that does `update` copy the files to the official UPDATES directory; see [P] **sysdir**. This process is all designed so that, should anything go wrong at any step along the way, no damage is done.

Updated binary executables, on the other hand, are just downloaded, and then you can either automatically or manually swap the executables. To automatically swap the executables, type `update swap`, and Stata will briefly quit to perform the copy before relaunching the newly downloaded executable. To manually swap the executables, (1) exit Stata, (2) rename the current executable, (3) rename the updated executable, (4) try Stata, and (5) erase the old executable.

`update` must have write access to both the STATA and UPDATES directories. You can obtain the names of these directories by typing `sysdir`. As system administrator, you must decide whether you are going to start Stata with such permissions (Unix users could first become superuser to ensure write access to the STATA and UPDATES directories) and trust Stata to do the right thing. That is what we recommend you do, but we provide the `into()` option for those who do not trust our recommendation.

To perform the final copying by hand, obtain the new executable by typing

```
. update executable, into(.)
```

That will place the new executable in the current directory. You need no special permissions to do this. Later you can copy the file into the appropriate place and give it the appropriate name. Type `update` without arguments; the default output will make it clear where the file goes and what its name must be. When you copy the file, be sure to make it executable by everybody.

To obtain the ado-file updates, make a new, empty directory, and then place the updates into it. For example, typing

```
. mkdir mydir
. update ado, into(mydir)
```

places the new, empty directory in the current directory under the name `mydir`. You need no special permissions to perform this step. Later, you can copy all the files in `mydir` to the official place. Type `update` without arguments; the default output will make it clear where the files go. When you copy the files, be sure to copy all of them and to make all of them readable by everybody.

Saved Results

update without a subcommand, update from, and update query save the following in r():

Scalars

r(inst_exe)	date of executable installed	(*)
r(avbl_exe)	date of executable available over web	(*) (**)
r(inst_ado)	date of ado-files installed	(*)
r(avbl_ado)	date of ado-files available over web	(*) (**)

Macros

r(dir_exe)	directory in which executable is stored
r(dir_ado)	directory in which ado-files are stored
r(name_exe)	name of the Stata executable

Notes:

* Dates are stored as integers counting the number of days since January 1, 1960; see [D] **dates and times**.

** These dates are not saved by update without a subcommand since update by itself reports information solely about the local computer and does not check what is available on the web.

Also See

[R] **adoupdate** — Update user-written ado-files

[R] **net** — Install and manage user-written additions from the Internet

[P] **sysdir** — Query and set system directories

[U] **28 Using the Internet to keep up to date**

[GSM] **20 Updating and extending Stata—Internet functionality**

[GSU] **20 Updating and extending Stata—Internet functionality**

[GSW] **20 Updating and extending Stata—Internet functionality**

Title

vce_option — Variance estimators

Syntax

estimation_cmd ... $\left[\, ,\ \text{vce}(\textit{vcetype})\ ...\right]$

vcetype	description
Likelihood based	
oim	observed information matrix (OIM)
opg	outer product of the gradient (OPG) vectors
Sandwich estimators	
<u>r</u>obust	Huber/White/sandwich estimator
<u>cl</u>uster *clustvar*	clustered sandwich estimator
Replication based	
<u>boot</u>strap $\left[\, ,\ \textit{bootstrap\_options}\right]$	bootstrap estimation
<u>jack</u>knife $\left[\, ,\ \textit{jackknife\_options}\right]$	jackknife estimation

Description

This entry describes the vce() option, which is common to most estimation commands. vce() specifies how to estimate the variance–covariance matrix (VCE) corresponding to the parameter estimates. The standard errors reported in the table of parameter estimates are the square root of the variances (diagonal elements) of the VCE.

Options

┌─── SE/Robust ──

vce(oim) is usually the default for models fitted using maximum likelihood. vce(oim) uses the observed information matrix (OIM); see [R] **ml**.

vce(opg) uses the sum of the outer product of the gradient (OPG) vectors; see [R] **ml**. This is the default VCE when the technique(bhhh) option is specified; see [R] **maximize**.

vce(robust) uses the robust or sandwich estimator of variance. This estimator is robust to some types of misspecification so long as the observations are independent; see [U] **20.15 Obtaining robust variance estimates**.

If the command allows pweights and you specify them, vce(robust) is implied; see [U] **20.17.3 Sampling weights**.

vce(cluster *clustvar*) specifies that the standard errors allow for intragroup correlation, relaxing the usual requirement that the observations be independent. That is, the observations are independent across groups (clusters) but not necessarily within groups. *clustvar* specifies to which group each observation belongs, e.g., vce(cluster personid) in data with repeated observations on individuals. vce(cluster *clustvar*) affects the standard errors and variance–covariance matrix of the estimators but not the estimated coefficients; see [U] **20.15 Obtaining robust variance estimates**.

vce(bootstrap [, *bootstrap_options*]) uses a bootstrap; see [R] **bootstrap**. After estimation with vce(bootstrap), see [R] **bootstrap postestimation** to obtain percentile-based or bias-corrected confidence intervals.

vce(jackknife [, *jackknife_options*]) uses the delete-one jackknife; see [R] **jackknife**.

Remarks

Remarks are presented under the following headings:

> *Prefix commands*
> *Passing options in* vce()

Prefix commands

Specifying vce(bootstrap) or vce(jackknife) is often equivalent to using the corresponding prefix command. Here is an example using jackknife with regress.

```
. use http://www.stata-press.com/data/r10/auto
(1978 Automobile Data)

. regress mpg turn trunk, vce(jackknife)
(running regress on estimation sample)

Jackknife replications (74)
——+—— 1 ——+—— 2 ——+—— 3 ——+—— 4 ——+—— 5
..................................................   50
......................
Linear regression                    Number of obs   =         74
                                     Replications    =         74
                                     F(  2,     73)  =      66.26
                                     Prob > F        =     0.0000
                                     R-squared       =     0.5521
                                     Adj R-squared   =     0.5395
                                     Root MSE        =     3.9260
```

mpg	Coef.	Jackknife Std. Err.	t	P>|t|	[95% Conf. Interval]	
turn	-.7610113	.150726	-5.05	0.000	-1.061408	-.4606147
trunk	-.3161825	.1282326	-2.47	0.016	-.5717498	-.0606152
_cons	55.82001	5.031107	11.09	0.000	45.79303	65.84699

```
. jackknife: regress mpg turn trunk
(running regress on estimation sample)

Jackknife replications (74)
   ────┼─── 1 ───┼─── 2 ───┼─── 3 ───┼─── 4 ───┼─── 5
..................................................    50
......................
```

Linear regression

Number of obs =	74
Replications =	74
F(2, 73) =	66.26
Prob > F =	0.0000
R-squared =	0.5521
Adj R-squared =	0.5395
Root MSE =	3.9260

mpg	Coef.	Jackknife Std. Err.	t	P>|t|	[95% Conf. Interval]	
turn	-.7610113	.150726	-5.05	0.000	-1.061408	-.4606147
trunk	-.3161825	.1282326	-2.47	0.016	-.5717498	-.0606152
_cons	55.82001	5.031107	11.09	0.000	45.79303	65.84699

Here it does not matter whether we specify option vce(jackknife) or instead use the jackknife prefix.

However, vce(jackknife) should be used in place of the jackknife prefix whenever available because they are not always equivalent. For example, to use the jackknife prefix with clogit properly, you must tell jackknife to omit whole groups rather than individual observations. Specifying vce(jackknife) does this automatically.

```
. use http://www.stata-press.com/data/r10/clogitid

. jackknife, cluster(id): clogit y x1 x2, group(id)
  (output omitted )
```

This extra information is automatically communicated to jackknife by clogit when the vce() option is specified.

```
. clogit y x1 x2, group(id) vce(jackknife)
(running clogit on estimation sample)

Jackknife replications (66)
   ────┼─── 1 ───┼─── 2 ───┼─── 3 ───┼─── 4 ───┼─── 5
..................................................    50
...............
```

Conditional (fixed-effects) logistic regression

Number of obs =	369
Replications =	66
F(2, 65) =	4.58
Prob > F =	0.0137
Log likelihood = -123.41386 Pseudo R2 =	0.0355

(Replications based on 66 clusters in id)

y	Coef.	Jackknife Std. Err.	t	P>|t|	[95% Conf. Interval]	
x1	.653363	.3010608	2.17	0.034	.052103	1.254623
x2	.0659169	.0487858	1.35	0.181	-.0315151	.1633489

Passing options in vce()

If you wish to specify more options to the bootstrap or jackknife estimation, you can include them within the `vce()` option. Below we request 300 bootstrap replications and save the replications in `bsreg.dta`:

```
. use http://www.stata-press.com/data/r10/auto
(1978 Automobile Data)
. regress mpg turn trunk, vce(bootstrap, nodots seed(123) rep(300) saving(bsreg))
Linear regression                     Number of obs   =        74
                                      Replications    =       300
                                      Wald chi2(2)    =    127.28
                                      Prob > chi2     =    0.0000
                                      R-squared       =    0.5521
                                      Adj R-squared   =    0.5395
                                      Root MSE        =    3.9260
```

mpg	Observed Coef.	Bootstrap Std. Err.	z	P>|z|	Normal-based [95% Conf. Interval]	
turn	-.7610113	.1361786	-5.59	0.000	-1.027916	-.4941062
trunk	-.3161825	.1145728	-2.76	0.006	-.540741	-.0916239
_cons	55.82001	4.69971	11.88	0.000	46.60875	65.03127

```
. bstat using bsreg
Bootstrap results                     Number of obs   =        74
                                      Replications    =       300

        command:  regress mpg turn trunk
```

mpg	Observed Coef.	Bootstrap Std. Err.	z	P>|z|	Normal-based [95% Conf. Interval]	
turn	-.7610113	.1361786	-5.59	0.000	-1.027916	-.4941062
trunk	-.3161825	.1145728	-2.76	0.006	-.540741	-.0916239
_cons	55.82001	4.69971	11.88	0.000	46.60875	65.03127

Methods and Formulas

By default, Stata's maximum likelihood estimators display standard errors based on variance estimates given by the inverse of the negative Hessian (second derivative) matrix. If `vce(robust)`, `vce(cluster clustvar)`, or `pweights` is specified, standard errors are based on the robust variance estimator (see [U] **20.15 Obtaining robust variance estimates**); likelihood-ratio tests are not appropriate here (see [SVY] **survey**), and the model χ^2 is from a Wald test. If `vce(opg)` is specified, the standard errors are based on the outer product of the gradients; this option has no effect on likelihood-ratio tests, though it does affect Wald tests.

If `vce(bootstrap)` or `vce(jackknife)` is specified, the standard errors are based on the chosen replication method; here the model χ^2 or F statistic is from a Wald test using the respective replication-based covariance matrix. The t distribution is used in the coefficient table when the `vce(jackknife)` option is specified. `vce(bootstrap)` and `vce(jackknife)` are also available with some commands that are not maximum likelihood estimators.

Also See

[R] **bootstrap** — Bootstrap sampling and estimation

[R] **jackknife** — Jackknife estimation

[XT] *vce_options* — Variance estimators

[U] **20 Estimation and postestimation commands**

Title

view — View files and logs

Syntax

Display file in Viewer

view [file] ["]*filename*["] [, asis adopath]

Bring up browser pointed to specified URL

view browse ["]*url*["]

Display help results in Viewer

view help [*topic_or_command_name*]

Display search results in Viewer

view search *keywords*

Display news results in Viewer

view news

Display net results in Viewer

view net [*netcmd*]

Display ado-results in Viewer

view ado [*adocmd*]

Display update results in Viewer

view update [*updatecmd*]

Programmer's analog to view file and view browse

view view_d

Programmer's analog to view help

view help_d

Programmer's analog to view search

view search_d

Programmer's analog to view net

```
view net_d
```

Programmer's analog to view ado

```
view ado_d
```

Programmer's analog to view update

```
view update_d
```

Description

view displays file contents in the Viewer.

view file displays the specified file. file is optional, so if you had a SMCL session log created by typing log using mylog, you could view it by typing view mylog.smcl. view file can properly display .smcl files (logs and the like), .sthlp files, and ASCII text files. view file's asis option specifies that the file be displayed as straight ASCII text, regardless of the *filename*'s extension.

view browse opens your browser pointed to *url*. Typing view browse http://www.stata.com would bring up your browser pointed to the http://www.stata.com web site.

view help does the same as the help command—see [R] **help**—but displays the result in the Viewer. For example, to review the help for Stata's print command, you could type view help print.

view search does the same as the search command—see [R] **search**—but displays the result in the Viewer. For instance, to search the online help for information on robust regression, you could type view search robust regression.

view news does the same as the news command—see [R] **news**—but displays the results in the Viewer. (news displays the latest news from *http://www.stata.com*.)

view net does the same as the net command—see [R] **net**—but displays the result in the Viewer. For instance, typing view net search hausman test would search the Internet for additions to Stata related to the Hausman test. Typing view net from http://www.stata.com would go to the Stata download site at *http://www.stata.com*.

view ado does the same as the ado command—see [R] **net**—but displays the result in the Viewer. For instance, typing view ado dir would show a list of files you have installed.

view update does the same as the update command—see [R] **update**—but displays the result in the Viewer. Typing view update would show the dates of what you have installed, and from there you could click to compare those dates with the latest updates available. Typing view update query would skip the first step and show the comparison.

The view *_d commands are more useful in programming contexts than they are interactively.

view view_d displays a dialog box from which you may type the name of a file or a URL to be displayed in the Viewer.

view help_d displays a help dialog box from which you may obtain interactive help on any Stata command.

view `search_d` displays a search dialog box from which you may obtain interactive help based on keywords.

view `net_d` displays a search dialog box from which you may search the Internet for additions to Stata (which you could then install).

view `ado_d` displays a dialog box from which you may search the user-written routines you have previously installed.

view `update_d` displays an update dialog box in which you may type the source from which updates are to be obtained.

Options

`asis`, allowed with view `file`, specifies that the file be displayed as ASCII text, regardless of the *filename*'s extension. view `file`'s default action is to display files ending in `.smcl` and `.sthlp` as SMCL.

`adopath`, allowed with view `file`, specifies that Stata search the `S_ADO` path for *filename* and display it, if found.

Remarks

Most users access the Viewer by selecting **File > View...** and proceeding from there. The view command allows you to skip that step. Some common interactive uses of view are

```
. view mysession.smcl
. view mysession.log

. view help print
. view help regress

. view news
. view browse http://www.stata.com

. view net search hausman test

. view net
. view ado

. view update query
```

Also, programmers find view useful for creating special effects.

Also See

[R] **help** — Display online help

[R] **net** — Install and manage user-written additions from the Internet

[R] **news** — Report Stata news

[R] **search** — Search Stata documentation

[R] **update** — Update Stata

[D] **type** — Display contents of a file

[GSM] **5 Using the Viewer**

[GSU] **5 Using the Viewer**

[GSW] **5 Using the Viewer**

Title

vwls — Variance-weighted least squares

Syntax

vwls *depvar indepvars* [*if*] [*in*] [*weight*] [, *options*]

options	description
Model	
<u>nocon</u>stant	suppress constant term
sd(*varname*)	variable containing estimate of conditional standard deviation
Reporting	
<u>l</u>evel(*#*)	set confidence level; default is level(95)

bootstrap, by, jackknife, rolling, statsby, and xi are allowed; see [U] **11.1.10 Prefix commands**.
Weights are not allowed with the bootstrap prefix.
fweights are allowed; see [U] **11.1.6 weight**.
See [U] **20 Estimation and postestimation commands** for more capabilities of estimation commands.

Description

vwls estimates a linear regression using variance-weighted least squares. It differs from ordinary least-squares (OLS) regression in that it does not assume homogeneity of variance, but requires that the conditional variance of *depvar* be estimated prior to the regression. The estimated variance need not be constant across observations. vwls treats the estimated variance as if it were the true variance when it computes the standard errors of the coefficients.

You must supply an estimate of the conditional standard deviation of *depvar* to vwls using the sd(*varname*) option, or you must have grouped data with the groups defined by the *indepvars* variables. In the latter case, vwls treats all *indepvars* as categorical variables, computes the mean and standard deviation of *depvar* separately for each subgroup, and computes the regression of the subgroup means on *indepvars*.

regress with analytic weights can be used to produce another kind of "variance-weighted least squares"; see the following remarks for an explanation of the difference.

Options

Model

noconstant; see [R] **estimation options**.

sd(*varname*) is an estimate of the conditional standard deviation of *depvar* (that is, it can vary observation by observation). All values of *varname* must be > 0. If you specify sd(), you cannot use fweights.

If sd() is not given, the data will be grouped by *indepvars*. Here *indepvars* are treated as categorical variables, and the means and standard deviations of *depvar* for each subgroup are calculated and used for the regression. Any subgroup for which the standard deviation is zero is dropped.

Reporting

level(#); see [R] estimation options.

Remarks

The vwls command is intended for use with two special—and different—types of data. The first contains data that consist of measurements from physical science experiments in which (1) all error is due solely to measurement errors and (2) the sizes of the measurement errors are known.

You can also use variance-weighted least-squares linear regression for certain problems in categorical data analysis, such as when all the independent variables are categorical and the outcome variable is either continuous or a quantity that can sensibly be averaged. If each of the subgroups defined by the categorical variables contains a reasonable number of subjects, then the variance of the outcome variable can be estimated independently within each subgroup. For the purposes of estimation, vwls treats each subgroup as one observation, with the dependent variable being the subgroup mean of the outcome variable.

The vwls command fits the model

$$y_i = \mathbf{x}_i \boldsymbol{\beta} + \varepsilon_i$$

where the errors ε_i are independent normal random variables with the distribution $\varepsilon_i \sim N(0, \nu_i)$. The independent variables $\mathbf{x}_i$ are assumed to be known without error.

As described above, vwls assumes that you already have estimates s_i^2 for the variances ν_i. The error variance is not estimated in the regression. The estimates s_i^2 are used to compute the standard errors of the coefficients; see *Methods and Formulas* below.

In contrast, weighted ordinary least-squares regression assumes that the errors have the distribution $\varepsilon_i \sim N(0, \sigma^2/w_i)$, where the w_i are known weights and σ^2 is an unknown parameter that is estimated in the regression. This is the difference from variance-weighted least squares: in weighted OLS, the magnitude of the error variance is estimated in the regression using all the data.

▷ Example 1

An artificial, but informative, example illustrates the difference between variance-weighted least squares and weighted OLS.

We measure the quantities x_i and y_i and estimate that the standard deviation of y_i is s_i. We enter the data into Stata:

```
. input x y s

            x           y           s
  1.        1         1.2         0.5
  2.        2         1.9         0.5
  3.        3         3.2         1
  4.        4         4.3         1
  5.        5         4.9         1
  6.        6         6.0         2
  7.        7         7.2         2
  8.        8         7.9         2
  9. end
```

Since we want observations with smaller variance to carry larger weight in the regression, we compute an OLS regression with analytic weights proportional to the inverse of the squared standard deviations:

```
. regress y x [aweight=s^(-2)]
(sum of wgt is   1.1750e+01)
```

Source	SS	df	MS		
Model	22.6310183	1	22.6310183		
Residual	.193355117	6	.032225853		
Total	22.8243734	7	3.26062477		

Number of obs =   8
F(1,   6) =   702.26
Prob > F   =   0.0000
R-squared   =   0.9915
Adj R-squared =   0.9901
Root MSE   =   .17952

y	Coef.	Std. Err.	t	P>|t|	[95% Conf. Interval]
x	.9824683	.0370739	26.50	0.000	.8917517 1.073185
_cons	.1138554	.1120078	1.02	0.349	-.1602179 .3879288

If we compute a variance-weighted least-squares regression using `vwls`, we get the same results for the coefficient estimates but very different standard errors:

```
. vwls y x, sd(s)
```

Variance-weighted least-squares regression   Number of obs   =   8
Goodness-of-fit chi2(6)   =   0.28   Model chi2(1)   =   33.24
Prob > chi2   = 0.9996   Prob > chi2   =   0.0000

y	Coef.	Std. Err.	z	P>|z|	[95% Conf. Interval]
x	.9824683	.170409	5.77	0.000	.6484728 1.316464
_cons	.1138554	.51484	0.22	0.825	-.8952124 1.122923

Although the values of y_i were nicely linear with x_i, the `vwls` regression used the large estimates for the standard deviations to compute large standard errors for the coefficients. For weighted OLS regression, however, the scale of the analytic weights has no effect on the standard errors of the coefficients—only the relative proportions of the analytic weights affect the regression.

If we are sure of the sizes of our error estimates for y_i, using `vwls` is valid. However, if we can only estimate the relative proportions of error among the y_i, `vwls` is not appropriate.

◁

▷ Example 2

Let us now consider an example of the use of `vwls` with categorical data. Suppose that we have blood pressure data for $n = 400$ subjects, categorized by gender and race (black or white). Here is a description of the data:

(Continued on next page)

```
. use http://www.stata-press.com/data/r10/bp, clear
. table gender race, c(mean bp sd bp freq) row col format(%8.1f)
```

		Race	
Gender	White	Black	Total
Female	117.1	118.5	117.8
	10.3	11.6	10.9
	100	100	200
Male	122.1	125.8	124.0
	10.6	15.5	13.3
	100	100	200
Total	119.6	122.2	120.9
	10.7	14.1	12.6
	200	200	400

Performing a variance-weighted regression using vwls gives

```
. vwls bp gender race
Variance-weighted least-squares regression      Number of obs   =      400
Goodness-of-fit chi2(1)      =      0.88          Model chi2(2)   =    27.11
Prob > chi2                  =    0.3486          Prob > chi2     =   0.0000
```

bp	Coef.	Std. Err.	z	P>|z|	[95% Conf. Interval]	
gender	5.876522	1.170241	5.02	0.000	3.582892	8.170151
race	2.372818	1.191683	1.99	0.046	.0371631	4.708473
_cons	116.6486	.9296297	125.48	0.000	114.8266	118.4707

By comparison, an OLS regression gives the following result:

```
. regress bp gender race
```

Source	SS	df	MS
Model	4485.66639	2	2242.83319
Residual	58442.7305	397	147.210908
Total	62928.3969	399	157.71528

```
Number of obs =      400
F(  2,   397) =    15.24
Prob > F      =   0.0000
R-squared     =   0.0713
Adj R-squared =   0.0666
Root MSE      =   12.133
```

bp	Coef.	Std. Err.	t	P>|t|	[95% Conf. Interval]	
gender	6.1775	1.213305	5.09	0.000	3.792194	8.562806
race	2.5875	1.213305	2.13	0.034	.2021938	4.972806
_cons	116.4862	1.050753	110.86	0.000	114.4205	118.552

Note the larger value for the race coefficient (and smaller p-value) in the OLS regression. The assumption of homogeneity of variance in OLS means that the mean for black men pulls the regression line higher than in the vwls regression, which takes into account the larger variance for black men and reduces its effect on the regression.

◁

Saved Results

vwls saves the following in e():

Scalars
  e(N)   number of observations
  e(df_m)   model degrees of freedom
  e(chi2)   model χ^2
  e(df_gf)   goodness-of-fit degrees of freedom
  e(chi2_gf)   goodness-of-fit χ^2

Macros
  e(cmd)   vwls
  e(cmdline)   command as typed
  e(depvar)   name of dependent variable
  e(properties) b V

Matrices
  e(b)   coefficient vector
  e(V)   variance–covariance matrix of the estimators

Functions
  e(sample)   marks estimation sample

Methods and Formulas

vwls is implemented as an ado-file.

Let $\mathbf{y} = (y_1, y_2, \ldots, y_n)'$ be the vector of observations of the dependent variable, where n is the number of observations. When sd() is specified, let s_1, s_2, ..., s_n be the standard deviations supplied by sd(). For categorical data, when sd() is not given, the means and standard deviations of y for each subgroup are computed, and n becomes the number of subgroups, $\mathbf{y}$ is the vector of subgroup means, and s_i are the standard deviations for the subgroups.

Let $\mathbf{V} = \mathrm{diag}(s_1^2, s_2^2, \ldots, s_n^2)$ denote the estimate of the variance of $\mathbf{y}$. Then the estimated regression coefficients are

$$\mathbf{b} = (\mathbf{X}'\mathbf{V}^{-1}\mathbf{X})^{-1}\mathbf{X}'\mathbf{V}^{-1}\mathbf{y}$$

and their estimated covariance matrix is

$$\widehat{\mathrm{Cov}}(\mathbf{b}) = (\mathbf{X}'\mathbf{V}^{-1}\mathbf{X})^{-1}$$

A statistic for the goodness of fit of the model is

$$Q = (\mathbf{y} - \mathbf{X}\mathbf{b})'\,\mathbf{V}^{-1}(\mathbf{y} - \mathbf{X}\mathbf{b})$$

where Q has a χ^2 distribution with $n - k$ degrees of freedom (k is the number of independent variables plus the constant, if any).

References

Gini, R., and J. Pasquini. 2006. Automatic generation of documents. *Stata Journal* 6: 22–39.

Grizzle, J. E., C. F. Starmer, and G. G. Koch. 1969. Analysis of categorical data by linear models. *Biometrics* 25: 489–504.

Press, W. H., S. A. Teukolsky, W. T. Vetterling, and B. P. Flannery. 1992. *Numerical Recipes in C: The Art of Scientific Computing*. 2nd ed. Cambridge: Cambridge University Press.

Also See

[R] **vwls postestimation** — Postestimation tools for vwls

[R] **regress** — Linear regression

[U] **11.1.6 weight**

[U] **20 Estimation and postestimation commands**

Title

vwls postestimation — Postestimation tools for vwls

Description

The following postestimation commands are available for vwls:

command	description
adjust	adjusted predictions of $\mathbf{x}\beta$
estat	VCE and estimation sample summary
estimates	cataloging estimation results
lincom	point estimates, standard errors, testing, and inference for linear combinations of coefficients
linktest	link test for model specification
mfx	marginal effects or elasticities
nlcom	point estimates, standard errors, testing, and inference for nonlinear combinations of coefficients
predict	predictions, residuals, influence statistics, and other diagnostic measures
predictnl	point estimates, standard errors, testing, and inference for generalized predictions
test	Wald tests for simple and composite linear hypotheses
testnl	Wald tests of nonlinear hypotheses

See the corresponding entries in the *Stata Base Reference Manual* for details.

Syntax for predict

predict [*type*] *newvar* [*if*] [*in*] [, xb stdp]

These statistics are available both in and out of sample; type predict ... if e(sample) ... if wanted only for the estimation sample.

Options for predict

xb, the default, calculates the linear prediction.

stdp calculates the standard error of the linear prediction.

Methods and Formulas

All postestimation commands listed above are implemented as ado-files.

Also See

[R] **vwls** — Variance-weighted least squares

[U] **20 Estimation and postestimation commands**

Title

> **which** — Display location and version for an ado-file

Syntax

> which *fname*[.*ftype*] [, all]

Description

which looks for *fname*.*ftype* along the S_ADO path. If Stata finds the file, which displays the full path and filename, along with, if the file is ASCII, all lines in the file that begin with "*!" in the first column. If Stata cannot find the file, which issues the message "file not found along adopath" and sets the return code to 111. *ftype* must be a file type for which Stata usually looks along the ado-path to find. Allowable *ftype*s are

> .ado, .class, .dlg, .idlg, .sthlp, .ihlp, .hlp, .key, .mata, .mlib, .mo, .mnu, .plugin, .scheme, and .style

If *ftype* is omitted, which assumes .ado. When searching for .ado files, if Stata cannot find the file, Stata then checks to see if *fname* is a built-in Stata command, allowing for valid abbreviations. If it is, the message "built-in command" is displayed; if not, the message "command not found as either built-in or ado-file" is displayed, and the return code is set to 111.

For information about internal version control, see [P] **version**.

Option

all forces which to report the location of all files matching the *fname*.*ftype* found along the search path. The default is to report just the first one found.

Remarks

If you write programs, you know that you make changes to the programs over time. If you are like us, you also end up with multiple versions of the program stored on your disk, perhaps in different directories. You may even have given copies of your programs to other Stata users, and you may not remember which version of a program you or your friends are using. The which command helps you solve this problem.

▷ Example 1

The which command displays the path for *filename*.ado and any lines in the code that begin with "*!". For example, we might want information about the notes command, described in [D] **notes**, which is an ado-file written by StataCorp. Here is what happens when we type which notes:

```
. which notes
C:\Program Files\Stata10\ado\base\n\notes.ado
*! version 2.0.0  13mar2007
```

which displays the path for the notes.ado file and also a line beginning with "*!" that indicates the version of the file. This is how we, at StataCorp, do version control—see [U] **18.11.1 Version** for an explanation of our version control numbers.

We do not need to be so formal. which will display anything typed after lines that begin with '*!'. For instance, we might write myprog.ado:

```
. which myprog
.\myprog.ado
*! first written 1/03/2007
*! bug fix on 1/05/2007 (no variance case)
*! updated 1/24/2007 to include noconstant option
*! still suspicious if variable takes on only two values
```

It does not matter where in the program the lines beginning with *! are—which will list them (in particular, our "still suspicious" comment was buried about 50 lines down in the code). All that is important is that the *! marker appear in the first two columns of a line.

◁

▷ Example 2

If we type which *command*, where *command* is a built-in command rather than an ado-file, Stata responds with

```
. which summarize
built-in command:  summarize
```

If *command* was neither a built-in command nor an ado-file, Stata would respond with

```
. which junk
command junk not found as either built-in or ado-file
r(111);
```

◁

Also See

[P] **findfile** — Find file in path

[P] **version** — Version control

[U] **17 Ado-files**

[U] **18.11.1 Version**

Title

xi — Interaction expansion

Syntax

xi [, prefix(*string*) noomit] *term(s)*

xi [, prefix(*string*) noomit] : *any_stata_command varlist_with_terms* ...

where a *term* has the form

i.*varname*	or	I.*varname*
i.*varname*$_1$*i.*varname*$_2$		I.*varname*$_1$*I.*varname*$_2$
i.*varname*$_1$**varname*$_3$		I.*varname*$_1$**varname*$_3$
i.*varname*$_1$|*varname*$_3$		I.*varname*$_1$|*varname*$_3$

varname, *varname*$_1$, and *varname*$_2$ denote numeric or string categorical variables. *varname*$_3$ denotes a continuous, numeric variable.

Description

xi expands terms containing categorical variables into indicator (also called dummy) variable sets by creating new variables, and, in the second syntax (xi: *any_stata_command*), executes the specified command with the expanded terms. The dummy variables created are

i.*varname*	Creates dummies for categorical variable *varname*
i.*varname*$_1$*i.*varname*$_2$	Creates dummies for categorical variables *varname*$_1$ and *varname*$_2$: all interactions and main effects
i.*varname*$_1$**varname*$_3$	Creates dummies for categorical variable *varname*$_1$ and continuous variable *varname*$_3$: all interactions and main effects
i.*varname*$_1$|*varname*$_3$	Creates dummies for categorical variable *varname*$_1$ and continuous variable *varname*$_3$: all interactions and main effect of *varname*$_3$, but no main effect of *varname*$_1$

Options

prefix(*string*) allows you to choose a prefix other than _I for the newly created interaction variables. The prefix cannot be longer than four characters. By default, xi will create interaction variables starting with _I. When you use xi, it drops all previously created interaction variables starting with the prefix specified in the prefix(*string*) option or with _I by default. Therefore, if you want to keep the variables with a certain prefix, specify a different prefix in the prefix(*string*) option.

noomit prevents xi from omitting groups. This option provides a way to generate an indicator variable for every category having one or more variables, which is useful when combined with the noconstant option of an estimation command.

Remarks

Remarks are presented under the following headings:

> *Background*
> *Indicator variables for simple effects*
> *Controlling the omitted dummy*
> *Categorical variable interactions*
> *Interactions with continuous variables*
> *Using xi: Interpreting output*
> *How xi names variables*
> *xi as a command rather than a command prefix*
> *Warnings*

xi provides a convenient way to include dummy or indicator variables when fitting a model (say, with regress, logistic, etc.). For instance, assume that the categorical variable agegrp contains 1 for ages 20–24, 2 for ages 25–39, 3 for ages 40–44, etc. Typing

    . xi: logistic outcome weight i.agegrp bp

estimates a logistic regression of outcome on weight, dummies for each agegrp category, and bp. That is, xi searches out and expands terms starting with "i." or "I." but ignores the other variables. xi will expand both numeric and string categorical variables, so if you had a string variable race containing "white", "black", and "other", typing

    . xi: logistic outcome weight bp i.agegrp i.race

would include indicator variables for the race group as well.

The i. indicator variables xi expands may appear anywhere in the *varlist*, so

    . xi: logistic outcome i.agegrp weight i.race bp

would fit the same model.

You can also create interactions of categorical variables; typing

    xi: logistic outcome weight bp i.agegrp*i.race

fits a model with indicator variables for all agegrp and race combinations, including the agegrp and race main-effect terms (i.e., the terms that are created when you just type i.agegrp i.race).

You can interact dummy variables with continuous variables; typing

    xi: logistic outcome bp i.agegrp*weight i.race

fits a model with indicator variables for all agegrp categories interacted with weight, plus the main-effect terms weight and i.agegrp.

You can get the interaction terms without the agegrp main effect (but with the weight main effect) by typing

    xi: logistic outcome bp i.agegrp|weight i.race

You can also include multiple interactions:

    xi: logistic outcome bp i.agegrp*weight i.agegrp*i.race

We will now back up and describe the construction of dummy variables in more detail.

Background

The terms *continuous, categorical,* and *indicator* or *dummy* variables are used below. Continuous variables measure something—such as height or weight—and at least conceptually can take on any real number over some range. Categorical variables, on the other hand, take on a finite number of values, each denoting membership in a subclass—for example, excellent, good, and poor, which might be coded 0, 1, 2 or 1, 2, 3 or even "Excellent", "Good", and "Poor". An indicator or dummy variable—the terms are used interchangeably—is a special type of two-valued categorical variable that contains values 0, denoting false, and 1, denoting true. The information contained in any k-valued categorical variable can be equally well represented by k indicator variables. Instead of one variable recording values representing excellent, good, and poor, you can have three indicator variables, indicating the truth or falseness of "result is excellent", "result is good", and "result is poor".

xi provides a convenient way to convert categorical variables to dummy or indicator variables when you fit a model (say with regress, logistic, etc.).

▷ Example 1

For instance, assume that the categorical variable agegrp contains 1 for ages 20–24, 2 for ages 25–39, and 3 for ages 40–44. (There is no one over 44 in our data.) As it stands, agegrp would be a poor candidate for inclusion in a model even if we thought age affected the outcome. The reason is that the coding would restrict the effect of being in the second age group to be twice the effect of being in the first, and, similarly, the effect of being in the third to be three times the first. That is, if we fitted the model,

$$y = \beta_0 + \beta_1 \, \texttt{agegrp} + X\beta_2$$

the effect of being in the first age group is β_1, the second $2\beta_1$, and the third $3\beta_1$. If the coding 1, 2, and 3 is arbitrary, we could just as well have coded the age groups 1, 4, and 9, making the effects β_1, $4\beta_1$, and $9\beta_1$.

The solution is to convert the categorical variable agegrp to a set of indicator variables a_1, a_2, and a_3, where a_i is 1 if the individual is a member of the ith age group and 0 otherwise. We can then fit the model:

$$y = \beta_0 + \beta_{11}a_1 + \beta_{12}a_2 + \beta_{13}a_3 + X\beta_2$$

The effect of being in age group 1 is now β_{11}; 2, β_{12}; and 3, β_{13}; and these results are independent of our (arbitrary) coding. The only difficulty at this point is that the model is unidentified in the sense that there are an infinite number of $(\beta_0, \beta_{11}, \beta_{12}, \beta_{13})$ that fit the data equally well.

To see this, pretend that $(\beta_0, \beta_{11}, \beta_{12}, \beta_{13}) = (1, 1, 3, 4)$. The predicted values of y for the various age groups are

$$y = \begin{cases} 1 + 1 + X\beta_2 = 2 + X\beta_2 & \text{(age group 1)} \\ 1 + 3 + X\beta_2 = 4 + X\beta_2 & \text{(age group 2)} \\ 1 + 4 + X\beta_2 = 5 + X\beta_2 & \text{(age group 3)} \end{cases}$$

Now pretend that $(\beta_0, \beta_{11}, \beta_{12}, \beta_{13}) = (2, 0, 2, 3)$. The predicted values of y are

$$y = \begin{cases} 2 + 0 + X\beta_2 = 2 + X\beta_2 & \text{(age group 1)} \\ 2 + 2 + X\beta_2 = 4 + X\beta_2 & \text{(age group 2)} \\ 2 + 3 + X\beta_2 = 5 + X\beta_2 & \text{(age group 3)} \end{cases}$$

These two sets of predictions are indistinguishable: for age group 1, $y = 2 + X\beta_2$ regardless of the coefficient vector used, and similarly for age groups 2 and 3. This arises because we have three equations and four unknowns. Any solution is as good as any other, and, for our purposes, we merely need to choose one of them. The popular selection method is to set the coefficient on the first indicator variable to 0 (as we have done in our second coefficient vector). This is equivalent to fitting the model

$$y = \beta_0 + \beta_{12}a_2 + \beta_{13}a_3 + X\beta_2$$

How we select a particular coefficient vector (identifies the model) does not matter. It does, however, affect the *interpretation* of the coefficients.

For instance, we could just as well choose to omit the second group. In our artificial example, this would yield $(\beta_0, \beta_{11}, \beta_{12}, \beta_{13}) = (4, -2, 0, 1)$ instead of $(2, 0, 2, 3)$. These coefficient vectors are the same in the sense that

$$y = \begin{cases} 2 + 0 + X\beta_2 = 2 + X\beta_2 = 4 - 2 + X\beta_2 & \text{(age group 1)} \\ 2 + 2 + X\beta_2 = 4 + X\beta_2 = 4 + 0 + X\beta_2 & \text{(age group 2)} \\ 2 + 3 + X\beta_2 = 5 + X\beta_2 = 4 + 1 + X\beta_2 & \text{(age group 3)} \end{cases}$$

But what does it mean that β_{13} can just as well be 3 or 1? We obtain $\beta_{13} = 3$ when we set $\beta_{11} = 0$, so $\beta_{13} = \beta_{13} - \beta_{11}$ and β_{13} measures the difference between age groups 3 and 1.

In the second case, we obtain $\beta_{13} = 1$ when we set $\beta_{12} = 0$, so $\beta_{13} - \beta_{12} = 1$ and β_{13} measures the difference between age groups 3 and 2. There is no inconsistency. According to our $\beta_{12} = 0$ model, the difference between age groups 3 and 1 is $\beta_{13} - \beta_{11} = 1 - (-2) = 3$, the same result we got in the $\beta_{11} = 0$ model.

◁

▷ Example 2

The issue of interpretation is important because it can affect the way we discuss results. Imagine that we are studying recovery after a coronary bypass operation. Assume that the age groups are (1) children under 13 (we have two of them), (2) young adults under 25 (we have a handful of them), (3) adults under 46 (of which we have even more), (4) mature adults under 56, (5) older adults under 65, and (6) elderly adults. We follow the prescription of omitting the first group, so all our results are reported relative to children under 13. While there is nothing statistically wrong with this, readers will be suspicious when we make statements like "compared with young children, older and elder adults ...". Moreover, we will probably have to end each statement with "although results are not statistically significant" because we have only two children in our comparison group. Of course, even with results reported in this way, we can do reasonable comparisons (say, with mature adults), but we will have to do extra work to perform the appropriate linear hypothesis test using Stata's `test` command.

Here it would be better to force the omitted group to be more reasonable, such as mature adults. There is, however, a generic rule for automatic comparison group selection that, although less popular, tends to work better than the omit-the-first-group rule. That rule is to omit the most prevalent group. The most prevalent is usually a reasonable baseline.

◁

In any case, the prescription for categorical variables is

1. Convert each k-valued categorical variable to k indicator variables.

2. Drop one of the k indicator variables; any one will do, but dropping the first is popular, dropping the most prevalent is probably better in terms of having the computer guess at a reasonable interpretation, and dropping a specified one often eases interpretation the most.

3. Fit the model on the remaining $k - 1$ indicator variables.

xi automates this procedure.

We will now consider each of xi's features in detail.

Indicator variables for simple effects

When you type i.*varname*, xi internally tabulates *varname* (which may be a string or a numeric variable) and creates indicator (dummy) variables for each observed value, omitting the indicator for the smallest value. For instance, say that agegrp takes on the values 1, 2, 3, and 4. Typing

```
xi: logistic outcome i.agegrp
```

creates indicator variables named _Iagegrp_2, _Iagegrp_3, and _Iagegrp_4. (xi chooses the names and tries to make them readable; xi guarantees that the names are unique.) The expanded logistic model is

```
. logistic outcome _Iagegrp_2 _Iagegrp_3 _Iagegrp_4
```

Afterward, you can drop the new variables xi leaves behind by typing 'drop _I*' (note the capitalization).

xi provides the following features when you type i.*varname*:

1. *varname* may be string or numeric.

2. Dummy variables are created automatically.

3. By default, the dummy-variable set is identified by dropping the dummy corresponding to the smallest value of the variable (how to specify otherwise is discussed below).

4. The new dummy variables are left in your dataset. By default, the names of the new dummy variables start with _I; therefore, you can drop them by typing 'drop _I*'. You do not have to do this; each time you use xi, any automatically generated dummies with the same prefix as the one specified in the prefix(*string*) option, or _I by default, are dropped and new ones are created.

5. The new dummy variables have variable labels so that you can determine what they correspond to by typing 'describe'.

6. xi may be used with any Stata command (not just logistic).

Controlling the omitted dummy

By default, i.*varname* omits the dummy corresponding to the smallest value of *varname*; for a string variable, this is interpreted as dropping the first in an alphabetical, case-sensitive sort. xi provides two alternatives to dropping the first: xi will drop the dummy corresponding to the most prevalent value of *varname*, or xi will let you choose the particular dummy to be dropped.

To change xi's behavior to dropping the most prevalent dummy, type

```
. char _dta[omit] prevalent
```

although whether you type "prevalent" or "yes" or anything else does not matter. Setting this characteristic affects the expansion of all categorical variables in the dataset. If you resave your dataset, the prevalent preference will be remembered. If you want to change the behavior back to the default drop-the-first rule, type

```
. char _dta[omit]
```

to clear the characteristic.

Once you set _dta[omit], i.*varname* omits the dummy corresponding to the most prevalent value of *varname*. Thus the coefficients on the dummies have the interpretation of change from the most prevalent group. For example,

```
. char _dta[omit] prevalent
. xi: regress y i.agegrp
```

might create _Iagegrp_1 through _Iagegrp_4, resulting in _Iagegrp_2 being omitted if agegrp = 2 is most common (as opposed to the default dropping of _Iagegrp_1). The model is then

$$y = b_0 + b_1 \text{ \_Iagegrp\_1} + b_3 \text{ \_Iagegrp\_3} + b_4 \text{ \_Iagegrp\_4} + u$$

Then

Predicted y for agegrp 1 = $b_0 + b_1$ Predicted y for agegrp 3 = $b_0 + b_3$

Predicted y for agegrp 2 = b_0 Predicted y for agegrp 4 = $b_0 + b_4$

Thus the model's reported t or Z statistics are for a test of whether each group is different from the most prevalent group.

Perhaps you wish to omit the dummy for agegrp 3 instead. You do this by setting the variable's omit characteristic:

```
. char agegrp[omit] 3
```

This overrides _dta[omit] if you have set it. Now when you type

```
. xi: regress y i.agegrp
```

_Iagegrp_3 will be omitted, and you will fit the model:

$$y = b_0' + b_1' \text{ \_Iagegrp\_1} + b_2' \text{ \_Iagegrp\_2} + b_4' \text{ \_Iagegrp\_4} + u$$

Later if you want to return to the default omission, type

```
. char agegrp[omit]
```

to clear the characteristic.

In summary, i.*varname* omits the first group by default, but if you define

```
. char _dta[omit] prevalent
```

the default behavior changes to dropping the most prevalent group. Either way, if you define a characteristic of the form

```
. char varname[omit] #
```

or, if *varname* is a string,

```
. char varname[omit] string-literal
```

the specified value will be omitted.

Examples: . char agegrp[omit] 1

. char race[omit] White   (for race a string variable)

. char agegrp[omit]   (to restore default for agegrp)

Categorical variable interactions

i.*varname*$_1$*i.*varname*$_2$ creates the dummy variables associated with the interaction of the categorical variables *varname*$_1$ and *varname*$_2$. The identification rules—which categories are omitted—are the same as those for i.*varname*. For instance, assume that `agegrp` takes on four values and `race` takes on three values. Typing

```
. xi: regress y i.agegrp*i.race
```

results in

model :	dummies for:

$$y = a + b_2 \text{ \_Iagegrp\_2} + b_3 \text{ \_Iagegrp\_3} + b_4 \text{ \_Iagegrp\_4} \quad \text{(agegrp)}$$
$$+c_2 \text{ \_Irace\_2} + c_3 \text{ \_Irace\_3} \quad \text{(race)}$$
$$+d_{22} \text{ \_IageXrac\_2\_2} + d_{23} \text{ \_IageXrac\_2\_3}$$
$$+d_{32} \text{ \_IageXrac\_3\_2} + d_{33} \text{ \_IageXrac\_3\_3} \quad \text{(agegrp*race)}$$
$$+d_{42} \text{ \_IageXrac\_4\_2} + d_{43} \text{ \_IageXrac\_4\_3}$$
$$+u$$

That is, typing

```
. xi: regress y i.agegrp*i.race
```

is the same as typing

```
. xi: regress y i.agegrp i.race i.agegrp*i.race
```

Although there are many other ways the interaction could have been parameterized, this method has the advantage that you can test the joint significance of the interactions by typing

```
. testparm _IageXrac*
```

When you perform the estimation step, whether you specify `i.agegrp*i.race` or `i.race*i.agegrp` makes no difference (other than in the names given to the interaction terms; in the first case, the names will begin with `_IageXrac`; in the second, `_IracXage`). Thus

```
. xi: regress y i.race*i.agegrp
```

fits the same model.

You may also include multiple interactions simultaneously:

```
. xi: regress y i.agegrp*i.race i.agegrp*i.sex
```

The model fitted is

model :	dummies for:

$$y = a + b_2 \text{ \_Iagegrp\_2} + b_3 \text{ \_Iagegrp\_3} + b_4 \text{ \_Iagegrp\_4} \quad \text{(agegrp)}$$
$$+c_2 \text{ \_Irace\_2} + c_3 \text{ \_Irace\_3} \quad \text{(race)}$$
$$+d_{22} \text{ \_IageXrac\_2\_2} + d_{23} \text{ \_IageXrac\_2\_3}$$
$$+d_{32} \text{ \_IageXrac\_3\_2} + d_{33} \text{ \_IageXrac\_3\_3} \quad \text{(agegrp*race)}$$
$$+d_{42} \text{ \_IageXrac\_4\_2} + d_{43} \text{ \_IageXrac\_4\_3}$$
$$+e_2 \text{ \_Isex\_2} \quad \text{(sex)}$$
$$+f_{22} \text{ \_IageXsex\_2\_2} + f_{23} \text{ \_IageXsex\_2\_3} + f_{24} \text{ \_IageXsex\_2\_4} \quad \text{(agegrp*sex)}$$
$$+u$$

The `agegrp` dummies are (correctly) included only once.

Interactions with continuous variables

i.*varname*$_1$**varname*$_2$ (as distinguished from i.*varname*$_1$**i.*varname*$_2$—note the second i.) specifies an interaction of a categorical variable with a continuous variable. For instance,

```
. xi: regress y i.agegr*wgt
```

results in the model:

$$y = a + b_2 \_\text{Iagegrp}\_2 + b_3 \_\text{Iagegrp}\_3 + b_4 \_\text{Iagegrp}\_4 \qquad \text{(agegrp dummies)}$$
$$+ c\,\text{wgt} \qquad\qquad\qquad\qquad\qquad\qquad\qquad\qquad \text{(continuous wgt effect)}$$
$$+ d_2 \_\text{IageXwgt}\_2 + d_3 \_\text{IageXwgt}\_3 + d_4 \_\text{IageXwgt}\_4 \quad \text{(agegrp*wgt interactions)}$$
$$+ u$$

A variation on this notation, using | rather than *, omits the agegrp dummies. Typing

```
. xi: regress y i.agegrp|wgt
```

fits the model

$$y = a' + c'\,\text{wgt} \qquad\qquad\qquad\qquad\qquad\qquad\qquad\qquad \text{(continuous wgt effect)}$$
$$+ d_2' \_\text{IageXwgt}\_2 + d_3' \_\text{IageXwgt}\_3 + d_4' \_\text{IageXwgt}\_4 \quad \text{(agegrp*wgt interactions)}$$
$$+ u'$$

The predicted values of y are

agegrp*wgt model	agegrp|wgt model	
$y = a + c\,\text{wgt}$	$a' + c'\,\text{wgt}$	if agegrp $= 1$
$a + c\,\text{wgt} + b_2 + d_2\,\text{wgt}$	$a' + c'\,\text{wgt} + d_2'\,\text{wgt}$	if agegrp $= 2$
$a + c\,\text{wgt} + b_3 + d_3\,\text{wgt}$	$a' + c'\,\text{wgt} + d_3'\,\text{wgt}$	if agegrp $= 3$
$a + c\,\text{wgt} + b_4 + d_4\,\text{wgt}$	$a' + c'\,\text{wgt} + d_4'\,\text{wgt}$	if agegrp $= 4$

That is, typing

```
. xi: regress y i.agegrp*wgt
```

is equivalent to typing

```
. xi: regress y i.agegrp i.agegrp|wgt
```

In either case, you do not need to specify separately the continuous variable wgt; it is included automatically.

Using xi: Interpreting output

```
. xi: regress mpg i.rep78
i.rep78              _Irep78_1-5    (naturally coded; _Irep78_1 omitted)
    (output from regress appears )
```

Interpretation: i.rep78 expanded to the dummies _Irep78_1, _Irep78_2, ..., _Irep78_5. The numbers on the end are "natural" in the sense that _Irep78_1 corresponds to rep78 $= 1$, _Irep78_2 to rep78 $= 2$, and so on. Finally, the dummy for rep78 $= 1$ was omitted.

```
. xi: regress mpg i.make
i.make               _Imake_1-74    (_Imake_1 for make==AMC Concord omitted)
    (output from regress appears )
```

Interpretation: i.make expanded to _Imake_1, _Imake_2, ..., _Imake_74. The coding is not natural because make is a string variable. _Imake_1 corresponds to one make, _Imake_2 another, and so on. You can find out the coding by typing describe. _Imake_1 for the AMC Concord was omitted.

How xi names variables

By default, xi assigns to the dummy variables it creates names having the form

$$\_Istub\_groupid$$

You may subsequently refer to the entire set of variables by typing '_Istub*'. For example,

name	= _I +	stub	+ _ +	groupid	Entire set
_Iagegrp_1	_I	agegrp	_	1	_Iagegrp*
_Iagegrp_2	_I	agegrp	_	2	_Iagegrp*
_IageXwgt_1	_I	ageXwgt	_	1	_IageXwgt*
_IageXrac_1_2	_I	ageXrac	_	1_2	_IageXrac*
_IageXrac_2_1	_I	ageXrac	_	2_1	_IageXrac*

If you specify a prefix in the prefix(*string*) option, say, _S, xi will name the variables starting with the prefix

$$\_Sstub\_groupid$$

xi as a command rather than a command prefix

xi can be used as a command prefix or as a command by itself. In the latter form, xi merely creates the indicator and interaction variables. Typing

```
. xi: regress y i.agegrp*wgt
i.agegrp                 _Iagegrp_1-4  (naturally coded; _Iagegrp_1 omitted)
i.agegrp*wgt             _IageXwgt_1-4  (coded as above)
  (output from regress appears )
```

is equivalent to typing

```
. xi i.agegrp*wgt
i.agegrp                 _Iagegrp_1-4  (naturally coded; _Iagegrp_1 omitted)
i.agegrp*wgt             _IageXwgt_1-4  (coded as above)
. regress y _Iagegrp* _IageXwgt*
  (output from regress appears )
```

Warnings

1. xi creates new variables in your dataset; most are bytes, but interactions with continuous variables will have the storage type of the underlying continuous variable. You may get the message "insufficient memory". If so, you will need to increase the amount of memory allocated to Stata's data areas; see [U] **6 Setting the size of memory**.

2. When using xi with an estimation command, you may get the message "matsize too small". If so, see [R] **matsize**.

Saved Results

xi saves the following characteristics:

$$
\begin{aligned}
&\_\text{dta}[\_\_\text{xi}\_\_\text{Vars}\_\_\text{Prefix}\_\_] \quad \text{prefix names} \\
&\_\text{dta}[\_\_\text{xi}\_\_\text{Vars}\_\_\text{To}\_\_\text{Drop}\_\_] \quad \text{variables created}
\end{aligned}
$$

Methods and Formulas

xi is implemented as an ado-file.

References

Hendrickx, J. 1999. dm73: Using categorical variables in Stata. *Stata Technical Bulletin* 52: 2–8. Reprinted in *Stata Technical Bulletin Reprints*, vol. 9, pp. 51–59.

——. 2000. dm73.1: Contrasts for categorical variables: Update. *Stata Technical Bulletin* 54: 7. Reprinted in *Stata Technical Bulletin Reprints*, vol. 9, pp. 60–61.

——. 2001a. dm73.2: Contrasts for categorical variables: Update. *Stata Technical Bulletin* 59: 2–5. Reprinted in *Stata Technical Bulletin Reprints*, vol. 10, pp. 9–14.

——. 2001b. dm73.3: Contrasts for categorical variables: Update. *Stata Technical Bulletin* 61: 5. Reprinted in *Stata Technical Bulletin Reprints*, vol. 10, pp. 14–15.

Also See

[U] **11.1.10 Prefix commands**

[U] **20 Estimation and postestimation commands**

Title

zinb — Zero-inflated negative binomial regression

Syntax

zinb *depvar* [*indepvars*] [*if*] [*in*] [*weight*] ,

    <u>inf</u>late(*varlist*[, <u>off</u>set(*varname*)] | _cons) [*options*]

options	description
Model	
* <u>inf</u>late( )	equation that determines whether the count is zero
<u>nocon</u>stant	suppress constant term
<u>expo</u>sure(*varname_e*)	include ln(*varname_e*) in model with coefficient constrained to 1
<u>off</u>set(*varname_o*)	include *varname_o* in model with coefficient constrained to 1
<u>constr</u>aints(*constraints*)	apply specified linear constraints
<u>coll</u>inear	keep collinear variables
probit	use probit model to characterize excess zeros; default is logit
SE/Robust	
vce(*vcetype*)	*vcetype* may be oim, <u>r</u>obust, <u>cl</u>uster *clustvar*, opg, <u>boot</u>strap, or <u>jack</u>knife
Reporting	
<u>l</u>evel(#)	set confidence level; default is level(95)
irr	report incidence-rate ratios
vuong	perform Vuong test
zip	perform ZIP likelihood-ratio test
Max options	
maximize_options	control maximization process; seldom used

* <u>inf</u>late(*varlist*[, <u>off</u>set(*varname*)] | _cons) is required.
bootstrap, by, jackknife, rolling, statsby, svy, and xi are allowed; see [U] **11.1.10 Prefix commands**.
Weights are not allowed with the bootstrap prefix.
vce(), vuong, zip, and weights are not allowed with the svy prefix.
fweights, iweights, and pweights are allowed; see [U] **11.1.6 weight**.
See [U] **20 Estimation and postestimation commands** for more capabilities of estimation commands.

Description

    zinb estimates a zero-inflated negative binomial regression of *depvar* on *indepvars*, where *depvar* is a nonnegative count variable.

Options

_____ Model _____

inflate(*varlist*[, offset(*varname*)] | _cons) specifies the equation that determines whether the observed count is zero. Conceptually, omitting inflate() would be equivalent to fitting the model with nbreg.

  inflate(*varlist*[, offset(*varname*)]) specifies the variables in the equation. You may optionally include an offset for this *varlist*.

  inflate(_cons) specifies that the equation determining whether the count is zero contains only an intercept. To run a zero-inflated model of *depvar* with only an intercept in both equations, type zinb *depvar*, inflate(_cons).

noconstant, exposure(*varname_e*), offset(*varname_o*), constraints(*constraints*), collinear; see [R] **estimation options**.

probit requests that a probit, instead of logit, model be used to characterize the excess zeros in the data.

_____ SE/Robust _____

vce(*vcetype*) specifies the type of standard error reported, which includes types that are derived from asymptotic theory, that are robust to some kinds of misspecification, that allow for intragroup correlation, and that use bootstrap or jackknife methods; see [R] *vce_option*.

_____ Reporting _____

level(*#*); see [R] **estimation options**.

irr reports estimated coefficients transformed to incidence-rate ratios, that is, e^{β_i} rather than β_i. Standard errors and confidence intervals are similarly transformed. This option affects how results are displayed, not how they are estimated or stored. irr may be specified at estimation or when replaying previously estimated results.

vuong specifies that the Vuong (1989) test of ZINB versus negative binomial be reported. This test statistic has a standard normal distribution with large positive values favoring the ZINB model and large negative values favoring the negative binomial model.

zip requests that a likelihood-ratio test comparing the zero-inflated negative binomial model with the zero-inflated Poisson model be included in the output.

_____ Max options _____

maximize_options: difficult, technique(*algorithm_spec*), iterate(*#*), [no]log, trace, gradient, showstep, hessian, shownrtolerance, tolerance(*#*), ltolerance(*#*), gtolerance(*#*), nrtolerance(*#*), nonrtolerance, from(*init_specs*); see [R] **maximize**. These options are seldom used.

Setting the optimization type to technique(bhhh) resets the default *vcetype* to vce(opg).

Remarks

See Long (1997, 242–247) and Greene (2003, 749–752) for a discussion of zero-modified count models. For information about the test developed by Vuong (1989), see Greene (2003) and Long (1997). Greene (1994) applied the test to zero-inflated Poisson and negative binomial models, and there is a description of that work in Greene (2003).

Negative binomial regression fits models of the number of occurrences (counts) of an event. You could use nbreg for this (see [R] **nbreg**), but in some count-data models, you might want to account for the prevalence of zero counts in the data.

For instance, you could count how many fish each visitor to a park catches. Many visitors may catch zero, as they do not fish (as opposed to being unsuccessful). You may be able to model whether a person fishes depending on several covariates related to fishing activity and model how many fish a person catches depending on several covariates having to do with the success of catching fish (type of lure/bait, time of day, temperature, season, etc.). This is the type of data for which the zinb command is useful.

The zero-inflated (or zero-altered) negative binomial model allows overdispersion through the splitting process that models the outcomes as zero or nonzero.

▷ Example 1

We have data on the number of fish caught by visitors to a national park. Some of the visitors do not fish, but we do not have the data on whether a person fished or not; we have data merely on how many fish were caught, together with several covariates. As our data have a preponderance of zeros (142 of 250), we use the zinb command to model the outcome.

```
. use http://www.stata-press.com/data/r10/fish

. zinb count persons livebait, inf(child camper) vuong

Fitting constant-only model:

Iteration 0:   log likelihood = -519.33992
  (output omitted )
Iteration 8:   log likelihood = -442.66299

Fitting full model:

Iteration 0:   log likelihood = -442.66299  (not concave)
  (output omitted )
Iteration 8:   log likelihood = -401.54776
```

Zero-inflated negative binomial regression			Number of obs	=		250
			Nonzero obs	=		108
			Zero obs	=		142
Inflation model = logit			LR chi2(2)	=		82.23
Log likelihood = -401.5478			Prob > chi2	=		0.0000

	Coef.	Std. Err.	z	P>|z|	[95% Conf. Interval]	
count						
persons	.9742984	.1034938	9.41	0.000	.7714543	1.177142
livebait	1.557523	.4124424	3.78	0.000	.7491503	2.365895
_cons	-2.730064	.476953	-5.72	0.000	-3.664874	-1.795253
inflate						
child	3.185999	.7468551	4.27	0.000	1.72219	4.649808
camper	-2.020951	.872054	-2.32	0.020	-3.730146	-.3117567
_cons	-2.695385	.8929071	-3.02	0.003	-4.44545	-.9453189
/lnalpha	.5110429	.1816816	2.81	0.005	.1549535	.8671323
alpha	1.667029	.3028685			1.167604	2.380076

```
Vuong test of zinb vs. standard negative binomial: z =    5.59  Pr>z = 0.0000
```

In general, Vuong test statistics that are significantly positive favor the zero-inflated models, whereas those that are significantly negative favor the non–zero-inflated models. Thus in the above model, the zero inflation is significant.   ◁

Saved Results

zinb saves the following in e():

Scalars

e(N)	number of observations
e(k)	number of parameters
e(k_eq)	number of equations
e(k_eq_model)	number of equations in model Wald test
e(k_aux)	number of auxiliary parameters
e(k_dv)	number of dependent variables
e(N_zero)	number of zero observations
e(df_m)	model degrees of freedom
e(ll)	log likelihood
e(ll_0)	log likelihood, constant-only model
e(df_c)	degrees of freedom for comparison test
e(N_clust)	number of clusters
e(chi2)	χ^2
e(chi2_c)	χ^2 for comparison test
e(p)	significance of model test
e(chi2_p)	χ^2 test against Poisson model
e(chi2_cp)	χ^2 for test of $\alpha = 0$
e(vuong)	Vuong test statistic
e(rank)	rank of e(V)
e(ic)	number of iterations
e(rc)	return code
e(converged)	1 if converged, 0 otherwise

Macros

e(cmd)	zinb
e(cmdline)	command as typed
e(depvar)	name of dependent variable
e(inflate)	logit or probit
e(wtype)	weight type
e(wexp)	weight expression
e(title)	title in estimation output
e(clustvar)	name of cluster variable
e(offset1)	offset
e(offset2)	offset for inflate()
e(chi2type)	Wald or LR; type of model χ^2 test
e(chi2_ct)	Wald or LR; type of model χ^2 test corresponding to e(chi2_c)
e(chi2_cpt)	Wald or LR; type of model χ^2 test corresponding to e(chi2_cp)
e(chi2_pt)	Wald or LR; type of model χ^2 test corresponding to e(chi2_p)
e(vce)	*vcetype* specified in vce()
e(vcetype)	title used to label Std. Err.
e(diparm#)	display transformed parameter #
e(opt)	type of optimization
e(ml_method)	type of ml method
e(user)	name of likelihood-evaluator program
e(technique)	maximization technique
e(crittype)	optimization criterion
e(properties)	b V
e(predict)	program used to implement predict

Matrices

e(b)	coefficient vector
e(V)	variance–covariance matrix of the estimators
e(gradient)	gradient vector
e(ilog)	iteration log (up to 20 iterations)

Functions

e(sample)	marks estimation sample

Methods and Formulas

`zinb` is implemented as an ado-file.

Several models in the literature are (correctly) described as zero inflated. The `zinb` command maximizes the log likelihood $\ln L$, defined by

$$m = 1/\alpha$$
$$p_j = 1/(1 + \alpha\mu_j)$$
$$\xi_j^\beta = \mathbf{x}_j\boldsymbol{\beta} + \text{offset}_j^\beta$$
$$\xi_j^\gamma = \mathbf{z}_j\boldsymbol{\gamma} + \text{offset}_j^\gamma$$
$$\mu_j = \exp(\xi_j^\beta)$$
$$\ln L = \sum_{j \in S} w_j \ln \left[F(\xi_j^\gamma) + \left\{ 1 - F(\xi_j^\gamma) \right\} p_j^m \right]$$
$$+ \sum_{j \notin S} w_j \left[\ln\left\{ 1 - F(\xi_j^\gamma) \right\} + \ln\Gamma(m + y_j) - \ln\Gamma(y_j + 1) \right.$$
$$\left. - \ln\Gamma(m) + m \ln p_j + y_j \ln(1 - p_j) \right]$$

where w_j are the weights, F is the logit link (or probit link if `probit` was specified), and S is the set of observations for which the outcome $y_j = 0$.

References

Greene, W. H. 1994. Accounting for excess zeros and sample selection in Poisson and negative binomial regression models. Working Paper no. EC-94-10, Department of Economics, Stern School of Business, New York University.

——. 2003. *Econometric Analysis*. 5th ed. Upper Saddle River, NJ: Prentice Hall.

Long, J. S. 1997. *Regression Models for Categorical and Limited Dependent Variables*. Thousand Oaks, CA: Sage.

Long, J. S., and J. Freese. 2001. Predicted probabilities for count models. *Stata Journal* 1: 51–57.

——. 2006. *Regression Models for Categorical Dependent Variables Using Stata*. 2nd ed. College Station, TX: Stata Press.

Mullahy, J. 1986. Specification and testing of some modified count-data models. *Journal of Econometrics* 33: 341–365.

Vuong, Q. 1989. Likelihood ratio tests for model selection and non-nested hypotheses. *Econometrica*: 57: 307–334.

Also See

Title

zinb postestimation — Postestimation tools for zinb

Description

The following postestimation commands are available for zinb:

command	description
adjust	adjusted predictions of $\mathbf{x}\beta$ or $\exp(\mathbf{x}\beta)$
estat	AIC, BIC, VCE, and estimation sample summary
estat (svy)	postestimation statistics for survey data
estimates	cataloging estimation results
lincom	point estimates, standard errors, testing, and inference for linear combinations of coefficients
lrtest[1]	likelihood-ratio test
mfx	marginal effects or elasticities
nlcom	point estimates, standard errors, testing, and inference for nonlinear combinations of coefficients
predict	predictions, residuals, influence statistics, and other diagnostic measures
predictnl	point estimates, standard errors, testing, and inference for generalized predictions
suest	seemingly unrelated estimation
test	Wald tests for simple and composite linear hypotheses
testnl	Wald tests of nonlinear hypotheses

[1] lrtest is not appropriate with svy estimation results.

See the corresponding entries in the *Stata Base Reference Manual* for details, but see [SVY] **estat** for details about estat (svy).

Syntax for predict

predict [*type*] *newvar* [*if*] [*in*] [, *statistic* <u>nooff</u>set]

predict [*type*] { *stub** | *newvar*$_{\text{reg}}$ *newvar*$_{\text{inflate}}$ *newvar*$_{\text{lnalpha}}$ } [*if*] [*in*] , <u>sc</u>ores

statistic	description
Main	
n	number of events; the default
ir	incidence rate
<u>pr</u>	probability of a zero outcome
xb	linear prediction
stdp	standard error of the linear prediction

These statistics are available both in and out of sample; type predict ... if e(sample) ... if wanted only for the estimation sample.

Options for predict

> ⌐ Main └───

n, the default, calculates the predicted number of events, which is $(1 - p_j)\exp(\mathbf{x}_j\boldsymbol{\beta})$ if neither offset() nor exposure() was specified when the model was fitted, where p_j is the predicted probability of a zero outcome; $(1 - p_j)\exp((\mathbf{x}_j\boldsymbol{\beta}) + \text{offset}_j)$ if offset() was specified; or $(1 - p_j)(\exp(\mathbf{x}_j\boldsymbol{\beta}) \times \text{exposure}_j)$ if exposure() was specified.

ir calculates the incidence rate $\exp(\mathbf{x}_j\boldsymbol{\beta})$, which is the predicted number of events when exposure is 1. This is equivalent to specifying both n and nooffset options.

pr calculates the probability of a zero outcome $F(\mathbf{z}_j\boldsymbol{\gamma})$, where $F()$ is the logit or probit link function. If offset() was specified within the inflate() option, then $F(\mathbf{z}_j\boldsymbol{\gamma} + \text{offset}_j^\gamma)$ is calculated.

xb calculates the linear prediction, which is $\mathbf{x}_j\boldsymbol{\beta}$ if neither offset() nor exposure() was specified; $\mathbf{x}_j\boldsymbol{\beta} + \text{offset}_j$ if offset() was specified; or $\mathbf{x}_j\boldsymbol{\beta} + \ln(\text{exposure}_j)$ if exposure() was specified; see nooffset below.

stdp calculates the standard error of the linear prediction.

nooffset is relevant only if you specified offset() or exposure() when you fitted the model. It modifies the calculations made by predict so that they ignore the offset or exposure variable; the linear prediction is treated as $\mathbf{x}_j\boldsymbol{\beta}$ rather than as $\mathbf{x}_j\boldsymbol{\beta} + \text{offset}_j$ or $\mathbf{x}_j\boldsymbol{\beta} + \ln(\text{exposure}_j)$. Specifying predict ..., nooffset is equivalent to specifying predict ..., ir.

scores calculates equation-level score variables.

The first new variable will contain $\partial \ln L/\partial(\mathbf{x}_j\boldsymbol{\beta})$.

The second new variable will contain $\partial \ln L/\partial(\mathbf{z}_j\boldsymbol{\gamma})$.

The third new variable will contain $\partial \ln L/\partial \ln\alpha$.

Methods and Formulas

All postestimation commands listed above are implemented as ado-files.

Also See

[R] **zinb** — Zero-inflated negative binomial regression

[U] **20 Estimation and postestimation commands**

Title

> **zip** — Zero-inflated Poisson regression

Syntax

> zip *depvar* [*indepvars*] [*if*] [*in*] [*weight*] ,
>
> inflate(*varlist*[, offset(*varname*)]|_cons) [*options*]

options	description
Model	
*inflate()	equation that determines whether the count is zero
noconstant	suppress constant term
exposure(*varname_e*)	include ln($varname_e$) in model with coefficient constrained to 1
offset(*varname_o*)	include $varname_o$ in model with coefficient constrained to 1
constraints(*constraints*)	apply specified linear constraints
collinear	keep collinear variables
probit	use probit model to characterize excess zeros; default is logit
SE/Robust	
vce(*vcetype*)	*vcetype* may be oim, robust, cluster *clustvar*, opg, bootstrap, or jackknife
Reporting	
level(*#*)	set confidence level; default is level(95)
irr	report incidence-rate ratios
vuong	perform Vuong test
Max options	
maximize_options	control maximization process; seldom used

*inflate(*varlist*[, offset(*varname*)]|_cons) is required.

bootstrap, by, jackknife, rolling, statsby, svy, and xi are allowed; see [U] **11.1.10 Prefix commands**.

Weights are not allowed with the bootstrap prefix.

vce(), vuong, and weights are not allowed with the svy prefix.

fweights, iweights, and pweights are allowed; see [U] **11.1.6 weight**.

See [U] **20 Estimation and postestimation commands** for more capabilities of estimation commands.

Description

zip estimates a zero-inflated Poisson regression of *depvar* on *indepvars*, where *depvar* is a nonnegative count variable.

Options

Model

inflate(*varlist*[, offset(*varname*)] | _cons) specifies the equation that determines whether the observed count is zero. Conceptually, omitting inflate() would be equivalent to fitting the model with poisson.

inflate(*varlist*[, offset(*varname*)]) specifies the variables in the equation. You may optionally include an offset for this *varlist*.

inflate(_cons) specifies that the equation determining whether the count is zero contains only an intercept. To run a zero-inflated model of *depvar* with only an intercept in both equations, type zip *depvar*, inflate(_cons).

noconstant, exposure(*varname_e*), offset(*varname_o*), constraints(*constraints*), collinear; see [R] **estimation options**.

probit requests that a probit, instead of logit, model be used to characterize the excess zeros in the data.

SE/Robust

vce(*vcetype*) specifies the type of standard error reported, which includes types that are derived from asymptotic theory, that are robust to some kinds of misspecification, that allow for intragroup correlation, and that use bootstrap or jackknife methods; see [R] **vce_option**.

Reporting

level(*#*); see [R] **estimation options**.

irr reports estimated coefficients transformed to incidence-rate ratios, that is, e^b rather than b. Standard errors and confidence intervals are similarly transformed. This option affects how results are displayed, not how they are estimated or stored. irr may be specified at estimation or when replaying previously estimated results.

vuong specifies that the Vuong (1989) test of ZIP versus Poisson be reported. This test statistic has a standard normal distribution with large positive values favoring the ZIP model and large negative values favoring the Poisson model.

Max options

maximize_options: difficult, technique(*algorithm_spec*), iterate(*#*), [no]log, trace, gradient, showstep, hessian, shownrtolerance, tolerance(*#*), ltolerance(*#*), gtolerance(*#*), nrtolerance(*#*), nonrtolerance, from(*init_specs*); see [R] **maximize**. These options are seldom used.

Setting the optimization type to technique(bhhh) resets the default *vcetype* to vce(opg).

Remarks

See Long (1997, 242–247) and Greene (2003, 749–752) for a discussion of zero-modified count models. For information about the test developed by Vuong (1989), see Greene (2003) and Long (1997). Greene (1994) applied the test to zero-inflated Poisson and negative binomial models, as described in Greene (2003).

Poisson regression fits models of the number of occurrences (counts) of an event. You could use `poisson` for this (see [R] **poisson**), but in some count-data models, you might want to account for the prevalence of zero counts in the data.

For instance, you might count how many fish each visitor to a park catches. Many visitors may catch zero, as they do not fish (as opposed to being unsuccessful). You may be able to model whether a person fishes depending on several covariates related to fishing activity and model how many fish a person catches depending on several covariates having to do with the success of catching fish (type of lure/bait, time of day, temperature, season, etc.). This is the type of data for which the `zip` command is useful.

The zero-inflated (or zero-altered) Poisson model allows overdispersion through the splitting process that models the outcomes as zero or nonzero.

▷ Example 1

We have data on the number of fish caught by visitors to a national park. Some of the visitors do not fish, but we do not have the data on whether a person fished or not; we merely have data on how many fish were caught together with several covariates. As our data have a preponderance of zeros (142 of 250), we use the `zip` command to model the outcome.

```
. use http://www.stata-press.com/data/r10/fish

. zip count persons livebait, inf(child camper) vuong

Fitting constant-only model:

Iteration 0:   log likelihood =  -1347.807
Iteration 1:   log likelihood = -1305.3245
  (output omitted )
Iteration 4:   log likelihood = -1103.9425

Fitting full model:

Iteration 0:   log likelihood = -1103.9425
Iteration 1:   log likelihood =  -896.2346
  (output omitted )
Iteration 5:   log likelihood = -850.70142
```

Zero-inflated Poisson regression

				Number of obs	=	250
				Nonzero obs	=	108
				Zero obs	=	142

Inflation model = logit
Log likelihood = -850.7014

				LR chi2(2)	=	506.48
				Prob > chi2	=	0.0000

	Coef.	Std. Err.	z	P>|z|	[95% Conf. Interval]	
count						
persons	.8068853	.0453288	17.80	0.000	.7180424	.8957281
livebait	1.757289	.2446082	7.18	0.000	1.277866	2.236713
_cons	-2.178472	.2860289	-7.62	0.000	-2.739078	-1.617865
inflate						
child	1.602571	.2797719	5.73	0.000	1.054228	2.150913
camper	-1.015698	.365259	-2.78	0.005	-1.731593	-.2998038
_cons	-.4922872	.3114562	-1.58	0.114	-1.10273	.1181558

Vuong test of zip vs. standard Poisson: z =   3.95 Pr>z = 0.0000

In general, Vuong test statistics that are significantly positive favor the zero-inflated models, while those that are significantly negative favor the non–zero-inflated models. Thus in the above model, the zero inflation is significant.

◁

Saved Results

zip saves the following in e():

Scalars

e(N)	number of observations
e(k)	number of parameters
e(k_eq)	number of equations
e(k_eq_model)	number of equations in model Wald test
e(k_dv)	number of dependent variables
e(N_zero)	number of zero observations
e(df_m)	model degrees of freedom
e(ll)	log likelihood
e(ll_0)	log likelihood, constant-only model
e(ll_c)	log likelihood, comparison model
e(df_c)	degrees of freedom for comparison test
e(N_clust)	number of clusters
e(chi2)	χ^2
e(chi2_c)	χ^2 for comparison test
e(p)	significance of model test
e(vuong)	Vuong test statistic
e(rank)	rank of e(V)
e(ic)	number of iterations
e(rc)	return code
e(converged)	1 if converged, 0 otherwise

Macros

e(cmd)	zip
e(cmdline)	command as typed
e(depvar)	name of dependent variable
e(inflate)	logit or probit
e(wtype)	weight type
e(wexp)	weight expression
e(title)	title in estimation output
e(clustvar)	name of cluster variable
e(offset1)	offset
e(offset2)	offset for inflate()
e(chi2type)	Wald or LR; type of model χ^2 test
e(chi2_ct)	Wald or LR; type of model χ^2 test corresponding to e(chi2_c)
e(vce)	*vcetype* specified in vce()
e(vcetype)	title used to label Std. Err.
e(opt)	type of optimization
e(ml_method)	type of ml method
e(user)	name of likelihood-evaluator program
e(technique)	maximization technique
e(crittype)	optimization criterion
e(properties)	b V
e(predict)	program used to implement predict

Matrices

e(b)	coefficient vector
e(V)	variance–covariance matrix of the estimators
e(gradient)	gradient vector
e(ilog)	iteration log (up to 20 iterations)

Functions

e(sample)	marks estimation sample

Methods and Formulas

`zip` is implemented as an ado-file.

Several models in the literature are (correctly) described as zero inflated. The `zip` command maximizes the log-likelihood $\ln L$, defined by

$$\xi_j^\beta = \mathbf{x}_j \boldsymbol{\beta} + \text{offset}_j^\beta$$
$$\xi_j^\gamma = \mathbf{z}_j \boldsymbol{\gamma} + \text{offset}_j^\gamma$$
$$\ln L = \sum_{i \in S} w_j \ln \left[F(\xi_j^\gamma) + \left\{ 1 - F(\xi_j^\gamma) \right\} \exp(-\lambda_j) \right] +$$
$$\sum_{i \notin S} w_j \left[\ln \left\{ 1 - F(\xi_j^\gamma) \right\} - \lambda_j + \xi_j^\beta y_j - \ln(y_j!) \right]$$

where w_j are the weights, F is the logit link (or probit link if `probit` was specified), and S is the set of observations for which the outcome $y_j = 0$.

References

Greene, W. H. 1994. Accounting for excess zeros and sample selection in Poisson and negative binomial regression models. Working Paper no. EC-94-10, Department of Economics, Stern School of Business, New York University.

———. 2003. *Econometric Analysis.* 5th ed. Upper Saddle River, NJ: Prentice Hall.

Lambert, D. 1992. Zero-inflated Poisson regression, with an application to defects in manufacturing. *Technometrics* 34: 1–14.

Long, J. S. 1997. *Regression Models for Categorical and Limited Dependent Variables.* Thousand Oaks, CA: Sage.

Long, J. S., and J. Freese. 2001. Predicted probabilities for count models. *Stata Journal* 1: 51–57.

———. 2006. *Regression Models for Categorical Dependent Variables Using Stata.* 2nd ed. College Station, TX: Stata Press.

Mullahy, J. 1986. Specification and testing of some modified count-data models. *Journal of Econometrics* 33: 341–365.

Vuong, Q. 1989. Likelihood ratio tests for model selection and non-nested hypotheses. *Econometrica*: 57: 307–334.

Also See

[R] **zip postestimation** — Postestimation tools for zip

[R] **zinb** — Zero-inflated negative binomial regression

[R] **ztnb** — Zero-truncated negative binomial regression

[R] **ztp** — Zero-truncated Poisson regression

[R] **nbreg** — Negative binomial regression

[R] **poisson** — Poisson regression

[XT] **xtnbreg** — Fixed-effects, random-effects, & population-averaged negative binomial models

[U] **20 Estimation and postestimation commands**

Title

zip postestimation — Postestimation tools for zip

Description

The following postestimation commands are available for zip:

command	description
adjust	adjusted predictions of $\mathbf{x}\beta$ or $\exp(\mathbf{x}\beta)$
estat	AIC, BIC, VCE, and estimation sample summary
estat (svy)	postestimation statistics for survey data
estimates	cataloging estimation results
lincom	point estimates, standard errors, testing, and inference for linear combinations of coefficients
lrtest[1]	likelihood-ratio test
mfx	marginal effects or elasticities
nlcom	point estimates, standard errors, testing, and inference for nonlinear combinations of coefficients
predict	predictions, residuals, influence statistics, and other diagnostic measures
predictnl	point estimates, standard errors, testing, and inference for generalized predictions
suest	seemingly unrelated estimation
test	Wald tests for simple and composite linear hypotheses
testnl	Wald tests of nonlinear hypotheses

[1] lrtest is not appropriate with svy estimation results.

See the corresponding entries in the *Stata Base Reference Manual* for details, but see [SVY] **estat** for details about estat (svy).

Syntax for predict

predict [*type*] *newvar* [*if*] [*in*] [, *statistic* nooffset]

predict [*type*] { *stub** | *newvar*$_{reg}$ *newvar*$_{inflate}$ } [*if*] [*in*] , scores

statistic	description
Main	
n	number of events; the default
ir	incidence rate
pr	probability of a zero outcome
xb	linear prediction
stdp	standard error of the linear prediction

These statistics are available both in and out of sample; type predict ... if e(sample) ... if wanted only for the estimation sample.

Options for predict

⌐ Main ⌐

n, the default, calculates the predicted number of events, which is $(1 - p_j) \exp(\mathbf{x}_j \boldsymbol{\beta})$ if neither offset() nor exposure() was specified when the model was fitted, where p_j is the predicted probability of a zero outcome; $(1 - p_j) \exp((\mathbf{x}_j \boldsymbol{\beta}) + \text{offset}_j)$ if offset() was specified; or $(1 - p_j)(\exp(\mathbf{x}_j \boldsymbol{\beta}) \times \text{exposure}_j)$ if exposure() was specified.

ir calculates the incidence rate $\exp(\mathbf{x}_j \boldsymbol{\beta})$, which is the predicted number of events when exposure is 1. This is equivalent to specifying both n and nooffset options.

pr calculates the probability of a zero outcome $F(\mathbf{z}_j \boldsymbol{\gamma})$, where $F()$ is the logit or probit link function. If offset() was specified within the inflate() option, then $F(\mathbf{z}_j \boldsymbol{\gamma} + \text{offset}_j^{\gamma})$ is calculated.

xb calculates the linear prediction, which is $\mathbf{x}_j \boldsymbol{\beta}$ if neither offset() nor exposure() was specified; $\mathbf{x}_j \boldsymbol{\beta} + \text{offset}_j$ if offset() was specified; or $\mathbf{x}_j \boldsymbol{\beta} + \ln(\text{exposure}_j)$ if exposure() was specified; see nooffset below.

stdp calculates the standard error of the linear prediction.

nooffset is relevant only if you specified offset() or exposure() when you fitted the model. It modifies the calculations made by predict so that they ignore the offset or exposure variable; the linear prediction is treated as $\mathbf{x}_j \boldsymbol{\beta}$ rather than as $\mathbf{x}_j \boldsymbol{\beta} + \text{offset}_j$ or $\mathbf{x}_j \boldsymbol{\beta} + \ln(\text{exposure}_j)$. Specifying predict ... , nooffset is equivalent to specifying predict ... , ir.

scores calculates equation-level score variables.

The first new variable will contain $\partial \ln L / \partial(\mathbf{x}_j \boldsymbol{\beta})$.

The second new variable will contain $\partial \ln L / \partial(\mathbf{z}_j \boldsymbol{\gamma})$.

Methods and Formulas

All postestimation commands listed above are implemented as ado-files.

Also See

[R] **zip** — Zero-inflated Poisson regression

[U] **20 Estimation and postestimation commands**

Title

ztnb — Zero-truncated negative binomial regression

Syntax

ztnb *depvar* [*indepvars*] [*if*] [*in*] [*weight*] [, *options*]

options	description
Model	
<u>nocon</u>stant	suppress constant term
dispersion(<u>mean</u>)	parameterization of dispersion; dispersion(mean) is the default
dispersion(<u>c</u>onstant)	constant dispersion for all observations
<u>e</u>xposure(*varname*_e)	include ln(*varname*_e) in model with coefficient constrained to 1
<u>off</u>set(*varname*_o)	include *varname*_o in model with coefficient constrained to 1
<u>cons</u>traints(*constraints*)	apply specified linear constraints
<u>col</u>linear	keep collinear variables
SE/Robust	
vce(*vcetype*)	*vcetype* may be oim, <u>r</u>obust, <u>cl</u>uster *clustvar*, opg, <u>boot</u>strap, or <u>jack</u>knife
Reporting	
<u>l</u>evel(#)	set confidence level; default is level(95)
<u>nolr</u>test	suppress likelihood-ratio test
<u>irr</u>	report incidence-rate ratios
Max options	
maximize_options	control the maximization process; seldom used

bootstrap, by, jackknife, rolling, statsby, svy, and xi are allowed; see [U] **11.1.10 Prefix commands**.
Weights are not allowed with the bootstrap prefix.
vce() and weights are not allowed with the svy prefix.
fweights, iweights, and pweights are allowed; see [U] **11.1.6 weight**.
See [U] **20 Estimation and postestimation commands** for more capabilities of estimation commands.

Description

ztnb fits a zero-truncated negative binomial regression model of *depvar* on *indepvars*, where *depvar* is a positive count variable.

Options

<u>Model</u>

noconstant; see [R] **estimation options**.

dispersion(mean | constant) specifies the parameterization of the model. dispersion(mean), the default, yields a model with dispersion equal to $1 + \alpha \exp(\mathbf{x}_j\boldsymbol{\beta} + \text{offset}_j)$; that is, the dispersion is a function of the expected mean: $\exp(\mathbf{x}_j\boldsymbol{\beta} + \text{offset}_j)$. dispersion(constant) has dispersion equal to $1 + \delta$; that is, it is a constant for all observations.

exposure(*varname_e*), offset(*varname_o*), constraints(*constraints*), collinear; see [R] **estimation options**.

_____ SE/Robust _____

vce(*vcetype*) specifies the type of standard error reported, which includes types that are derived from asymptotic theory, that are robust to some kinds of misspecification, that allow for intragroup correlation, and that use bootstrap or jackknife methods; see [R] *vce_option*.

_____ Reporting _____

level(*#*); see [R] **estimation options**.

nolrtest suppresses fitting the Poisson model. Without this option, a comparison Poisson model is fitted, and the likelihood is used in a likelihood-ratio test of the null hypothesis that the dispersion parameter is zero.

irr reports estimated coefficients transformed to incidence-rate ratios, i.e., e^b rather than b. Standard errors and confidence intervals are similarly transformed. This option affects how results are displayed, not how they are estimated or stored. You can specify irr at estimation or when you replay previously estimated results.

_____ Max options _____

maximize_options: difficult, technique(*algorithm_spec*), iterate(*#*), [no]log, trace, gradient, showstep, hessian, shownrtolerance, tolerance(*#*), ltolerance(*#*), gtolerance(*#*), nrtolerance(*#*), nonrtolerance, from(*init_specs*); see [R] **maximize**. These options are seldom used.

Setting the optimization type to technique(bhhh) resets the default *vcetype* to vce(opg).

Remarks

Grogger and Carson (1991) showed that overdispersion causes inconsistent estimation of the mean in the truncated Poisson model, so they proposed the zero-truncated negative binomial model as an alternative. If the data exhibit zero truncation but not overdispersion, the zero-truncated Poisson model is appropriate; see [R] **ztp**. See Long (1997, chap. 8) for an introduction to negative binomial and zero-truncated negative binomial regression.

ztnb fits two different parameterizations of zero-truncated negative binomial models, namely the mean-dispersion and constant-dispersion models. They are equivalent to those modeled by nbreg; see [R] **nbreg**.

▷ Example 1

We illustrate the zero-truncated negative binomial model by using the HCFA's 1997 MedPar dataset (Hilbe 1999). The data represent 475 patients in Arizona who were assigned a diagnostic-related group (DRG) of patients having a ventilator. Length of stay, the dependent variable, is a positive integer; it cannot have zero values. The variables used in the study include

provnum	Provider number (Hospital identifier)
died	Patient died while in hospital
hmo	Patient is a member of an HMO
los	Length of stay
type1	Emergency admission
type2	Urgent admission (first available bed)
type3	Elective admission

```
. use http://www.stata-press.com/data/r10/medpar
. ztnb los died hmo type2-type3, vce(cluster provnum) nolog
```

Zero-truncated negative binomial regression				Number of obs	=	1495
Dispersion = mean				Wald chi2(4)	=	36.01
Log likelihood = -4737.535				Prob > chi2	=	0.0000

(Std. Err. adjusted for 54 clusters in provnum)

los	Coef.	Robust Std. Err.	z	P>|z|	[95% Conf. Interval]	
died	-.2521884	.061533	-4.10	0.000	-.3727908	-.1315859
hmo	-.0754173	.0533132	-1.41	0.157	-.1799091	.0290746
type2	.2685095	.0666474	4.03	0.000	.137883	.3991359
type3	.7668101	.2183505	3.51	0.000	.338851	1.194769
_cons	2.224028	.034727	64.04	0.000	2.155964	2.292091
/lnalpha	-.630108	.0764019			-.779853	-.480363
alpha	.5325343	.0406866			.4584734	.6185588

Likelihood-ratio test of alpha=0:   chibar2(01) = 4218.24 Prob>=chibar2 = 0.000

Because observations within the same hospital (provnum) are likely to be correlated, we specified the vce(cluster provnum) option. The results show that whether the patient died in the hospital and the type of admission have significant effects on the patient's length of stay.

◁

▷ Example 2

Here we compare the zero-truncated Poisson model with the zero-truncated negative binomial model. The following data appeared in Rodríguez (1993).

```
. use http://www.stata-press.com/data/r10/rod93
. quietly tab cohort, gen(coh)
. ztp deaths coh2 coh3, exp(exposure) nolog
```

Zero-truncated Poisson regression				Number of obs	=	21
				LR chi2(2)	=	49.16
				Prob > chi2	=	0.0000
Log likelihood = -2159.5098				Pseudo R2	=	0.0113

deaths	Coef.	Std. Err.	z	P>|z|	[95% Conf. Interval]	
coh2	-.3020296	.0573367	-5.27	0.000	-.4144076	-.1896517
coh3	.074248	.0589754	1.26	0.208	-.0413417	.1898377
_cons	-3.899522	.0411385	-94.79	0.000	-3.980152	-3.818892
exposure	(exposure)					

```
. ztnb deaths coh2 coh3, exp(exposure) nolog
Zero-truncated negative binomial regression          Number of obs   =        21
                                                     LR chi2(2)      =      0.32
Dispersion       = mean                              Prob > chi2     =    0.8535
Log likelihood = -130.05845                          Pseudo R2       =    0.0012
```

deaths	Coef.	Std. Err.	z	P>|z|	[95% Conf. Interval]	
coh2	-.2628645	.817489	-0.32	0.748	-1.865114	1.339384
coh3	-.4587405	.8167563	-0.56	0.574	-2.059553	1.142072
_cons	-2.181593	.5831368	-3.74	0.000	-3.32452	-1.038666
exposure	(exposure)					
/lnalpha	.8916014	.4062167			.0954313	1.687772
alpha	2.439032	.9907757			1.100133	5.407417

```
Likelihood-ratio test of alpha=0:  chibar2(01) = 4058.90 Prob>=chibar2 = 0.000
```

Comparing the zero-truncated Poisson and zero-truncated negative binomial models, we see that ztnb has a much larger log likelihood. To test $\alpha = 0$, ztnb performs a likelihood-ratio test. The large χ^2 value of 4058.90 strongly suggests that the zero-truncated negative binomial model is more appropriate.

◁

❑ Technical Note

The usual Gaussian test of $\alpha = 0$ is omitted since this test occurs on the boundary, invalidating the usual theory associated with such tests. However, the likelihood-ratio test of $\alpha = 0$ has been modified to be valid on the boundary. In particular, the null distribution of the likelihood-ratio test statistic is not the usual χ_1^2, but rather a $50:50$ mixture of a χ_0^2 (point mass at zero) and a χ_1^2, denoted as $\overline{\chi}_{01}^2$. See Gutierrez, Carter, and Drukker (2001) for more details.

❑

Saved Results

ztnb saves the following in e():

Scalars

e(N)	number of observations	e(alpha)	the value of alpha
e(k)	number of parameters	e(N_clust)	number of clusters
e(k_eq)	number of equations	e(chi2)	χ^2
e(k_eq_model)	number of equations in model	e(chi2_c)	χ^2 for comparison test
	Wald test	e(p)	significance
e(k_aux)	number of auxiliary parameters	e(rank)	rank of e(V)
e(k_dv)	number of dependent variables	e(rank0)	rank of e(V) for constant-only
e(df_m)	model degrees of freedom		model
e(r2_p)	pseudo-R-squared	e(ic)	number of iterations
e(ll)	log likelihood	e(rc)	return code
e(ll_0)	log likelihood, constant-only model	e(converged)	1 if converged, 0 otherwise
e(ll_c)	log likelihood, comparison model		

Macros

e(cmd)	ztnb	e(chi2_ct)	Wald or LR; type of model χ^2 test
e(cmdline)	command as typed		corresponding to e(chi2_c)
e(depvar)	name of dependent variable	e(diparm#)	display transformed parameter #
e(wtype)	weight type	e(diparm_opt2)	options for displaying transformed
e(wexp)	weight expression		parameters
e(title)	title in estimation output	e(opt)	type of optimization
e(clustvar)	name of cluster variable	e(ml_method)	type of ml method
e(offset)	offset	e(user)	name of likelihood-evaluator program
e(dispers)	mean or constant	e(technique)	maximization technique
e(vce)	*vcetype* specified in vce()	e(crittype)	optimization criterion
e(vcetype)	title used to label Std. Err.	e(properties)	b V
e(chi2type)	Wald or LR; type of model χ^2 test	e(predict)	program used to implement predict

Matrices

e(b)	coefficient vector	e(V)	variance–covariance matrix of
e(ilog)	iteration log (up to 20 iterations)		the estimators
e(gradient)	gradient vector		

Functions

e(sample)	marks estimation sample

Methods and Formulas

ztnb is implemented as an ado-file.

Mean-dispersion model

A negative binomial distribution can be regarded as a gamma mixture of Poisson random variables. The number of times something occurs, y_j, is distributed as Poisson$(\nu_j \mu_j)$. That is, its conditional likelihood is

$$f(y_j \mid \nu_j) = \frac{(\nu_j \mu_j)^{y_j} e^{-\nu_j \mu_j}}{\Gamma(y_j + 1)}$$

where $\mu_j = \exp(\mathbf{x}_j \boldsymbol{\beta} + \text{offset}_j)$ and ν_j is an unobserved parameter with a gamma$(1/\alpha, \alpha)$ density:

$$g(\nu) = \frac{\nu^{(1-\alpha)/\alpha}e^{-\nu/\alpha}}{\alpha^{1/\alpha}\Gamma(1/\alpha)}$$

This gamma distribution has mean 1 and variance α, where α is the ancillary parameter.

The unconditional likelihood for the jth observation is therefore

$$f(y_j) = \int_0^\infty f(y_j \mid \nu)g(\nu)\,d\nu = \frac{\Gamma(m+y_j)}{\Gamma(y_j+1)\Gamma(m)}\,p_j^m(1-p_j)^{y_j}$$

where $p_j = 1/(1+\alpha\mu_j)$ and $m = 1/\alpha$. Solutions for α are handled by searching for $\ln\alpha$ since α must be greater than zero. The conditional probability of observing y_j events given that $y_j > 0$ is

$$\Pr(Y = y_j \mid y_j > 0, \mathbf{x}_j) = \frac{f(y_j)}{1 - (1+\alpha\mu_j)^{-\alpha^{-1}}}$$

The log likelihood (with weights w_j and offsets) is given by

$$m = 1/\alpha \qquad p_j = 1/(1+\alpha\mu_j) \qquad \mu_j = \exp(\mathbf{x}_j\boldsymbol{\beta} + \text{offset}_j)$$

$$\ln L = \sum_{j=1}^n w_j \Bigg[\ln\{\Gamma(m+y_j)\} - \ln\{\Gamma(y_j+1)\}$$
$$- \ln\{\Gamma(m)\} + m\ln(p_j) + y_j\ln(1-p_j) - \ln(1-p_j^m) \Bigg]$$

Constant-dispersion model

The constant-dispersion model assumes that y_j is conditionally distributed as $\text{Poisson}(\mu_j^*)$, where $\mu_j^* \sim \text{gamma}(\mu_j/\delta, \delta)$ for some dispersion parameter δ (by contrast, the mean-dispersion model assumes $\mu_j^* \sim \text{gamma}(1/\alpha, \alpha\mu_j)$). The log likelihood is given by

$$m_j = \mu_j/\delta \qquad p = 1/(1+\delta)$$

$$\ln L = \sum_{j=1}^n w_j \Bigg[\ln\{\Gamma(m_j+y_j)\} - \ln\{\Gamma(y_j+1)\}$$
$$- \ln\{\Gamma(m_j)\} + m_j\ln(p) + y_j\ln(1-p) - \ln(1-p^{m_j}) \Bigg]$$

with everything else defined as before in the calculations for the mean-dispersion model.

Acknowledgment

We gratefully acknowledge the previous work by Joseph Hilbe, Arizona State University, USA; see Hilbe (1999).

References

Cameron, A. C., and P. K. Trivedi. 1998. *Regression Analysis of Count Data*. Cambridge: Cambridge University Press.

Grogger, J. T., and R. T. Carson. 1991. Models for truncated counts. In *Journal of Applied Econometrics* 6: 225–238.

Gutierrez, R. G., S. L. Carter, and D. M. Drukker. 2001. On boundary-value likelihood-ratio tests. *Stata Technical Bulletin* 60: 15–18. Reprinted in *Stata Technical Bulletin Reprints*, vol. 10, pp. 269–273.

Hilbe, J. 1998. sg91: Robust variance estimators for MLE Poisson and negative binomial regression. *Stata Technical Bulletin* 45: 26–28. Reprinted in *Stata Technical Bulletin Reprints*, vol. 8, pp. 177–180.

——. 1999. sg102: Zero-truncated Poisson and negative binomial regression. *Stata Technical Bulletin* 47: 37–40. Reprinted in *Stata Technical Bulletin Reprints*, vol. 8, pp. 233–236.

Long, J. S. 1997. *Regression Models for Categorical and Limited Dependent Variables*. Thousand Oaks, CA: Sage.

Rodríguez, G. 1993. sbe10: An improvement to poisson. *Stata Technical Bulletin* 11: 11–14. Reprinted in *Stata Technical Bulletin Reprints*, vol. 2, pp. 94–98.

Simonoff, J. S. 2003 *Analyzing Categorical Data*. New York: Springer.

Also See

Title

ztnb postestimation — Postestimation tools for ztnb

Description

The following postestimation commands are available for `ztnb`:

command	description
adjust	adjusted predictions of $\mathbf{x}\beta$ or $\exp(\mathbf{x}\beta)$
estat	AIC, BIC, VCE, and estimation sample summary
estat (svy)	postestimation statistics for survey data
estimates	cataloging estimation results
lincom	point estimates, standard errors, testing, and inference for linear combinations of coefficients
lrtest[1]	likelihood-ratio test
mfx	marginal effects or elasticities
nlcom	point estimates, standard errors, testing, and inference for nonlinear combinations of coefficients
predict	predictions, residuals, influence statistics, and other diagnostic measures
predictnl	point estimates, standard errors, testing, and inference for generalized predictions
suest	seemingly unrelated estimation
test	Wald tests for simple and composite linear hypotheses
testnl	Wald tests of nonlinear hypotheses

[1] `lrtest` is not appropriate with `svy` estimation results.

See the corresponding entries in the *Stata Base Reference Manual* for details, but see
[SVY] **estat** for details about `estat` (svy).

Syntax for predict

predict [*type*] *newvar* [*if*] [*in*] [, *statistic* <u>nooff</u>set]

predict [*type*] { *stub** | *newvar*_{reg} *newvar*_{disp} } [*if*] [*in*] , <u>sc</u>ores

statistic	description
Main	
n	number of events; the default
ir	incidence rate
cm	estimate of the conditional mean, $E(y_j \mid y_j > 0)$
xb	linear prediction
stdp	standard error of the linear prediction

These statistics are available both in and out of sample; type `predict ... if e(sample) ...` if wanted only for the estimation sample.

Options for predict

Main

n, the default, calculates the predicted number of events, which is $\exp(\mathbf{x}_j\boldsymbol{\beta})$ if neither offset() nor exposure() was specified when the model was fitted; $\exp(\mathbf{x}_j\boldsymbol{\beta} + \text{offset}_j)$ if offset() was specified; or $\exp(\mathbf{x}_j\boldsymbol{\beta}) \times \text{exposure}_j$ if exposure() was specified.

ir calculates the incidence rate $\exp(\mathbf{x}_j\boldsymbol{\beta})$, which is the predicted number of events when exposure is 1. This is equivalent to specifying both n and nooffset options.

cm calculates the estimate of the conditional mean.

$$E(y_j \mid y_j > 0) = \frac{E(y_j)}{\Pr(y_j > 0)}$$

xb calculates the linear prediction, which is $\mathbf{x}_j\boldsymbol{\beta}$ if neither offset() nor exposure() was specified; $\mathbf{x}_j\boldsymbol{\beta} + \text{offset}_j$ if offset() was specified; or $\mathbf{x}_j\boldsymbol{\beta} + \ln(\text{exposure}_j)$ if exposure() was specified; see nooffset below.

stdp calculates the standard error of the linear prediction.

nooffset is relevant only if you specified offset() or exposure() when you fitted the model. It modifies the calculations made by predict so that they ignore the offset or exposure variable; the linear prediction is treated as $\mathbf{x}_j\boldsymbol{\beta}$ rather than as $\mathbf{x}_j\boldsymbol{\beta} + \text{offset}_j$ or $\mathbf{x}_j\boldsymbol{\beta} + \ln(\text{exposure}_j)$. Specifying predict ..., nooffset is equivalent to specifying predict ..., ir.

scores calculates equation-level score variables.

The first new variable will contain $\partial \ln L / \partial(\mathbf{x}_j\boldsymbol{\beta})$.

The second new variable will contain $\partial \ln L / \partial(\ln\alpha)$ for dispersion(mean).

The second new variable will contain $\partial \ln L / \partial(\ln\delta)$ for dispersion(constant).

Methods and Formulas

All postestimation commands listed above are implemented as ado-files.

In the following, we use the same notation as in [R] **ztnb**.

Mean-dispersion model

The equation-level scores are given by

$$\text{score}(\mathbf{x}\boldsymbol{\beta})_j = p_j(y_j - \mu_j) - \frac{p_j^{(m+1)}\mu_j}{1 - p_j^m}$$

$$\text{score}(\tau)_j = -m\left\{\frac{\alpha(\mu_j - y_j)}{1 + \alpha\mu_j} - \ln(1 + \alpha\mu_j) + \psi(y_j + m) - \psi(m)\right\}$$

$$- \frac{p_j^m}{1 - p_j^m}\left\{m\ln(p_j) + \mu_j p_j\right\}$$

where $\tau_j = \ln\alpha_j$ and $\psi(z)$ is the digamma function.

Constant-dispersion model

The equation-level scores are given by

$$\text{score}(\mathbf{x}\boldsymbol{\beta})_j = m_j \left\{ \psi(y_j + m_j) - \psi(m_j) + \ln(p) + \frac{p^{m_j} \ln(p)}{1 - p^{m_j}} \right\}$$

$$\text{score}(\tau)_j = y_j - (y_j + m_j)(1 - p) - \text{score}(\mathbf{x}\boldsymbol{\beta})_j - \frac{\mu_j p}{1 - p^{m_j}}$$

where $\tau_j = \ln\delta_j$.

Also See

[R] **ztnb** — Zero-truncated negative binomial regression

[U] **20 Estimation and postestimation commands**

Title

ztp — Zero-truncated Poisson regression

Syntax

ztp *depvar* [*indepvars*] [*if*] [*in*] [*weight*] [, *options*]

options	description
Model	
<u>nocon</u>stant	suppress constant term
<u>exp</u>osure(*varname_e*)	include ln(*varname_e*) in model with coefficient constrained to 1
<u>off</u>set(*varname_o*)	include *varname_o* in model with coefficient constrained to 1
<u>constr</u>aints(*constraints*)	apply specified linear constraints
<u>col</u>linear	keep collinear variables
SE/Robust	
vce(*vcetype*)	*vcetype* may be oim, <u>r</u>obust, <u>cl</u>uster *clustvar*, opg, <u>boot</u>strap, or <u>jack</u>knife
Reporting	
<u>l</u>evel(#)	set confidence level; default is level(95)
<u>irr</u>	report incidence-rate ratios
Max options	
maximize_options	control the maximization process; seldom used

depvar and *indepvars* may contain time-series operators; see [U] **11.4.3 Time-series varlists**.
bootstrap, by, jackknife, rolling, statsby, svy, and xi are allowed; see [U] **11.1.10 Prefix commands**.
Weights are not allowed with the bootstrap prefix.
vce() and weights are not allowed with the svy prefix.
fweights, iweights, and pweights are allowed; see [U] **11.1.6 weight**.
See [U] **20 Estimation and postestimation commands** for more capabilities of estimation commands.

Description

ztp fits a zero-truncated Poisson regression of *depvar* on *indepvars*, where *depvar* is a positive count variable.

Options for ztp

⌐ **Model** ⌐

noconstant, exposure(*varname_e*), offset(*varname_o*), constraints(*constraints*), collinear; see [R] **estimation options**.

⌐ **SE/Robust** ⌐

vce(*vcetype*) specifies the type of standard error reported, which includes types that are derived from asymptotic theory, that are robust to some kinds of misspecification, that allow for intragroup correlation, and that use bootstrap or jackknife methods; see [R] *vce_option*.

┌─ Reporting └

level(*#*); see [R] **estimation options**.

irr reports estimated coefficients transformed to incidence-rate ratios, that is, e^{β_i} rather than β_i. Standard errors and confidence intervals are similarly transformed. This option affects how results are displayed, not how they are estimated or stored. irr may be specified at estimation or when replaying previously estimated results.

┌─ Max options └

maximize_options: <u>diff</u>icult, <u>tech</u>nique(*algorithm_spec*), <u>iter</u>ate(*#*), [<u>no</u>]<u>log</u>, <u>tr</u>ace, gradient, showstep, <u>hess</u>ian, <u>shownr</u>tolerance, <u>tol</u>erance(*#*), <u>ltol</u>erance(*#*), <u>gtol</u>erance(*#*), <u>nrtol</u>erance(*#*), <u>nonrtol</u>erance, from(*init_specs*); see [R] **maximize**. These options are seldom used.

Setting the optimization type to technique(bhhh) resets the default *vcetype* to vce(opg).

Remarks

Zero-truncated Poisson regression is used to model the number of occurrences of an event when that number is restricted to be positive. If zero is an admissible value for the dependent variable, then standard Poisson regression may be more appropriate; see [R] **poisson**. Zero-truncated poisson regression was first proposed by Grogger and Carson (1991). See Long (1997, chap. 8) for an introduction to Poisson and zero-truncated Poisson regression.

Suppose that you counted the number of times a patient has medical treatments, and those patients who have at least one treatment are randomly selected into the sample. You might examine the length of hospital stay of patients, which has to be positive. In those cases, the ztp command should be used instead of poisson.

▷ Example 1

We will illustrate the zero-truncated Poisson model by using the dataset of running shoes for a sample of runners who registered an online running log, which is used by Simonoff (2003). A running-shoe marketing executive is interested in knowing how the number of running shoes purchased related to other factors such as gender, marital status, age, education, income, typical number of runs per week, average miles run per week, and the preferred type of running. These data are naturally truncated. We fit a zero-truncated Poisson regression model based on distance running, runs per week, miles run per week, gender, and age.

```
. use http://www.stata-press.com/data/r10/runshoes

. ztp shoes distance rpweek mpweek male age
Iteration 0:   log likelihood = -86.267921
Iteration 1:   log likelihood = -84.096471
Iteration 2:   log likelihood = -84.079524
Iteration 3:   log likelihood = -84.079517
```

Zero-truncated Poisson regression

Number of obs = 60
LR chi2(5) = 27.10
Prob > chi2 = 0.0001

Log likelihood = -84.079517

Pseudo R2 = 0.1388

shoes	Coef.	Std. Err.	z	P>|z|	[95% Conf. Interval]	
distance	.5101724	.2336767	2.18	0.029	.0521746	.9681703
rpweek	.1822472	.1146021	1.59	0.112	-.0423688	.4068631
mpweek	.0143156	.0099102	1.44	0.149	-.0051081	.0337393
male	.1314745	.2480547	0.53	0.596	-.3547039	.6176528
age	.011772	.010266	1.15	0.252	-.0083489	.0318929
_cons	-1.174196	.6524461	-1.80	0.072	-2.452967	.1045745

◁

Saved Results

ztp saves the following in e():

Scalars

e(N)	number of observations	e(ll_0)	log likelihood, constant-only model
e(k)	number of parameters	e(N_clust)	number of clusters
e(k_eq)	number of equations	e(rc)	return code
e(k_eq_model)	number of equations in model Wald test	e(chi2)	χ^2
		e(p)	significance
e(k_dv)	number of dependent variables	e(ic)	number of iterations
e(df_m)	model degrees of freedom	e(rank)	rank of e(V)
e(r2_p)	pseudo-R-squared	e(converged)	1 if converged, 0 otherwise
e(ll)	log likelihood		

Macros

e(cmd)	ztp	e(vce)	*vcetype* specified in vce()
e(cmdline)	command as typed	e(vcetype)	title used to label Std. Err.
e(depvar)	name of dependent variable	e(opt)	type of optimization
e(wtype)	weight type	e(ml_method)	type of ml method
e(wexp)	weight expression	e(user)	name of likelihood-evaluator program
e(title)	title in estimation output	e(technique)	maximization technique
e(clustvar)	name of cluster variable	e(crittype)	optimization criterion
e(offset)	offset	e(properties)	b V
e(chi2type)	Wald or LR; type of model χ^2 test	e(predict)	program used to implement predict

Matrices

e(b)	coefficient vector	e(V)	variance–covariance matrix of the estimators
e(ilog)	iteration log (up to 20 iterations)		
e(gradient)	gradient vector		

Functions

e(sample)	marks estimation sample

Methods and Formulas

`ztp` is implemented as an ado-file.

The conditional probability of observing y_j events given that $y_j > 0$ is

$$\Pr(Y = y_j \mid y_j > 0, \mathbf{x}_j) = \frac{e^{-\lambda}\lambda^{y_j}}{y_j!(1 - e^{-\lambda})}$$

The log likelihood (with weights w_j and offsets) is given by

$$\xi_j = \mathbf{x}_j\boldsymbol{\beta} + \text{offset}_j$$

$$f(y_j) = \frac{e^{-\exp(\xi_j)}e^{\xi_j y_j}}{y_j!\{1 - e^{-\exp(\xi_j)}\}}$$

$$\ln L = \sum_{j=1}^{n} w_j \left[-e^{\xi_j} + \xi_j y_j - \ln(y_j!) - \ln\{1 - e^{-\exp(\xi_j)}\} \right]$$

Acknowledgment

We gratefully acknowledge the previous work by Joseph Hilbe, Arizona State University, USA; see Hilbe (1999).

References

Cameron, A. C., and P. K. Trivedi. 1998. *Regression Analysis of Count Data.* Cambridge: Cambridge University Press.

Grogger, J. T., and R. T. Carson. 1991. Models for truncated counts. In *Journal of Applied Econometrics* 6: 225–238.

Hilbe, J. 1998. sg91: Robust variance estimators for MLE Poisson and negative binomial regression. *Stata Technical Bulletin* 45: 26–28. Reprinted in *Stata Technical Bulletin Reprints*, vol. 8, pp. 177–180.

———. 1999. sg102: Zero-truncated Poisson and negative binomial regression. *Stata Technical Bulletin* 47: 37–40. Reprinted in *Stata Technical Bulletin Reprints*, vol. 8, pp. 233–236.

Hilbe, J., and D. H. Judson. 1998. sg94: Right, left, and uncensored Poisson regression. *Stata Technical Bulletin* 46: 18–20. Reprinted in *Stata Technical Bulletin Reprints*, vol. 8, pp. 186–189.

Long, J. S. 1997. *Regression Models for Categorical and Limited Dependent Variables.* Thousand Oaks, CA: Sage.

Simonoff, J. S. 2003 *Analyzing Categorical Data.* New York: Springer.

Also See

Title

ztp postestimation — Postestimation tools for ztp

Description

The following postestimation commands are available for `ztp`:

command	description
adjust	adjusted predictions of $\mathbf{x}\beta$ or $\exp(\mathbf{x}\beta)$
estat	AIC, BIC, VCE, and estimation sample summary
estat (svy)	postestimation statistics for survey data
estimates	cataloging estimation results
lincom	point estimates, standard errors, testing, and inference for linear combinations of coefficients
lrtest[1]	likelihood-ratio test
mfx	marginal effects or elasticities
nlcom	point estimates, standard errors, testing, and inference for nonlinear combinations of coefficients
predict	predictions, residuals, influence statistics, and other diagnostic measures
predictnl	point estimates, standard errors, testing, and inference for generalized predictions
suest	seemingly unrelated estimation
test	Wald tests for simple and composite linear hypotheses
testnl	Wald tests of nonlinear hypotheses

[1] `lrtest` is not appropriate with `svy` estimation results.

See the corresponding entries in the *Stata Base Reference Manual* for details, but see
[SVY] **estat** for details about `estat` (svy).

Syntax for predict

predict [*type*] *newvar* [*if*] [*in*] [, *statistic* <u>nooff</u>set]

statistic	description
Main	
n	number of events; the default
ir	incidence rate
cm	estimate of the conditional mean, $E(y_j \mid y_j > 0)$
xb	linear prediction
stdp	standard error of the linear prediction
<u>score</u>	first derivative of the log likelihood with respect to $\mathbf{x}_j\beta$

These statistics are available both in and out of sample; type `predict ... if e(sample) ...` if wanted
only for the estimation sample.

603

Options for predict

n, the default, calculates the predicted number of events, which is $\exp(\mathbf{x}_j\boldsymbol{\beta})$ if neither offset() nor exposure() was specified when the model was fitted; $\exp(\mathbf{x}_j\boldsymbol{\beta} + \text{offset}_j)$ if offset() was specified; or $\exp(\mathbf{x}_j\boldsymbol{\beta}) \times \text{exposure}_j$ if exposure() was specified.

ir calculates the incidence rate $\exp(\mathbf{x}_j\boldsymbol{\beta})$, which is the predicted number of events when exposure is 1. This is equivalent to specifying both n and nooffset options.

cm calculates the estimate of the conditional mean.

$$E(y_j \mid y_j > 0) = \frac{E(y_j)}{\Pr(y_j > 0)}$$

xb calculates the linear prediction, which is $\mathbf{x}_j\boldsymbol{\beta}$ if neither offset() nor exposure() was specified; $\mathbf{x}_j\boldsymbol{\beta} + \text{offset}_j$ if offset() was specified; or $\mathbf{x}_j\boldsymbol{\beta} + \ln(\text{exposure}_j)$ if exposure() was specified; see nooffset below.

stdp calculates the standard error of the linear prediction.

score calculates the equation-level score, $\partial \ln L / \partial(\mathbf{x}_j\boldsymbol{\beta})$.

nooffset is relevant only if you specified offset() or exposure() when you fitted the model. It modifies the calculations made by predict so that they ignore the offset or exposure variable; the linear prediction is treated as $\mathbf{x}_j\boldsymbol{\beta}$ rather than as $\mathbf{x}_j\boldsymbol{\beta} + \text{offset}_j$ or $\mathbf{x}_j\boldsymbol{\beta} + \ln(\text{exposure}_j)$. Specifying predict ..., nooffset is equivalent to specifying predict ..., ir.

Methods and Formulas

All postestimation commands listed above are implemented as ado-files.

In the following, we use the same notation as in [R] **ztp**.

The equation-level scores are given by

$$\text{score}(\mathbf{x}\boldsymbol{\beta})_j = y_j - e^{\xi_j} - \frac{e^{-\exp(\xi_j)}e^{\xi_j}}{1 - e^{-\exp(\xi_j)}}$$

Also See

[R] **ztp** — Zero-truncated Poisson regression

[U] **20 Estimation and postestimation commands**

Author Index

This is the author index for the 3-volume *Stata Base Reference Manual*.

A

Abramowitz, M., [R] **orthog**
Abrams, K. R., [R] **meta**
Abramson, J. H., [R] **kappa**, [R] **meta**
Abramson, Z. H., [R] **kappa**, [R] **meta**
Achen, C. H., [R] **scobit**
Acock, A. C., [R] **alpha**, [R] **anova**, [R] **nestreg**,
    [R] **oneway**, [R] **prtest**, [R] **ranksum**
Afifi, A. A., [R] **anova**, [R] **stepwise**
Agresti, A., [R] **ci**, [R] **expoisson**, [R] **tabulate twoway**
Aigner, D. J., [R] **frontier**
Aitchison, J., [R] **ologit**, [R] **oprobit**
Aitken, A. C., [R] **reg3**
Aivazian, S. A., [R] **ksmirnov**
Akaike, H., [R] **BIC note**, [R] **estat**, [R] **glm**
Aldrich, J. H., [R] **logit**, [R] **mlogit**, [R] **probit**
Alexandersson, A., [R] **regress**
Alf, E., [R] **rocfit**
Alldredge, J. R., [R] **pk**, [R] **pkcross**, [R] **pkequiv**
Allen, M. J., [R] **alpha**
Allison, P. D., [R] **rologit**, [R] **testnl**
Altman, D. G., [R] **anova**, [R] **fracpoly**, [R] **kappa**,
    [R] **kwallis**, [R] **meta**, [R] **mfp**, [R] **nptrend**,
    [R] **oneway**
Ambler, G., [R] **fracpoly**, [R] **mfp**, [R] **regress**
Amemiya, T., [R] **cnreg**, [R] **glogit**, [R] **intreg**,
    [R] **ivprobit**, [R] **nlogit**, [R] **tobit**
Andersen, E. B., [R] **clogit**
Anderson, J. A., [R] **ologit**, [R] **slogit**
Anderson, R. E., [R] **rologit**
Anderson, T. W., [R] **ivregress postestimation**
Andrews, D. F., [R] **rreg**
Andrews, R. W. K., [R] **ivregress**
Anscombe, F. J., [R] **binreg postestimation**, [R] **glm**,
    [R] **glm postestimation**
Arbuthnott, J., [R] **signrank**
Archer, K. J., [R] **logistic**, [R] **logistic postestimation**,
    [R] **logit**, [R] **logit postestimation**
Arminger, G., [R] **suest**
Armitage, P., [R] **ameans**, [R] **expoisson**
Armstrong, R. D., [R] **qreg**
Arthur, B. S., [R] **symmetry**
Atkinson, A. C., [R] **boxcox**, [R] **nl**
Azen, S. P., [R] **anova**

B

Babiker, A., [R] **sampsi**
Baker, R. J., [R] **glm**
Baker, R. M., [R] **ivregress postestimation**
Bakker, A., [R] **mean**

Balanger, A., [R] **sktest**
Baltagi, B. H., [R] **hausman**
Bamber, D., [R] **roc**, [R] **rocfit**
Bancroft, T. A., [R] **stepwise**
Barnard, G. A., [R] **spearman**
Barnow, B., [R] **treatreg**
Barrison, I. G., [R] **binreg**
Bartlett, M. S., [R] **oneway**
Bartus, T., [R] **mfx**
Basmann, R. L., [R] **ivregress**, [R] **ivregress**
    postestimation
Bassett, G., Jr., [R] **qreg**
Basu, A., [R] **glm**
Baum, C. F., [R] **heckman**, [R] **heckprob**,
    [R] **ivregress**, [R] **mfx**, [R] **net**, [R] **net**
    search, [R] **regress postestimation**, [R] **regress**
    postestimation time series, [R] **ssc**
Bayart, D., [R] **qc**
Beale, E. M. L., [R] **stepwise**, [R] **test**
Beaton, A. E., [R] **rreg**
Becketti, S., [R] **fracpoly**, [R] **runtest**, [R] **spearman**
Begg, C. B., [R] **meta**
Beggs, S., [R] **rologit**
Bellocco, R., [R] **meta**
Belsley, D. A., [R] **estat**, [R] **regress postestimation**
Bendel, R. B., [R] **stepwise**
Benedetti, J. K., [R] **tetrachoric**
Beniger, J. R., [R] **cumul**
Bera, A. K., [R] **sktest**
Beran, R. J., [R] **regress postestimation time series**
Berk, K. N., [R] **stepwise**
Berk, R. A., [R] **rreg**
Berkson, J., [R] **logit**, [R] **probit**
Bern, P. H., [R] **nestreg**
Bernasco, W., [R] **tetrachoric**
Berndt, E. K., [R] **glm**
Berndt, E. R., [R] **treatreg**, [R] **truncreg**
Bernstein, I. H., [R] **alpha**
Berry, G., [R] **ameans**, [R] **expoisson**
Bewley, R., [R] **reg3**
Beyer, W. H., [R] **qc**
Bickel, P. J., [R] **rreg**
Bickenböller, H., [R] **symmetry**
Birdsall, T. G., [R] **logistic postestimation**
Black, W. C., [R] **rologit**
Blackwell, J. L., III, [R] **areg**
Bland, M., [R] **ranksum**, [R] **signrank**
Bleda, M. J., [R] **alpha**
Bliss, C. I., [R] **probit**
Bloch, D. A., [R] **brier**
Bloomfield, P., [R] **qreg**
Blundell, R., [R] **ivprobit**
BMDP, [R] **symmetry**
Boice, J. D., [R] **bitest**
Boland, P. J., [R] **ttest**
Bolduc, D., [R] **asmprobit**
Bollen, K. A., [R] **regress postestimation**

Subject Index

This is the subject index for the 3-volume *Stata Base Reference Manual*. Readers may also want to consult the combined subject index (and the combined author index) in the *Stata Quick Reference and Index*. The combined index indexes the *Getting Started* manuals, the *User's Guide*, and all the Reference manuals except the *Mata Reference Manual*.

Semicolons set off the most important entries from the rest. Sometimes no entry will be set off with semicolons; this means all entries are equally important.

A

about command, [R] **about**
absorption in regression, [R] **areg**
acprplot command, [R] **regress postestimation**
added-variable plots, [R] **regress postestimation**
adjust command, [R] **adjust**
adjusted partial residual plot, [R] **regress postestimation**
ado command, [R] **net**
ado describe command, [R] **net**
ado dir command, [R] **net**
ado-files,
  editing, [R] **doedit**
  installing, [R] **net**, [R] **sj**, [R] **ssc**
  location of, [R] **which**
  official, [R] **update**
ado uninstall command, [R] **net**
ado, update subcommand, [R] **update**
ado, view subcommand, [R] **view**
ado_d, view subcommand, [R] **view**
adosize, set subcommand, [R] **set**
adoupdate command, [R] **adoupdate**
agreement, interrater, [R] **kappa**
AIC, [R] **BIC note**, [R] **estat**, [R] **estimates stats**, [R] **glm**
Akaike information criterion, *see* AIC
all, update subcommand, [R] **update**
alpha coefficient, Cronbach's, [R] **alpha**
alpha command, [R] **alpha**
alternative-specific multinomial probit regression, [R] **asmprobit**
alternatives, estat subcommand, [R] **asclogit postestimation**, [R] **asmprobit postestimation**, [R] **asroprobit postestimation**
ameans command, [R] **ameans**
analysis of covariance, *see* ANCOVA
analysis of variance, *see* ANOVA
analysis-of-variance test of normality, [R] **swilk**
ANCOVA, [R] **anova**
ANOVA, [R] **anova**, [R] **loneway**, [R] **oneway**
  Kruskal–Wallis, [R] **kwallis**
  repeated measures, [R] **anova**
anova command, [R] **anova**, [R] **anova postestimation**, *also see* postestimation command

ARCH effects, testing for, [R] **regress postestimation time series**
archlm, estat subcommand, [R] **regress postestimation time series**
areg command, [R] **areg**, [R] **areg postestimation**, *also see* postestimation command
asclogit command, [R] **asclogit**, [R] **asclogit postestimation**, *also see* postestimation command
asmprobit command, [R] **asmprobit**, [R] **asmprobit postestimation**, *also see* postestimation command
asroprobit command, [R] **asroprobit**, [R] **asroprobit postestimation**, *also see* postestimation command
association, measures of, [R] **tabulate twoway**
asymmetry, [R] **lnskew0**, [R] **lv**, [R] **sktest**, [R] **summarize**
attribute tables, [R] **table**, [R] **tabulate twoway**, [R] **tabulate, summarize()**
AUC, [R] **pk**
augmented
  component-plus-residual plot, [R] **regress postestimation**
  partial residual plot, [R] **regress postestimation**
autocorrelation, [R] **regress postestimation time series**
autoregressive conditional heteroskedasticity, testing for, [R] **regress postestimation time series**
autotabgraphs, set subcommand, [R] **set**
averages, *see* means
avplot and avplots commands, [R] **regress postestimation**

B

Bartlett's test for equal variances, [R] **oneway**
Bayesian information criterion, *see* BIC
bcskew0 command, [R] **lnskew0**
Berndt–Hall–Hall–Hausman algorithm, [R] **ml**
beta coefficients, [R] **ivregress**, [R] **nlsur**, [R] **regress**, [R] **regress postestimation**
BFGS algorithm, [R] **ml**
bgodfrey, estat subcommand, [R] **regress postestimation time series**
BHHH algorithm, [R] **ml**
bias corrected and acelerated, [R] **bootstrap postestimation**
BIC, [R] **BIC note**, [R] **estat**, [R] **estimates stats**, [R] **glm**
binary outcome models, *see* dichotomous outcome models
binomial
  distribution, confidence intervals, [R] **ci**
  family regression, [R] **binreg**
  probability test, [R] **bitest**
binreg command, [R] **binreg**, [R] **binreg postestimation**, *also see* postestimation command
bioequivalence tests, [R] **pk**, [R] **pkequiv**
biopharmaceutical data, *see* pk (pharmacokinetic data)
biprobit command, [R] **biprobit**, [R] **biprobit postestimation**, *also see* postestimation command
bitest and bitesti commands, [R] **bitest**

M

P

S